Werner Bonrath, Jonathan Medlock, Marc-André Müller, Jan Schütz
Catalysis for Fine Chemicals

Also of interest

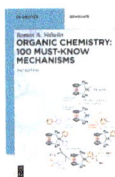

Organic Chemistry: 100 Must-Know Mechanisms
Valiulin, 2023
ISBN 978-3-11-078682-8, e-ISBN (PDF) 978-3-11-078683-5

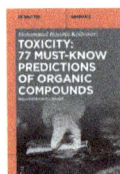

Toxicity: 77 Must-Know Predictions of Organic Compounds.
Including Ionic Liquids
Keshavarz, 2023
ISBN 978-3-11-118912-3, e-ISBN (PDF) 978-3-11-119092-1

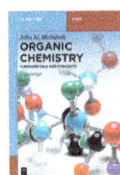

Organic Chemistry.
Fundamentals and Concepts
McIntosh, 2022
ISBN 978-3-11-077820-5, e-ISBN (PDF) 978-3-11-077831-1

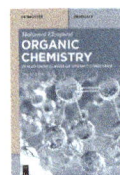

Organic Chemistry.
25 Must-Know Classes of Organic Compounds
Elzagheid, 2025
ISBN 978-3-11-138199-2, e-ISBN (PDF) 978-3-11-138275-3

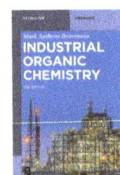

Industrial Organic Chemistry
Benvenuto, 2024
ISBN 978-3-11-132991-8, e-ISBN (PDF) 978-3-11-133035-8

Werner Bonrath, Jonathan Medlock,
Marc-André Müller, Jan Schütz

Catalysis for Fine Chemicals

——

2nd, Revised and Extended Edition

DE GRUYTER

Authors

Werner Bonrath
DSM-Firmenich
P.O. Box 2676
CH-4002 Basel
Switzerland
werner.bonrath@dsm-firmenich.com

Jonathan Medlock
DSM-Firmenich
Switzerland
jonathan.medlock@dsm-firmenich.com

Marc-André Müller
Solvias AG
Switzerland
marc-andre.mueller@solvias.com

Jan Schütz
DSM-Firmenich
Switzerland
jan.schuetz@dsm-firmenich.com

ISBN 978-3-11-109609-4
e-ISBN (PDF) 978-3-11-110267-2
e-ISBN (EPUB) 978-3-11-110285-6

Library of Congress Control Number: 2024931480

Bibliographic information published by the Deutsche Nationalbibliothek
The Deutsche Nationalbibliothek lists this publication in the Deutsche Nationalbibliografie;
detailed bibliographic data are available on the Internet at http://dnb.dnb.de.

© 2024 Walter de Gruyter GmbH, Berlin/Boston
Cover image: alex.pin/Adobe Stock
Typesetting: Integra Software Services Pvt. Ltd.
Printing and binding: CPI books GmbH, Leck

www.degruyter.com

For our children Anna, Charlotte, Edward, Elise, Jakob, Mia, Rebecca, Sophie and Uwe.

Preface to the second edition

The development in the field of industrial process chemistry and in particular industrial catalysis never stands still. The desire for continuous improvement is driven by the demand for more efficient processes producing high yields and selectivities in shorter times with lower energy consumption and reduced waste generation. In addition, new technologies and transformations become ripe for implementation and existing processes are improved upon or replaced. We have tried to capture some of these improvements, alternative processes and new technologies in this second edition.

In almost every chapter, we have included new examples of catalytic processes for the production of fine chemicals for example homogeneous and heterogeneous hydrogenation processes to aroma compounds and agrochemicals; the application of solid acids and bases to prepare key fine chemical building blocks and new bio-transformations to form pharmaceutical intermediates. Furthermore, due to growing interest in the "electrification of chemistry", using green electricity to improve the sustainability of chemical processes, we have included a completely new chapter on electrochemistry containing background information and example processes. This is certainly an area that will continue to expand in the coming years.

We hope you continue to share our enjoyment and fascination in the huge success industrial catalysis has brought modern life.

Werner Bonrath, Jonathan Medlock, Marc-André Müller and Jan Schütz
Basel, April 2024

https://doi.org/10.1515/9783111102672-202

Preface

Today, unbeknownst to most consumers, chemistry and, in particular, catalysis play a fundamental role in the efficient manufacture of all manner of products used in our daily life. Polymers to agrochemicals, nutritional ingredients to flavour and fragrance compounds, dyestuffs and adhesives to pharmaceuticals, all use catalytic processes on production scales ranging from several 100 kg to several 100,000 tons per annum (and even higher). With increasing demands for more environmentally friendly processes in larger volumes at cheaper prices, catalysts and catalytic processes are ideally placed to facilitate this change, now and in the future. New catalytic methodologies developed in both industry and academia are finding application on an industrial scale for the synthesis of (poly)functional molecules, and this will only increase in the future. This makes the field of catalysis not only a challenging field to work in but also an incredibly rewarding one.

Our goal in writing this book is to provide an overview of current catalytic processes that are applied on an industrial scale for the manufacture of fine chemicals. We hope it will highlight the many examples of excellent chemical processes that have been developed by numerous scientists throughout the world and inspire current and future students and practitioners of chemistry and chemical engineering to work in the field of catalysis. We start with a basic review of important catalytic concepts before moving on to give instructive examples of the key technologies currently applied. The following chapters group the examples into overarching themes or topics such as oxidation and hydrogenation reactions, bond forming and rearrangement reactions, and specific technologies such as acid–base and phase transfer catalysis, biotransformations and reactions in the gas phase. Where helpful, we have also attempted to provide background information to provide context to the discussed applications and also discuss mechanistic aspects where they provide additional insight.

The examples chosen are based on our experience of working in the field for a number of years and also work published by industrial scientists in journal articles and patents. It would be an almost impossible task to compile a comprehensive account of all industrial catalytic processes for fine chemicals, so we have restricted ourselves to a limited number of examples for each topic or transformation. We are sure that many other excellent examples could have been chosen, and we apologise in advance for any omissions that we have made. Where possible, we have attempted to provide cross-references between chapters when different catalytic steps are used in the production of the same compound, or even where one technology has superseded the use of another. Several of the examples chosen could have been discussed in multiple chapters, and we have made a decision where best to place them.

In preparing the manuscript, we have attempted to provide references to literature and patent publications so that the interested reader may find further, more specific details. Unfortunately, to avoid assisting (potential) competitors, the exact details of all industrialised catalytic process are not available in the open literature, so some

https://doi.org/10.1515/9783111102672-203

details given are restricted to pilot-scale or even laboratory-scale experiments. We have tried to ensure that all the examples given here are accurate; however, given the large amount of information condensed into a relatively small book, it is possible that some errors have been introduced. These are the responsibility of the authors, and we apologise in advance for them. We would welcome all feedback and errors that need correcting.

Finally, we would like to thank our numerous former and present colleagues for interesting challenges, fruitful discussions and developing innovative solutions. We especially thank our families for their support whilst we prepared this overview. We hope that you enjoy and are inspired by the field of industrial catalysis.

Werner Bonrath, Jonathan Medlock, Marc-André Müller and Jan Schütz
Basel, August 2020

Contents

1 Introduction and fundamental aspects

Today, the application of catalysts is fundamental for efficient synthetic methods and production processes. During the last decades in the fine chemicals industry, and partly in the pharmaceutical industry, many chemical conversions which require stoichiometric amounts of reagents have been replaced by catalytic processes. In the bulk chemical industry, the application of catalytic methods is already well established. This difference can be partly explained by the higher complexity of pharmaceuticals and fine chemicals which makes catalysis more demanding and process development more expensive and complex.

In this chapter, we review some of the general concepts, techniques and principles that are important for the discussion of the examples in the later chapters. It is not supposed to be a comprehensive introduction to these topics, rather a refresher/reminder of important points to remember. For more detailed information and discussion of the points in this chapter, please refer to a standard textbook on catalysis, for example, [1–6].

The definition of the term *catalyst* is based on the fundamental work of Wilhelm Ostwald (Nobel Prize, 1909).

> *A catalyst is a compound which influences the reaction rate of a chemical reaction without being consumed and without the ability to influence the thermodynamic equilibrium of the catalysed reaction [7–12].*

Catalysts influence reactions that are thermodynamically possible (see also *Hess'* law); the function of a catalyst is to decrease the activation energy (E_A) of the reaction by providing an alternative pathway or mechanism for the reaction to follow (Figure 1.1).

Wilhelm Ostwald had a tremendous influence on the field of physical chemistry and catalytic concepts. His work was world-leading for decades and resulted in various industrial applications. For instance, the Ostwald process for nitric acid later became the basis of the work of Haber, Bosch and Mittasch for the ammonia process from hydrogen and nitrogen. Nowadays, catalysts have significant impact on industrial production processes. It is assumed that 90% of all chemically obtained industrial goods come in contact with at least one catalytical production step during their manufacture [13].

The main targets of catalytic reactions are:
1) Mild reaction conditions,
2) Decreased waste formation,
3) Increased selectivity,
4) Increased efficiency.

From a physical chemistry point of view, catalytic reactions can be divided into homogeneous and heterogeneous reactions. In heterogeneous catalysis, the catalyst and the

https://doi.org/10.1515/9783111102672-001

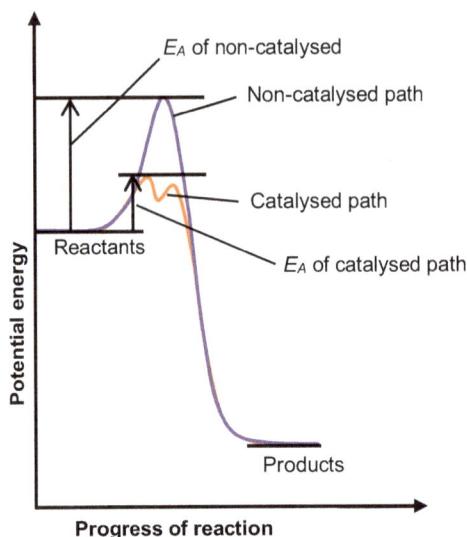

Figure 1.1: Energy–reaction coordinate of catalysed and non-catalysed reactions.

substrate are in separate phases, for example liquid/solid, solid/gas. In homogeneous catalysis, the catalyst and substrate are in one phase. The general advantages and disadvantages of homo- and heterogeneous catalysed reactions are summarised in Table 1.1. In particular, because of their easy separation and recovery, heterogeneous catalysts are more frequently found with increasing product volumes and decreasing product prices. In contrast, for asymmetric reactions, homogeneous catalysts are clearly superior to heterogeneous systems.

Table 1.1: Advantages and disadvantages of heterogeneous and homogeneous catalysed reactions.

	Heterogeneous	Homogeneous
Activity	Variable	High
Selectivity	Variable	High
Reaction conditions	Harsh	Mild
Sensitivity to impurities/deactivation	Low	High
Diffusion problems	Possible	None
Catalyst separation/recycling	Easy	Difficult/expensive
Active site	Poorly defined	Well-defined
Heat transfer	Can be problematic	Easy

1.1 General concepts of catalysis

1.1.1 Basics of thermodynamics and kinetics for chemical reactions

For the comprehension of catalysed chemical reactions, an understanding of thermodynamics and kinetics is essential. The Gibbs free energy (G) is a thermodynamic state function that is used to describe the driving force of a chemical reaction. It is defined as:

$$\Delta G^0 = \Delta H^0 - T\Delta S^0$$

ΔG^0 = Gibbs free energy; ΔH^0 = Reaction enthalpy; ΔS^0 = Entropy;
T = Absolute temperature (K)

The reaction enthalpy (ΔH) is the energy that is absorbed or released during the course of the reaction. *Hess'* law states that the change of enthalpy in a chemical reaction (i.e. the heat of reaction at constant pressure) is independent of the pathway between the initial and final states [14]. This means that for each reaction, the change of standard enthalpy is identical in the catalysed and non-catalysed reactions.

In other words, if a chemical change takes place by several different pathways, the overall enthalpy change is the same, regardless of the route by which the chemical change occurs (provided the initial and final conditions are the same). From this point of view, a catalyst does not influence the equilibrium of a reaction, but only the kinetics of the reaction.

Hess' law allows the enthalpy change (ΔH) for a reaction to be calculated even when it cannot be measured directly. This is accomplished by performing basic algebraic operations based on the chemical equations of the reactions using previously determined values for the enthalpies of formation:

$$\Delta H^0 = \sum \Delta H^0_{f(products)} - \sum \Delta H^0_{f(reactants)}$$

For as simple reaction such as the hydrogenation of ethene, it is necessary to formally break a C=C bond and an H–H bond. In contrast, two C–H bonds and a C–C bond are formed (Scheme 1.1). An amount of energy of 1,038 kJ/mol (602 + 436 kJ/mol) is required for bond breaking and −1,172 kJ/mol (2x − 413 + (−346 kJ/mol)) is released through the bond formations. So, the enthalpy of the reaction is −134 kJ/mol (1,038 kJ/mol + (−1,172 kJ/mol)).

If the net enthalpy change is negative ($\Delta H_{net} < 0$), the reaction is exothermic (e.g. hydrogenation of ethene) and is more likely to be spontaneous; positive ΔH values correspond to endothermic reactions (Figure 1.2).

The concepts of *Hess'* law can be expanded to include changes in entropy and Gibbs free energy. The *Bordwell* thermodynamic cycle is an example of such an extension which takes advantage of easily measured equilibria and redox potentials to de-

Bond	Energy [kJ/mol]
C=C	602
C−C	346
C−H	413
H−H	436

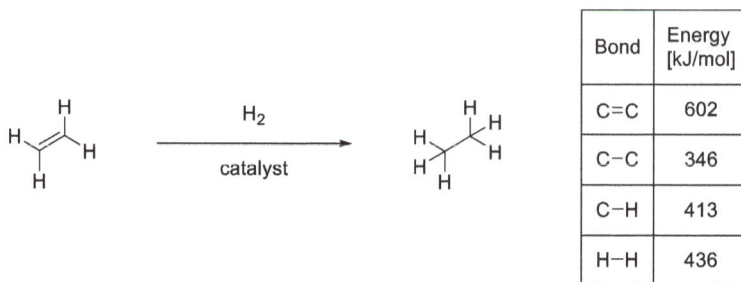

Scheme 1.1: Hydrogenation of ethene to ethane.

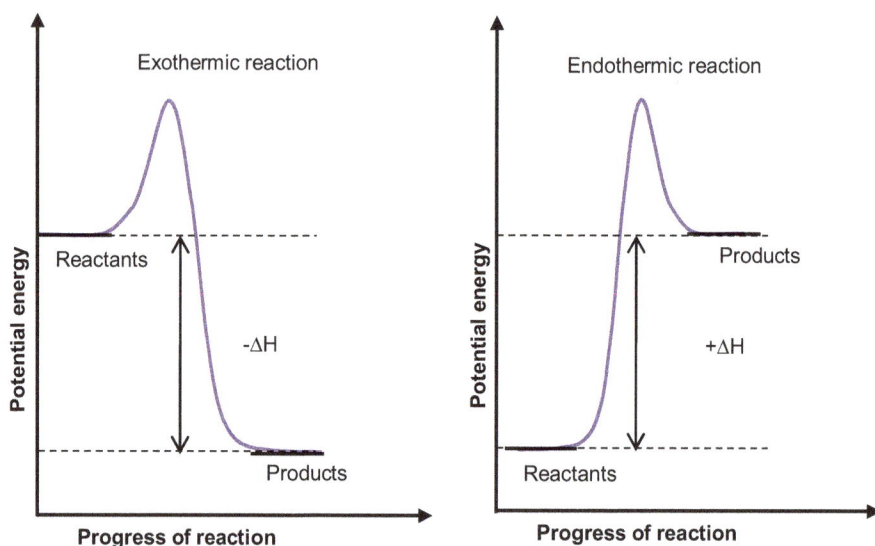

Figure 1.2: Difference between an endothermic and exothermic reaction.

termine experimentally inaccessible Gibbs free energy values. Combining ΔG^0 values from *Bordwell* thermodynamic cycles and ΔH^0 values found with *Hess'* law can be helpful in determining entropy values which are not measured directly, and therefore must be calculated through alternative routes.

For free energy:

$$\Delta G^0_{reaction} = \sum \Delta G^0_{f(products)} - \sum \Delta G^0_{f(reactants)}$$

If $\Delta G^0 < 0$, the reaction is favourable or spontaneous whereas for $\Delta G^0 > 0$, the reaction is unfavourable or non-spontaneous.

Entropy also plays a key role in determining spontaneity. Chemical reactions that are exothermic and result in an increased disorder are always spontaneous, whereas endothermic reactions that result in an increase of order are never spontaneous.

Some reactions with a positive enthalpy change are nevertheless spontaneous, for example dissolving ammonium nitrate in water which consumes heat from its surrounding. The change in enthalpy is positive, but the change in entropy is large enough to make the Gibbs free energy negative and the reaction spontaneous.

The entropy (S) is a measurement for the state of disorder of a system. The better a system is organised, the lower is the entropy; in nature there is an inherent tendency towards disorder. In a closed system, entropy never decreases. Because entropy can be measured as an absolute value, not relative to those of the elements in their reference states (as with ΔH^0 and ΔG^0), there is no need to use the entropy of formation; one simply uses the absolute entropies for the products and reactants:

$$\sum S^0_{\text{reaction}} = \sum S^0_{(\text{products})} - \sum S^0_{(\text{reactants})}$$

The picture drawn of a chemical reaction until now has been rather simplified. The energy profiles depicted so far only show reactions with one transition state (e.g. S_N2 reaction). However, reactions involving multiple steps are more usual (Figure 1.3). This is especially true for catalysed reactions, which proceed via more transition states and intermediates due to the interaction of the catalyst with the reactants. This factor makes the understanding of the reaction mechanism for catalysed reactions more challenging.

The transition state is defined by the IUPAC (International Union of Pure and Applied Chemistry) as follows:

> In the formalism of "transition state theory" the transition state of an elementary reactions is that set of states (each characterised by its own geometry and energy) in which an assembly of atoms, when randomly placed there, would have an equal probability of forming the reactants or of forming the products of that elementary reactions. [15]

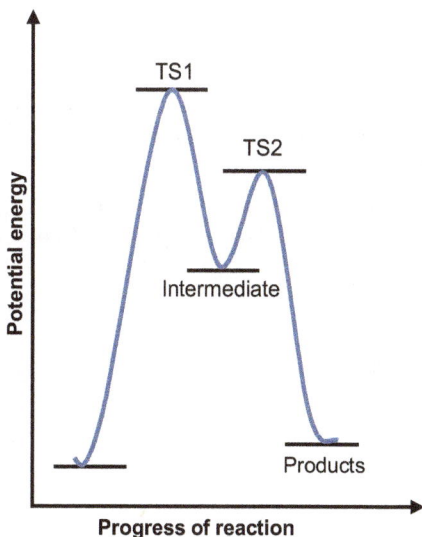

Figure 1.3: Reaction profile with two transition states (TS1 and TS2).

In a chemical reaction involving more elementary steps, there are more than one energy barriers and consequently multiple transition states. The step involving the highest barrier in going to the transition state is the rate-determining step of the reaction ($\Delta G_1^{\ddagger}$, Figure 1.4).

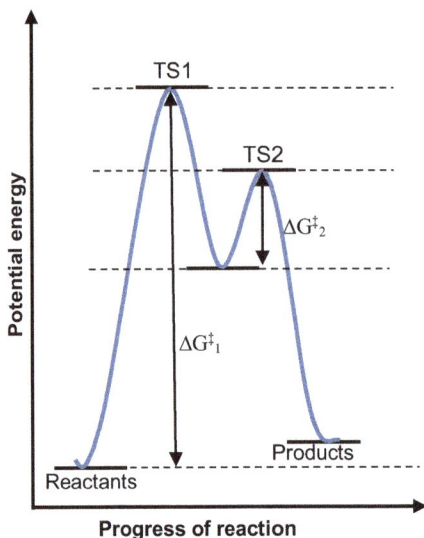

Figure 1.4: Energy profile of a reaction involving two elementary steps with $\Delta G_1^{\ddagger}$ being the rate determining step.

The understanding of chemical reactions and processes, especially reactions influenced by a catalyst, requires the discussion of other fundamental physical principles. The *Arrhenius* equation for monomolecular reactions [16] describes the dependency of the reaction rate k on the temperature T:

$$k = A \cdot e^{-\frac{E_A}{R \cdot T}}$$

where E_A is the activation energy (kJ/mol[1]), R is the gas constant (8.314 J $\cdot$ K^{-1}/mol), T is the absolute temperature (K), A is the frequency factor. The *Arrhenius* equation is also used to describe diffusion phenomena in solid-state systems.

The activation energy (E_A) of a reaction can be determined by measuring two reaction rates at two different temperatures for the same reaction:

(1) $\quad k_1 = A \cdot e^{-\frac{E_A}{R \cdot T_1}}$

(2) $\quad k_2 = A \cdot e^{-\frac{E_A}{R \cdot T_2}}$

After applying the natural logarithm (ln) to both sides:

(1') $\quad \ln(k_1) = \ln(A) - \frac{E_A}{R T_1}$,

(2') $\quad \ln(k_2) = \ln(A) - \frac{E_A}{R T_2}$,

Subtracting $\ln(k_1)$ from $\ln(k_2)$ results in:

$$\ln(k_2) - \ln(k_1) = -\frac{E_A}{R \cdot T_2} + \frac{E_A}{R \cdot T_1}$$

Regrouping and rearranging to give the activation energy E_A, followed by multiplication with R and division $(1/T_1-1/T_2)$ results in:

$$E_A = R \cdot \frac{\ln\left(\frac{k_2}{k_1}\right)}{\frac{1}{T_1} - \frac{1}{T_2}} = R \cdot \ln\left(\frac{k_2}{k_1}\right) \cdot \frac{T_1 \cdot T_2}{T_2 - T_1}$$

The rate of a chemical reaction depends on the concentration of the compounds in the system. If there is a catalytic reaction, and the reaction cycle involves several elementary steps, the rate-determining step is the slowest step in the reaction.

The *Eyring–Polanyi* equation is used in chemical kinetics to describe how the rate of a chemical reaction varies with temperature. It was developed almost simultaneously in 1935 by Henry Eyring, Meredith Gwynne Evans and Michael Polanyi. This equation is derived from transition state theory (also known as activated complex theory) and is equivalent to the empirical *Arrhenius* equation, both readily derived from statistical thermodynamics in the kinetic theory of gases [17–19].

The general form of the *Eyring–Polanyi* equation is similar to the *Arrhenius* equation:

$$k = \frac{\kappa\, k_B T}{h} e^{-\frac{\Delta G^{\ddagger}}{R \cdot T}}$$

where $\Delta G^{\ddagger}$ is the Gibbs energy of activation, κ = transmission factor, k = reaction rate constant, T = absolute temperature, R = gas constant, k_B = *Boltzmann* constant, h = *Planck*'s constant.

The transmission factor (κ) describes what proportion of the transition state molecules proceed to the product. This is usually assumed to equal to one, which means that all molecules at the transition state continue to form the product.

The free energy of activation ($\Delta G^{\ddagger}$) (available thermal energy) at 25 °C is 20 kcal/mol (84 kJ/mol). This means that reactions with similar or lower activation energy take place at room temperature, whereas reactions with higher activation energy often require heating.

The fundamental parameter, when looking at reaction kinetics, is the rate of reaction which defines the amount of compound that is converted per unit time and volume (mol/s × m^3). Considering the reaction kinetics for a simple bimolecular reaction with a rate constant k, this could be a dimerisation reaction where two molecules of the same species react together (eq. (1.1)), or a reaction where two different molecules react together (eq. (1.2)), for example an alkene chlorination or a hydrogenation reaction. The reaction rates are second order; that is, they depend on the concentration of both reacting species.

$$2A \rightarrow P \tag{1.1}$$

$$A + B \rightarrow P \tag{1.2}$$

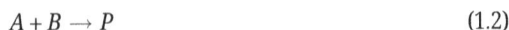

If we consider eq. (1.1), the rate of reaction v is:

$$v = -\frac{1}{2}\frac{d[A]}{dt} = \frac{d[P]}{dt} = k[A]^2$$

with ½ being the stoichiometric factor of the reaction.

If we consider eq. (1.2), the rate of reaction v is:

$$v = -\frac{d[A]}{dt} = -\frac{d[B]}{dt} = \frac{d[P]}{dt} = k[A][B]$$

This equation applies to most bimolecular reactions. However, if one reactant is present in a huge excess (e.g. B), for example if the solvent is also a reactant, the reaction kinetics is pseudo-first order with respect to the minor component (e.g. A) and the reaction rate simplifies to:

$$v = -\frac{d[A]}{dt} = \frac{d[P]}{dt} = k'[A]$$

1.1.1.1 *Michaelis–Menten* kinetics

In homogeneous catalysis, where the substrate interacts with a catalyst, we have a special case of reaction kinetics. For example, the two-step conversion of a substrate into a product involving the reversible binding of the substrate (S) to a catalyst (Cat), forming a catalyst–substrate complex ($Cat–S$); this then reacts irreversibly to form the product (P) and the catalyst, which can continue in the catalytic cycle. This is known as *Michaelis–Menten* kinetics (Scheme 1.2) [20] and was first applied for enzyme-catalysed reactions [21].

$$Cat + S \underset{k_{-1}}{\overset{k_1}{\rightleftharpoons}} Cat\text{-}S \overset{k_2}{\longrightarrow} Cat + P$$

Scheme 1.2: Model reaction to explain *Michaelis–Menten* kinetics.

The rate of formation of the product depends on the concentration of the catalyst–substrate complex and the rate constant k_2:

$$v = \frac{d[P]}{dt} = k_2[Cat - S]$$

The rate of formation of the catalyst–substrate complex is dependent upon the rate constants k_1 and k_{-1} in the initial equilibrium of catalyst, substrate and complex:

$$v_1 = \frac{d[Cat - S]}{dt} = v_{+1} - v_{-1} = k_1[Cat][S] - k_{-1}[Cat - S]$$

However, the total amount of catalyst in the system is constant and identical to the initial catalyst concentration ($[Cat]_0$), so:

$$[Cat]_0 = [Cat] + [Cat - S]$$

The quasi-steady-state approximation is made with the assumption that the concentration of the catalyst species is small compared to the amount of starting materials and products ($[S]$ and $[P] \gg [Cat]$ and $[Cat–S]$), and their change in concentration in negligible:

$$\frac{d[Cat - S]}{dt} = -\frac{d[Cat]}{dt} = 0 = k_1[Cat][S] - k_{-1}[Cat - S] - k_2[Cat - S]$$

Therefore, substituting $[Cat]$ using the formula above:

$$k_1[Cat][S] = k_1[S]([Cat]_0 - [Cat - S]) = [Cat - S](k_{-1} + k_2)$$

Rearranging gives:

$$[Cat - S] = \frac{[Cat]_0[S]}{\left(\frac{k_{-1}+k_2}{k_1}\right) + [S]}$$

Returning to the original equation for the formation of the product, this gives us the *Michaelis–Menten* rate expression:

$$v = \frac{d[P]}{dt} = k_2[Cat - S] = \frac{k_2[Cat]_0[S]}{\left(\frac{k_{-1}+k_2}{k_1}\right) + [S]} = v_{max}\frac{[S]}{K_M + [S]}$$

where K_M is known as the *Michaelis* constant.

1.1.2 General principles of heterogeneous catalysis

In heterogeneous reactions, the catalyst and reactants are in two or more different phases (e.g. solid and liquid or gas phase); the reactants need to be transported to the catalyst, before they can react. Mass transfer limitations between these phases may significantly influence the rate of reactions. Mass transfer effects are usually discussed in systems with a solid catalyst and a liquid or gaseous reactant. Depending on the system, the transport to the outer surface of a catalyst particle and, in certain cases, also the transport inside the particle have to be considered.

The concept of mass transport effects in heterogeneous catalysis is based on Ostwald's ideas, later extended by Walther Nernst, and can be summarised as follows [22]:

1. Diffusion of the reactants through a thin layer at the interface of the solid and fluid phases (film diffusion)
2. Diffusion of the reactants through the pores of the catalyst particle (pore diffusion)
3. Adsorption of the reactants at the inner surface of the pores
4. Chemical reaction at the specific active sites on the catalyst surface
5. Desorption of the products from the inner surface of the pores
6. Diffusion of the products through the pores of the catalyst particle (pore diffusion)
7. Diffusion of the products through a thin layer at the interface of solid and fluid phase into the bulk phase (film diffusion)

Three diffusion mechanisms for liquid or gaseous transport in porous media can be observed:

1. When the mean free path λ (*Knudsen* diffusion) of the diffusing molecule is much less than the average pore diameter d_{pore}, molecule–molecule collisions dominate and bulk diffusion occurs ($d_{pore} \gg \lambda$).
2. When molecule–pore wall interactions dominate, then $d_{pore} < \lambda$. In microporous solids (e.g. zeolites), the pore diameter is close to the size of the reactant molecule, the latter can diffuse within the pores only by remaining constantly in contact with the pore walls (surface diffusion).
3. When the mass transport in or on the catalyst is determined by diffusion the following phenomena occur:
    a. If the pore volume is large compared to the mean free path, normal diffusion effects are dominant, *Fick*'s first law of diffusion dominates.
    b. In cases of small pores (compared to mean free path), the *Knudsen* diffusion predominates. The interaction of the molecules with the wall is much higher than their interaction with each other.

Walther Nernst (winner of the 1920 Nobel Prize in Chemistry) significantly influenced physical chemistry with his work on electrochemistry, dissociation and thermodynamics. In 1905, he formulated the *Nernst* theorem, now known as the third law of thermodynamics (the entropy of a system approaches a constant value at absolute zero), which means that absolute zero cannot be reached. Walther Nernst enlarged the view of chemical reactions, with his work on kinetic reaction rates, which was fundamental for an understanding of diffusion phenomena.

In chemical reactions, the correlation between the equilibrium constant and temperature at constant pressure is described by the *van't Hoff* equation:

$$\left(\frac{d\ln k}{dT}\right)_p = \frac{\Delta H_m^o(T)}{RT^2}$$

where k = equilibrium constant, T = temperature, $\Delta H_m^0(T)$ = standard enthalpy (molar) at the particular temperature T, R = gas constant.

Assuming that the reaction enthalpy, H_m^o, is constant at different temperatures:

$$\left(\frac{d\ln k}{d\frac{1}{T}}\right)_p = -\frac{\Delta H_m^o}{R}$$

This equation was introduced by Jacobus Henricus van't Hoff (Nobel Prize, 1901). The *van't Hoff* equation results in the Q_{10} temperature coefficient (Q_{10}). It is useful to describe the dependency of the reaction rate on the temperature in biological, chemical and ecological reaction systems, and was introduced in 1884. In the following years, the work of van't Hoff was the foundation of the *Arrhenius* equation which was introduced in 1889.

$$Q_{10} = \left(\frac{k_2}{k_1}\right)^{\frac{10\,K}{T_2 - T_1}}$$

1.1.3 Mechanisms and kinetics on catalyst surfaces

1.1.3.1 Concept of active sites
The concept of active sites was introduced by H. S. Taylor in the 1920s [23]. Taylor worked on calorimetry and developed the concept that catalysed chemical reactions are not catalysed over the whole surface of the catalyst; the reactions are only catalysed at certain active sites (Figure 1.5) [24]. In his theory, the concentration of active sites is rate-determining for catalytic reaction steps, rather than the total concentration of available surface sites.

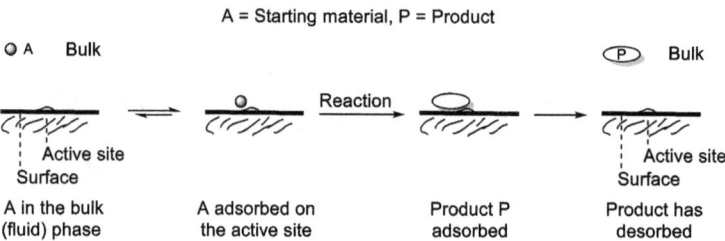

Figure 1.5: Concept of active site.

The fundamental concept of active sites was further developed by several scientists. The concept now includes the structure of active sites and the electronic environment to explain the catalyst performance [25–30]. These researchers recognised that the surface structure of the catalyst and the surface orbitals have a fundamental role in the catalyst activity (the electronic theory of active sites).

1.1.3.2 Langmuir–Hinshelwood

The adsorption of particles on a surface was described by I. Langmuir (Nobel Prize in Chemistry, 1932). The so-called *Langmuir* isotherm allows the calculation of the equilibrium concentration and surface coverage for a species. The following assumptions are made: adsorption takes place in a monomolecular layer on a homogeneous surface, the adsorption is consistent in all places and there is no interaction between the adsorbed species [31].

Langmuir then went on to develop a model for a bi-molecular reaction taking place on a catalyst surface that was later refined by C. Hinshelwood. This theory can be used to explain reactions for which the absorption of all reactants on the surface is required before a reaction may occur. The catalytic reaction taking place on the surface is divided into a number of individual steps: adsorption, diffusion on the surface, reaction and desorption (Figure 1.6).

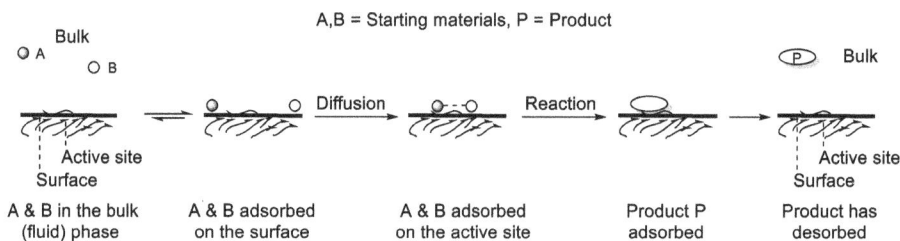

Figure 1.6: *Langmuir–Hinshelwood* mechanism.

In the first step, the starting materials A and B are adsorbed on the surface from the gas phase.

$$A_g \rightarrow A_{ads}$$

$$B_g \rightarrow B_{ads}$$

The adsorbed compounds A and B undergo diffusion on the surface and then react to form the product P, which then undergoes desorption to the gas phase.

$$A_{ads} + B_{ads} \rightarrow P_{ads}$$

$$P_{ads} \rightarrow P_g$$

The rate constants k_1, k_{-1}, k_2 and k_{-2} stand for the adsorption and desorption of compound A, B and k for the reaction. The rate law is:

$$r = k\theta_A\theta_B C_S^2$$

where C_S is the total number of active sites, θ is the surface coverage of A and B, r is the reaction rate.

Assuming steady-state conditions and that the rate-limiting step is the reaction of A and B:

$$\frac{dc}{dt} = 0$$

$$\theta_A + \theta_B + \theta_E = 1, \text{ and } \theta_A = K_1 C_A \theta_E$$

$$r = k[A]^2 \frac{K_1 K_2[A][B]}{(1 + K_1[A] + K_2[B])^2}$$

where dc is the change of concentration of all compounds (A, B and P), θ_E is the fraction of empty sites, K_1 and K_2 are adsorption constants.

This rate expression is rather complicated, but can be simplified by considering different situations depending on how strongly the reactants bind to the surface and also their relative concentrations.

If both compounds are very weakly adsorbed:

$$1 \gg K_1[A], \ K_2[B]$$

$$r = kC_S^2 K_1 K_2[A][B]$$

The reaction order is 1 for A and B, in total 2.

The following equation is valid for *the low adsorption of one compound:*

$$K_1[A] \gg K_2[B]$$

$$r = kC_S^2 \frac{K_1 K_2[A][B]}{(1 + K_1[A])^2}$$

The reaction order is 1 with respect to B. Considering the reaction order with respect to A, there are two extreme possibilities. At low concentrations of compound A, the equation above becomes:

$$r = kC_S^2 K_1 K_2[A][B]$$

The reaction order is 1 with respect to A. However, if there are high concentrations of A, the rate expression simplifies to:

$$r = kC_S^2 \frac{K_2[B]}{K_1[A]}$$

Therefore, the reaction order is −1 with respect to A.

And if one compound is available in high concentrations, for example A, then:
The reaction order is 1 with respect to B and −1 with respect to A:

$$K_1[A] \gg K_2[B]$$

$$r = kC_S^2 \frac{K_2[B]}{K_1[A]}$$

1.1.3.3 Eley–Rideal

The *Eley–Rideal* mechanism was introduced in the 1930s and follows the concept of the adsorption of compound A from the gas phase on the surface and reaction with a compound B from the gas phase (without being adsorbed) to form product P on the surface. Product P undergoes desorption from the surface into the gas phase (Figure 1.7).

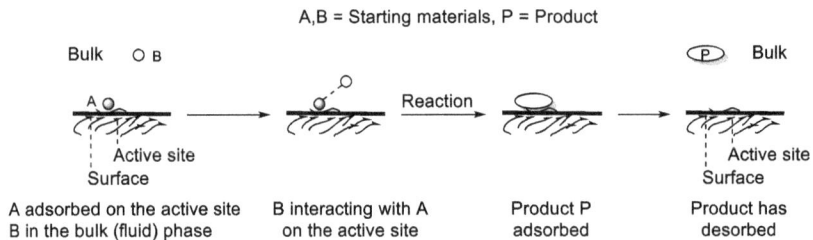

Figure 1.7: *Eley–Rideal* mechanism.

The steps in the *Eley–Rideal* mechanism are:

$$A_g \rightarrow A_{ads}$$

$$A_{ads} + B_g \rightarrow P_{ads}$$

$$P_{ads} \rightarrow P_g$$

With $r = kC_S\theta_A[B]$ and assuming steady-state conditions (of A bound to the catalyst surface):

$$r = \frac{K_1[A]}{K_1[A] + 1} kC_S[B]$$

The reaction order is 1 with respect to B. To determine the reaction order with respect to A, we again consider two extreme conditions. At *high concentrations of A*:
$r = kC_S[B]$ and the reaction order is 0 with respect to A,
and *low concentrations of A*:
$r = kC_S[B]K_1[A]$ and the reaction order is 1 with respect to A.

The *Langmuir–Hinshelwood* mechanism is a good model for reactions such as the Pt-catalysed oxidation of CO to CO_2, the formation of methanol (CH_3OH) from CO and H_2 and some hydrogenation reactions. The *Eley–Rideal* mechanism is a good model for reactions such as the Pt-catalysed oxidation of ammonia (NH_3) to N_2 and water (H_2O), the oxidation of ethylene (C_2H_4) to ethylene oxide and some hydrogenation reaction (e.g. acetylene (C_2H_2) to ethylene (C_2H_4) on nickel or iron catalysts).

The interpretation of the kinetics of heterogeneous catalysed reactions by the *Langmuir–Hinshelwood* or *Eley–Rideal* mechanisms is based on the assumed homogeneity of the surface and induced surface interactions. We must consider that a real or working catalyst, due to its heterogeneity, is significantly more complicated than the above-mentioned models.

1.1.3.4 *Mars–van Krevelen*

In oxidation reactions, another mechanism is often discussed which assumes that the oxygen used in the oxidation reaction comes from the lattice of the catalyst. The formed vacancy is replenished in a later step by oxygen from the gas phase [32]. The kinetic equations of the *Mars–van Krevelen* mechanism are as follows:

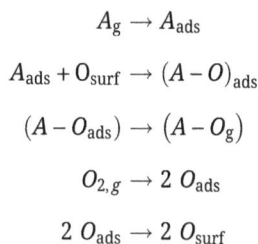

$$A_g \rightarrow A_{ads}$$

$$A_{ads} + O_{surf} \rightarrow (A - O)_{ads}$$

$$(A - O_{ads}) \rightarrow (A - O_g)$$

$$O_{2,g} \rightarrow 2\, O_{ads}$$

$$2\, O_{ads} \rightarrow 2\, O_{surf}$$

An example of a reaction following the *Mars–van Krevelen* mechanism is the formation of propene by oxidative propane dehydration reactions on vanadium catalysts. The *Mars–van Krevelen* mechanism has the feature that lattice components are involved in the product formation, resulting in a universal character of this mechanism [33].

1.1.4 Characterisation of solid catalysts

In developing and improving catalysts and catalytic processes for the production of fine chemicals, an understanding of the catalyst, how it operates and the factors that affect performance is a huge benefit, but can also be a very difficult undertaking. Catalyst characterisation can also help to understand the mechanism of the reaction and how the substrate and product interact with the catalyst. A wide variety of techniques are available, and usually no single technique provides all the answers; a combination of techniques is required. The ideal situation would be to investigate each key step in the catalytic cycle *in operando* (under working conditions); however, this can be very difficult to achieve. Adapting the process to be compatible with the requirements of the measuring instrument can result in the data obtained not being completely representative of the actual catalyst and process. Nevertheless, this is the "holy grail" of catalyst characterisation.

Due to these difficulties, the majority of catalyst characterisation, especially of heterogeneous catalysts, takes place *ex situ* with both fresh and used catalysts. A range of techniques are available which can provide significant insight (including structure–activity relationships) and permit improvements in both catalysts and processes. An introduction to several important methods for the characterisation of solid catalysts is described below and for more detailed information, suitable books and review articles should be consulted [34–37].

1.1.4.1 Catalyst surface area and pore size and distribution

In the field of solid catalysts, the surface area and pore size/distribution of the catalyst are important factors in catalytic activity. These properties are usually measured by physisorption techniques which involve the non-dissociative interactions between the catalyst surface and various probe molecules, which do not chemically react with the surface. Typically, nitrogen is used at its boiling point (77 K) to measure the surface area and pore volume; alternative probe molecules include argon, carbon dioxide and water. Mercury can be used to measure pore sizes and distribution. Inorganic materials such as mesoporous silica and layered clay minerals have high surface areas of several hundred m^2/g, indicating the possibility of their application as efficient catalytic materials.

In the measurement of porosity, a known amount of sample is pre-treated to remove adsorbed water and air that might be in the pores of the material. Then a known amount of probe material is dosed into the vessel containing the sample and the amount of probe adsorbed is calculated from the pressure change after equilibrium has been reached. The values are usually plotted as an adsorption isotherm of the amount adsorbed probe against relative pressure (p/p_o, where p_o is the saturation pressure of the probe molecule at the measured temperature). At a first approximation, a *Langmuir* isotherm is formed where the surface is saturated with a monolayer of the probe material. However, this is an over-simplification and does not provide the correct value for the monolayer capacity. Therefore, a modification of the *Langmuir* theory is used, the *Brunauer–Emmett–Teller* (BET) theory.

BET theory explains the physical adsorption of gas molecules on a solid surface. It was first described in 1938 in *the Journal of the American Chemical Society* by Stephen Brunauer, Paul Hugh Emmett and Edward Teller [38, 39]. It is widely used in surface science for the calculation of surface areas of solids by physical adsorption of gas molecules. The BET theory considers both monolayer and multilayer adsorption of the probe molecule on the surface, with the following assumptions.

1. Gas molecules physically adsorb on a solid in layers.
2. Gas molecules only interact with adjacent layers.
3. The *Langmuir* theory can be applied to each layer.
4. The gas molecules in the first layer interact with the enthalpy of adsorption.
5. The gas molecules in further layers interact with the enthalpy of condensation.

The BET equation is an approximation and only fits part of the observed isotherm curve because of pore condensation. Usually, values in the range of $0.05 < p/p_0 < 0.35$ are used. The total surface area, pore volume and pore diameter can be calculated.

1.1.4.2 Size and shape of the catalyst particle and active phase

The size of a catalyst particle can be measured with a variety of techniques that include optical microscopy, light scattering or laser diffraction, usually using a suspension of the particles in an inert medium. In both of the latter techniques, the particle-size distribution is calculated using an optical model based on the observed scattering, which can yield variable results depending on the catalyst particle shape. These methods are suitable to measure sizes down to the sub-micrometre range.

For doped materials, more interesting and usually more relevant for catalyst performance is the size and shape of the active phase of a catalyst, such as a metal deposited on a carrier. These particles are usually significantly smaller, in the range of nanometres so other techniques are required, such as electron microscopy and chemisorption.

In electron microscopy, a beam of accelerated electrons is used to illuminate the sample of interest. Since the wavelength of an electron is significantly smaller than that of light, a significantly higher resolution image is achieved. Transmission electron microscopy (TEM) is generally used to study the bulk material of the catalyst whereas scanning electron microscopy (SEM) is mainly used to characterise the surface of the material. SEM can also be combined with energy dispersive X-ray spectroscopy (EDX) to determine the chemical composition of the catalyst surface and investigate the atoms present. This technique can also be used to "map the surface" of the material and determine regions of the solid with different compositions.

Chemisorption uses pulses of probe gases in equipment similar to that used for physisorption and is typically used to investigate the active sites of a catalyst. The quantity of adsorbed gas can be determined and then linked to the (relative) catalyst performance. For example, hydrogen or carbon monoxide is used to investigate the active phase of a supported metal catalyst. It provides quantitative information about the metal surface area, metal dispersion and metal crystallite size. Alternatively, other probe molecules such as ammonia can also be used to investigate the number and strength of acidic sites on a catalyst.

1.1.4.3 Chemical composition (and structure)

The bulk chemical composition of a catalyst can be measured by traditional techniques such as inductively coupled plasma (ICP)-optical emission spectrometry and ICP-mass spectrometry to determine contents down to the ppm level. This is a destructive technique that requires the entire sample to be dissolved/digested to form a homogeneous solution before measurement.

To determine the composition of the catalyst surface, SEM-EDX can be used, as described earlier, but only in a semi-quantitative manner. In addition, X-ray photoelectron spectroscopy (XPS) can be used, where a beam of X-rays is used to irradiate the sample and then the number and energy of electrons emitted from the surface are measured. The accuracy range is in the parts-per-thousand level and XPS has the additional benefit of determining the oxidation state of the surface material. Because XPS only deals with the surface of the catalyst, it can be used to determine the dispersion of particles over a support material by looking at the intensity of the signals originating from both the support material and the doped particles.

X-ray diffraction (XRD) uses the elastic scattering of X-ray photons to determine the crystalline phases that are present in a catalyst and can also be used to estimate particle sizes (based on the width/shape of the diffraction peaks). XRD can also be used *in situ* in specially designed reactors to monitor solid-state reactions or reactions on solid surfaces such as oxidation and reduction. In homogeneous catalysis and biocatalysis, the structure of single crystals can be determined by XRD which can give significant information about the special geometry of the active site of the catalyst.

1.1.4.4 Electronic properties

The electronic properties of a catalyst, in particular the oxidation state of the active sites, can be important for understanding the mechanism of the reaction and the catalyst's performance. One technique to investigate this is XPS, as described earlier. Temperature-programmed techniques such as temperature-programmed reduction and temperature-programmed oxidation can be used to determine the oxidation state of the catalyst sample. In this technique, the catalyst sample is subjected to a flow of gas (usually H_2 or O_2) whilst undergoing a programmed temperature rise. These measurements allow the investigation of the reducibility/oxidisability of a catalyst formulation, which is of particular importance when deciding the calcination temperature of a catalyst sample.

All the above-mentioned techniques can be considered "standard techniques" that are widely available. In addition, there are a number of specialist techniques to give more detailed information about the catalyst. These include: extended X-ray absorption fine structure, X-ray absorptions near-edge spectroscopy, *Mössbauer* spectroscopy, ion spectroscopy, secondary ion mass spectrometry, low-energy ion scattering, atomic force microscopy and electron spin resonance. In addition, infrared, *Raman* and ultraviolet spectroscopy can be used in combination with probe molecules to investigate bindings to the catalyst surface, which is also possible *in situ/in operando*. A discussion of these techniques is beyond the scope of this chapter.

1.1.5 Terms and definitions

For an understanding chemical reactions, especially catalytic reactions, there are several important definitions and these are discussed further in the chapter.

1.1.5.1 Yield, conversion, selectivity, space time yield

The conversion (X_i) of a starting material is the ratio of consumed material to the initial amount of starting material [40, 41]:

$$X_i = \frac{n_{i0} - n_i}{n_{i0}}$$

where n_{i0} is the mol of the material at $t = 0$ (start of reaction) and n_i is the mol of the materials at a given reaction time (usually end of reaction) [42].

The yield of a reaction is defined as the amount of a specific product formed per mol of reactant consumed:

$$\text{Yield}(Y_{Pi}) = \frac{n_P v_i}{n_{i0} v_p}$$

where n_P is the mol of product P, and v_i, and v_P are stoichiometric coefficients of each species in the reaction.

The selectivity (S_{Pi}) is the total amount of reactants that have formed the product P divided by the total amount of reactant consumed:

$$S_{Pi} = \frac{n_P v_i}{(n_{i0} - n_i) v_p}$$

For a reactor system without the recycle of unconverted starting material, the selectivity is the ratio of yield and conversion, which gives the same final expression:

$$S = \frac{Y_{Pi}}{X_i} = \frac{n_P v_i}{(n_{i0} - n_i) v_p}$$

It must be stated that the selectivity is measured at a specific point of time. As the reaction proceeds (conversion increases), the selectivity can decrease whereas the yield can increase; a hypothetical example in shown in Figure 1.8. The yield at 100% conversion may be lower than at partial conversion; for some industrial processes the selectivity can be more important than the yield or conversion. In these cases, the process is conducted at partial conversion and the unreacted starting material is recycled; overall, this can result in a more economical process than running to complete conversion.

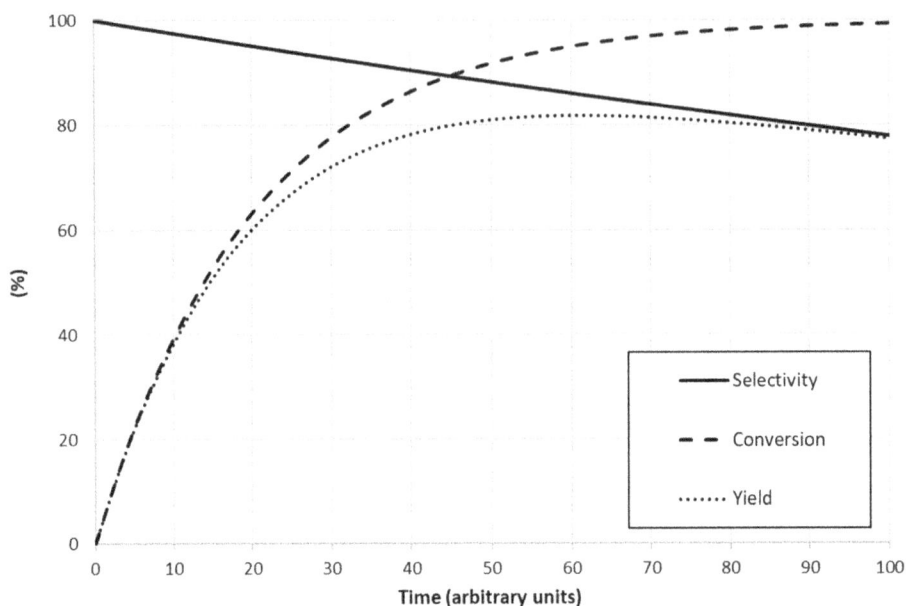

Figure 1.8: An example selectivity–conversion–yield chart for a hypothetical catalytic reaction.

Another important parameter from an economical point of view is the space time yield. It is the amount of product produced in a given time at a given reactor volume. The unit is mol l^{-1} s^{-1} or kg l^{-1} s^{-1}. This parameter determines the throughput of a process and will depend on the exact reaction conditions chosen.

1.1.5.2 Catalyst lifetime, turnover number, turnover frequency

As catalysts are a crucial part of an efficient industrial process, various parameters are available to define the catalyst performance. The catalyst lifetime is the length of time in which the catalyst operates at an acceptable activity or selectivity. The catalyst lifetime will depend on many parameters such as the reaction conditions and the purity of the feed material.

The turnover number (TON) is a number showing the catalyst performance:

$$TON = consumed \; mol \; starting \; material/mol \; catalyst$$

Especially, in homogeneous catalysis, the catalyst loading (s/c) is also commonly quoted. This is usually as a ratio of number of moles of starting material divided by the number of moles of catalyst. It is similar to but different from the TON, as one relates the amount of catalyst to the starting conditions (catalyst loading) and the other relates the catalyst amount to the actual catalyst performance, the amount of starting material consumed (TON).

The turnover frequency (TOF) describes the activity of a catalyst, and is defined as the TON per h:

$$TOF = TON\left(h^{-1}\right)$$

1.1.5.3 Fine chemicals, E-factor, atom economy, energy balance

Based on the fact that there exists no universal definition for fine chemicals, as well as for specialty chemicals or bulk chemicals, we follow the definition of R. Sheldon and H. v. Bekkum. Fine chemicals, according to Sheldon's classification, are compounds which are manufactured in production volumes of about 100 to 10,000 t/a (Table 1.2) [43–48]. An alternative definition is given by R. Pollak, which also uses other descriptors to differentiate between commodity, fine and specialty chemicals (Table 1.3) [49]. As can be seen, the definition is not exact and encompasses many different aspects.

Table 1.2: *Sheldon* classification of industry segments.

Industry segment	Production volume (t/a)	kg by-products/waste per kg product
Oil refining	10^6–10^8	<0.1
Bulk chemicals	10^4–10^6	<1–5
Fine chemicals	10^2–10^4	5–50
Pharmaceuticals	10^1–10^3	25–>100

Table 1.3: *Pollak's* definition of commodity, fine and specialty chemicals.

Commodity chemicals	Fine chemicals	Speciality chemicals
Single pure chemical substance	Single pure chemical substance	Mixtures
Produced in dedicated plants	Produced in multi-purpose plants	Formulated products
High volume/low price	Low volume (<1,000 t/a) and high price (>10 $/kg)	Undifferentiated
Many applications	Few applications	Undifferentiated
Sold on specification	Sold on specification "What they are"	Sold on performance "What they can do"

As can be seen in Table 1.2, Sheldon uses in his definition a measure of the amount of waste generated in producing a compound. This is known as the E-factor, and is defined as kg waste/kg product; it is a criterion for the efficiency of a process:

$$\text{E-factor} = \frac{\text{waste (kg)}}{\text{product (kg)}}$$

The E-factor encompasses significantly more information than the term *atom economy*, that calculates which atoms or mass of the starting material is found in the final product [50]:

$$\text{Atom economy} = \frac{\text{MW product}}{\sum_{i=1}^{n} \text{MWS}_i} \times 100\%$$

S_i = starting materials, MW = molecular weight

The focus in this book is, in general, on molecules with a price >10 \$/kg and a production volume (worldwide per annum) of around 10,000 t/a. To give a good overview of the range of industrial catalysis, also some processes that operate with higher or lower production volumes are discussed, because they produce multi-functional compounds. Furthermore, the trend during the last decade has been a general increase in the production volumes of most chemical products.

During the last two decades, the application of catalytic methods in the field of fine chemicals has increased, mainly due to the price pressure on those products. Other driving forces such as the reduction of waste, use of less toxic reagents and solvents, improvement of energy efficiency, recycle of catalysts and reagents and combination of unit operations to reduce costs and achieve more sustainable processes are in agreement with the 12 principles of "green chemistry" (Table 1.4) [51]. These principles can be seen as guidelines to maximise efficiency and minimise hazardous effects to human health and the environment.

Table 1.4: The 12 principles of green chemistry.

No.	Principle	Explanation
1	Prevention	Try to prevent waste to reduce purification and waste treatment.
2	Atom economy	All material used should end up in the final product.
3	Less hazardous chemical synthesis	Use and generate little or no toxic compounds.
4	Designing safer chemicals	Take the toxicity of a chemical into account already in the designing process.
5	Safer solvents and auxiliaries	Reduce solvents and auxiliaries and avoid if possible.
6	Design for energy efficiency	Use ambient temperature and pressure where possible.
7	Use of renewable feedstocks	Use feedstocks from renewables rather than those derived from fossil fuel whenever it is technically feasible.

Table 1.4 (continued)

No.	Principle	Explanation
8	Reduce derivatives	Avoid protection groups to reduce number of steps.
9	Catalysis	Avoid stoichiometric reactions.
10	Design for degradation	Design compounds to reduce toxic breakdown products.
11	Real-time analysis for pollution prevention	In-process monitoring to react prior to the formation of hazardous substances.
12	Inherently safer chemistry for accident prevention	Choose safer chemical processes to minimise chemical accidents, explosions, fires, *etc.*

Furthermore, unit operations, which means the number and type of unit operations and the size of the operation itself, need to be considered. All these factors influence not only the cost of a process, but are also responsible for the process efficiency and the energy balance. Choosing right operations and size of operation needs co-operation between engineers and chemists. In a more detailed view, the size of a unit operation affects the mass and energy transfer. Therefore, the prediction and determination of flow patterns, the stirrer speed and mixing are challenging for scale-up and will not be discussed in depth in this book. Nevertheless, these factors are also important for the energy balance, which can be calculated by:

$$WER = \frac{\sum waste\ energy}{\sum energy\ input}$$

WER = Waste energy ratio

In general, for processes, especially biotechnological processes, heat and heat transfer is a critical reaction parameter. Removal of heat is important in exothermic processes and usually a balance is sought where the produced heat is used to maintain the reaction temperature. If the rate of heat loss and applied cooling is not in balance with the generated heat then additional energy would be required. Other important roles for heat and heat transfer in biotechnological processes are procedures for sterilisation of the reaction vessel and the reaction media, and drying of the products.

For chemical and biotechnological processes, the energy balance is important to calculate economic aspects such as energy consumption or production during the process. Today, and in all future processes currently in development, energy balances are additionally important for the design of all types of equipment (columns, reactors, heat exchangers). Furthermore, it is important to design "green processes" with a reduced eco-footprint, for example neutral in emission of carbon dioxide equivalents by 2050 at the latest. Depending on the process, especially for processes where electricity is applied, for example electrolysis, arc processes or electrothermic processes, the source of the electrical energy is important, since this can have a significantly different environmental footprint. This field of industrial processing is a typical area where chemistry and biotechnology have a strong interaction with process engineering.

The energy balance can be expressed by:

$$Q_{met} + Q_{ag} + Q_{gas} = Q_{acc} + Q_{exch} + Q_{evap} + Q_{sen}$$

where Q_{met} is the rate of heat generation (per unit volume), Q_{ag} is the rate of heat generation during agitation (per unit volume), Q_{gas} is the rate of heat gassing, Q_{acc} is the rate of heat accumulation in the system, Q_{exch} is the rate of heat transfer per unit volume to the surroundings, Q_{evap} is the rate of heat loss by evaporation (per unit volume), Q_{sen} is the rate of sensible enthalpy gains by flow streams (exit – inlet, per unit volume).

In biotechnological processes, heat evaluation during metabolism is a consequence of the thermodynamics of microbial activity. During fermentation processes, the amount of heat produced is usually high, and if not removed it raises the temperature of the contents of the fermenter to a level beyond the optimum range for the system. In systems where the stoichiometry of the reaction is known, the rate of heat generation per unit volume due to the microbial metabolism (Q_{met}) can be calculated from the standard enthalpies of substrates, biomass and products:

$$Q_{met} = \sum_{l=1}^{m} biH_{p,i} - \sum_{j=1}^{n} ajH_{s,j}$$

where Q_{met} is the total heat generation, $H_{p,i}$ is the standard enthalpies of products, $H_{s,j}$ is the standard enthalpies of substrates, aj and bi are reaction specific variables.

1.1.5.4 Chirality, asymmetric synthesis and non-linear effects

The term *chirality* was first used by Lord Kelvin [52] and is defined by IUPAC as follows: "The geometric property of a rigid object (or spatial arrangement of points or atoms) of being non-superposable on its mirror image; such an object has no symmetry elements of the second kind (a mirror plane, $\sigma = S_1$, a centre of inversion, $i = S_2$, a rotation-reflection axis, S_{2n}). If the object is superposable on its mirror image the object is described as achiral" [53].

On a molecular level, chirality originates most frequently from carbon atoms having four different substituents directly attached. However, also other forms such as, axial chirality, planar chirality or inherent chirality are known. If a molecule contains one stereogenic centre such as a carbon atom with four different substituents, it occurs as a pair of enantiomers. These enantiomers have identical properties in an achiral environment; however, in a chiral environment such as the human body, enantiomers can show different effects. For example, (R)-penicillamine is highly toxic whereas (S)-penicillamine can be used for the treatment of Wilson's disease a genetic disorder in which copper builds up in the body (Figure 1.9) [54].

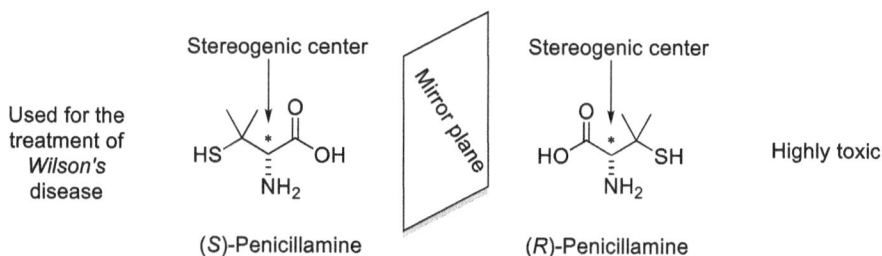

Figure 1.9: The two enantiomers of penicillamine and their effect on the human body.

One property of chiral molecules is optical rotation and its value is given the symbol $[\alpha]_D$. This is the ability to rotate the orientation of plane polarised light and can be clockwise or anti-clockwise depending on which enantiomer is dominant/present in excess. The magnitude of the rotation is dependent on many factors including concentration, solvent, temperature and trace impurities. Chiral molecules were originally given the designation D- or L-depending on if they rotated the polarised light clockwise (dextrorotatory) or anti-clockwise (levorotatory) respectively. However, it is not possible (except by complicated computational chemistry) to relate the absolute stereochemical descriptor ((R) or (S)) to the sense of optical rotation, so nowadays the absolute descriptor is preferred, but D- and L- are still used for older, often naturally derived molecules.

Based on these inherent properties, most pharmaceutical ingredients containing one or more stereogenic centres are supplied as one pure enantiomer [55]. This is not only important in the pharmaceutical industry, but also in the manufacture of fine and bulk chemicals. Various approaches to achieve the synthesis of one enantiomer over the other can be envisioned, such as starting from the chiral pool [56], the use of chiral auxiliaries [57], separation of the enantiomers by crystallisation or chromatography (a chiral stationary phase is required) or asymmetric synthesis. Amongst these options, asymmetric synthesis and catalysis is usually the most economical method [58].

If products formed differ in their absolute configuration, these products can be characterised in their enantiomeric excess (*ee*) and/or enantiomer ratio (*er*). The enantiomeric excess describes the excess of one enantiomer in a mixture of enantiomers. The term *ee* was introduced in chemistry by J. D. Morrison and H. S. Mosher and has high importance in synthetic chemistry [59].

The *ee* is defined as:

$$ee = \frac{|m_1 - m_2|}{m_1 + m_2} \times 100\%$$

m_1 = mass of enantiomer 1, m_2 = mass of enantiomer 2

In a racemic mixture, 1:1 mixture of enantiomers, the *ee* = 0.

The *ee*-value is losing its importance because this value is directly connected with the specific optical rotation, being the first way to measure the enantiomeric purity of a compound and the selectivity of an asymmetric reaction. The optical purity of an enantiomer is given as:

$$\text{Optical purity} = \frac{\text{Observed specific optical rotation}}{\text{Specific optical rotation of the pure enantiomer}} \times 100\%$$

As the amount of an enantiomer present in a mixture is now more commonly and exactly measured by other techniques, typically chromatographic methods and also NMR (nuclear magnetic resonance) techniques, it is more usual to report the ratio of enantiomers (*er*) [60–64]. In fact, the recommendation of IUPAC is that, the enantiomer ratio (*er*) should completely replace the *ee* value. The *er* can be calculated from the *ee* by

$$er = \frac{1+ee}{1-ee}$$

If a distribution of 95% of the (*S*)-enantiomer and 5% of the (*R*)-enantiomer is obtained, this would translate into an *ee* of 90% and an *er* of 95:5 = 19. In terms of energy differences, a $\Delta\Delta G^{\ddagger}$ of ~2.0 kcal/mol is needed to achieve 90% *ee* at room temperature. Whereas, for an enantiomeric excess of 98%, a $\Delta\Delta G^{\ddagger}$ of 2.60 kcal/mol is required. At −78 °C, this energy difference drops to only 1.80 kcal/mol.

Based on the above-mentioned discussion, it would be logical to assume that the enantiopurity of the product would directly correlate with enantiopurity of the catalyst/substrate (for a chiral ligand–metal complex or chiral auxiliary approach). However, whilst this is usually correct, it is not always the case and such a deviation is known as a non-linear effect.

Two scenarios are possible: a positive non-linear effect where the *ee/er* of the product is higher compared to the enantiopurity of the catalyst and a negative non-linear effect with the *ee/er* of the product lower compared to the enantiopurity of the catalyst (Figure 1.10).

Most frequently, this phenomenon arises by applying scalemic (non-racemic) mixture of ligands in a reaction. A simple case can be observed if two ligands are involved in the enantio-determining step, for example an ML_2 system with M for metal and L_S/L_R for ligand in its (*S*)- or (*R*)-form; three different metal complexes can be formed: $M(L_S)_2$, $M(L_R)_2$ and $M(L_SL_R)$, if only monometallic species are involved. Whereas, $M(L_S)_2$ and $M(L_R)_2$ are enantiomers, they result in the same *ee*-value for the product formed with the same reaction rate constant k^1. The third possibility $M(L_SL_R)$ is a heterochiral or meso complex which forms racemic product and has a different reaction rate constant k^2.

The equilibrium of $M(L_S)_2 + M(L_R)_2$ and $M(L_SL_R)$ is defined by the equilibrium constant K. If, for example, $M(L_SL_R)$ is significantly more stable compared to $M(L_S)_2 + M(L_R)_2$, all of the minor part of one ligand enantiomer is converted to the $M(L_SL_R)$ and only the chiral complex ($M(L_S)_2$ or $M(L_R)_2$) and the heterochiral chiral complex,

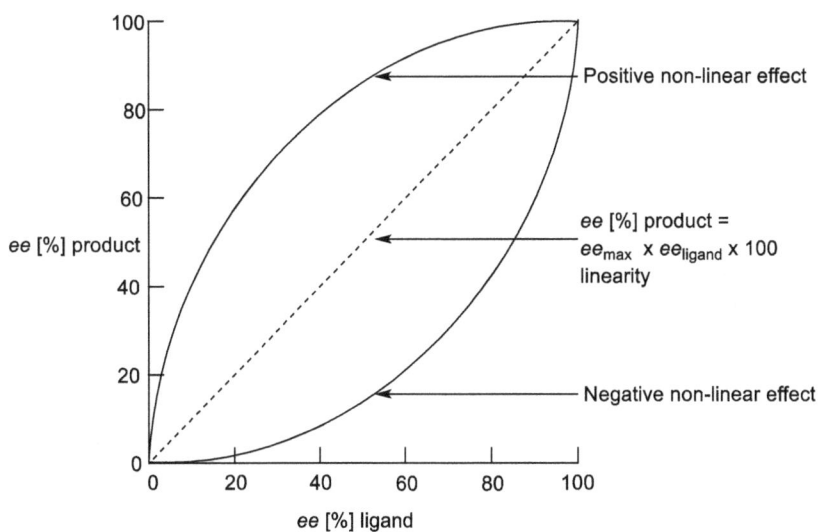

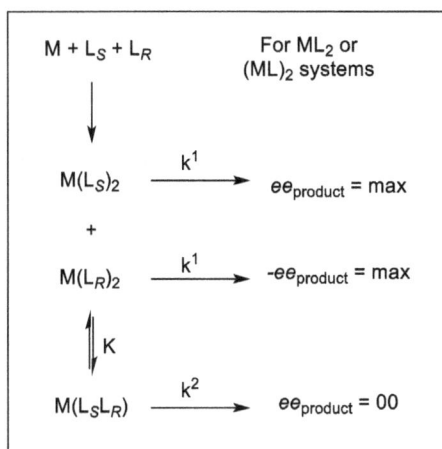

Figure 1.10: Non-linear effects.

$M(L_SL_R)$, are present in the reaction mixture. If then k^1 is significantly larger than k^2, then a strong positive linear effect is observed and the product has a higher ee/er than the applied ligand. If k^2 is greater than k^1, the opposite is true and a negative non-linear effect is observed [65].

Additionally, dynamic non-linear effects are known which describe the catalyst systems exhibiting various levels of enantioselectivity over the reaction course induced by catalyst/product or catalyst/substrate interactions [66]. In general, non-linear effects are also used as tool to elucidate the mechanism of a reaction.

1.2 Catalyst preparation

The application of catalysts in the chemical industry requires catalyst properties such as a high performance (high selectivity and high conversion), long catalyst lifetime, insensitivity to variations in the quality of the starting materials and knowledge about catalyst deactivation/regeneration. The scale-up of a catalyst preparation from laboratory scale into industrial scale and testing of these catalysts should be carried out under reaction conditions as close as possible to its industrial environment [67]. Furthermore, the configuration of the reactor is also important, for example fixed- or fluidised-beds or slurry phase reactors. After the reaction, the spent catalyst must be separated from the product, and, for efficient and economically viable processes, usually it is reactivated and reused. The reactivation procedure must be applicable on industrial scale and knowledge of the mechanism of catalyst deactivation is essential. Deactivation often occurs due to impurities in the starting materials resulting in catalyst poisoning, or sintering/ripening processes of the catalyst.

The preparation of catalysts for industrial processes in a multidisciplinary endeavour requiring specialists from several different fields to work together, these include: organic, inorganic, analytical and surface chemists, chemical engineers and specialists in kinetics, physical properties, corrosion and safety aspects. Whilst many companies have all the required experience in-house, in other cases two or more companies will work together to develop a new catalyst and catalytic process.

The development of such a catalyst and process is time-consuming and often starts with small amount of a catalyst being prepared and tested in a laboratory environment. Given the requirements to test many different formulations in a short period of time, high-throughput screening is often used at the beginning of a research and development program. Then as the key parameters are identified and specified, smaller variations are investigated to improve key requirements such as activity, selectivity and lifetime. Even when a suitable catalyst has been developed and scaled-up, the development usually does not stop as additional modifications and tweaks are tested to improve the performance further and develop second and later generations of catalyst. Depending on the application, variables that affect the mechanical stability of the catalyst may need to be optimised.

For a typical heterogeneous catalyst involving an active phase and a support material, the most comment preparation technique is impregnation [68]. A simplified block-flow diagram is shown in Scheme 1.3. A solution of the active catalyst (or a precursor) is generally added to a suspension of the carrier material, often in aqueous solution and after drying the catalyst is activated by reduction. Shaping of the catalyst can then take place if a particular form is required. Alternative methods of depositing an active phase on a carrier include vapour phase deposition or adsorption of a precursor dissolved in a solvent and evaporation of this solvent.

An alternative method for the preparation of heterogeneous catalysts is where the precipitation of the active component or its precursor is induced by a change in condi-

Precursor →
Carrier →
| Impregnation | → | Drying | → | Reduction/ Activation | → | Shaping | → Supported Catalyst

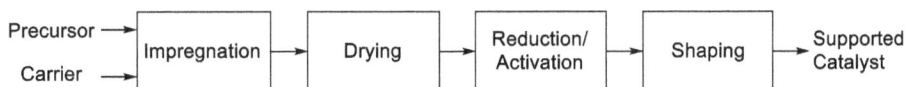

Scheme 1.3: Simplified block-flow diagram of the catalyst preparation by impregnation.

tions such as temperature or concentration (e.g. by evaporation). pH control, for example by addition of a base, is also often used for preparing catalysts by precipitation (Scheme 1.4). In some cases, co-precipitation can be used where the active and inactive species precipitate at the same time. This is a convenient way to obtain high metal loadings whilst ensuring small particle sizes. After filtration, the catalyst is activated (often by calcination) and can then be shaped into the final form. Further details of catalyst preparation and development can be found in the following references [69–72].

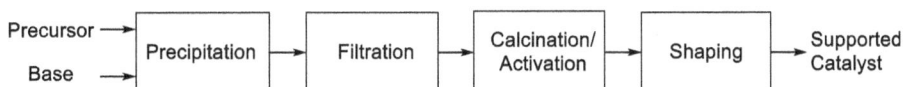

Precursor →
Base →
| Precipitation | → | Filtration | → | Calcination/ Activation | → | Shaping | → Supported Catalyst

Scheme 1.4: Catalyst preparation by precipitation (simplified).

After preparation, it has to be determined if the catalyst meets the required specifications for use. This is especially important when the catalyst is to be used on an industrial scale and key parameters for acceptance are usually defined during the development phase; the actual parameters chosen depend on the application. In addition to one or more of the characterisation above-mentioned techniques, the performance of the catalyst is usually determined by carrying out a use test, either using the exact substrate or a model compound. Here parameters such as activity, selectivity and lifetime are important. It should be remembered that not only the initial selectivity is important, but also how the selectivity varies as the reaction proceeds (see Figure 1.8).

In addition to developing and synthesising catalysts, the catalytic process has to be optimised in combination with reactor development. Factors that have to be considered are the reaction kinetics and the rate-limiting step, mass transport effects, impurity profiles of starting materials and products and economic factors. These parameters can often be in competition with each other, so a compromise needs to be found to result in the optimal catalytic process.

1.3 Chemical reactors for catalytic processes

In addition to discussing fundamental aspects of catalysis, for the synthesis of fine chemicals, not only does a catalyst and a catalytic process have to be developed, but it also has to be scaled up and operated in a chemical plant. Each chemical production plant is different, depending on the type of product produced, the scale and type of unit operations and whether it is a dedicated or multi-purpose plant. The inside of typical production plants from the twentieth century can be seen in figures later in the chapter. Figure 1.11 shows one level of a chemical production plant with a range of apparatus visible and the associated pipework. In Figure 1.12 (left), the inside of a chemical production plant is visible, showing the different levels with production equipment arranged above/below each other for optimal transfer of material between the different process stages. A closer view of, relatively small individual, pieces of production equipment can be seen in Figures 1.12 (right) and 1.13.

Figure 1.11: A chemical production plant in Dalry, Scotland for the synthesis of vitamins (photo courtesy of The Roche Historical Collection and Archive).

In particular, for catalytic reactions, a wide variety of reactor designs are utilised. A few examples are provided in this chapter and further examples can be found in the subsequent chapters. Industrial hydrogenation reactions are key transformations in the petrochemical and fine chemical industries. Hydrogenation reactions are often performed in the liquid phase (substrate) with catalysts such as nickel-alloy types or platinum group metals supported on a carrier. For liquid phase hydrogenation reactions, typical reactor types are fixed-bed reactors, trickle bed reactors, slurry-type reactors, or bubble column reactors; both batchwise and continuous operation modes are applied. For reactions using suspensions of catalyst, often batchwise operation is the preferred option due to the limited production volumes. Another reason for favouring batch reactions is their high flexibility (e.g. multipurpose plants) where a number of different re-

Figure 1.12: (Left) Vitamin production (Right) centrifuge, vitamin production, Roche Basel (photo courtesy of The Roche Historical Collection and Archive).

actions can be performed under similar (but not identical) reaction conditions. If harsh reaction conditions are required (high temperatures and pressures) continuous operation is preferred for reasons of process safety and product quality. Examples for possible reactor set-ups in batch operation are seen in Scheme 1.5 and for continuous operation (cascade of stirred tank reactors) in Scheme 1.6.

1.4 Outlook

Industrial catalytic processes have been known for many decades, especially in the area of bulk chemicals. The application of catalysis for products of smaller production volumes, but higher complexity and value has steadily been increasing. This is due not only to new targets in synthetic chemistry, but also the desire to replace old, stoichiometric processes, with more efficient processes, avoiding drawbacks such as hazard chemicals, protecting groups and high waste generation. In addition, during the last two decades, the target has changed in the direction of an efficient energy balance, use of renewable feedstocks, continuous processing, less hazardous reagents

Figure 1.13: Small-scale parallel batch chemical reactors, Roche Basel (photo courtesy of The Roche Historical Collection and Archive).

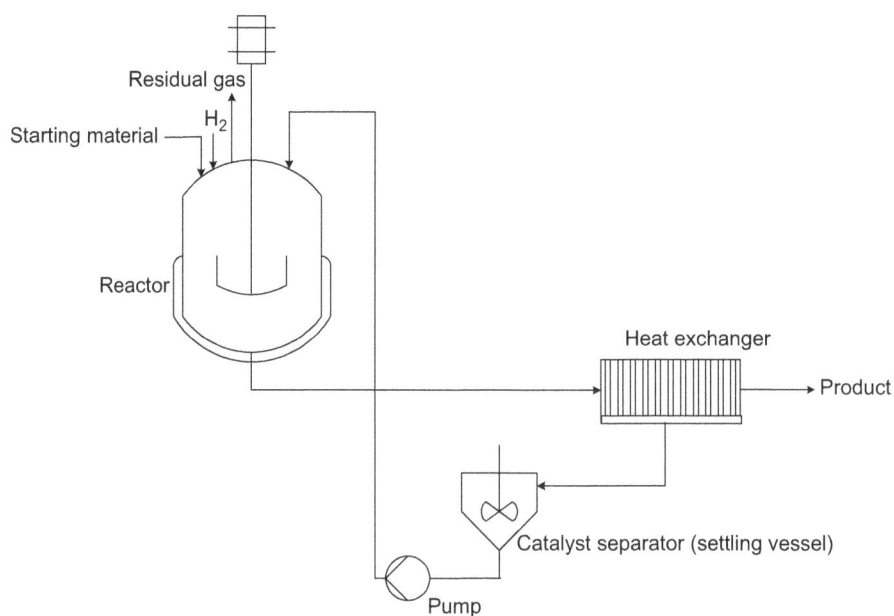

Scheme 1.5: Schematic of a batch hydrogenation set-up with a heterogeneous catalyst.

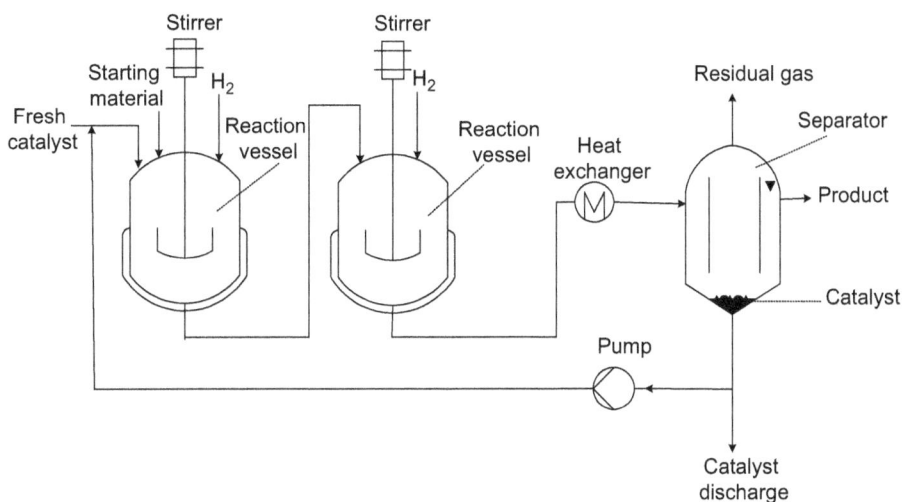

Scheme 1.6: Continuous hydrogenation set-up (cascade of stirred tank reactors).

and, of course, more catalysis. Therefore, nowadays even standard chemistry text books are dealing with these new aspects of sustainable chemistry [73].

In the following chapters, we highlight the application of a wide range of catalysts for the synthesis of not only fine chemicals, but also bulk chemicals and pharmaceuticals. Molecules such as flavour and fragrance compounds, lubricants, monomers, vitamins, amino acids, agrochemicals and drug substances can all be produced in high yield in efficient catalytic processes. In addition to describing the syntheses, we also try to highlight fundamental aspects of industrial catalysis. Whilst the vast majority of examples were and still are in active production, several examples have not realised on commercial scale for a number of reasons (e.g. an improved synthesis was later developed or the development of a particular drug substance was halted), but all would have been suitable for industrial application.

References

[1]   Levenspieler O. Chemical Reaction Engineering. New York, USA, John Wiley & Sons, 1999.
[2]   Thomas J M, Tomas W J. Principles and Practice of Heterogeneous Catalysis. Weinheim, Germany, Wiley-VCH, 2014.
[3]   Baerns M, Behr A, Brehm A, Gmehling J, Hofmann H, Onken U, Renken A, Hinrichs K-O, Palkovits R. Technische Chemie. Weinheim, Germany, Wiley-VCH, 2013.
[4]   Sheldon R A, van Bekkum H. Fine Chemicals through Heterogeneous Catalysis. Weinheim, Germany, Wiley-VCH, 2000.
[5]   Jacobsen E N, Pfaltz A, Yamamoto H. Comprehensive Asymmetric Catalysis I-III. Berlin Heidelberg, Germany, Springer-Verlag, 2004.

[6]   Fitzer E, Fritz W, Emig G. Technische Chemie, Einführung in Die Chemische Reaktionstechnik. Berlin Heidelberg, Germany, Springer, 1995.

[7]   Ostwald W. Über Katalysatoren. Z Angew Chem 1907, 49, 265–267.

[8]   Ostwald W. Über Katalyse. Chemiker Zeitung 1910, 34, 397–399.

[9]   Ostwald W. Grundriss der allgemeinen Chemie, 5th ed. Dresden, Verlag Theodor Steinkopf, 1917, 327–332.

[10]   Ostwald W. Über Katalysatoren. Angew Chem 1907, 20, 2113–2115.

[11]   Ostwald W. Ueber Katalyse. Z Elektrochem und Angew Phys Chem 1901, 7, 955–1004.

[12]   Dunsch L. Wilhelm Ostwald. Chem Unserer Zeit 1982, 16, 186–196.

[13]   Röper M. Homogene Katalyse in der Chemischen Industrie. Selektivität, Aktivität und Standzeit. Chem Unserer Zeit 2006, 40, 126–135.

[14]   Henry M. Germain Henri Hess and the Foundations of Thermochemistry. J Chem Educ 1951, 28, 581–583.

[15]   IUPAC. Compendium of Chemical Terminology, 2nd ed. (the "Gold Book"). Compiled by McNaught A D, Wilkinson. A. Oxford, Blackwell Scientific Publications, 1997, Online version (2019–) created by S. J. Chalk. ISBN 0-9678550-9-8. https://doi.org/10.1351/goldbook.

[16]   Arrhenius S. Über die Reaktionsgeschwindigkeit bei der Inversion von Rohrzucker durch Säuren. Z Phys Chem 1889, 4, 226–248.

[17]   Evans M G, Polanyi M. Some Applications of the transition state method to the calculation of reaction velocities; especially in solution. Trans Faraday Soc 1935, 31, 875–894.

[18]   Eyring H. The activated complex in chemical reactions. J Chem Phys 1935, 3, 107–115.

[19]   Eyring H, Polanyi M. Über einfache Gasreaktionen. Z Phys Chem B 1931, 12, 279–311.

[20]   Michaelis L, Menten M L. Die Kinetik der Invertierung. Biochem Z 1913, 49, 333–369.

[21]   Chakraborty S. Microfluidics and Microfabrication, ed. Springer, 2009, 1542–1549.

[22]   Nernst W. Theorie der Reaktionsgeschwindigkeit in heterogenen Systemen. Z Phys Chem 1904, 47, 52–55.

[23]   Taylor H S. A Theory of the Catalytic Surface. Proc Roy Soc A 1925, 108, 105–111.

[24]   Taylor H S. Fourth report of the committee on contact catalysis. J Phys Chem 1926, 30, 145–171.

[25]   Boreskov G K. Heterogeneous Catalysis. New York, Nova Science Publisher, 2003.

[26]   Dawden D A. Heterogeneous catalysis, part I. Theoretical basis. J Chem Soc 1950, 242–265.

[27]   Boudart M. Heterogeneous catalysis by metals. J Mol Catal 1985, 30, 27–38.

[28]   Pisarzhevsky L V. Selected works in Catalysis Izd. Acad Nauk USSR Kiev 1955.

[29]   Schwab G-M. Katalyse Vom Standpunkt der Chemischen Kinetik. Berlin, Springer, 1931.

[30]   Boudart M. Pauling's theory of metals in catalysis. J Am Chem Soc 1950, 72, 1040–1040.

[31]   Langmuir I. The adsoption of gases on plane surface of glass, mica and platinum. J Am Chem Soc 1918, 40, 1361–1403.

[32]   Mars-van K. Oxidations carried out by means of vanadium oxide catalysts. Chem Eng Sci 1954, 3, 41–59.

[33]   Doornkamp D, Ponec V. The universal character of the Mars van Krevelen mechanism. J Mol Catalysis A: General 2000, 162, 19–32.

[34]   Che M, Védrine J C. Characterization of Solid Materials and Heterogeneous Catalysts: From Structure to Surface Reactivity. Wiley-VCH, 2012.

[35]   Mul G, De Groot F, Mojet-Mol B, Tromp M. Chapter 7: Characterization of Catalysts. In: Hanefeld L, eds. Catalysis: An Integrated Textbook for Students, Wiley-VCH, 2018, 271–313.

[36]   Leofanti G, Tozzola G, Padovan M, Petrini G, Bordiga S, Zecchina A. Catalyst characterization: Characterization techniques. Catal Today 1997, 34, 307–327.

[37]   Leofanti G, Tozzola G, Padovan M, Petrini G, Bordiga S, Zecchina A. Catalyst characterization: Applications. Catal Today 1997, 34, 329–352.

[38] Brunauer S, Emmett P H, Teller E. Adsorption of gases in multimolecular layers. J Am Chem Soc 1938, 60, 309–319.

[39] Rouquerol J, Llewellyn P L, Rouquerol F. Is the BET equation applicable to microporous adsorbents? Stud Surf Sci Catal 2007, 160, 49–56.

[40] Fitzer E, Fritz W. Technische Chemie. 3rd ed. Berlin, Springer, 1982.

[41] Levenspiel O. Chemical Reaction Engineering, Ed. 3rd ed. New York, Wiley, 1999.

[42] Baerns M, Behr A, Brehm A, Gmehling J, Hoffmann H, Onken U, Renken A. Technisch Chemie, Ed. Weinheim, Wiley-VCH, 2006.

[43] Sheldon R A, Bekkum H V. Fine Chemicals through Heterogeneous Catalysis. Weinheim, Wiley-VCH, 2001, 1.

[44] Sheldon R A. Atom efficiency and catalysis in organic synthesis. Pure App Chem 1233, 200, 1233–1246.

[45] Sheldon R A. Selective catalytic synthesis of fine chemicals: Opportunities and trends. J Mol Catal A Chemical 1996, 107, 75–83.

[46] Sheldon R A. Organic synthesis- past, present and future. Chem Ind (London) 1992, 7, 903–906.

[47] Sheldon R A. Catalysis: The key to waste minimization. J Chem Tech Biotechnol 1997, 68, 381–38.

[48] Sheldon R A. Consider the environmental quotient. Chemtech 1994, 38, 38–47.

[49] Pollak P. Fine Chemicals–The Industry and the Business. New York, NY, USA, Wiley, 2007.

[50] Trost B M. Atomökonomische Synthesen- eine Herausforderung in der Organischen Chemie, die Homogen-katalyse als wegweisende Methode. Angew Chem 1995, 107, 285–307.

[51] Anastas P T, Warner J. Green Chemistry: Theory and Practice. New York, Oxford University Press, 1998.

[52] Lord K. The Molecular Tactics of a Crystal, Ed. Oxford, 1894.

[53] IUPAC. Compendium of Chemical Terminology, 2nd ed. (the "Gold Book"). Compiled by McNaught A D, Wilkinson. A. Oxford, Blackwell Scientific Publications, 1997, Online version (2019–) created by S. J. Chalk. ISBN 0-9678550-9-8. https://doi.org/10.1351/goldbook.

[54] Weigert W M, Offermanns H, Scherbich P. D-penicillamine-production and properties. Angew Chem Int Ed 1975, 14, 330–336.

[55] Murakami H. From racemates to single enantiomers – Chiral synthetic drugs over the last 20 years. Top Curr Chem 2007, 269, 273–299.

[56] Blaser H U. The chiral pool as a source of enantioselective catalysts and auxiliaries. Chem Rev 1992, 92, 935–952.

[57] Gnas Y, Glorius F. Chiral auxiliaries – principles and recent applications. Synthesis 2006, 1899–1930.

[58] Trost B M. The atom economy–a search for synthetic efficiency. Science 1991, 254, 1471–1477.

[59] Morrison J D, Mosher H S. Asymmetric Organic Reactions. Prentice-Hall, ed. Englewood Cliff, New Jersey, 1971.

[60] Eliel E L, Wilen S H. Stereochemistry of Organic Compounds, Ed. John Wiley & Sons, 1994, 221–240.

[61] Schurig V. Contributions to the theory and practice of the chromatographic separation of enantiomers. Chirality 2005, 17, 205–226.

[62] Kowalska T, Shaerma J. Thin Layer Chromatography in Chiral Separations and Analysis (Chromatographic Science Series. Band 98). CRC Press Taylor & Francis Group, 2007.

[63] Günther K, Martens J, Schickedanz M. Dünnschichtchromatographische Enantiomerentrennung mittels Ligandenaustausch. Angew Chem 1984, 96, 515–515.

[64] Günther K, Martens J, Messerschmidt M. Gas chromatographic separation of enantiomers: Determination of the optical purity of the chiral auxiliaries (R)- and (S)-1-Amino-2-methoxymethylpyrrolidine. J Chromatography A 1984, 203–205.

[65] Satyanarayana T, Abraham S, Kagan H B. Nonlinear effects in asymmetric catalysis. Angew Chem Int Ed 2009, 48, 456–494.

[66] Bryliakov K P. Dynamic nonlinear effects in asymmetric catalysis. ACS Catal 2019, 9, 5418–5438.

[67]   Armor J. Do you have really a better catalyst? Appl Catal A Gen 2005, 282, 1–4.
[68]   Munnik P, de Jongh P E, de Jong K P. Recent developments in the synthesis of supported catalysts. Chem Rev 2015, 115, 6687–6718.
[69]   Ozkan U S. Design of Heterogeneous Catalysts. Weinheim, Germany, Wiley-VCH Verlag GmbH & Co, 2009.
[70]   deJong K P. Synthesis of Solid Catalysts. Weinheim, Germany, Wiley-VCH Verlag GmbH & Co, 2009.
[71]   Thomas J M, Thomas W J. Principles and Practice of Heterogeneous Catalyst. Weinheim, Germany, Wiley-VCH Verlag GmbH & Co, 1997.
[72]   Perego C, Villa P. Catalyst preparation methods. Catal Today 1997, 34, 281–305.
[73]   Li C-J. Green Processes, Handbook of Green Chemistry, Vol. 7. Anastase P T ed. Wiley-VCH, 2012.

2 Heterogeneous hydrogenations

2.1 Introduction

Hydrogenation is probably the most widely used single transformation for the synthesis of fine chemicals. It is one of the most atom-economic reactions with every atom of the reacting molecules incorporated into the final product, except the catalyst. In addition to the high atom economy, hydrogenation reactions are usually very selective. Many fine chemicals are molecules that contain a number of different reactive moieties and functional groups; the correct choice of the metal catalyst (active metal, support material and other characteristics) and the reaction conditions allows almost exclusive reaction at the desired position in the molecule.

Within hydrogenation reactions, the use of heterogeneous catalysts is the most prevalent due to their many advantages, including: ease of catalyst separation by filtration, reuse, stability/long lifetime, high activity, high selectivity, recyclability and resistance to catalyst poisons. A wide variety of metal catalysts can be used, usually from the d-block of the periodic table. In addition, hydrogenation reactions are easy to workup, usually involving release of the hydrogen atmosphere (which can be recycled) and replacement with an inert atmosphere. This is followed by removal of the catalyst, often by filtration for heterogeneous systems. If the substrate is liquid, reactions can be performed neat, or at high concentrations with limited amounts of solvents.

One of the first people to react a metal with hydrogen gas was Johann Döbereiner in 1823, who observed that exposing hydrogen to powdered platinum followed by air resulted in the formation of water and the platinum became initially red-hot and then white-hot. Whilst not used for the production of chemicals, this discovery resulted in the development of the first gas lighters which were still in use at the start of the twentieth century [1].

By far the cheapest and most widely available hydrogen source is hydrogen gas, although other reducing agents (or hydrogen sources) are available and used in specific cases. On an industrial scale, the safe handling of hydrogen gas can easily be achieved. In addition, hydrogen gas has the advantages of being cheap, widely available, easily transported (and low volume per mole) and it leaves no other side product after use. Alternatively, for specialist applications, other hydrogen donors such as isopropanol, hydrazine or formic acid (including formate salts) can be used. These reagents have the benefit that the exact stoichiometry between substrate and reducing agent can be controlled and that the reduction does not need to take place under elevated pressure. Downsides include the higher cost of the hydrogen source and having to deal with side products (e.g. acetone from isopropanol and carbon dioxide from formic acid). A selection of reducing agents is listed in Table 2.1, with their approximate price per mole – hydrogen gas is significantly cheaper than all others.

Currently, most hydrogen is produced from fossil fuels such as oil or natural gas, either as the main product, or as a by-product in petroleum crackers and must be pu-

https://doi.org/10.1515/9783111102672-002

Table 2.1: Approximate prices for various reducing agents, based on quantities of >1,000 tons.

Reducing agent		Approximate price (€/mol)
Gas	Hydrogen	0.01
Alcohols	Ethanol	0.04
	2-Propanol	0.07
	Glycerol	0.07
Others	Formic acid	0.08
	Hydrazine	0.19
	Sodium borohydride (NaBH$_4$)	0.80

rified before use [2]. Other methods to produce hydrogen include: the conversion of carbon and water to CO and H$_2$ (syngas), photocatalysis, electrolysis and other electrochemical processes. Electrochemical water splitting has the advantage of producing very pure hydrogen for specialist applications like fuel cells. The main driving force for using fossil fuels is the lower energy demand, which is 20–25% of that required from the generation of hydrogen from water. However, with the drive for more environmentally friendly chemical processes with the reduction of greenhouse gas emissions (such as CO$_2$) and the increasing availability (and falling price) of renewable energy, "green hydrogen" is becoming more widely available and is likely to make up a significant fraction of future hydrogen usage by the chemical industry.

Although hydrogen is a potentially explosive gas, this is no problem on an industrial scale; with the right equipment it can be handled very safely. In particular, when a heterogeneous hydrogenation catalyst is used, it is important to ensure good mixing of the most common gas–liquid–solid three-phase mixture. On a small (laboratory) and pilot-plant scale, this was previously performed in a shaken reactor system. For example, Figure 2.1 shows the mini-/pilot-plant hydrogenation laboratory at Roche (Basel) in the middle of the twentieth century. A stirred reactor can be observed in the centre, and two shaker reactors (one small, one larger) can be seen on the left-hand side.

Nowadays, reactions with hydrogen gas are performed predominantly in the liquid phase in standard batch reactors at slightly elevated pressures (1–3 bar above ambient pressure) (Figure 2.2). However, hydrogenation reactions at 10 bar, 50 bar or even higher pressures are also possible in suitable equipment. The dosing of the gas can be above or below the surface of the reaction mixture, depending on the reaction requirements and reaction rate. Depending on the application and the lifetime of the (heterogeneous) catalyst, at the end of the reaction, the used catalyst can be sent for disposal/refining or returned to the reactor for use in subsequent batches.

In addition to typical batch reactors, (heterogeneous) hydrogenation reactions can be carried out in the liquid or gas phase and in continuous mode, using a cascade

Figure 2.1: Photo of Roche (Basel, Switzerland) hydrogenation mini-/pilot-plant (1947–1968). Photo courtesy of the Roche Historical Archive and used with permission.

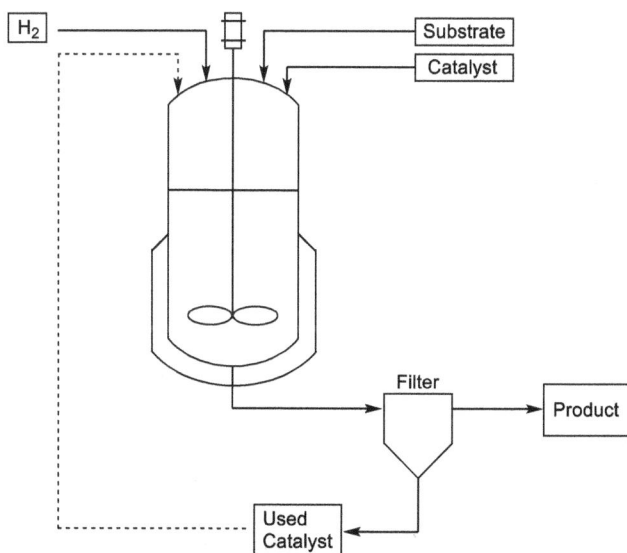

Figure 2.2: Schematic of a typical hydrogenation batch reactor set-up.

of stirred-tank reactors, fixed-bed or fluidised-bed reactors or slurry-based loop reactors. For high productivity, fast and continuous hydrogenation reactions that require high mass-transfer rates, an Advanced Buss Loop® reactor is often used (Figure 2.3).

Here, a suspension of heterogeneous catalyst in the reaction mixture circulates at high speed through a loop containing a reactor section, a pump, a heat exchanger and a crossflow filter. The system is constantly fed with substrate and hydrogen, and there is a constant outflow of product via the filter unit. To ensure excellent three-phase mixing, the reaction mixture passes through a nozzle at high speed and enters the gas–liquid ejector which produces small and well-dispersed gas bubbles. The gas is taken from both the main supply and also the reactor headspace. The excellent mixing means that baffles are not needed in the reactor. This reactor system can also be used for other reactions involving gases such as oxidations.

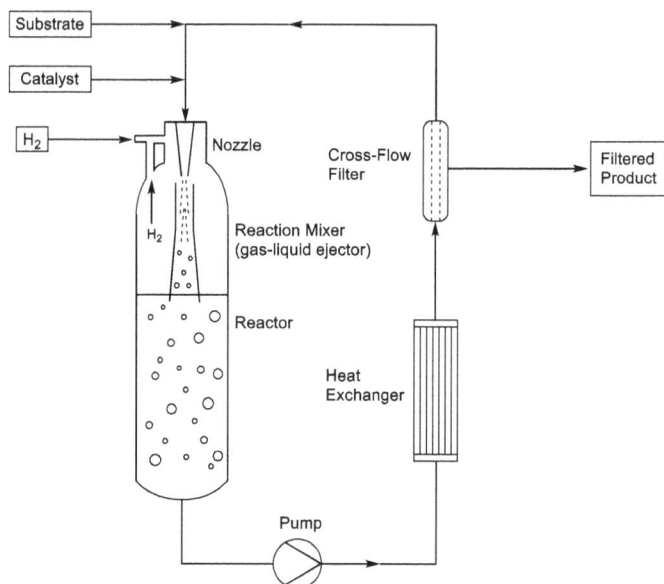

Figure 2.3: Schematic of an Advanced Buss Loop® reactor.

The choice of reactor depends on many factors, including whether the plant is designed for a single or multiple products, the choice of catalysts and reaction conditions, formation of by-products/impurities, mass-transfer effects and economic reasons. Within the list of reaction conditions, temperature, pressure, reaction/residence time (space-time yield), solvent and reaction exotherm have the most influence.

Nickel catalysts were some of the first to be used in hydrogenation reactions, following the discovery by Paul Sabatier (Nobel Prize in Chemistry in 1912) that traces of nickel catalysed the addition of hydrogen to organic molecules [3]. Shortly afterwards, Wilhelm Norman developed the nickel-catalysed hydrogenation of liquid edible oils and fats to produce solid materials, for example the "hardening" of oleic acid into stearic acid (Scheme 2.1). This technology is the predecessor to the current production of margarines. Nowadays, the main metals used in industrial hydrogenation are palladium, platinum, rhodium, ruthenium, cobalt and copper, in addition to nickel. The precious metals (platinum group metals (PGMs): Pd, Pt, Rh, Ru) are generally highly active and operate under milder conditions with small amounts of active metal (usually deposited on a carrier). The base metals (Co, Cu, Ni) are generally less active and their use requires more severe reaction conditions (generally higher temperatures and pressures) and higher loadings. The choice of metal for a hydrogenation reaction is based on many factors, including reaction selectivity, conditions required and cost. The price of metals (especially PGMs) can vary considerably depending on the metal; an overview of prices at the start of 2020 is shown in Table 2.2 [4, 5]. In addition, the cost of certain metals can vary widely over time due to outside influences, since their use in chemical production processes is generally only a minor proportion of the total worldwide use (e.g. the use of in catalytic converters/autocatalysts, electrical/electrochemical industries, in fuel cells, in jewellery and as an investment). Figure 2.4 shows the variation in price of palladium and platinum in the time period 2000–2019, with a sharp fall in prices after the financial crisis in 2008 and a recovery thereafter [4]. PGMs (from chemical catalysts and other sources) are always recycled/refined, but the supply of new metals is based on mining and the extraction of mixed ores from places such as South Africa, Russia, North America and China.

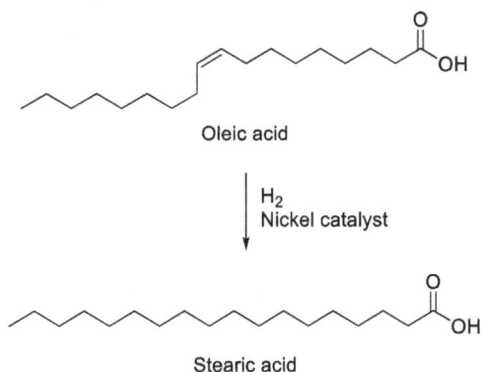

Oleic acid

H_2
Nickel catalyst

Stearic acid

Scheme 2.1: Hydrogenation of oleic acid to stearic acid.

In the detailed discussion further in the chapter, we have highlighted industrially important catalytic hydrogenations for the production of fine chemicals. There are, of course, many large-scale industrial hydrogenation processes, such as the *Haber–Bosch* process for the synthesis of ammonia, the *Fischer–Tropsch* process for formation of hydrocarbons and the hydrogenation of acetylene from ethylene streams (for polymerisation) that are extremely important; however, they are beyond the scope of this chapter. The examples of hydrogenation reactions are first ordered by the metal catalyst used and then later by the type of chemical transformation.

Table 2.2: Approximate metal prices at start of 2023.

Metal	€/kg
Rh	370,000
Pd	53,000
Ir	140,000
Pt	32,000
Ru	14,000
Co	30
Ni	19
Cu	7

Pt and Pd Prices (EUR/kg)

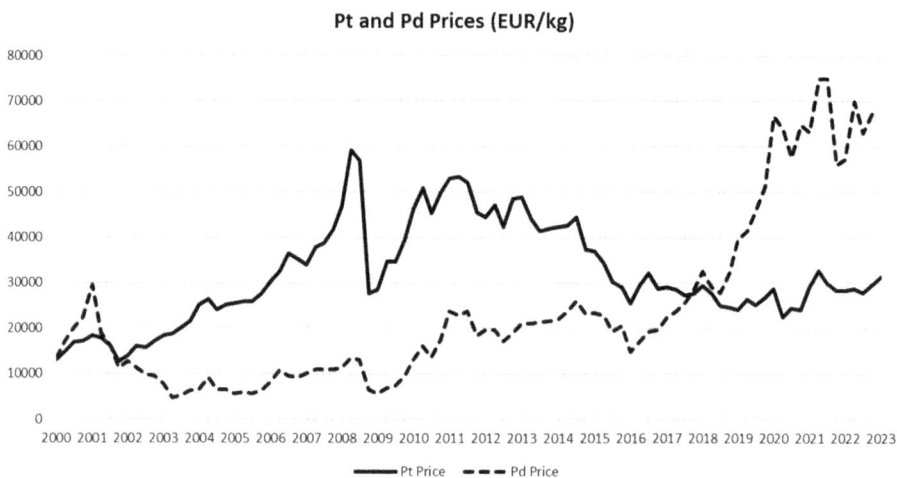

Figure 2.4: Change in palladium and platinum prices over time.

2.2 Nickel catalysts

The most widely used nickel catalysts are "sponge" or "skeletal" metal catalysts that are unsupported and were developed by Murray Raney in 1925 and are widely known as Raney® Ni.[1] This type of catalyst was developed originally for the hydrogenation of cottonseed oil into shortening, a fat related to margarine that is solid at room temperature and used in many food products [3]. One of the main challenges of heteroge-

---

1 Raney® is a trademark of the W.R. Grace and Co, USA.

neous metal hydrogenation catalysts is obtaining a highly active, highly dispersed metal surface for binding of the substrate for an efficient and cost-effective process. In supported metal catalysts, active metal nanoparticles are distributed on a generally inert support material. In contrast, in sponge metal catalysts, a large metal surface area is produced by an alternative process.

Initially, a base metal (such as nickel) is alloyed with aluminium at high temperature (>1,200 °C). This alloy is crushed and ground into pellets or a coarse powder, depending on the application. These are then treated with a sodium hydroxide solution which leaches out a proportion of the aluminium leaving behind an activated skeletal metal surface covered in adsorbed hydrogen. A schematic of the process is shown in Figure 2.5. The catalysts are highly active and must be stored under water as they are pyrophoric when dried. Catalysts can be supplied in forms for use in slurry reactors or fixed-beds. Several manufacturers also offer a non-pyrophoric version where the water has been displaced by an aliphatic amine, which releases the active metal catalyst on addition to the substrate/solvent.

The catalysts prepared by this method generally have a particle size of 10–80 μm and a surface area of 50–150 m^2g^{-1}. Nickel metal forms approximately 80–90% of the total mass. For certain applications, other metals such as Mo, Cr, and Fe can be added to the initial alloy as promotors in the hydrogenation reaction. These generally make up 3–10% of the total mass, with the remainder being un-leached aluminium. The amount of leached/remaining aluminium is critical to the activity and stability of the hydrogenation catalyst.

2.2.1 Hydrogenation of nitriles

In the synthesis of fine chemicals, one of the most widely used applications of sponge nickel catalysts is the selective reduction of nitriles to primary amines. This hydrogenation proceeds via the stepwise reduction of the nitrile to the primary imine, followed by subsequent reduction to the primary amine. However, if significant amounts of the intermediate primary imine accumulate, then this can react with the desired primary amine to produce a secondary imine in an equilibrium reaction. This secondary imine can be reduced to the secondary amine, which could further react to yield the tertiary amine (Scheme 2.2). These secondary and tertiary amines sometimes have an intrinsic value or may just be "lost yield". In the worst case, it can be difficult to remove the impurities which negatively affects the downstream chemistry. Therefore, a significant amount of effort is put into optimising the catalyst and the reaction conditions to minimise these side reactions. One of the most common ways is to conduct the hydrogenation in the presence of ammonia. This affects the equilibrium between primary imine/ amine and secondary imine, resulting in less of the unwanted side reactions. However, as will be seen, other impurities can also be formed. Generally, high selectivities for the amine products can be achieved; however, one area where sponge nickel is unsuccess-

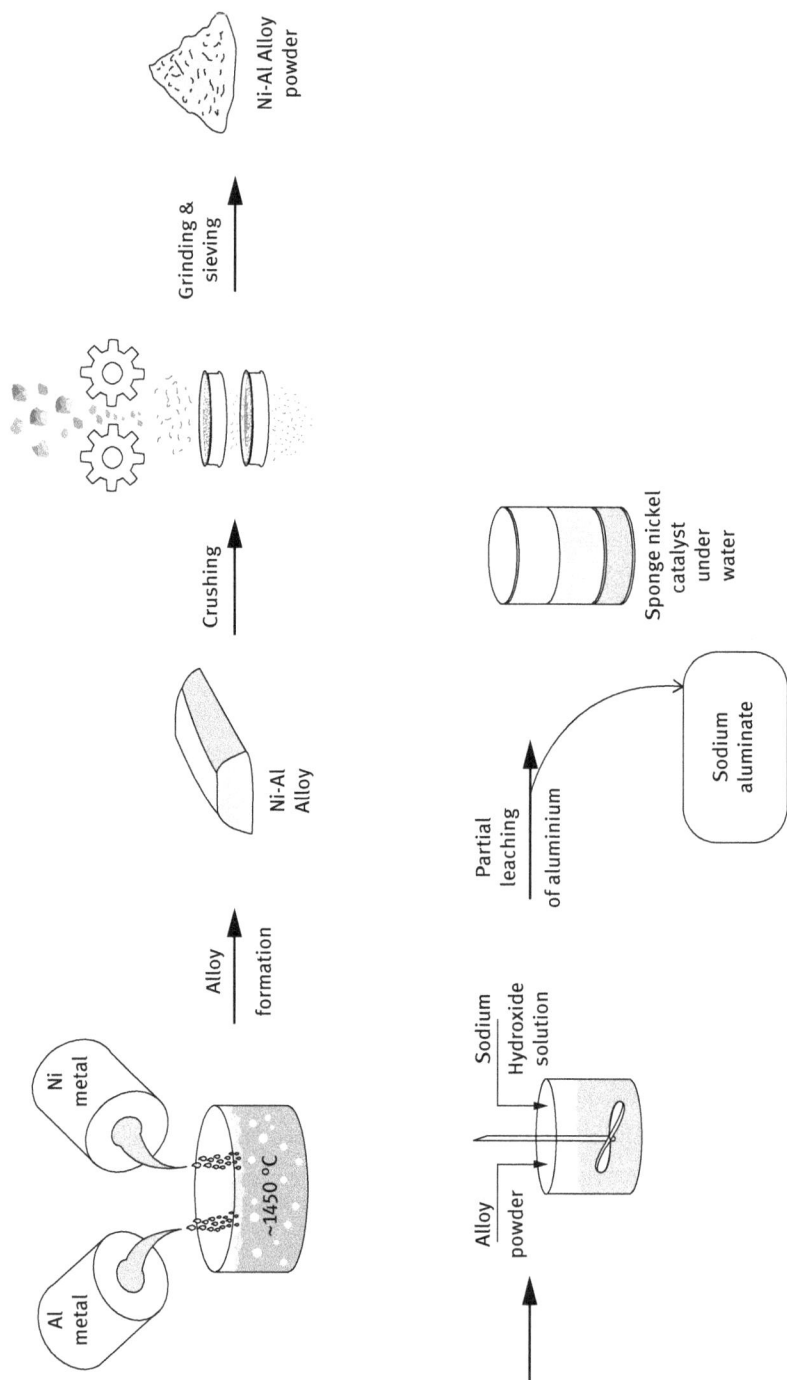

Figure 2.5: Manufacture of "sponge" metal catalysts such as Raney® nickel.

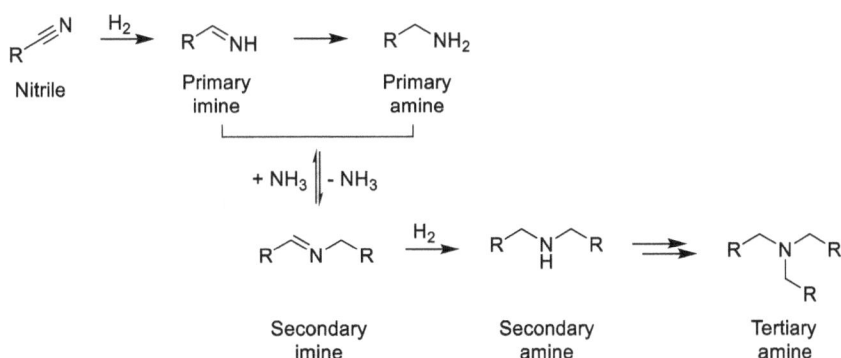

Scheme 2.2: General reaction pathway for nitrile hydrogenation and side product formation.

ful is the selective reduction of unsaturated nitriles – in this case, the fully saturated amine is obtained. A partial solution to this problem is the use of sponge cobalt catalysts which show moderate to good selectivity for the unsaturated amine [6].

One of the largest-scale nitrile hydrogenation reactions with nickel catalysts is the reduction of adiponitrile to hexamethylene diamine, one of the two components of nylon 6,6 (with adipic acid) and produced on over 1 million tonnes per year (Scheme 2.3). The reaction can be performed with either a reduced cobalt oxide or iron oxide catalyst in a packed bed reactor at high temperatures and pressures (>100 °C, >200 bar) or in a bubble column with a Fe/Cr-promoted nickel catalyst under milder conditions (50–100 °C, 15–35 bar) [7, 8] (Scheme 2.4). The control of side products is absolutely critical as high-purity diamine is required for polymerisation. In contrast to the process with iron or cobalt oxide, the nickel process does not use ammonia, but sodium hydroxide to maintain the catalyst activity and reduce the formation of impurities. In addition to formation of

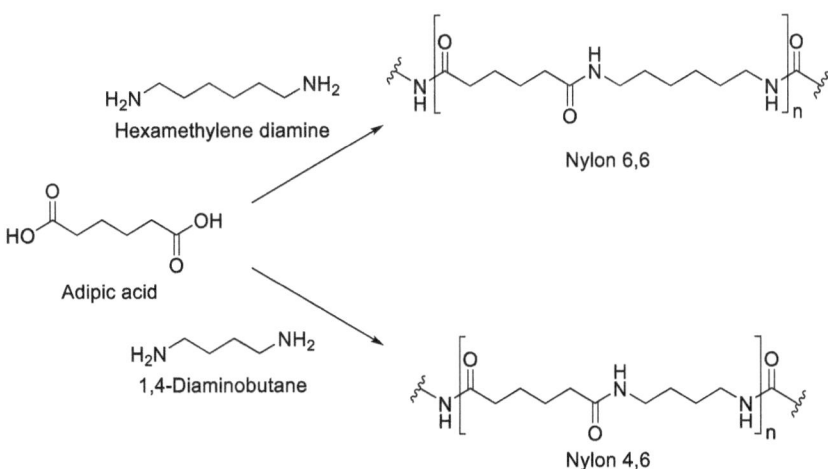

Scheme 2.3: Formation of nylon 6,6 and nylon 4,6 from adipic acid.

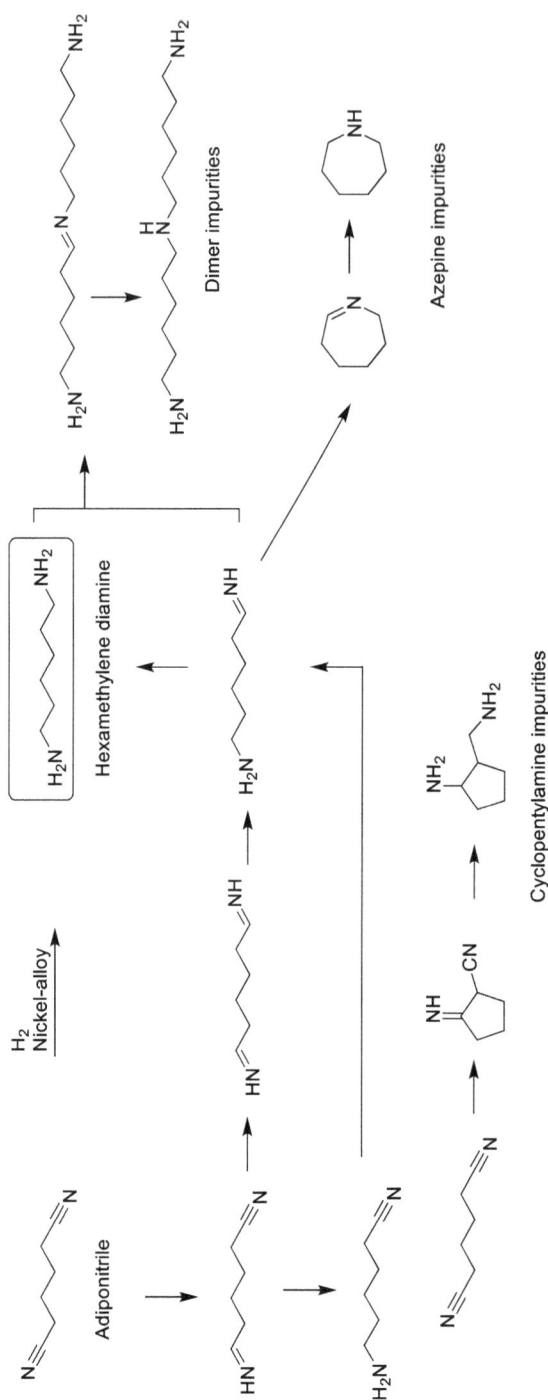

Scheme 2.4: Hydrogenation of adiponitrile to hexamethylene diamine showing by-products and their formation.

dimer imines and amines, cyclic by-products can be formed from either the starting dini-trile (cyclopentyl amine derivatives via addition of α-acidic carbon to the nitrile group) or the intermediate amino-imine forming tetrahydro- and hexahydro-azepines.

A related process, performed on a smaller scale, is the production of 1,4-diaminobutane by DSM (Scheme 2.5). This is used to make the polymer nylon 4,6 (Stanyl®) which has a higher melting point than traditional polyamides and higher crystallinity. This allows its use in high-temperature applications and low friction and high wear per-formance. 1,4-Diaminobutane is produced by skeletal nickel hydrogenation of succinoni-trile [9, 10] in the presence of ammonia at elevated temperatures and hydrogen pressures of >50 bar. The production of the by-product pyrrolidine is thus minimised.

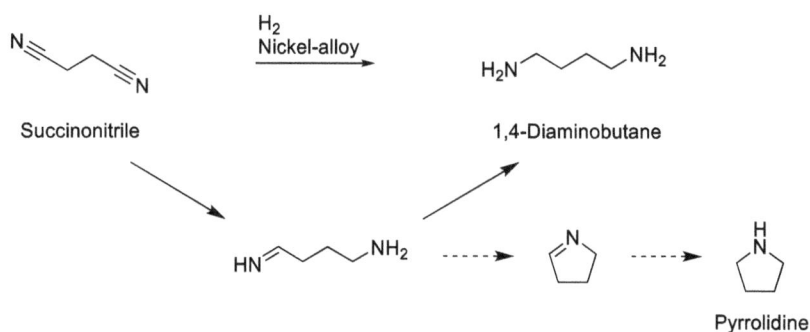

Scheme 2.5: Hydrogenation of succinonitrile to 1,4-diaminobutane.

Another nitrile hydrogenation, where the control of impurities is critical, is the pro-duction of vitamin B_1 (thiamine), which is produced worldwide on a scale of approxi-mately 9,000 t/a. One of the two key intermediates for the synthesis of the thiazolium ring is *Grewe* diamine. This compound can readily be prepared from pynitril (which in turn comes from the condensation of malononitrile with orthoformate and an ace-timidate hydrochloride) (Scheme 2.6). The hydrogenation of pynitril to *Grewe* diamine is performed with a skeletal nickel catalyst [11]. However, if significant amounts of primary imine intermediate are produced (see Scheme 2.2), then this can react with the product *Grewe* diamine to produce the secondary imine by-product (and after sub-sequent hydrogenation, the secondary amine by-product). The reaction takes place at elevated temperatures (>100 °C) and pressures (>25 bar H_2). To reduce the amount of secondary amine/imine by-products, ammonia gas is added to the reaction mixture to shift the equilibrium to the primary imine/amide. Even so, significant amounts (at least 2%) of by-products are formed. Whilst this constitutes a loss in yield, an addi-tional problem is that the secondary imine/amine impurities cause problems with the following purification and downstream chemistry. Therefore, the impurities had to be reduced further. One solution was the pretreatment of the skeletal nickel catalyst with a partial-poisoning agent – in this way, the relative rates of the two hydrogenation steps are more similar and there is a limited build-up of the intermediate primary imine. It was found that aldehydes, ketones, carbon monoxide and carbon dioxide were suitable, reducing the amount of secondary amine to less than 1% and resulting in a *Grewe* diamine selectivity of >99.5% [12, 13].

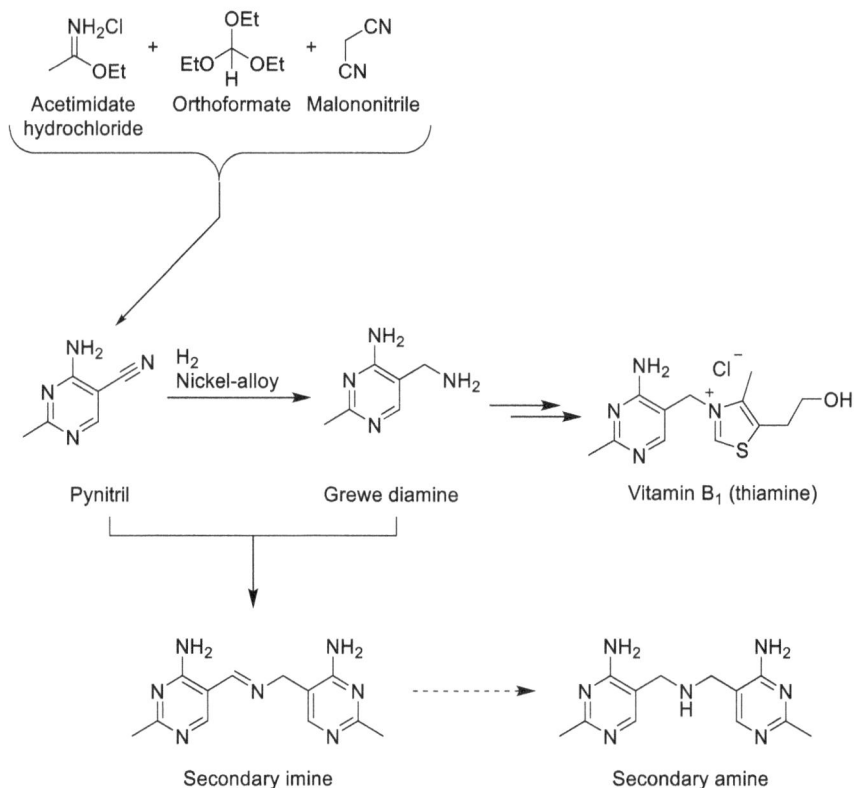

Scheme 2.6: Hydrogenation of Pynitril to *Grewe* diamine in the synthesis of vitamin B$_1$.

Skeletal nickel catalysts have also been used to produce pharmaceuticals and one example is the synthesis of pregabalin, a γ-aminobutyric acid analogue that was developed by Parke-Davis/Warner-Lambert (later becoming Pfizer), for the treatment of several central nervous system disorders. In the first process synthesis, a racemic diester was decarboxylated and then the nitrile group was hydrogenated to the corresponding amine with a nickel catalyst under approximately 3 bar hydrogen pressure at ambient temperature. The resulting racemic pregabalin was then resolved using mandelic acid [14]. An improved route with significantly lower usage of chemicals and solvents was later developed where the stereogenic centre was introduced by an enzyme-catalysed kinetic resolution to form the chiral monoester (see Chapter 10: Biocatalysis) (Scheme 2.7) [15]. The efficient nickel hydrogenation was retained from the

First generation process route:

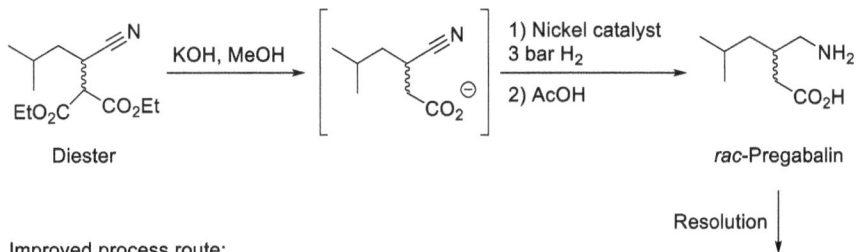

Improved process route:

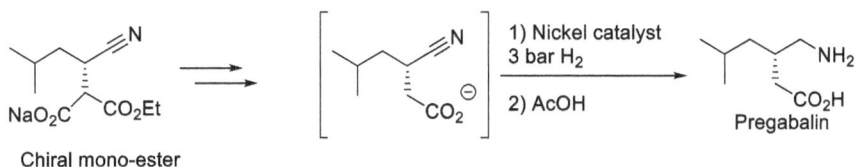

Scheme 2.7: Synthesis of pregabalin via a nickel-catalysed nitrile hydrogenation.

original route resulting in high purity of the final drug substance with no loss in enantiomeric excess of the intermediate.

Although skeletal nickel catalysts are the most widely used for nitrile hydrogenation, sometimes they are not effective enough, supported nickel catalysts can be an alternative. Tranexamic acid is a widely used antifibrinolytic drug to prevent excessive loss of blood, for example during surgery. It was first produced in 1962 and is included in the World Health Organization's List of Essential Medicines. A common route is to hydrogenate the benzylamine with platinum oxide to give (racemic) tranexamic acid (for more details, see Section 2.3.2.2). However, the production of the starting benzylamine uses a number of toxic reagents that should be avoided (Scheme 2.8). An alternative, high-yielding synthesis has been developed by a group in China in which two separate hydrogenation steps are used [16]. The acetyl-protected benzylamine was synthesised by the hydrogenation of the corresponding nitrile with a nickel on alumina catalyst. The hydrogenation takes place in the presence of acetic anhydride at 50 °C and 10 bar hydrogen pressure. As soon as the benzylamine is produced, it is immediately acetylated, thus stopping the formation of a range of impurities. To complete the synthesis, the aromatic ring could be reduced with ruthenium on alumina at 70 bar hydrogen. Isomerisation, saponification and removal of the acetyl protecting group were achieved with barium hydroxide. The combined yield over the three steps was 70%. It is currently not known if this new route has been commercialised.

Common Route

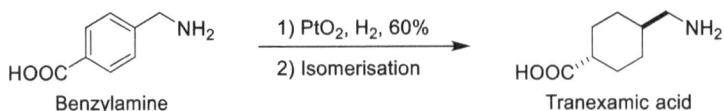

Benzylamine

1) PtO₂, H₂, 60%

2) Isomerisation

Tranexamic acid

New Route

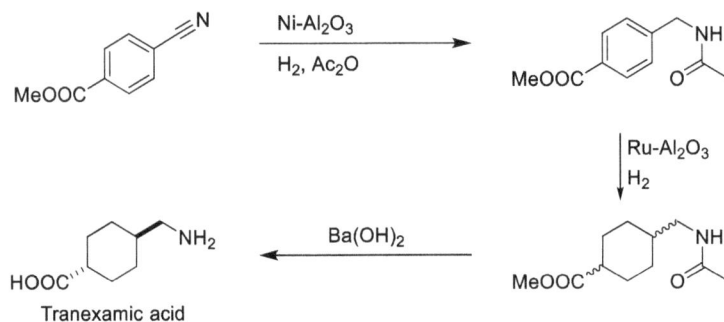

Ni-Al₂O₃

H₂, Ac₂O

Ru-Al₂O₃

H₂

Ba(OH)₂

Tranexamic acid

Scheme 2.8: Synthesis of tranexamic acid.

2.2.2 Hydrogenation of carbonyl groups

One of the large-scale uses of nickel catalysts in the reduction of carbonyl groups is the hydrogenation of glucose to sorbitol (Scheme 2.9). Sorbitol itself can be used as a sugar substitute, in toothpaste and mouthwash, as a thickener in cosmetics, and it is also the key intermediate in the preparation of *L*-ascorbic acid (vitamin C). Although other reduction methods now exist, including microbiological and electrochemical methods, the nickel-catalysed hydrogenation of glucose to sorbitol used to dominate production. The reaction is carried out at elevated temperatures (>100 °C) and high pressure (>25 bar) and can be run batchwise or in continuous operation. Sorbitol is obtained in high yield and selectivity with only small amounts of other sugars [17, 18].

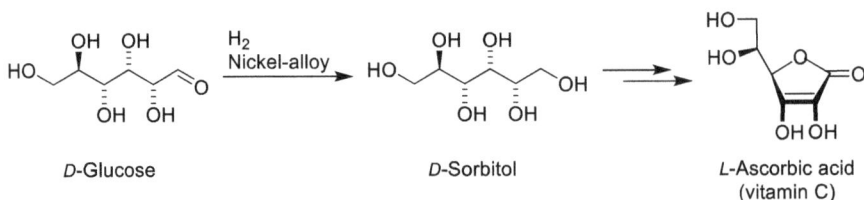

D-Glucose

H₂
Nickel-alloy

D-Sorbitol

L-Ascorbic acid
(vitamin C)

Scheme 2.9: Hydrogenation of *D*-glucose to *D*-sorbitol for vitamin C production.

Another example of nickel catalysts being used for the reduction of a carbonyl group is the preparation of intermediates in the synthesis of resveratrol [19]. Resveratrol is a natural polyphenol that is found in grapes, blueberries and several other fruits and is commonly used in dietary supplements. It can be extracted from natural sources, or produced chemically (Scheme 2.10). One chemical route requires the reduction of a diacetoxyacetophone to the secondary alcohol. A number of precious metal catalysts (e.g. platinum or palladium on carbon) were tested, but these resulted in lower yields and the formation of impurities (such as the "des-OAc" or the "ethyl impurity") that are difficult to remove. Using nickel catalysts in a variety of solvents at 70 °C and low pressure (2 bar) resulted in high yields (>95%) and good selectivities. In addition, the catalyst was stable and could be reused multiple times.

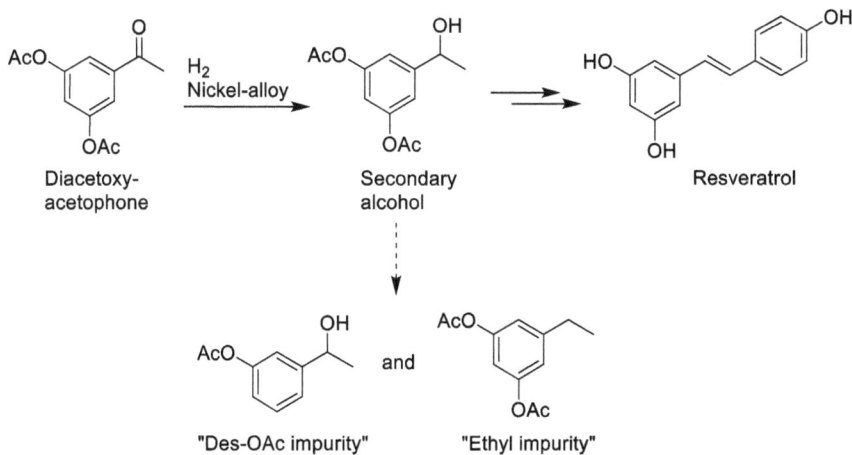

Scheme 2.10: Hydrogenation of an acetophenone derivative in the synthesis of resveratrol.

2.2.3 Hydrogenation of other functional groups

In addition to the reduction of nitriles and carbonyl groups, nickel catalysts can reduce other functionalities. One example is the production of 1,4-butanediol which is produced worldwide at a scale of approximately 4 million tonnes per year. It is used as an industrial solvent, in some plastics and also to prepare tetrahydrofuran and γ-butyrolactone. Several manufacturing processes exist and recently a bio-based 1,4-butanediol has been commercialised using the fermentation of glucose [20]. However, the *Reppe* technology starting from acetylene and formaldehyde is still the most common (Scheme 2.11). Butynediol as an aqueous solution is mixed with hydrogen and is passed over a fixed-bed catalyst containing nickel. These processes operate at 80–170 °C and 250–300 bar hydrogen. Skeletal nickel can also be used in a slurry form at 20 bar hydrogen pressure and 120–140 °C, or as a fixed-bed catalyst.

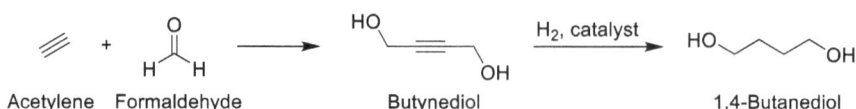

Scheme 2.11: Hydrogenation of butynediol to 1,4-butanediol.

Nitro groups can also be reduced using nickel catalysts. One large-scale example is the reduction of 2,4-dinitrotoluene to 2,4-diaminotoluene, which is used to produce toluenediisocyanate, used in the flexible foam and elastomer industries (Scheme 2.12). A range of temperatures, pressures and solvents can be used. Suitable catalysts are palladium on carbon or nickel catalysts (supported or skeletal metal). Since the starting, nitroarene can act as a catalyst poison to obtain an efficient and cost-effective process; the concentration of dinitrotoluene is kept as low as possible [21, 22].

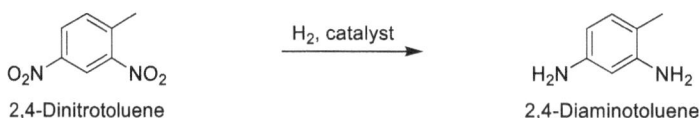

Scheme 2.12: Hydrogenation of 2,4-dinitrotoluene to 2,4-diaminotoluene.

A further example of the use of nickel catalysts is in the synthesis of the drug conivaptan, which was approved in 2004 for use in the treatment of low blood sodium levels. In the initial drug discovery route, an aromatic nitro group was reduced with a palladium on carbon catalyst [23]. For the process chemistry route, a related intermediate was chosen (Scheme 2.13), and this required the use of a nickel catalyst. The reaction took place under mild conditions and in greater than 90% yield.

Nickel catalysts can also be used for the removal of protecting groups. One example is the preparation of a dipeptide amidine for the use in personal care (skin) applications (Scheme 2.14) [24]. Although several methods to prepare this compound were already known, problems occurred due to the solubility of intermediates and the large number of steps required. Several approaches to an improved process were investigated, with the removal of various protecting groups by hydrogenation as the last step. In order to be compatible with the available production equipment, hydrogenation reactions had to be performed at less than 2 bar hydrogen and as concentrated as possible to obtain a high space-time efficiency. Numerous catalysts (including precious metals) were tested on a variety of different substrates with a nickel-catalysed reduction of the hydroxy- or acetoxy-imidamide being optimal. For solubility reasons, tetrahydrofuran (THF)–water or acetic acid–water mixtures were used. Although only produced on a relatively small scale so far, further scale-up and development is planned.

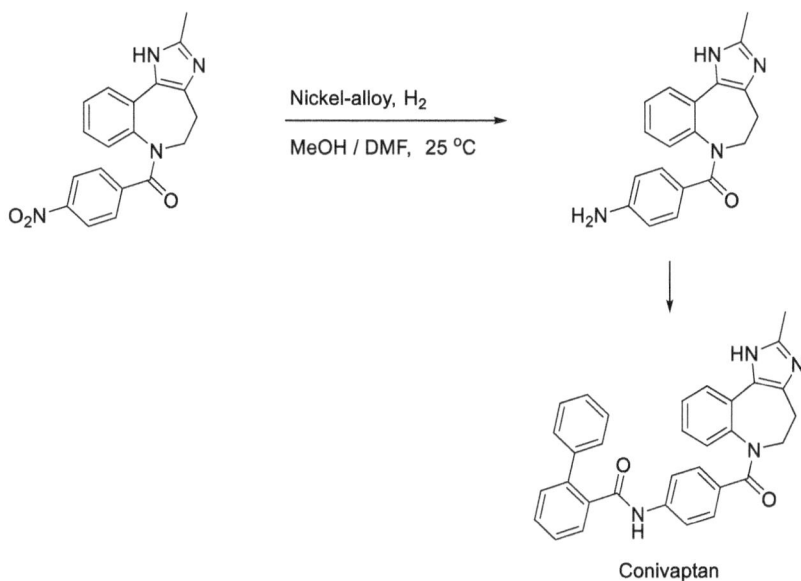

Nickel-alloy, H$_2$

MeOH / DMF, 25 °C

O$_2$N—

H$_2$N—

Conivaptan

Scheme 2.13: Nickel-catalysed hydrogenation of a nitro group in the synthesis of conivaptan.

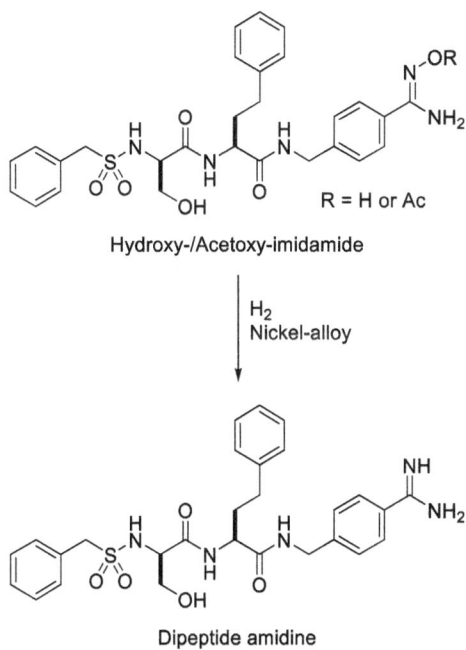

N–OR

NH$_2$

OH

R = H or Ac

Hydroxy-/Acetoxy-imidamide

H$_2$
Nickel-alloy

NH

NH$_2$

OH

Dipeptide amidine

Scheme 2.14: Protecting group cleavage of a dipeptide.

2.2.4 Reduction of multiple functional groups in the same reaction

As can be seen, nickel catalysts are incredibly versatile in the type of functional groups that they can hydrogenate. This can allow very efficient chemical processes by combining multiple steps into one reaction. One example of this is the synthesis of tetrahydrogeraniol, which is a fragrance molecule which has a fresh aroma with hints of rose and citrus. There are many possible ways to produce this compound; however, a very efficient one is the hydrogenation of citral, itself an aroma compound that is produced in bulk quantities (Scheme 2.15). A number of different catalysts were screened with supported precious metal catalysts giving mixtures of partially hydrogenated compounds [25, 26]. Skeletal nickel catalysts proved to be very effective producing tetrahydrogeraniol in high yield and selectively (both >95%) at 40–80 °C and 5–20 bar hydrogen pressure. In addition, the catalyst could be reused multiple times, making the process even more cost effective.

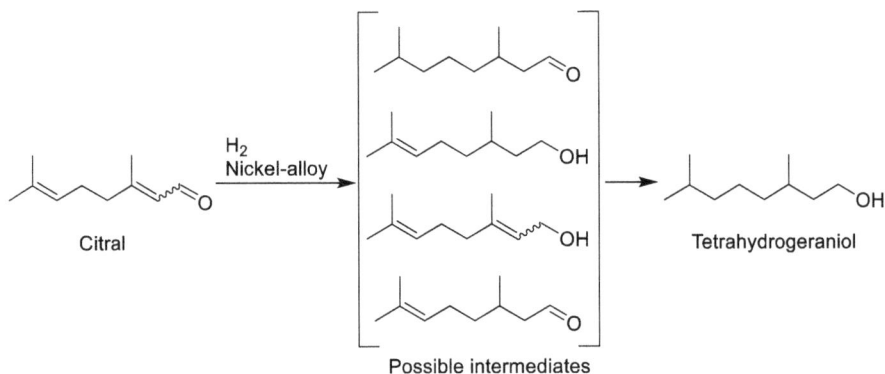

Scheme 2.15: Hydrogenation of citral to tetrahydrogeraniol.

A related example of a skeletal nickel catalyst being used for both the reduction of C=C and C=O double bonds comes from the pharmaceutical industry. Researchers at Wyeth needed to prepare a bicyclic amine for a pharmaceutical compound under development [27]. A key intermediate was the *cis*-diol (Scheme 2.16). A simple route to this would be hydrogenation of the bulk, renewable feedstock hydroxymethylfurfural. Due to the range of equipment available in production, reaction conditions had to be developed using a maximum pressure of 5 bar hydrogen. A variety of palladium catalysts were tested; however, the reaction was very sluggish with often remaining starting material and also significant amounts of the partially reduced intermediate. The switch to skeletal nickel allowed complete hydrogenation in a range of solvents giving 93–95% yield of the desired *cis*-diol with the remaining material being the *trans*-diol. After removal of the catalyst and a solvent exchange, the product could be used in the next step without any purification.

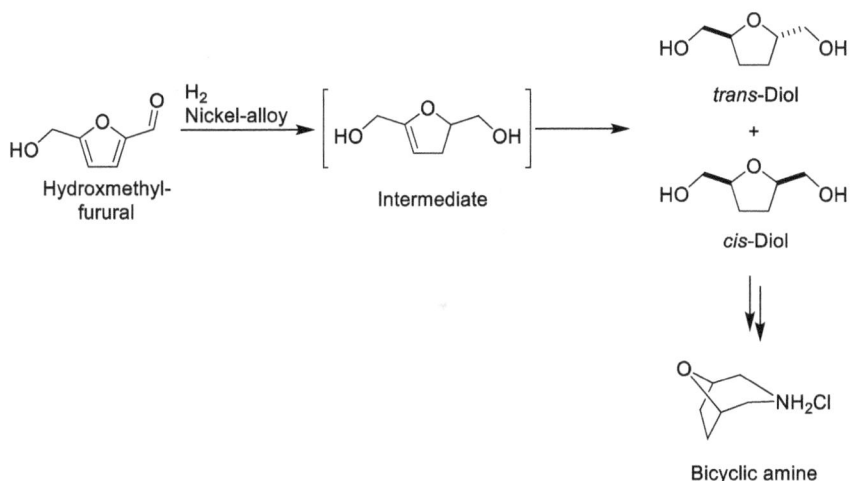

Scheme 2.16: Hydrogenation of hydroxymethylfurfural.

2.3 Platinum Group Metal (PGM) based catalysts

PGMs are a group of metallic elements from the second and third rows of the d-block in the periodic table. The six PGMs are ruthenium, rhodium, palladium, osmium, iridium and platinum. They have similar chemical and physical properties and are found as mixtures in mineral deposits. Due to their scarcity and the fact that they have to be separated from each other, they are expensive (see Table 2.2), but also highly active in catalytic reactions.

Due to the cost and activity of the metals, catalysts derived from PGMs are usually applied as supported catalysts, that is, a small amount of metal (generally 1–10%) deposited on a carrier. This allows for a better distribution of the metal, a higher available surface area for reactions and more efficient processes. In addition, supported PGM catalysts are often reused multiple times (or have a long lifetime in continuous processes) and when the catalyst is spent, it is collected and sent for refining to recover the metal. Although small losses are inevitable, recovery rates of 99% are possible for catalysts such as palladium supported on charcoal. In fact, due to the high cost of the metal, the small losses in use, recovery and refining can make up a significant part of the total catalyst cost.

The support material used for metal catalysts is not just an inert material, it can have a significant influence on the reaction performance, it determines the metal dispersion, distribution and surface area. In addition, it provides shape, size and mechanical strength to the catalyst particle which can be critical, especially in continuous fixed-bed reactors. The support can also have secondary effects influencing the binding of substrates/products and also supressing unwanted side reactions [28].

For powder-based PGM catalysts, a list of common catalyst supports and their typical properties are listed in Table 2.3. In addition to the materials shown in the table, ceria (CeO_2), titania (TiO_2), zirconia (ZrO_2) graphite and barium sulphate ($BaSO_4$) can be used. As can be seen, activated carbon has the highest specific surface area and is often used for the preparation of hydrogenation catalysts. Other supports are used for specific applications, for example when a basic support is required (e.g. alumina) or a non-porous support (e.g. calcium carbonate or barium sulphate). The reaction conditions to be used can also influence the choice of catalyst support, since it must be stable (chemically, thermally and mechanically) during operation. In addition to the support material (and its origin/crystal form), there are many other factors that influence the performance of a supported PGM catalyst; a summary of some of these is shown in Table 2.4.

Table 2.3: Common catalyst supports for powder PGM catalysts.

Support material	Specific surface area (m^2 g^{-1})	Typical pore volume (m^3 g^{-1})	Average pore diameter (nm)
Activated carbon	800–1,200	0.2–2.0	1–4
Silica	400–800	0.4–0.8	2–8
Silica–alumina	200–600	0.5–0.7	3–15
Activated alumina	100–350	0.4	4–9
Calcium carbonate	<10	–	–

Table 2.4: Factors that can influence the performance of a PGM catalyst.

Impurities	Support
	Metal
	Reaction mixture
Reaction mixture	Solvent
	Polarity
	Solubility of reactive gas
	Substrate concentration
Surface effects	Cleaning
	Binding groups

Table 2.4 (continued)

Active metal	Metal and metal precursor used
	Mono- or bi-metallic systems
	Metal nanoparticles produced:
	Size
	Shape
	Oxidation state
	Acidity
	Location of metal on the support
Dopants	Deliberate
	Non-deliberate
Treatment/activation	Washing
	Drying
	Calcination
	Pre-reduction
	Formulation/shaping

Regarding the metal nanoparticle size and distribution, a significant variation in performance can be observed depending on the location of the metal particles, substrate size and reaction conditions. These parameters can all be varied depending on the method used to produce the catalyst. When the metal is deposited mainly on the outer surface of the support and not in the pores, this results in an "egg-shell" catalyst which has a lower metal dispersion, but may be more suitable for larger substrates or those where mass-transfer effects are significant. Uniform catalysts have a much larger surface area and are suitable for smaller molecules and/or reactions at higher pressures since the metal in the support pores is more accessible. An intermediate catalyst is somewhere between the two extremes; a simple schematic to illustrate this is shown in Figure 2.6. The location of the metal can also affect the selectivity, if side reactions are possible. The distribution of metal nanoparticles can additionally affect the leaching (loss of metal) from the catalyst, which can be important for economic reasons and also reasons of safety/product quality if a metal-catalysed step is the last one in a multi-step sequence.

The handling of precious metal salts and the preparation of supported catalysts is usually not performed by the company using them in the preparation of fine chemicals. There are a number of catalyst manufactures who have specialised in this area and can offer a wide range of catalysts for testing and commercial use. For small-scale applications, an "off the shelf" catalyst using an existing catalyst recipe is often used. "Catalyst kits" are available to allow a first screening to identify a lead catalyst for the desired transformation. This "hit" can be further refined (and further samples provided) in discussion with the manufacturer.

For larger volume products, or those with specific selectivity requirements, often a catalyst will be developed for the particular molecule or process, between the catalyst

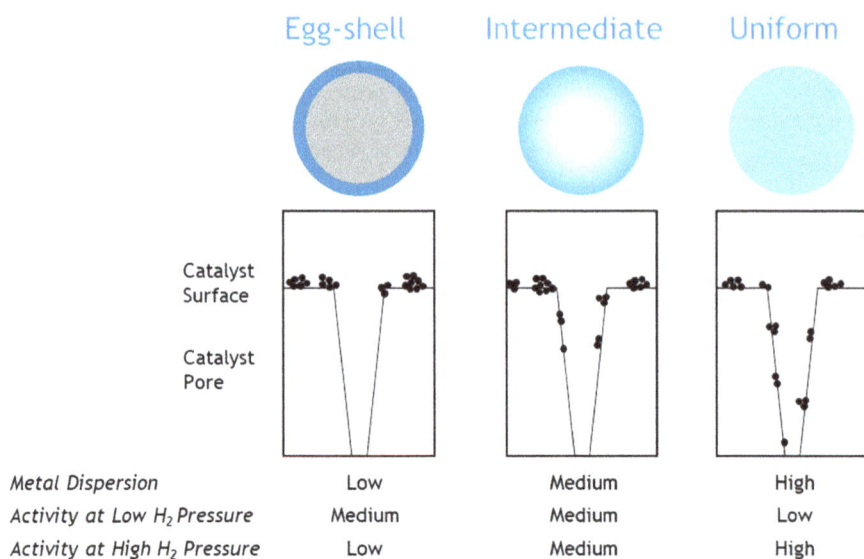

Figure 2.6: Possible distribution of metal particles in supported catalysts.

	Egg-shell	Intermediate	Uniform
Metal Dispersion	Low	Medium	High
Activity at Low H_2 Pressure	Medium	Medium	Low
Activity at High H_2 Pressure	Low	Medium	High

user and the catalyst manufacturer. This often results from multiple cycles of semi-empirical catalyst design and synthesis followed by use-testing. In rarer cases, a specific catalyst recipe may have been developed by the fine chemical manufacturer itself and then given to a catalyst manufacturer to produce (or "toll") the catalyst.

2.3.1 Palladium catalysts

2.3.1.1 Reduction of carbon-carbon multiple bonds

In general, the rate of hydrogenation of alkynes and alkenes to alkanes over supported PGM catalysts is in the following order [28]:

$$Pd > Rh > Pt \gg Ru$$

For this reason, supported palladium catalysts are some of the most widely used for the reduction of carbon-carbon double bonds, with usually a 5% palladium on carbon catalyst being used. However, a 1–10% Pd palladium loading is also known and *Pearlman*'s catalyst (10% or 20% palladium hydroxide on carbon) [29] can also be used. In certain situations, *Pearlman*'s catalyst can give superior performance. One disadvantage of palladium-based catalysts is that if there are other substituted C=C double bonds that should not be reduced in a molecule, use of a palladium catalyst can result in Z/E isomerisation or double bond migration. In this case, an alternative metal for the hydrogenation reaction needs to be chosen. Discounting electronic effects, the more substituted a C=C multiple bond is, the slower its rate of hydrogenation. This means that if there are multiple possible reduction sites, the least hindered will, generally, be reduced first.

One of the large-volume applications of palladium on carbon catalysts is the reduction of intermediates in the side chain of α-tocopherol derivatives. α-Tocopherol is the most important member of the group of vitamin E compounds and has antioxidant properties. α-Tocopherol (and its derivatives) is produced on a scale of >70,000 tonnes per year for application in human and animal nutrition, as well as use in the pharmaceutical, food and cosmetic industries [30]. The final steps involve a condensation of trimethylhydroquinone with isophytol (see Chapter 8: Acid–base-catalysed reactions). Isophytol is produced in a multi-step sequence involving the hydrogenation of multiple C=C double bonds (Scheme 2.17). One of the key intermediates is hexahydro-pseudoionone, a C_{13}-ketone. This compound can be prepared by the hydrogenation of either pseudoionone (hydrogenation of 3 × C=C) or geranylacetone (hydrogenation of 2 × C=C). Both of these compounds can be prepared using catalytic processes (see Chapter 6: C–C-bond and C–N-bond forming reactions). These hydrogenation reactions take place with a palladium on carbon catalyst at elevated temperatures (up to 90 °C) and hydrogen pressures (<10 bar, ideally < 5 bar). Hexahydro-pseudoionone then undergoes a C_5-chain extension to a C_{18}-allene-ketone, which is also hydrogenated with a palladium on carbon catalyst to give a C_{18}-ketone [31]. Due to the high production volumes of α-tocopherol, each of these steps is performed on >10,000 tonnes per year scale. In addition to these batchwise slurry processes, continuous hydrogenation using a fixed-bed catalyst has also been reported [32].

Related examples are the preparation of the aroma compounds tetrahydrolinalool and 3,7-dimethyloct-1-en-3-ol. Both molecules are related to the aroma compound linalool (see Scheme 2.24 and Section 2.3.2) with similar aromas. Both syntheses use the same raw materials and in both cases, all the unsaturated C=C bonds are reduced. For tetrahydrolinalool synthesis, methylhept-5-en-2-one undergoes catalytic addition of acetylene and then both the C=C and C $\equiv$ C bonds are completely hydrogenated with a palladium on carbon catalyst (Scheme 2.18). 3,7-Dimethyloct-1-en-3-ol is produced in a similar way, but with a change in order of the chemical steps. First, methylheptan-2-one is produced by a standard hydrogenation with a palladium on carbon catalyst. Then the catalytic addition of acetylene takes place. To produce the final aroma compound, a selective semi-hydrogenation of the triple bond to the double bond takes place – this is described more in Section 2.3.1.2.1 [30].

Hydrogenation of C=C double bonds can also be diastereoselective. A good example of this is the preparation of (+)-biotin, a vitamin that has three stereogenic centres (and therefore eight possible stereoisomers). Only one isomer has the full biological activity; so, a stereoselective synthesis is critical. Biotin is an essential co-factor for many enzymes in the body, and a deficiency can result in dermatitis and central nervous system abnormalities; biotin also has beneficial effects in the hair and in nails. In the industrial synthesis of biotin, a key intermediate containing two of the required stereogenic centres is the D-lactone. This compound can be prepared by a diastereose-

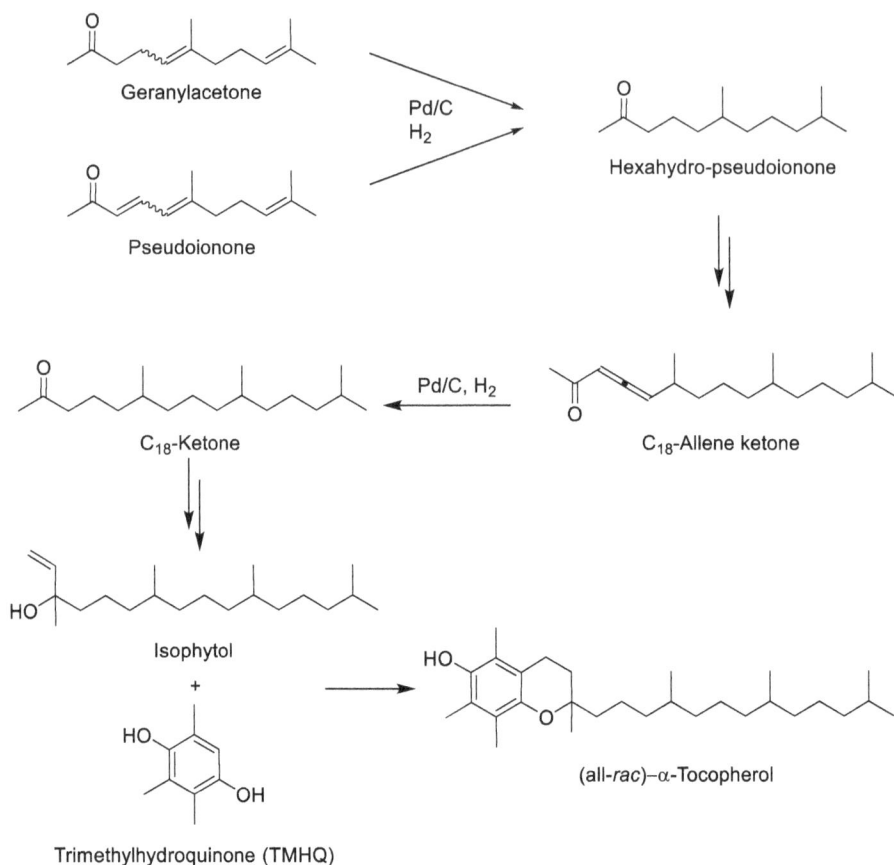

Scheme 2.17: Hydrogenation reactions used in the production of (all-*rac*)-α-tocopherol (and derivatives).

lective ring opening of the corresponding anhydride [11, 33] or by a homogeneous asymmetric hydrogenation of the same anhydride [34] (see Chapter 3: Homogeneous hydrogenations). The *D*-lactone is converted to the thiolactone, and then the side-chain is added via a *Grignard* reaction and elimination forming a C=C double bond. This is then hydrogenated with a palladium on carbon catalyst to give the final stereo-genic centre in the correct configuration (Scheme 2.19). The diastereoselectivity is very high due to the fused 5–5 ring system and the fact that only one face of the C=C double bond can bind to the catalyst surface. Under the conditions used for the hydro-genation reaction, the benzyl protecting groups on the nitrogen are stable. The other noteworthy feature of this reaction is that the reaction proceeds all, since sulphur-containing compounds are known to strongly deactivate palladium catalysts. Higher catalyst loadings (compared to other C=C bond reductions) are required, but an effi-cient, cost-effective process is still possible.

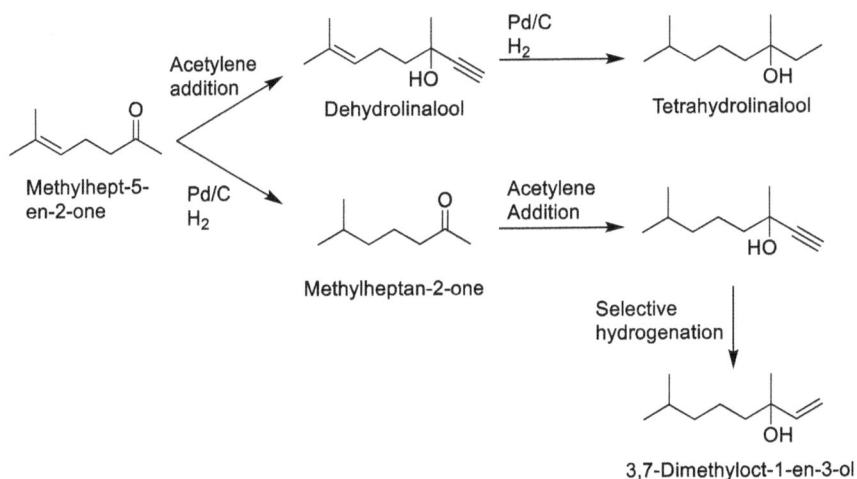

Scheme 2.18: Hydrogenation reactions used in the production of two aroma compounds.

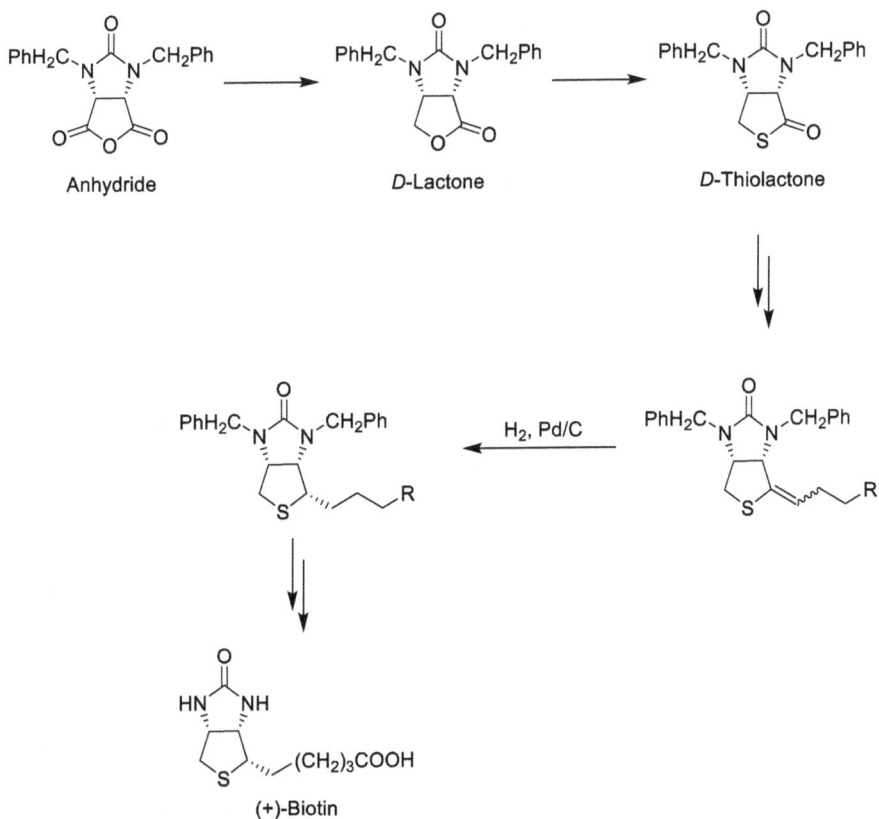

Scheme 2.19: Synthesis of (+)-biotin.

An example of using a palladium catalyst on a non-standard support and using alternative solvents is the hydrogenation of isophorone (Scheme 2.20). The product, dihydroisophorone (3,3,5-trimethylcyclohexanone), is used as a solvent for paints, varnishes and vinyl resins. Many previous hydrogenation approaches only showed high selectivity at low conversions; if reactions were run at high conversions, lower selectivities were observed. High selectivities at high conversions are important since the separation of the starting material and product (and potential side products) is difficult and costly. The best results were obtained in a flow reactor using super-critical carbon dioxide ($scCO_2$) as solvent with palladium supported on an acidic ion-exchange resin (Deloxan®). Full conversion and excellent selectivity (>99%) was observed at 140–200 °C at 25–40 weight% concentration of isophorone in $scCO_2$. This combination of catalyst, solvent and flow chemistry has also been applied to a number of other hydrogenation substrates [35].

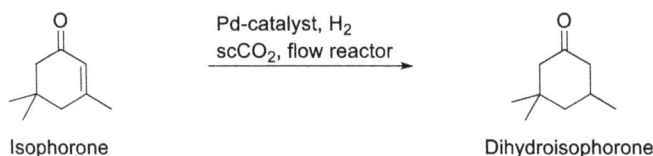

Isophorone

Pd-catalyst, H_2
$scCO_2$, flow reactor

Dihydroisophorone

Scheme 2.20: Hydrogenation of isophorone in supercritical CO_2.

Musk compounds are one of the most important classes of aroma compounds (Figure 2.7). These macrocyclic compounds are relative expensive based on their difficult synthesis, mainly because the cyclisation reaction to form a macrocycle is thermodynamically unfavoured. Typical industrial routes to musk compounds are polymerisation or oligomerisation reactions, acyloin condensation reactions, ring expansion reactions and transesterification reactions [36, 37]. Improvements in the synthesis of macrocyclic ketones lead to the *Stoll-Hansly-Prelog* process for the manufacture of 15-pentadecanolide (tradenames Exaltone®, Pentalide®, Cyclopentadecanolid®) which allows the synthesis of this class of compounds at higher concentration [38].

The real breakthrough in macrocycle synthesis of lactones and ketones was the *Wilke* trimerisation of butadiene followed by further functionalisation of the resulting cyclododecatriene [39, 40], which can undergo a ring extension applying appropriate reaction conditions. The production volume of musk aroma compounds is assumed to be several t/a, and of cyclododecane 10,000 t/a (see below) [36, 41].

In the field of industrial muscone synthesis an important intermediate is muscenone®, a mixture of isomers ((E/Z)-3-methylcyclopentadeca-4-en-1-one and (E/Z)-3-methylcyclopentadeca-5-en-1-one). Starting from cyclododecanone, a cyclisation reaction with dimethyl methylsuccinate in presence of a Brønsted acid, preferred phosphoric acid, followed by hydrogenation and elimination results in the formation of methylbicyclo[10.3.0]-pentadec-1(12)-ene. Ozonolysis and reductive workup followed by hydrogenation of the corresponding ketone in presence of a

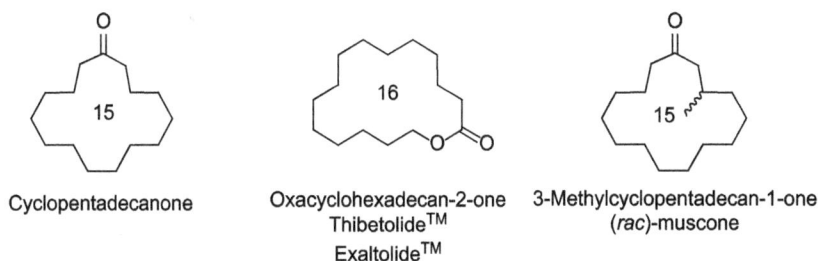

Cyclopentadecanone

Oxacyclohexadecan-2-one
Thibetolide™
Exaltolide™

3-Methylcyclopentadecan-1-one
(*rac*)-muscone

Figure 2.7: Macrocyclic aroma compounds.

Ni-alloy catalyst, and an acid treatment muscenone® is formed. The hydrogenation of the mixture of isomers in presence of Pd/C results in the formation of (*rac*)-muscone (Scheme 2.21) [36, 37, 40, 42, 43].

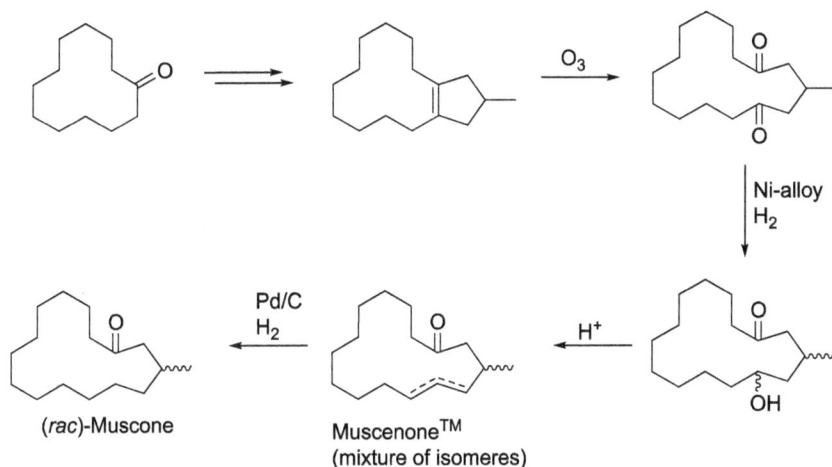

(*rac*)-Muscone

Muscenone™
(mixture of isomeres)

Scheme 2.21: Approach to muscenone® and (*rac*)-muscone.

2.3.1.2 Selective hydrogenations

Selective hydrogenation reactions can be the reduction of one functional group, leaving one or multiple other functional groups untouched (e.g. the reduction of a C=C double bond in the presence of a carbonyl group), or a semi-hydrogenation where the functional group is only partially reduced (e.g. an alkyne to an alkene, or a carboxylic acid/acid chloride to an aldehyde).

2.3.1.2.1 Reduction of alkynes to alkenes

The selective reduction of alkynes to alkenes is widely applied in the synthesis of vitamins and aroma compounds. Probably the largest application of this technology is in the synthesis of isophytol, the key intermediate in the vitamin E (α-tocopherol) and vitamin K_1 side chain (Scheme 2.22). Here a specially modified palladium catalyst is used to ensure that the over-hydrogenation to the alkane (dihydroisophytol) is minimised. The catalyst used is known as the "*Lindlar* catalyst", which is palladium supported on calcium carbonate, doped with lead [44, 45]. Before discussing in detail the selective semi-hydrogenation of terminal alkynes, we will first discuss the original application of the *Lindlar* catalyst in the preparation of vitamin A derivatives (retinol/ retinyl acetate).

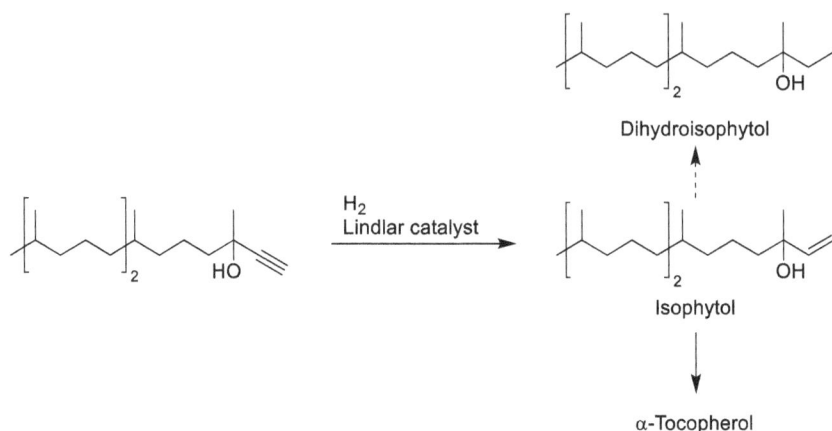

Scheme 2.22: Semi-hydrogenation of an alkyne to an alkene for the preparation of isophytol.

The first industrial synthetic route to vitamin A acetate (retinyl acetate) was developed by Otto Isler and co-workers at F. Hoffmann-La Roche. This involved the coupling of a C_{14}-building block derived from β-ionone with a C_6-building block to produce a C_{20}-compound called oxenin (Scheme 2.23). The following steps to vitamin A acetate involved the semi-hydrogenation of the carbon-carbon triple bond, elimination of water and acylation [46]. Originally, the partial reduction of the alkyne was performed by using a poisoned palladium on carbon catalyst. This was successful, but lead to some over-hydrogenated impurities that were difficult to remove; these could only be reduced by careful monitoring the consumption of hydrogen gas and stopping the reaction at the correct point. A significant improvement was made by Herbert Lindlar; a palladium on calcium carbonate catalyst was used which had significantly lower support surface area (see Table 2.3) and thus lower activity. This was then improved upon by adding an additional metal (lead) to the catalyst preparation to further reduce the activity and reduce the over-hydrogenation. Further improvements could be made by adding an organic modifier (such as an amine; e.g. quinoline) which resulted in the semi-hydrogenation reaction stopping almost completely after the alkyne had been consumed. Since the reduction takes place on the surface of the catalyst, the hydrogen is delivered to the same side of the carbon-carbon triple bond (*syn*-addition). Therefore, when an internal alkyne is reduced with a heterogeneous catalyst, almost exclusively the (Z)-isomer is produced. Traces of the (E)-isomer can sometimes be observed, but these are generally formed by isomerisation of the product, after reduction.

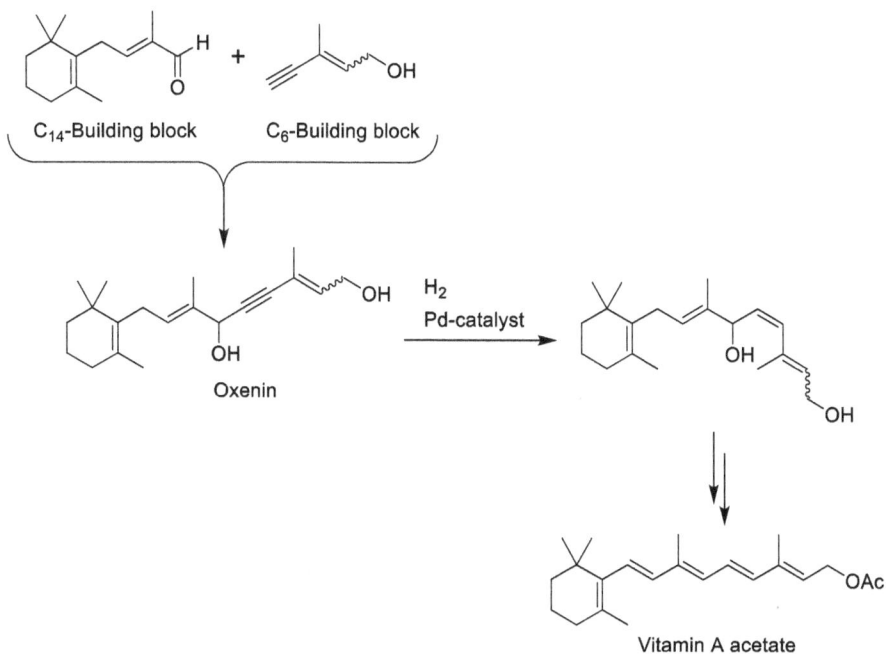

Scheme 2.23: Semi-hydrogenation of an internal alkyne in the preparation of vitamin A acetate.

There have been a number of investigations to understand why the *Lindlar* catalyst is so successful. In addition, there have been attempts to produce a better catalyst for the selective semi-hydrogenation of alkynes, but until now, none have surpassed the original in industrial application. In general, the hydrogenation of an alkyne with a supported palladium catalyst follows a sequential reaction pathway where the alkyne is first hydrogenated to the alkene, which then undergoes subsequent hydrogenation to the alkane; in an ideal situation, the reaction would be stopped as soon as all the alkyne is consumed. The alkyne binds more strongly to the catalyst surface than the alkene and alkane; as long as sufficient alkyne is present in the reaction mixture and surface exchange is fast enough, alkyne from the reaction mixture displaces the product alkene from the catalyst surface (Figure 2.8) [47]. This results in high selectivities at partial conversions; however as the concentration of alkyne decreases, hydrogenation of the desired alkene increases, generating the unwanted alkane and lowering the selectivity.

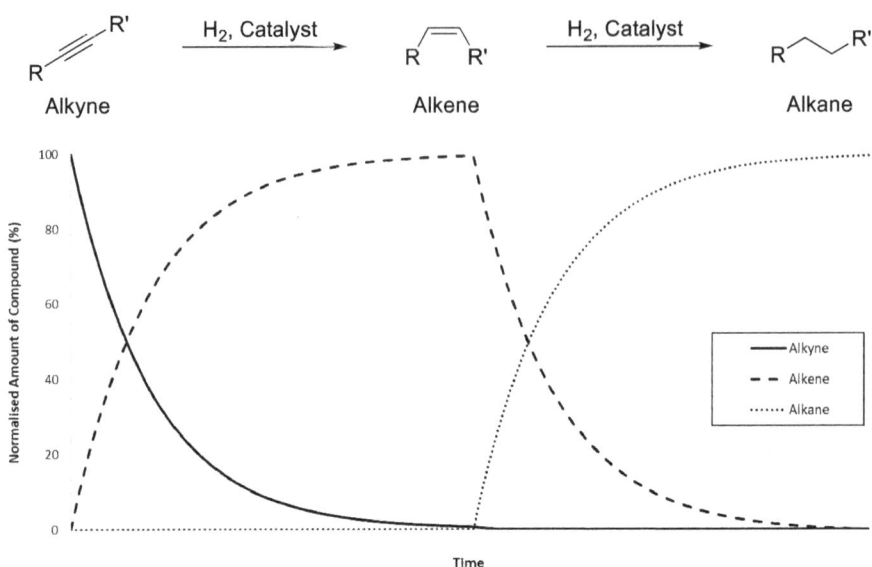

Figure 2.8: Schematic of the theoretical "perfect" semi-hydrogenation of an alkyne (with both rates of hydrogenation identical).

An interesting comparison of the relative rates of the first (alkyne–alkene) and second (alkene–alkane) hydrogenation steps can been seen in an example experiment testing different catalysts for the hydrogenation of a terminal alkyne. The reactions were run under identical conditions and the amount of consumed hydrogen was measured. The reaction with palladium on carbon proceeded very quickly and two equivalents of hydrogen were consumed, yielding the alkane with no obvious change in hydrogenation rate between the two sequential reaction steps (Figure 2.9). Using a palladium on calcium carbonate catalyst, the overall rate of hydrogenation (slope of the curve) was slightly slower, but again the reaction smoothly consumed two equivalents of hydrogen resulting in complete over-hydrogenation. When an organic catalyst modifier was added to the reaction with palladium on calcium carbonate, a clear difference in the two rates of hydrogenation was observed with a "corner" in the hydrogen consumption curve at 100–110% of the theoretical hydrogen amount (for a single hydrogenation). The subsequent rate of hydrogenation of the alkene to alkane was significantly lower. Using a *Lindlar* catalyst with the same catalyst modifier showed an even larger improvement with the rate of alkene hydrogenation reduced to almost zero [48]. The ideal catalyst would stop completely at the alkene stage giving 100% selectivity. In practice, using the *Lindlar* catalyst with an organic modifier and careful optimisation of the reaction conditions, selectivities of >95% are possible on an industrial scale.

As stated previously, the preparation of isophytol from dehydroisophytol by a selective semi-hydrogenation is widely used in the synthesis of α-tocopherol (vitamin E). The semi-hydrogenation uses a traditional *Lindlar* catalyst at elevated temperatures (<100 °C)

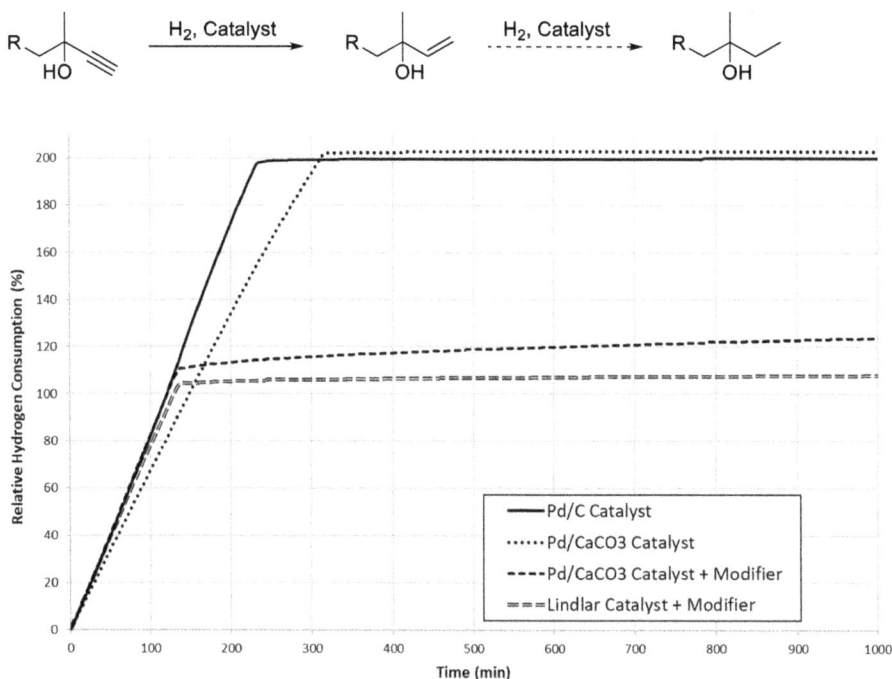

Figure 2.9: Hydrogenation of a terminal alkyne with various catalysts and conditions.

and slightly elevated pressures (Scheme 2.24). To meet the production demands for α-tocopherol, production volumes are >50,000 t/a. This C_{20}-building block can be made by a series of sequential chain elongations (see Chapter 6: C–C-bond and C–N-bond forming reactions), from hexahydropseudoione (Scheme 2.17). This C_{13}-building block can be produced from several sources, including a C_5-building block 2-methyl-but-3-yn-2-ol (MBY) used by DSM Nutritional Products. This compound undergoes a similar, but different, semi-hydrogenation to yield 2-methyl-but-3-en-2-ol (MBE). The selectivity is very high (>98%), and the catalyst can be reused multiple times [30]. The selective reduction of alkynols to enols is also used by a number of Chinese manufacturers of vitamin E intermediates and aroma compounds [49–51].

Three related semi-hydrogenations are used to produce aroma compounds with a lavender-like aroma. The parent compound is linalool, a C_{10}-molecule which is produced by selective hydrogenation with a *Lindlar* catalyst. Two derivatives, linalyl acetate and 3,7-dimethyloct-1-en-3-ol (Scheme 2.18), are related compounds with slightly different aromas and stabilities and are produced by comparable semi-hydrogenations. Although structurally similar and related to MBY and dehydroisophytol, each of these semi-hydrogenation reactions operates under different conditions to obtain the best possible selectivity (Scheme 2.25). Small variations in substrate structure can have a sig-

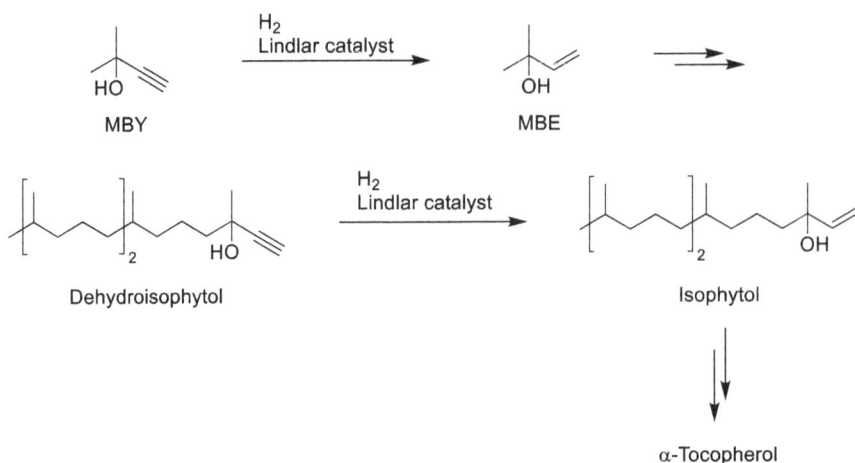

Scheme 2.24: Semi-hydrogenation of terminal alkynes for the synthesis of α-tocopherol.

nificant difference in semi-hydrogenation performance, and a cost-effective process requires individual optimisation of the reaction conditions [30].

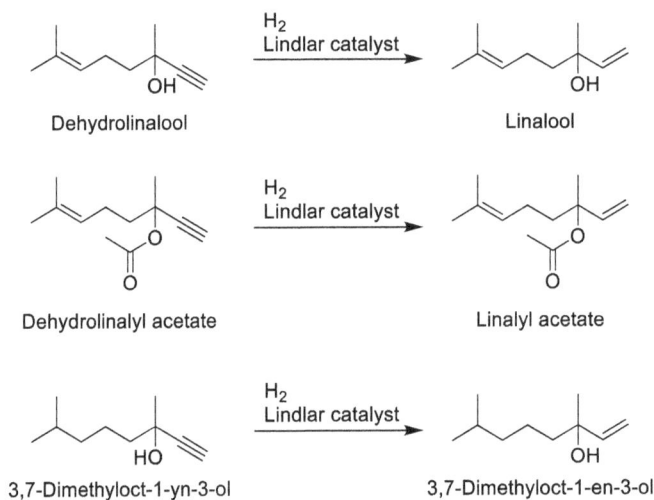

Scheme 2.25: Semi-hydrogenation of terminal alkynes producing various aroma compounds.

Alternative lavender aromas are important in the F + F industry, especially when oxidation products of molecules such as linalool/linalyl acetate may cause allergic dermatitis in small groups of people. An interesting alternative for linalyl acetate is Bergalin® (3,7-dimethylocten-3yl acetate) – like 3,7-dimethyloct-1-en-3-ol (above), it does not contain the trisubstituted C-C double bond and is less susceptible to oxidation. Bergalin can be readily synthesised from methylheptenone in excellent yield over all steps (>95%) (Scheme 2.26). This process involved hydrogenation of the C-C double bond with a palladium on carbon catalyst, followed by ethynylation and acylation. Then the C-C triple bond is hydrogenated in high selectivity with a *Lindlar* catalyst.

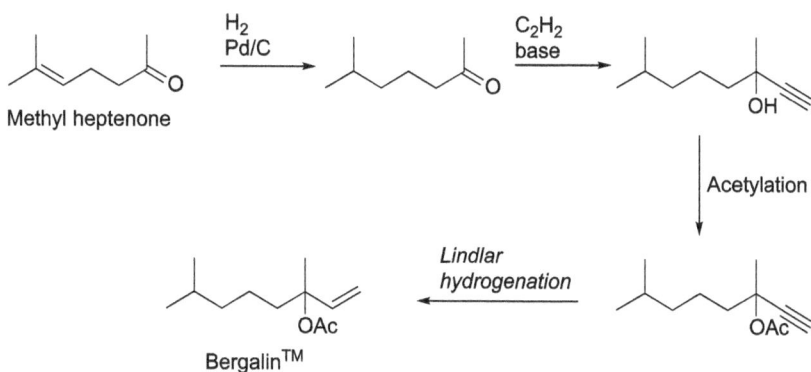

Scheme 2.26: Approach to 3,7-dimethylocten-3-yl acetate.

In addition to the selective semi-hydrogenation of terminal alkynes to alkenes, the *Lindlar* catalyst has also been used for the selective reduction of internal alkynes. One example is in the synthesis of vitamin A acetate described previously (Scheme 2.24). A further example of a related structure is the synthesis of astaxanthin, a carotenoid with a bright pink colour. It is a natural antioxidant and is also the molecule that is responsible for the pink-red colour in aquatic organisms, for example shrimps and salmon. It is produced by algae and then enters the aquatic food chain. However, since there is not enough natural material available to satisfy demand, astaxanthin is produced chemically for use in dietary supplements for humans and aquaculture.

A key step in the chemical synthesis is the selective mono-reduction of an internal carbon-carbon triple bond (Scheme 2.27). This was originally performed with zinc/acetic acid, and later by a semi-hydrogenation using a *Lindlar* catalyst [52]. In addition to optimising the reaction conditions, the preparation of the *Lindlar* catalyst was optimised. It was found that significantly increased selectivity for the mono-reduction could be achieved by varying the calcium carbonate particle size used in the *Lindlar* catalyst, with the highest selectivities being obtained when the calcium carbonate had a median diameter of more than 10 μm [53].

In the field of aroma compounds, in addition to the examples described previously, the *Lindlar* catalyst has been widely used for the reduction of internal alkynes in the synthesis of methyl jasmonate and derivatives. These are aroma compounds found in the jasmine plant and were and still are widely used in perfumes, "up to the middle of the twentieth century, ca. 80% of the marketed compositions contained a basic note extracted from this precious handpicked flower" [54]. Extracts of jasmine

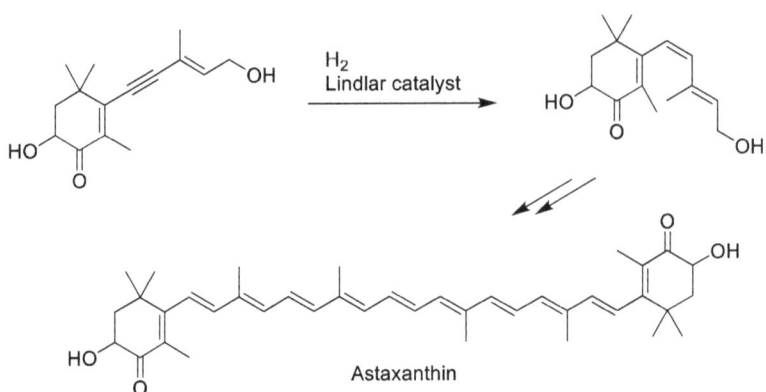

Scheme 2.27: Semi-hydrogenation of an internal alkyne in the preparation of astaxanthin.

could cost > 10,000 €/kg and supplies were severely limited, so it made perfect sense to synthesise some of the components of jasmine extracts.

The first synthetic route used the selective reduction of an alkyne with the *Lindlar* catalyst to directly produce methyl jasmonate (Scheme 2.28). A huge amount of work followed in a number of companies to develop better synthetic routes that could be commercialised. Two important routes developed by Büchi from Firmenich (Switzerland) [55] and Tsuji from Nippon Zeon (Japan) [56] used a semi-hydrogenation with a *Lindlar* catalyst on the same cyclopentenone intermediate, but the following chemical steps were different. The hydrogenation typically takes place around room temperature under a slightly elevated hydrogen pressure (<2 bar). All of these routes produce a mixture of relative stereochemistry in the five-membered ring and were initially racemic. Nowadays, a number of different qualities of methyl jasmonate and related compounds (e.g. Hedione®, Paradisone®, Magnoline® and Fleuramone®) are manufactured by chemical synthesis.

A simpler aroma compound that is also produced by the hydrogenation of an internal alkyne is known as "leaf alcohol" (*cis*-hexen-1-ol), and it has an intense fresh "green" odour similar to that of freshly cut grass. It is produced on a scale of approximately 10 t/a and uses a *Lindlar* catalyst for the reduction of the 3-hexyn-1-ol to the *cis*-alkene (Scheme 2.29) [57, 58]. In fact, this semi-hydrogenation reaction is often used as a test reaction when research groups are trying to develop a novel replacement for the *Lindlar* hydrogenation. Although there have been some interesting results, including a selective lead-free catalyst [59], it is unclear if these catalysts are currently used on an industrial scale for the production of leaf alcohol or even in other semi-hydrogenation reactions.

1,4-Butenediol is a key intermediate in the synthesis of vitamin B₆ (pyridoxine) and also of the insecticide endosulfan and is produced on a scale of several hundred tons per year. It is synthesised industrially by the partial hydrogenation of 1,4-butynediol with a palladium catalyst. Butynediol is prepared from acetylene and

First synthetic route:

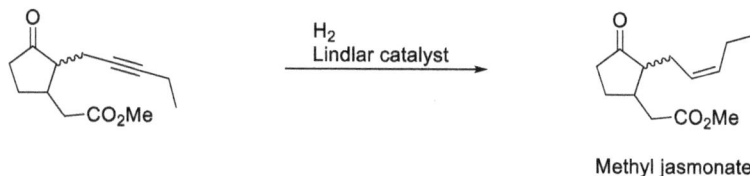

Methyl jasmonate

Firmenich and Nippon Zeon Routes:

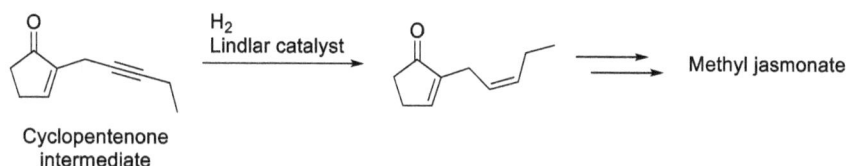

Cyclopentenone
intermediate

Scheme 2.28: Routes to methyl jasmonate using a selective alkyne–alkene reduction.

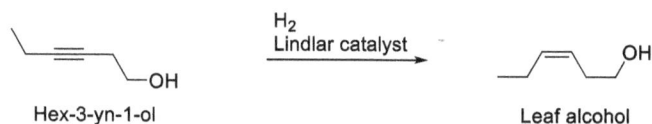

Hex-3-yn-1-ol

Leaf alcohol

Scheme 2.29: Preparation of leaf alcohol by semi-hydrogenation reaction.

formaldehyde (see Section 2.2.3, Scheme 2.11). The preferred hydrogenation catalyst is palladium, poisoned with zinc or other metals [20, 60]. The reaction is performed with an aqueous solution of butynediol at 60–80 °C with 4–13 bar hydrogen pressure (Scheme 2.30).

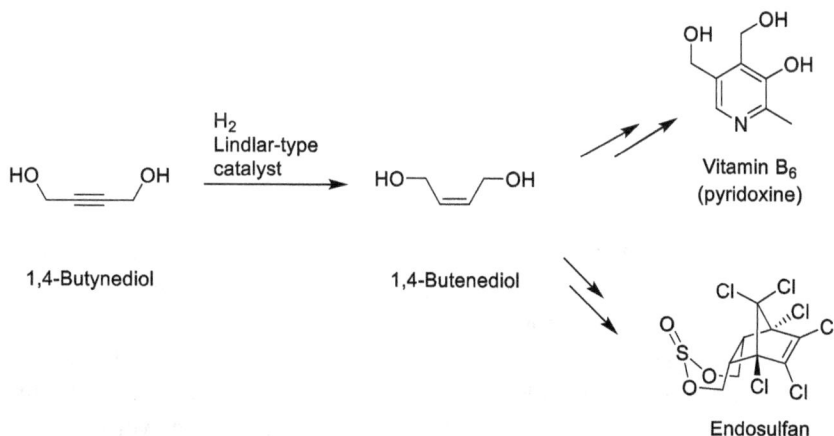

1,4-Butynediol

1,4-Butenediol

Vitamin B_6
(pyridoxine)

Endosulfan

Scheme 2.30: Semi-hydrogenation of an internal alkyne in the preparation of butenediol.

2.3.1.3 Hydrogenation of other functional groups and deprotection reactions

2.3.1.3.1 Hydrogenation reactions with aromatic groups

The hydrogenation of hydroquinones and related compounds are used in the production of vitamin E (α-tocopherol) and K_1 (phylloquinone). Here the quinone or derivative must be reduced to the aromatic hydroquinone, before condensation with the side chain can take place. In both cases, the starting material is a solid at room temperature, so the reactions are performed in a suitable solvent (Scheme 2.31). Palladium on carbon catalysts are used and the hydrogenations take place at low to medium pressure and at elevated temperatures to achieve high rates of reaction and also for solubility reasons. For α-tocopherol, trimethyl-1,4-benzoquinone (TMQ) is the starting material, which can be prepared from the oxidation of trimethylphenol (see Chapter 4: Oxidations). The resulting trimethylhydroquinone (TMHQ) can be condensed with the isophytol side chain (see Scheme 2.17). A similar sequence is used for the production of vitamin K_1, starting from menadione. The resulting menadiol undergoes further steps including condensation with an isomeric sidechain (phytol) to give the final product [30]. These reactions can also be performed in a continuous manner using a cascade of batch reactors or a loop reactor.

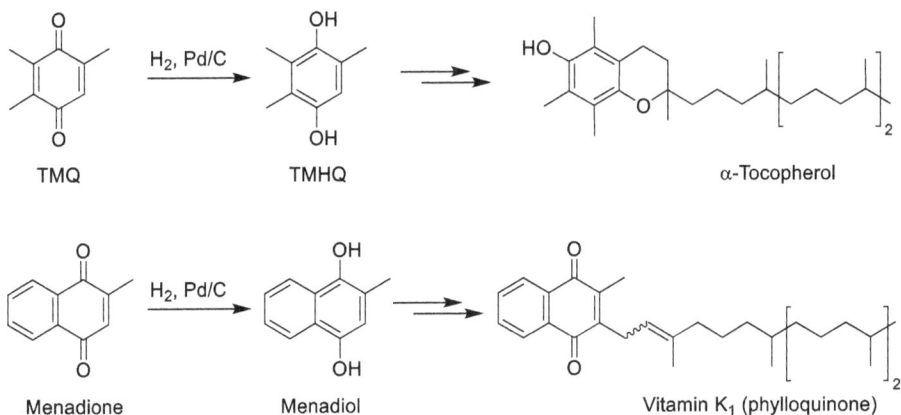

TMQ TMHQ α-Tocopherol

Menadione Menadiol Vitamin K_1 (phylloquinone)

Scheme 2.31: Hydrogenation of quinones (and derivatives) in the synthesis of α-tocopherol and vitamin K_1.

A different example, in this case where the aromatic ring itself is actually hydrogenated, is the production of cyclohexanone. As described before (Section 2.2, nickel), cyclohexanone is one of the intermediates of ε-caprolactam, the precursor to nylon 6. Phenol can be hydrogenated with a range of precious metal catalysts in the vapour phase [61]. Ambient pressure and high temperatures (>140 °C) are used and yields up to 95% are possible (at 100% conversion). Alternatively, a liquid-phase hydrogenation with palladium on carbon catalysts under milder conditions results in 99% yield at

90% conversion, based on recovered starting material (Scheme 2.32). If the reaction is performed with a nickel-alloy catalyst, then complete reduction to cyclohexanol is observed (in >99% selectivity).

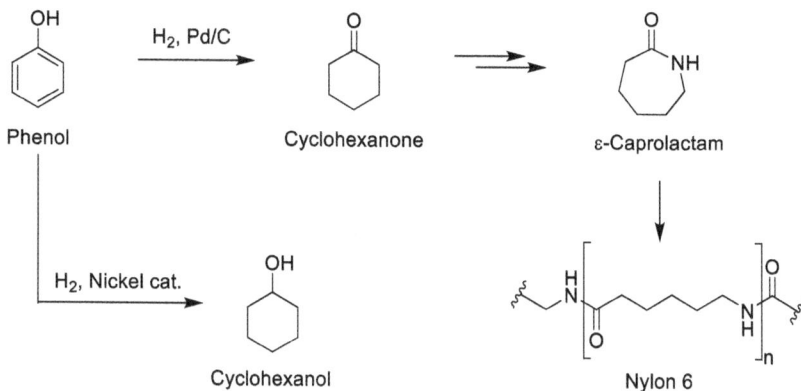

Scheme 2.32: Hydrogenation of phenol for the synthesis of caprolactam and nylon 6.

Heteroaromatic groups can also be reduced and an example is the triazole intermediate of sitagliptin, a new treatment for diabetes developed by Merck and launched in 2006. The key triazole fragment is prepared by hydrogenation of triazolopyrazine under moderate conditions (3.5 bar H_2, 45 °C) in 51% yield (Scheme 2.33) [62]. A hydrogenation reaction was also used to cleave a benzyl N-oxide in a later intermediate of sitagliptin (see Section 2.3.1.3.3).

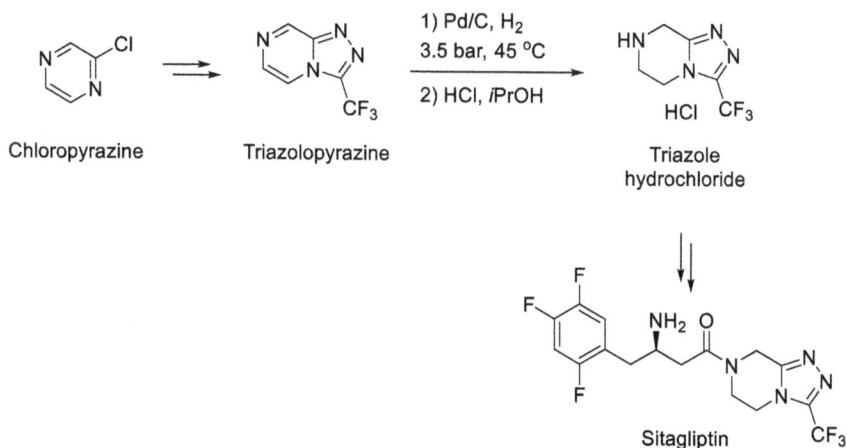

Scheme 2.33: Hydrogenation of a heteroaromatic group in the synthesis of sitagliptin.

2.3.1.3.2 Other functional groups

As seen previously, the production of ε-caprolactam uses the hydrogenation of phenol to produce cyclohexanone (Scheme 2.32); the other reacting species is hydroxylamine. This reacts with cyclohexanone to form the corresponding oxime, which then undergoes a sulphuric acid catalysed *Beckmann* rearrangement to yield caprolactam (see Chapter 8: Acid–base-catalysed reactions), the overall yield from cyclohexanone is about 98% (Scheme 2.34). The whole process ("HPO process") from ammonium nitrate to the oxime is fully integrated and uses multiple recycle streams for maximum efficiency. An ammonium nitrate/phosphoric acid buffer is reacted with hydrogen in the presence of a supported palladium catalyst in a three-phase bubble column reactor. Careful choice of reaction conditions and the metal-promoter for the palladium catalyst increases the activity and selectivity of the catalyst. Continuous process development with more concentrated process flows and reduced raw material and energy consumption allows individual production plants to reach over 200,000 t/a [63].

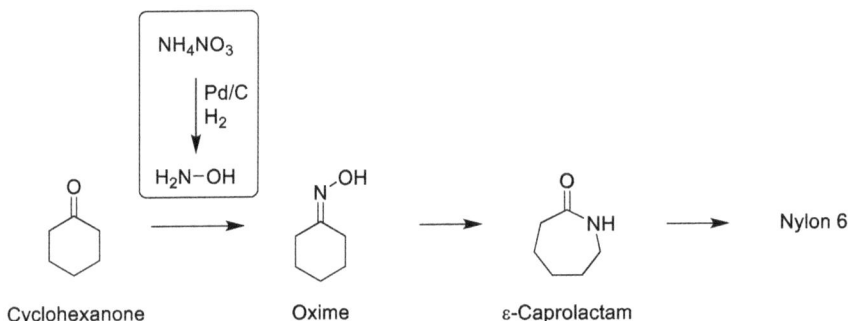

Scheme 2.34: Preparation of hydroxylamine for the synthesis of caprolactam and nylon 6.

A common functional group interconversion in the synthesis of fine chemicals is the reduction of a nitro group to an amine. This is usually performed by a metal-catalysed hydrogenation and can be achieved with nickel (see Section 2.2), platinum (see Section 2.3.2) or palladium. One example is the preparation of nintedanib, developed by Boehringer Ingelheim for the treatment of idiopathic pulmonary fibrosis (a condition in which the lungs become progressively scarred over time) and also non-small cell lung cancer. Here, two different aromatic nitro groups are reduced, both with a palladium on carbon catalyst (Scheme 2.35). In the first, the preparation of the aniline fragment, the hydrogenation takes place under very mild conditions with a 10% palladium on carbon catalyst [64] in >70% yield. In the second, slightly stronger conditions are required (3 bar H_2, 45 °C) giving the corresponding aniline in 87% yield; after the hydrogenation is complete, the reaction mixture is heated to induce decarboxylation and cyclisation to form the desired oxindole fragment [65]. These two components are subsequently combined to give the final drug substance.

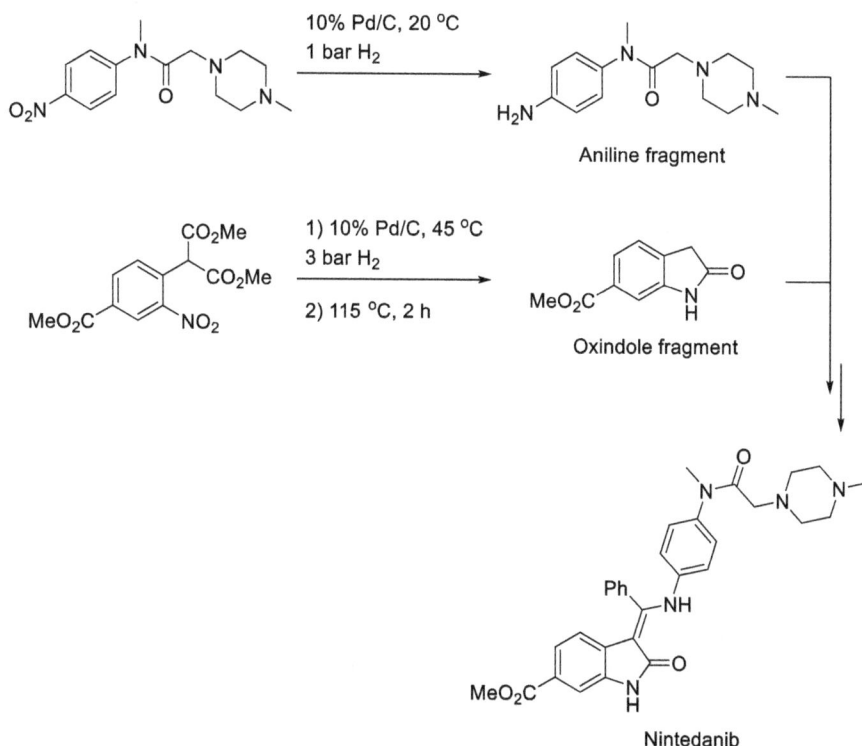

Scheme 2.35: Two palladium-catalysed hydrogenations of nitro groups in the synthesis of nintedanib.

An interesting combination of a nitro group hydrogenation and also the reduction of an aromatic group is the preparation of *trans*-4-aminocyclohexyl acetic acid, an intermediate in the synthesis of the drug cariprazine, an anti-psychotic agent used in the treatment of schizophrenia and bipolar disorder, which is sold as a racemate. Whilst the reaction can be performed using skeletal nickel catalysts, high temperatures and pressures are required and significant quantities of the unwanted *cis*-isomer are produced. It was discovered that significantly higher *trans*–*cis* ratio can be achieved if the reaction is performed under milder conditions with a palladium on carbon catalyst; in addition the reaction is performed in two stages (Scheme 2.36). First the nitro group is reduced at 45 °C with a small over-pressure (0.6 bar). Then the temperature and pressure are increased to reduce the aromatic ring [66]. The unwanted *cis*-isomer can then be removed by crystallisation, giving an overall yield of 40% of the pure *trans*-isomer.

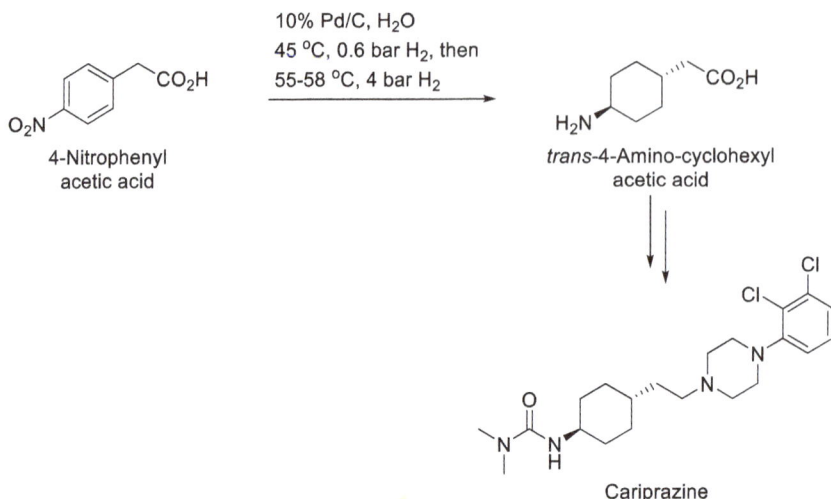

Scheme 2.36: Consecutive reduction of a nitro group and an aromatic ring in the synthesis of cariprazine.

2.3.1.3.3 Debenzylation reactions

In general, protecting groups are usually avoided in the preparation of fine chemicals since this approach increases the number of chemical steps by two (one for the protection and one for deprotection). However, there are cases where protection is required and palladium catalysts are excellent for the removal of benzyl protecting groups to produce carboxylic acids, alcohols and amines. One main use (albeit in small volumes) is in the field of peptide synthesis. Palladium-catalysed debenzylations are generally highly efficient processes giving only toluene (from the hydrogenolysis) as a side product.

One example of a debenzylation from the pharmaceutical industry is from a group at Wyeth [27]. They had previously used a nickel-alloy catalyst to produce the *cis*-diol from hydroxymethylfurural in the synthesis of a drug intermediate (see Section 2.2.4, Scheme 2.16). This was then converted into a tertiary amine bearing a benzyl group. However, the desired compound was the bicyclic secondary amine and this was synthesised by deprotection of the tertiary amine (Scheme 2.37). Optimal conditions required the use of *Pearlman*'s catalyst (palladium hydroxide on carbon) rather than palladium on carbon. Reactions were run at moderate temperature (60–65 °C) and 4–5 bar hydrogen pressure.

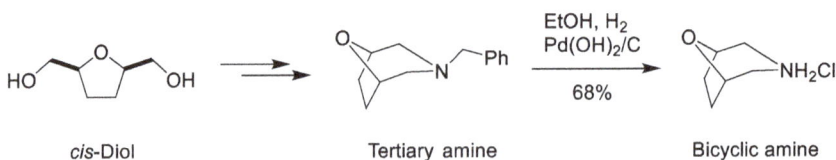

Scheme 2.37: Preparation of a bicyclic amine by debenzylation.

The cleavage of benzyl groups from amines is used twice in the synthesis of the drug delafloxacin, using the same catalyst but under different reaction conditions (Scheme 2.38) [67]. A diamino pyridine derivative was prepared from the dibenzyl precursor using *Pearlman's* catalyst with formic acid as the hydrogen source. This has the advantage that the reaction does not take place under pressure and the stoichiometry of the amount of hydrogen can be better controlled. A separate intermediate, the hydroxy azetidine hydrochloride was prepared by a similar debenzylation, but this time using hydrogen gas. In both cases, the benzyl protected substrates were prepared from benzylamine [68].

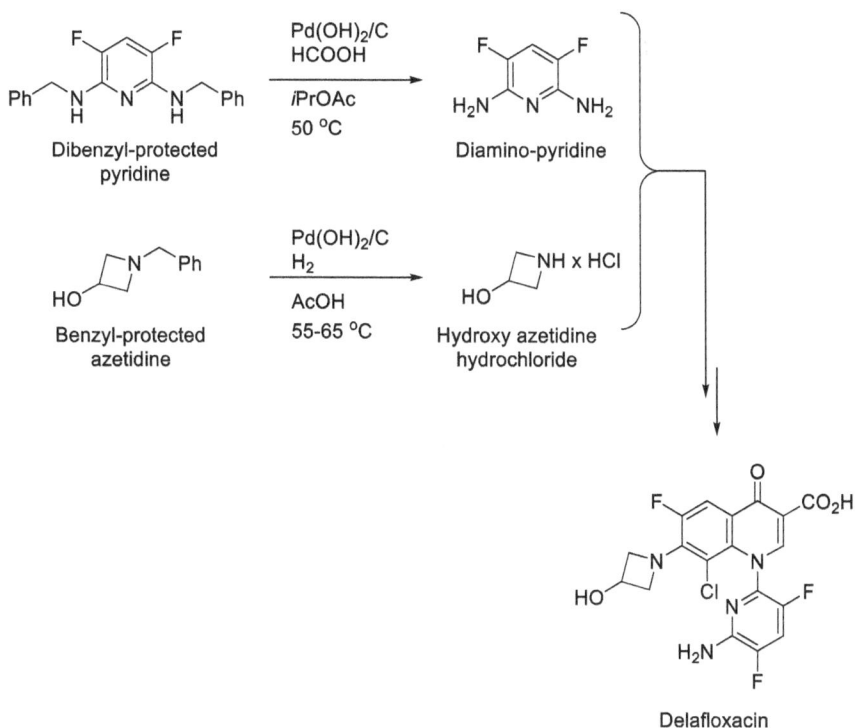

Scheme 2.38: Synthesis of delafloxacin utilising two different debenzylation reactions.

The cleavage of benzyl groups from oxygen atoms is also performed by heterogeneous hydrogenation with palladium catalysts. In the synthesis of sitagliptin, not only was a benzyl group cleaved from a hydroxy species (in this case a hydroxylamine), but also the N–O bond of the hydroxylamine was cleaved under the same reaction conditions to give the final drug substance (Scheme 2.39) [56]. This synthetic route was eventually replaced by an asymmetric hydrogenation approach (see Chapter 3: Homogeneous hydrogenations) and later by a biocatalytic process using a transaminase (see Chapter 10: Biocatalysis).

The carboxybenzyl-protecting group ($PhCH_2OCO-$), often written as Cbz- or Z-, is very commonly used as a protection group for nitrogen atoms, especially in the area

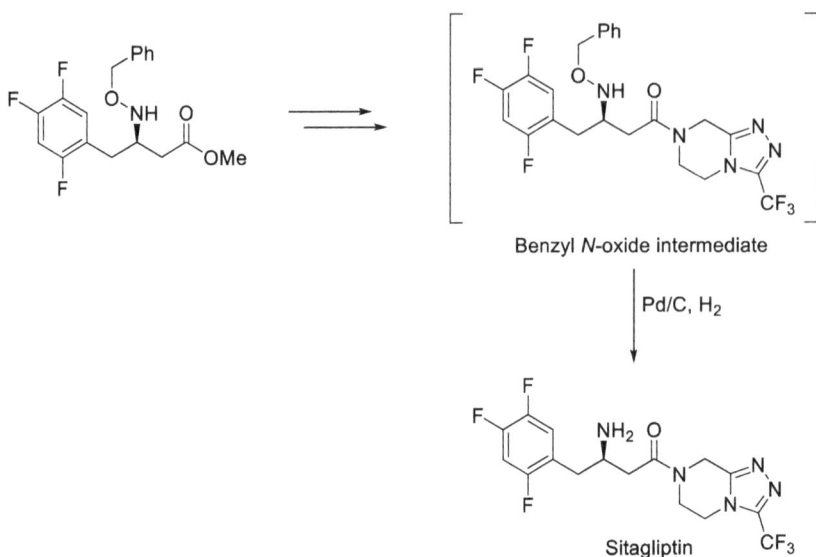

Scheme 2.39: Reductive cleavage of a benzyl-protected *N*-oxide in the synthesis of sitagliptin.

of peptide chemistry. It is readily removed by heterogeneous hydrogenation using a palladium on carbon catalyst. Reaction conditions are generally mild requiring temperatures and pressures at, or slightly above, ambient levels, and reactions are fast and selective. The Cbz-protecting group has the advantage that the deprotection is orthogonal (i.e. can be introduced/removed under different conditions) to other nitrogen protecting groups such as amides and other carbonates.

On a large scale, Cbz-deprotection has been utilised in the synthesis of the artificial sweetener aspartame which has a worldwide production volume of 15,000 t/a. The original synthesis used the condensation of an aspartic acid derivative with phenyl alanine methyl ester; however, this approach resulted in the formation of a bitter tasting byproduct. An improved process used a biocatalytic condensation of a Cbz-protected aspartic acid to produce the Cbz-protected aspartame (see Chapter 10: Biocatalysis). The Cbz group was then removed by catalytic hydrogenation with 5% palladium on charcoal at ambient pressure and temperatures of up to 60 °C (Scheme 2.40) [69].

The Cbz-protecting group is widely used in the pharmaceutical industry, and one example is the Merck synthesis of raltegravir, a HIV integrase inhibitor that was approved in 2007 for use in combination therapies with other antiretroviral drugs. A convergent synthesis was used with an amine building block as a key component for the final steps. This building block was synthesised from the Cbz-protected precursor with palladium on carbon at slightly elevated temperature and pressure (Scheme 2.41) [70, 71].

In all of the above-mentioned examples, the benzyl group is removed by catalytic hydrogenation. A related example, where the *aromatic group is actually the wanted*

Scheme 2.40: Removal of a Cbz-protecting group in the preparation of aspartame.

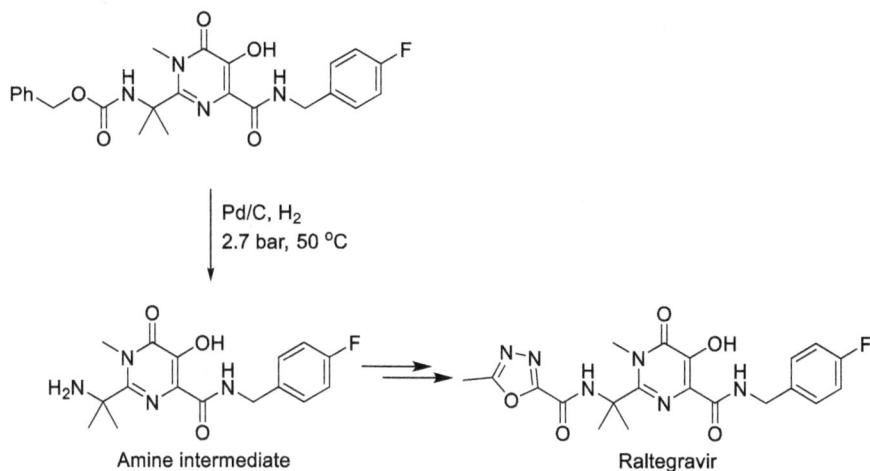

Scheme 2.41: Cbz deprotection in the synthesis of the HIV drug raltegravir.

product, is an alternative preparation of TMHQ, the aromatic core of α-tocopherol, as discussed in Section 2.3.1.3.1. Here, the synthesis starts from a dimethyl starting material; two palladium-catalysed hydrogenation steps are required [72]. First, the 2,6-dimethylquinone (2,6-DMQ) is reduced to the corresponding hydroquinone in up to 97% yield, in an analogous way to the trimethyl equivalent (Scheme 2.31). Then the missing carbon atom is introduced by an aminomethylation using morpholine and formaldehyde. Direct methylation of the aromatic ring is not used, to ensure that only the trisubstituted compound is produced. The morpholine group is then removed (and later recycled) by a further hydrogenation with palladium on carbon (Scheme 2.42) in >97% conversion and 91–94% yield.

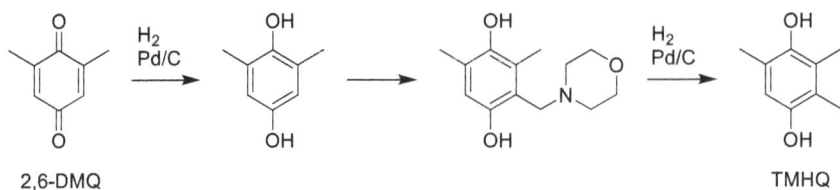

Scheme 2.42: An alternative route to trimethylhydroquinone (TMHQ).

2.3.2 Platinum catalysts

Compared to palladium, platinum-based catalysts are less widely used than their palladium counterparts. One of the reasons is for the hydrogenation reactions that they can catalyse – palladium is significantly more active and selective than platinum. For example, in the reduction of carbon-carbon double and triple bonds, debenzylation reactions and the fact that only benzylic ketones are generally reduced by palladium catalysts [28]. A further reason is that until recently, platinum was 2–5 times more expensive than palladium (see Figure 2.4). Whether this price difference will remain in the coming years is yet to be seen, but there are several applications where platinum is the preferred metal catalyst in the synthesis of fine chemicals. Most of these applications come from the field of the synthesis of pharmaceuticals, since these higher-priced molecules are better able to absorb the (previously) higher metal price of platinum.

2.3.2.1 Supported platinum catalysts

Platinum-based catalysts are often used for the reduction of nitro groups, especially when there are other functional groups in the molecule that could also be hydrogenated. A good example of this reaction is found in the preparation of the anticoagulant betrixaban, which was discovered by Millennium Pharmaceuticals and later developed by Merck Sharp & Dohme. The synthesis required the reduction of an aryl nitro group; however, the molecule also contained a chloro-pyridine moiety (Scheme 2.43). Use of a palladium catalyst would likely also cause partial dechlorination, so a sulphided platinum catalyst was chosen. Under mild conditions, 90% yield was obtained with less than 0.1% of the corresponding dechlorinated impurity [73].

Metolachlor is highly active grass herbicide, discovered in 1970, that is produced on >10,000 t/a and is used in weed control in crops such as maize, soybean, peanuts and cotton. It was originally prepared as a racemate, but since the (S)-isomer is significantly more active, it is now sold as a diastereomerically enriched mixture and produced by homogeneous asymmetric hydrogenation (see Chapter 3: Homogeneous hydrogenations). The original synthesis, for the racemic product was developed in the 1970s and was operated for approximately 20 years until the process for the single isomer was implemented [74]. Methyl ethylaniline undergoes a platinum-catalysed reductive alkylation with an aqueous solution of methoxyacetone in the presence of small

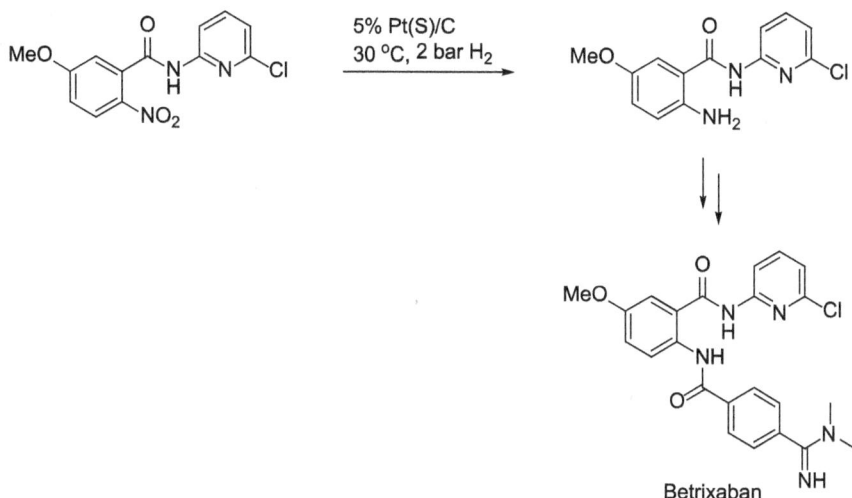

Scheme 2.43: Platinum catalysed nitro group reduction in the synthesis of betrixaban.

amounts of sulphuric acid (Scheme 2.44). The reaction takes place at slightly elevated temperatures and pressures (50 °C, 5 bar) with approximately 1 wt% of a 5% platinum on carbon catalyst. The catalyst can be reused more than 20 times without a significant loss in activity or selectivity, resulting in a very cost-effective process [75].

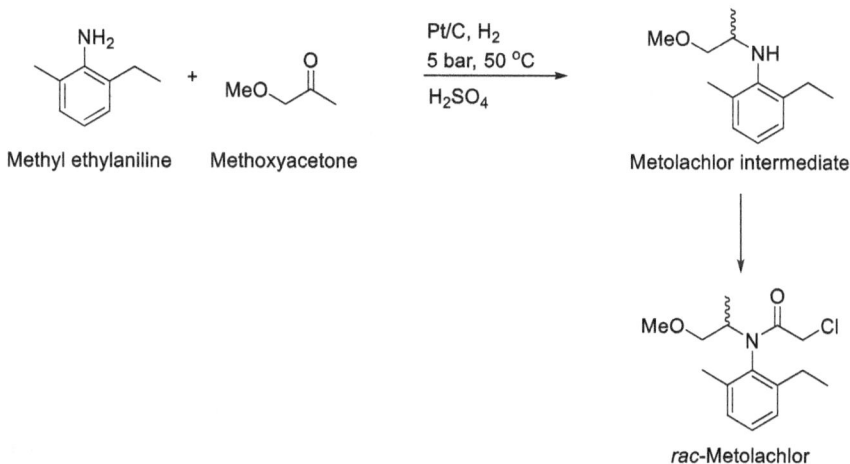

Scheme 2.44: The original industrial synthesis of the herbicide metolachlor.

A similar transformation was used in the synthesis of the drug safinamide which was approved in 2015 for the treatment of mid- and late-stage *Parkinson's* disease. The process chemistry route was similar to the original medicinal chemistry route, involving a reductive amination between *L*-alaninamide and a benzaldehyde derivative. The traditional synthetic chemistry conditions for a reductive amination (NaBH₃CN, MeOH) resulted in low yields and significant amounts of unwanted by-products. Use of a heterogeneous hydrogenation catalyst showed a significant improvement. Palladium on carbon resulted in 72% yield in a one-pot reaction where the imine was formed *in situ*. Switching to platinum on carbon catalyst resulted in a 92% yield after 5 h at 30–35 °C and 5 bar hydrogen pressure. The applied pressure could even be reduced to 1 bar with only a minor reduction in yield (90%) (Scheme 2.45) [76]. In all cases, the unwanted side products were formed in less than 0.01%. The process could also be used for the related drug substance ralfinamide, which is undergoing investigation for the treatment of neuropathic pain.

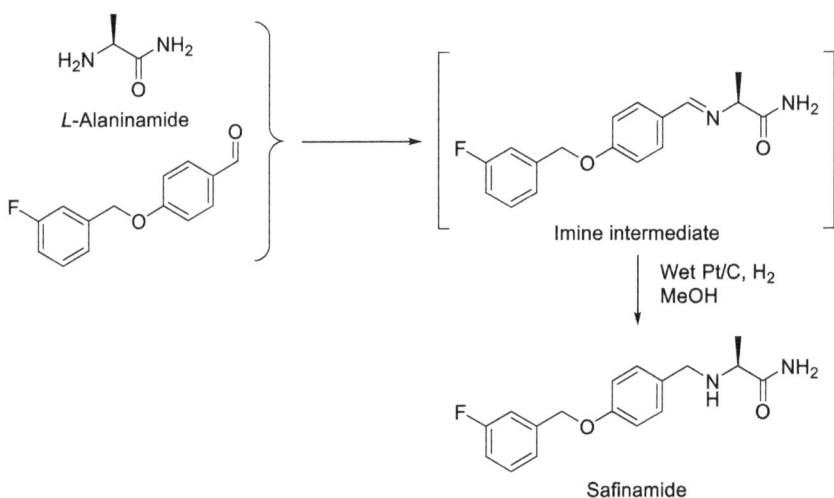

Scheme 2.45: Platinum-catalysed reductive amination in the synthesis of safinamide.

2.3.2.2 Unsupported platinum catalysts
In addition to supported platinum catalysts (i.e. platinum metal on a support), unsupported platinum oxide (known as the *Adams'* catalyst) [77] is also used for certain applications, in particular the reduction of aromatic and heteroaromatic groups. Two examples are given below.

As described in Section 2.2.1, tranexamic acid is a widely used as an antifibrinolytic drug. Its synthesis contains at least one heterogeneous hydrogenation step. The original route involves the hydrogenation of the aromatic ring of a benzyl amine using platinum oxide. The substrate is readily available in two steps in two different routes from simple and cheap aromatic starting materials. The hydrogenation reaction is performed under mild reaction conditions (1.5 bar H₂, 40 °C) [16]. A mixture of *cis*- and *trans*- isomers is produced from the hydrogenation, but the mixture can be isomerised to the wanted *trans*-isomer by heating with base. Although the yield is moderate, it is a short and effective route (Scheme 2.46).

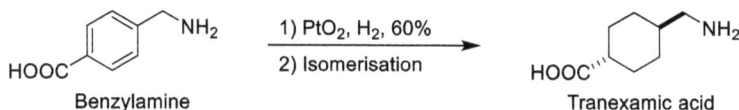

Scheme 2.46: Synthesis of tranexamic acid using a platinum oxide catalysed hydrogenation.

Ritalin (*syn/threo* methylphenidate hydrochloride) is a widely used stimulant drug that is used to treat attention deficit hyperactivity disorder in children. The (2 R,2′R) isomer is significantly more active than its enantiomer and a number of different synthetic routes have been developed. One of the most successful involves the reduction of a pyridine ring with a platinum oxide catalyst (Scheme 2.47) under mild conditions (room temperature, 4 bar hydrogen pressure) [78]. The reduction gives a 4:1 mixture of diastereoisomers in quantitative yield that is then subjected to an isomerisation and crystallisation to give the desired *rac–syn/threo*-isomer in 85% yield. This compound can be converted into the *rac*-ritalin, or first resolved and then converted into the more active (2 R,2′R)-isomer.

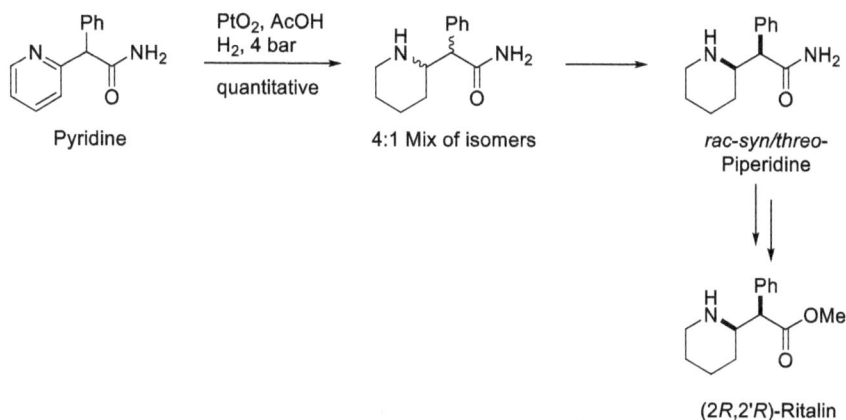

Scheme 2.47: Reduction of a pyridine group in the synthesis of Ritalin.

2.3.2.3 Asymmetric hydrogenations with platinum catalysts

Compared to homogeneous hydrogenation (see Chapter 3: Homogeneous hydrogenations), very few asymmetric hydrogenation reactions have been performed with heterogeneous catalysts. This is probably due to the difficulty in generating a uniform metal catalyst surface and a chiral environment on the metal surface in close proximity to the substrate binding site, without completely inhibiting the substrate binding.

The most successful reactions have been the asymmetric reduction of α-keto esters using a supported platinum catalyst modified with *cinchona* alkaloids such as quinine, cinchonidine, quinidine or cinchonine or their derivatives. Small changes to the structure of the chiral modifier can result in huge differences in enantioselectivity and choosing the correct modifier is usually a matter of trial and error [79, 80].

One successful application in the field of fine chemicals is the preparation of the (R)-hydroxyester intermediate of benazepril, a treatment for high blood pressure and kidney disease. The group at Solvias, working in collaboration with Ciba SC Life Science Molecules, investigated the asymmetric hydrogenation of several different α-keto esters before settling on the route shown in Scheme 2.48 [81]. The diketone (which is in equilibrium with its enol form) was successfully hydrogenated in >98% yield and up to 88% *ee* using a modified platinum on alumina catalyst. The best modifier was found to be the partially reduced 10,11-dihydrocinchonidine.

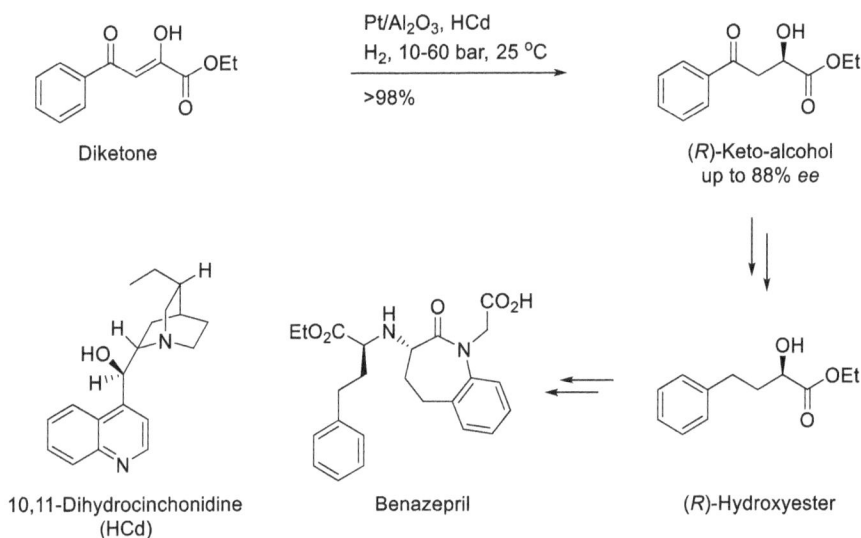

Scheme 2.48: Asymmetric heterogeneous hydrogenation in the synthesis of benazepril.

2.3.3 Other PGM catalysts

Compared to palladium and platinum catalysts, the use of other PGMs as heterogeneous hydrogenation catalysts for the synthesis of fine chemicals is significantly less advanced. There are a few reported examples of the use of ruthenium catalysts, and these are discussed below.

One example that has previously been mentioned is the synthesis of tranexamic acid. The traditional route uses a platinum oxide catalyst to reduce a phenyl ring (see Section 2.3.2.2, Scheme 2.46). An alternative high-yielding route (70% over three steps) was discussed previously (see Section 2.2.1, Scheme 2.8) involving two catalytic hydrogenation reactions (Scheme 2.49). Firstly a benzonitrile derivate was reduced to the corresponding benzylamine and *in situ* protected as an amide. Then a ruthenium on alumina catalyst was used for the reduction of the aromatic ring [16].

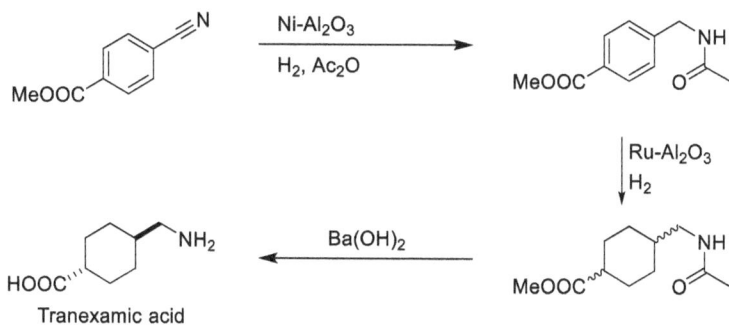

Scheme 2.49: Novel synthesis of tranexamic acid.

Heterogeneous ruthenium catalysts are also very selective for the reduction of carbonyl groups, especially aldehydes and ketones [82]. A very efficient and selective process has been developed in a trickle bed reactor for the reduction of glucose to sorbitol (Scheme 2.50) [83]. A ruthenium on carbon catalyst was used at 100 °C and 80 bar hydrogen pressure to give a selectivity of up to 99% at up to 99% conversion. However, the liquid flow rates had to be carefully monitored to minimise the epimerisation of the produced sorbitol to mannitol. Although the catalyst activity was stable over several weeks and no leaching of ruthenium was observed, it is not believed that this process has been implemented in production – the nickel catalysed process still dominates (see Section 2.2.2).

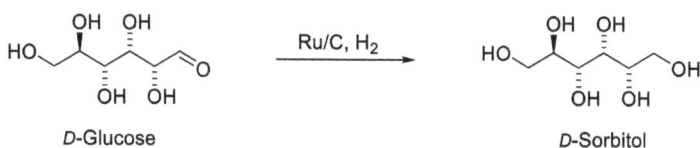

Scheme 2.50: Reduction of glucose to sorbitol.

Based on the growing importance of bio-based building blocks and the switch away from fossil-based feedstocks, it has been suggested that ruthenium-based catalysts could be ideal catalysts for the aqueous phase reduction/deoxygenation of these molecules in a high oxidation state [84]. This is especially true due to the high water solubility of these building blocks and also the accelerating effect that water can have on ruthenium catalysed hydrogenations. One possibility is the reduction of levulinic acid to γ-valerloactone, where promising results have been obtained and ruthenium catalysts are preferred (Scheme 2.51) [85].

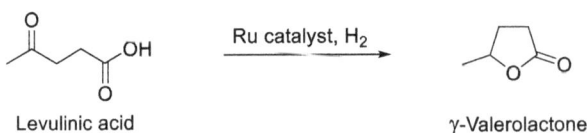

Scheme 2.51: Hydrogenation of levulinic acid to γ-valerolactone.

Currently, no industrial applications are known; only time will tell if heterogeneous ruthenium catalysts will become more widely used in hydrogenation reactions.

2.4 Other metal catalysts

As can be seen previously, nickel, palladium, platinum and ruthenium heterogeneous catalysts are widely used in hydrogenation reactions for the manufacture of fine chemicals. However, other catalytic metals can also be used. One example is the use of iron oxide or cobalt oxide catalysts in the hydrogenation of adiponitrile to hexamethylene diamine (Scheme 2.4) [7]. The disadvantages of such catalysts is that usually very high temperatures and pressures are required to obtain acceptable reaction rates.

An additional example of the use of a different metal catalyst is in the preparation of the aroma compound (−)menthol. Menthol is a terpenoid alcohol with three chiral centres but only the (−)-menthol isomer possesses the desired cool, clean, minty aroma/taste. It is one of the largest volume chiral compounds with an estimated production in 2010 of >20,000 tons per annum and volumes have increased since then. It is produced from natural sources and also synthesised chemically [86]. Three separate chemical routes have been commercialised – the Takasago route from myrcene, the BASF route from citral (see Chapter 3: Homogeneous hydrogenations) and the process originally developed by Haarmann and Reimer (now part of Symrise) in collaboration with Bayer AG (Scheme 2.52).

This process, which had an estimated production scale of >4,000 t/a in 2009, starts from *m*-cresol which undergoes a *Friedel–Crafts* alkylation to form thymol. Thymol is then hydrogenated in a continuous process at high temperatures (160–180 °C) and pressures (200–300 bar) with a fixed-bed catalyst consisting of mainly cobalt and manganese oxide [87–89]. The hydrogenation reaction produces a mixture of isomers, with approximately 60% of the desired diastereoisomer. The isomers are separated by distillation, and the unwanted ones are recycled. Racemic menthol is then resolved by crystallisation of the benzoate ester and seeding with the pure enantiomer. The unwanted enantiomer is racemised and recycled. Although a complicated process, the overall yield from thymol is greater than 90%.

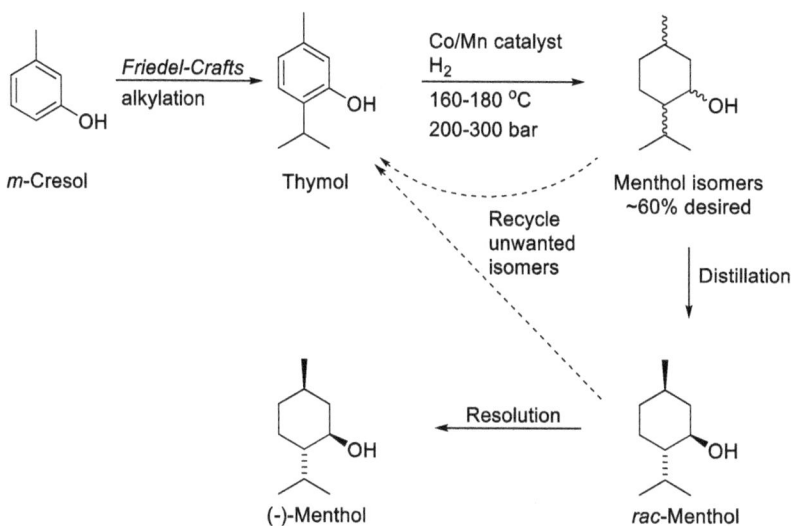

Scheme 2.52: Haarman & Reimer/Symrise preparation of menthol.

2.5 Summary

As can been seen from the wide variety of above-mentioned examples, heterogeneous hydrogenation is applicable for the reduction of a range of functional groups. It is used in multiple industries for the production of bulk chemicals, agrochemicals, vitamins and nutritional ingredients, aroma compounds, personal care applications, pharmaceuticals and many more. Hydrogenation is an efficient and atom-economical reaction that gives high selectivity and high productivity. A number of different active metals can be applied, and the catalyst can be easily recovered at the end of the reaction for reuse or refining. Hydrogenation has correctly been called "one of the most important weapons in the arsenal of the synthetic organic chemist" [90].

In this chapter, we have covered the application of heterogeneous catalysts for hydrogenation reactions, which make up the majority of the examples in the synthesis of fine chemicals (both in the number of reactions and in combined production volumes). However, there are still a few areas when heterogeneous hydrogenation is less successful, for example in asymmetric reductions, or certain cases in complicated/highly functionalised molecules where high levels of chemoselectivity are required. For these cases, homogeneous hydrogenation and biocatalytic reduction are viable alternatives and are covered in other chapters.

References

[1]   Kauffman G B. Johann Wolfgang Döbereiner's Feuerzeug. Platinum Metals Rev 1999, 43, 122–128.

[2]   Häussinger P, Lohmüller R, Watson A M. Hydrogen, 2. Production. In: Ullmann's Encyclopedia of Industrial Chemistry, Wiley-VCH, Weinheim, Germany, 2012, Vol. 18, 249–307.

[3]   Schmidt S R. The Raney Catalyst Legacy in Hydrogenation. In: Jackson S D, Ed Hydrogenation, Berlin, Germany, de Gruyter, 2018, 19–42.

[4]   Johnson Matthey PGM Prices (Accessed April 21 2020, at http://www.platinum.matthey.com/prices).

[5]   LME metal prices (Accessed April 21 2020, at https://www.lme.com/Metals).

[6]   Kukula P, Studer M, Blaser H-U. Chemoselective hydrogenation of α,β-unsaturated nitriles. Adv Synth Catal 2004, 346, 1487–1493.

[7]   Bartalini G, Giuggioli M. Process for the manufacture of hexamethylenediamine. US 3821305, Montedison Fibre October 22, 1970.

[8]   Allgeier A M, Sengupta S K. Nitrile Hydrogenation. In: Jackson S D, ed Hydrogenation, Berlin, Germany, de Gruyter, 2018, 107–154.

[9]   Hoffer B W, Moulijn J A. Hydrogenation of dinitriles on Raney-type Ni catalysts: Kinetic and mechanistic aspects. App Catal A 2009, 352, 193–201.

[10] Hoffer B W, Schoenmakers P H J, Mooijman P R M, et al. Mass transfer and kinetics of the three-phase hydrogenation of a dinitrile over a Raney-type nickel catalyst. Chem Eng Sci 2004, 59, 259–269.

[11] Eggersdorfer M, Laudert D, Letinois U, McClymont T, et al. One hundred years of vitamins – A success story of the natural sciences. Angew Chem Int Ed 2012, 51, 12960–12990.

[12] Degischer O G, Roessler F. Modifikation Eines Hydrierungskatalysators. EP 1108469, DSM, 2000.

[13] Ostgard D J, Roessler F, Karge R, Tacke T. The treatment of activated nickel catalysts for the selective hydrogenation of pynitrile. Chem Ind 2007, 115, 227–234.

[14] Hoekstra M S, Sobieray D M, Schwindt M A, et al. Chemical development of CI-1008, an enantiomerically pure anticonvulsant. Org Process Res Dev 1997, 1, 26–38.

[15] Martinez C A, Hu S, Dumond Y, Tao J, Kelleher P, Tully L. Development of a chemoenzymatic manufacturing process for pregabalin. Org Process Res Dev 2008, 12, 392–398.

[16] Li Z, Fang L, Wang J, Dong L, Guo Y, Xie Y. An improved and practical synthesis of tranexamic acid. Org Process Res Dev 2015, 19, 444–448.

[17] Schiweck H, Bär A, Vogel R, et al. Sugar Alcohols. In: Ullmann's Encyclopedia of Industrial Chemistry Online, Weinheim, Germany, Wiley-VCH, 2012.

[18] Arena B J. Deactivation of ruthenium catalysts in continuous glucose hydrogenation. App Catal A 1992, 87, 219–229.

[19] Bonrath W, Letinois U, Hugentobler M, Karge R, Lehmann H. Process for Resveratrol Intermediate. WO 2010079123, DSM Nutritional Products, 2009.

[20] Gräfje H, Körnig W, Weitz H-M, et al. Butanediols, Butenediol, and Butynediol. In: Ullmann's Encyclopedia of Industrial Chemistry Online, Weinheim, Germany, Wiley-VCH, 2019.

[21] Cartolano A R. Toluenediamine. In: Kirk-Othermer Encyclopedia of Chemical Technology Online, Hoboken, USA, John Wiley & Sons Inc, 2005.

[22] Wegener G, Brandt M, Duda L, et al. Trends in industrial catalysis in the polyurethane industry. App Catal A 2001, 221, 303–335.

[23] Tsunoda T, Yamazaki A, Iwamoto H, Sakamoto S. Practical Synthesis of N-{4-[(2-Methyl-4,5-dihydroimidazo[4,5-d][1]benzazepin-6 (1H)-yl)carbonyl]phenyl}biphenyl-2-carboxamide Monohydrochloride: An Arginine Vasopressin Antagonist. Org Process Res Dev 2003, 7, 883–887.

[24] Medlock J A, Wikstroem P. Process for the Production of a Dipeptide Derivative. WO 2015/197309, DSM, 2015.

[25] Bonrath W, Medlock J. Parallel Hydrogenation Experiments in the Fine Chemicals Industry. In: Hagemeyer A, Volpe A F, eds Modern Applications of High Throughput R&D in Heterogeneous Catalysis, Bentham Books, Bussum, The Netherlands, 2014, 341–351.

[26] no inventors stated. Hydrogenation of citral in a solvent. EP2765126, 2013.

[27] Connolly T J, Considine J L, Ding Z, et al. Efficient Synthesis of 8-Oxa-3-aza-bicyclo[3.2.1]octane Hydrochlo-ride. Org Process Res Dev 2010, 14, 459–465.

[28] Nishimura S. Handbook of Heterogeneous Catalytic Hydrogenation for Organic Synthesis, New York, NY, USA, John Wiley & Sons, Inc., 2001.

[29] Pearlman W M. Nobel metal hydroxides on carbon nonpyrophoric dry catalysts. Tetrahedron Lett 1967, 8, 1663–1664.

[30] Bonrath W, Wyss A, Litta G, Baldenius K-U, von dem Bussche-Hünnefeld L, Hilgemann E, Hoppe P, Stürmer R, Netscher T. Vitamin E (Tocopherols, Tocotrienols). In: Elvers B, ed. Ullmann's Encyclopedia of Industrial Chemistry Online, Weinheim, Germany, Wiley-VCH, 2021.

[31] Bonrath W, Medlock J, Schütz J, Wüstenberg W, Netscher T. Hydrogenation in the Vitamins and Fine Chemicals Industry – An Overview. In: Karamé I, ed. Hydrogenation, London, UK, InTech, 2012, 69–86.

[32] Bonrath W, Kircher T, Künzi R, Tschumi J. Process for the preparation of saturated aliphatic ketones. WO 2006029737, September 14, 2005.

[33] Seki M. Biological significance and development of practical synthesis of biotin. Med Res Rev 2006, 26, 434–482.

[34] Bonrath W, Karge R, Netscher T, Roessler F, Spindler F. Chiral Lactones by Asymmetric Hydrogenation – A Step Forward in (+)-biotin Production. In: Blaser H-U, Federsel H-J, ed. Asymmetric Catalysis on Industrial Scale: Challenges, Approaches, and Solutions. 2nd ed., Weinheim, Germany, Wiley-VCH, 2010, 27–39.

[35] Hitzler M G, Smail F R, Ross S K, Poliakoff M. Selective catalytic hydrogenation of organic compounds in supercritical fluids as a continuous process. Org Process Res Dev 1992, 2, 137.

[36] Fahlbusch K-G, Hammerschmidt F-J, Paten J, Pickenhagen W, Schatkowski D, Bauer K, Garbe D, Suburg H. Flavors and Fragrances. In: Elvers, ed. Ullmann's Encyclopedia of Industrial Chemistry, Online ed., Weinheim, Germany, Wiley-VCH Verlag GmbH & Co, 2012.

[37] Krafft P. AromaChemicals IV: Musks: In: Rowe ed. Chemistry and Technology of Flavours and Fragrances, Online ed. Backwell Publishing Ltd, 2005, 143–168.

[38] Ohloff G. Der Forscher Leopold Ruzicka. CHIMIA 1987, 41, 181–187.

[39] Wilke G, Bogdanovic B, Borner P, Breil H, Hardt P, Heimbach P, Herrmann G, Kaminsky H J, Keim W, Kröner M, Müller H, Müller E W, Oberkirch W, Schneider J, Stefeder J, Tanaka K, Weyer K. Cyclooligomerization of butadiene and transition metal π-complexes. Angew Chem Int Ed Engl 1963, 2, 105–115.

[40] Williams A S. The synthesis of macrocylic musks. Synthesis 1999, 1707–1723.

[41] Pletcher D, Walsh F C. Industrial Electrochemistry, 2nd ed. Springer Science + Business Media, 1990.

[42] Felix D, Schreiber J, Ohloff G, Eschenmoser A. α,β-epoxy-keton → alkinon-fragmentierung I: Synthese von exalton und rac-muscon aus cyclodecanon. Helv Chim Acta 1971, 54, 2896–2912.

[43] Becker J J, Schulte-Elte K-H, Ohloff G. Methods for the Manufacture of Macrocyclic Compounds. US 3778483, Firmenich, December 11, 1973.

[44] Lindlar H. Ein neuer Katalysator für selektive Hydrierungen. Helv Chim Acta 1952, 35, 446–450.

[45] Lindlar H, Dubuis R. Palladium catalyst for partial reduction of acetylenes. Org Synth 1966, 46, 89.

[46] Isler O, Huber W, Ronco A, Kofler M. Synthese des Vitamin A. Helv Chim Acta 1947, 30, 1911–1927.

[47] Molnar A, Sarkany A. Varga M Hydrogenation of C=C multiple bonds: Chemo-, regio- and stereo-selectivity. J Mol Catal A 2001, 173, 185–221.

[48] Lehmann H, Medlock J. Unpublished results.

[49]   Che L, Chen L, Fumin Z, et al. Catalyst for Selective Hydrogenation of Dehydroisophytol for Synthesizing Isophytol. CN 101869845, Zhejiang NHU Co. Ltd, 2012.

[50]   Bao Y, Li Y, Zhang Y. Method for Preparing Enol through Partial Hydrogenation of Alkynol. CN 109293472, Wanhua Chemical Group Co. Ltd, 2019.

[51]   Bao Y, Li J, Li Y, Song J, Song M, Zhang Y. Catalyst for Preparing Enol from Alkynol through Partial Hydrogenation, Preparation Method of the Catalyst, and Method for Preparing Enol by Using the Catalyst, CN 110124742, Wanhua Chemical Group Co. Ltd, 2019.

[52]   Widmer E, Zell R, Broger E A, et al. Technische Verfahren Zur Synthese von Carotinoiden Und Verwandten Verbindungen Aus 6-0x0-isophoron. 11. Ein Neues Konzept Für Die Synthese von (3*RS*, 3'*RS*)-Astaxanthin. Helv Chim Acta 1981, 64, 2436–2446.

[53]   Bonrath W, Buss A, Medlock J A, Mueller T. New Catalytic System. WO 2013/190076, DSM IP Assets, 2013.

[54]   Chapuis C. The Jubilee of Methyl Jasmonate and Hedione®. Helv Chim Acta 2012, 95, 1479–1511.

[55]   Buchi G H Process for the preparation of cyclic ketones. US 3941828, Firmenich SA, 1974.

[56]   Kataoka H, Yamada T, Goto K, Tsuji J. An efficient synthetic method of methyl (±)-jasmonate. Tetrahedron 1987, 43, 4107–4112.

[57]   Ohloff G, Pickenhagen W, Kraft P. Scent and Chemistry: The Molecular World of Odors. Zürich, Wiley-VCH, 2012.

[58]   Sondheimer F. Studies of compounds related to natural perfumes. Part I. Concerning cis- and trans-hex-3-en-1-ol. J Chem Soc 1950, 877–882.

[59]   Witte P T, de Groen M, de Rooij R M, et al. Highly active and selective precious metal catalysts by use of the reduction-deposition method. Stud Surf Sci Catal 2010, 175, 135–143.

[60]   Hoffmann H, Boettger G, Bör K, Wache H, Gräfje H, Körnig W. Katalysator zur Partiellen Hydrierung. DE 2431939, BASF 1974.

[61]   Musser M T. Cyclohexanol and Cyclohexanone. In: Ullmann's Encyclopedia of Industrial Chemistry, Wiley-VCH, Weinheim, Germany, 2012, 11, 49–60.

[62]   Hansen K B, Balsells J, Dreher S, et al. First generation process for the preparation of the DPP-IV inhibitor sitagliptin. Org Process Res Dev 2005, 9, 634–639.

[63]   Tinge J, Groothaert M, Veld H O P, et al. Caprolactam. In: Ullmann's Encyclopedia of Industrial Chemistry Online, Weinheim, Germany, Wiley-VCH, 2018.

[64]   Albrecht W, Fischer D, Janssen C. Intedanib Salts and Solid State Forms Thereof. WO 2012/068441, Teva Pharmaceuticals 2011.

[65]   Merten J, Linz G, Schaubelt J, et al. Process for the Manufacture of an Indolinone Derivative. WO 2009/071523 Boehringer Ingelheim, 2008.

[66]   Mathe T, Hegedus L, Czibula L, Juhasz B, Nagyne Bagdy J, Markos D. Process for the preparation of trans 4-Amino-cyclohexyl acetic acid ethyl ester. WO 2010/070368 to Richter Gedeon Nyrt, 2009.

[67]   Haight A, Barnes D, Zhang G. Preparation of Pyridonecarboxylic Acid Antibacterials. WO 2006/015194, Abbott Laboratories, 2005.

[68]   Flick A C, Leverett C A, Ding H X, et al. Synthetic approaches to the new drugs approved during 2017. J Med Chem 2019, 62, 7340–7382.

[69]   Higuchi C, Kitada I, Nagamoto A, Enomoto K, Ajioka M, Yamaguchi A. Preparation Process of alpha-aspartyl-L-phenylalanine Methyl Ester. EP 0510552, Mitsui Toatsu Chemicals, 1992.

[70]   Liu K K-C, Sakya S M, O'Donnell C J, Li J. Synthetic approaches to the 2007 new drugs. Mini-Rev Med Chem 2008, 8, 1526–1548.

[71]   Belyk K M, Morrison H G, Jones P, Summa V. Potassium salt of an HIV intergrase inhibitor. WO 2006/060712 Merck & Co and Istituto di Ricerche di Biologia Molecolare P. Angeletti, 2005.

[72]   Bonrath W, Netscher T, Schütz J, Wüstenberg B Process for the Manufacture of TMHQ. WO 2012/025587, DSM, 2011.

[73] Pandey A, Leitao E P T, Rato J, Song Z J. Methods for Synthesizing Factor XA Inhibitors. WO 2011/084519, Millennium Pharmaceuticals and Merck Sharp & Dohme, 2010.

[74] Blaser H-U. The chiral switch of (S)-Metolachlor: A personal account of an industrial odyssey in asymmetric catalysis. Adv Synth Catal 2002, 344, 17–31.

[75] Bader R, Flatt P, Radimerski P. Process for the Preparation of 2-alkyl-6-methyl-N-(1′-methoxy-2′-propyl)-aniline and a Process for the Preparation of Their Chloracetanilides. EP 0605363, Ciba-Geigy AG, 1993.

[76] Barbanti E, Caccia C, Calvati P, Velardi F, Rufilli T, Bogogna L. Process for the Production of 2-[4-(3- and 2-Fluorobenzyloxy)benzylamino]proanamides. WO 2007/147491, Newron Pharmaceuticals, 2007.

[77] (no authors listed). The Story of the Adams' Catalyst. Platinum Metals Rev 1962, 6, 150–152.

[78] Khetani V, Luo Y, Ramaswamy S. Processes and Intermediates for Resolving Piperidyl Acetamide Stereoisomers. WO 9852921, Celgene 1998.

[79] Blaser H-U, Jalett H-P, Studer M Enantioselective hydrogenation of α-ketoesters using cinchona modified platinum catalysts and related systems: A review. Catal Today 1997, 37, 441–463.

[80] Studer M, Blaser H-U, Exner C. Enantioselective hydrogenation using heterogeneous modified catalysts: An Update. Adv Synth Catal 2003, 345, 45–65.

[81] Blaser H-U, Studer M. Cinchona-Modified platinum catalysts: From ligand acceleration to technical processes. Acc Chem Res 2007, 40, 1348–1356.

[82] Kluson P, Cerveny L. Selective hydrogenation over ruthenium catalysts. App Catal A 1995, 128, 13–31.

[83] Gallezot P, Nicolaus N, Fleche G, Fuertes P, Perrard A. Glucose hydrogenation on ruthenium catalysts in a trickle-bed reactor. J Catal 1998, 180, 51–55.

[84] Michel C, Gallezot P. Why is ruthenium an efficient catalyst for the aqueous-phase hydrogenation of biosourced carbonyl compounds? ACS Catal 2015, 5, 4130–4132.

[85] Manzer L E. Catalytic synthesis of α-methylene-γ-valerolactone: A biomass-derived acrylic monomer. App Catal A 2004, 272, 249–256.

[86] Leffingwell J, Leffingwell D. Chiral chemistry in flavours & Fragrances. Specialty Chem Mag, 2011, 30–33.

[87] Hopp R, Lawrence B M. Natural and Synthetic Menthol. In: Lawrence B M ed. Mint: The Genus Mentha. Boca Raton, USA, CRC Press, 2006, 371–397.

[88] Biedermann W. Kontinuierliuches Verfahren Zur Herstellung von d,l-Menthol, DE 2314813, Bayer AG, 1973.

[89] Darstow G, Petruck G-M. Verfahren zur herstellung von d,l-Menthol, EP 0743296, Bayer AG, 1996.

[90] Rylander P N. Hydrogenation Methods (Best Synthetic Methods). London, Academic Press, 1985.

3 Homogeneous hydrogenations

3.1 Introduction

As described in the previous chapter (Chapter 2: Heterogeneous hydrogenations), due to its attractive features, heterogeneous hydrogenation is one of the most fundamental and frequently applied reactions in all chemical industries on a production scale. Although heterogeneous catalysts offer distinct advantages over homogeneous systems, they usually fail to introduce high levels of chiral information to prochiral substrates. In this case, homogeneous hydrogenation is clearly advantageous and therefore most of the homogeneous hydrogenations applied in industry are conducted in an asymmetric fashion. In metal-catalysed asymmetric reactions, the chiral information of a transition metal complex is transferred during the reaction to the substrate to form an enantiomerically enriched product. In this case, the generally higher costs for homogeneous transformations are justified by the value of the intermediate or product obtained. The enormous impact of this transformation was recognised by the Nobel Prize awarded to R. Noyori and W. S. Knowles in 2001 [1, 2] "for their work on chirally catalysed hydrogenation reactions" together with B. Sharpless "for his work on chirally catalysed oxidation reactions" [3]. The number of possible catalysts is higher compared to heterogeneous systems, with an almost inexhaustible variety of different ligands that can be potentially applied. Additionally, "Since achieving 95% *ee* only involves energy differences of about 2 kcal, which is no more than the barrier encountered in a simple rotation of ethane, it is unlikely that before the fact one can predict what kind of ligand structures will be effective" (W.S. Knowles, 1983) [4].

Therefore, high-throughput screening becomes more relevant in order to allow for a rapid testing of the many available ligands under different reaction conditions, with or without various additives. Additionally, it is not uncommon that the price of the ligand is higher than that of the transition metal, and thus becomes the main cost driver for a reaction.

This chapter will deal with different catalytic systems that are applied on an industrial scale in the fine chemical industry. However, for certain catalysts, examples for pharmaceutical products or examples that were performed only on kilogram scale are chosen to explain the key features of the particular catalyst, such as to highlight the potential of the catalysts or concepts for hydrogenations or in catalysis in general.

Predominately molecular hydrogen is used as the reducing agent on a production scale for fine and bulk chemicals since it is the cheapest option. Alternatively transfer hydrogenations, referring to the addition of hydrogen to a molecule from a source different to molecular hydrogen, is an option. The sections herein are divided by metal source for the catalyst applied starting with Rh-based catalysts, the first class that was successfully applied on an industrial scale. Usually the hydrogenation of certain structural groups tends to proceed best with a specific catalyst group. Hydrogenations using non-molecular hydrogen as reducing agent are described in a separate section.

https://doi.org/10.1515/9783111102672-003

3.2 Rh-catalysed hydrogenations

Rh-based catalysts are among the most commonly applied catalysts for asymmetric hydrogenations in academia and in industrial environments. Also, the first asymmetric catalytic step on industrial scale was a hydrogenation accomplished by using an Rh-based catalyst (see Section 3.2.1). Many examples of successful applications in gram, kilogram and production scale have been reported, showing high catalyst reactivity in combination with excellent enantioselectivity, especially for the asymmetric reduction of C=C double bonds. It emerged that different functional groups adjacent to the double bond significantly increase the possibility of achieving excellent catalyst performance. High selectivities and reactivities were obtained, for example, for enamides, α,β-unsaturated carboxylic acids, α,β-unsaturated aldehydes and ketones naming only the most relevant functional groups [5]. However, in terms of price, Rh is among the most expensive metals. This aspect makes is essential to either develop a very efficient process that runs at very low catalyst loadings or to ensure the recovery of the catalyst and ligand after the reaction at high rates.

3.2.1 (*S*)-3-(3,4-Dihydroxyphenyl)alanine, *L*-DOPA

The application of chiral Rh-based catalysts in asymmetric homogeneous hydrogenation was the first groundbreaking example, demonstrating the feasibility of an asymmetric transformation on an industrial scale. This methodology was used as the key step in the synthesis of (*S*)-3-(3,4-dihydroxyphenyl)alanine, better known as *L*-DOPA.

Coincidently, it was discovered that *L*-DOPA is a potent compound in the treatment of *Parkinson's* diseases [6]. This compound was manufactured by Hoffmann-LaRoche by applying a synthesis which involves a prochiral enamide being hydrogenated in an achiral fashion and the enantiomers being separated by a resolution. Consequently, more than 50% of the material is lost in this step (Scheme 3.1). Luckily, a derivative of the prochiral enamide applied in this synthesis was found to be an ideal target for the application of asymmetric hydrogenation technology.

The basis for the asymmetric hydrogenation in the Monsanto production process for *L*-DOPA (Scheme 3.2) was the discovery of the *Wilkinson's* catalyst [RhCl(PPh$_3$)$_3$], a square planar 16-electron complex with rhodium at the oxidation state of +1 [7] and the development of methods for the preparation of chiral phosphines by *Mislow* and *Horner* (Figure 3.1) [8, 9]. In these compounds the lone pair represents the fourth substituent and contributes to a pyramidal shape. In contrast to nitrogen, the inversion barrier of the lone pair in phosphines is higher and half-life times of several hours at 120 °C are common [10].

The idea of using *Wilkinson's* catalyst in combination with a chiral phosphine as a catalyst for an asymmetric hydrogenation was then independently followed by the groups of Horner [11] and Knowles [12], although the first results revealed only mod-

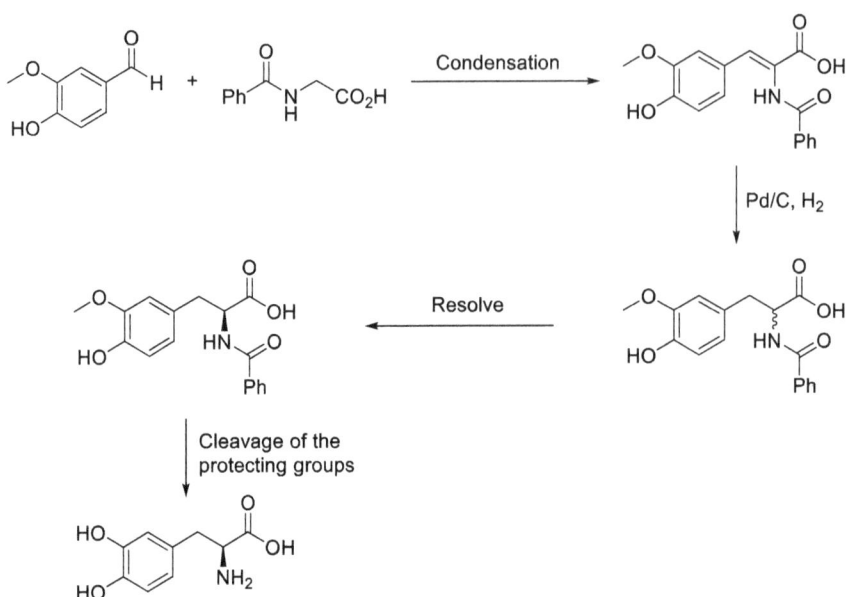

Scheme 3.1: *L*-DOPA synthesis by Roche before the implementation of the asymmetric hydrogenation.

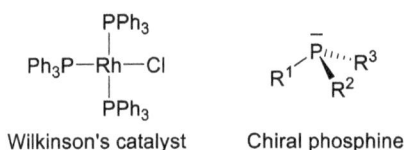

Wilkinson's catalyst   Chiral phosphine

Figure 3.1: *Wilkinson's* catalyst and a chiral phosphine, two major parts of the puzzle for the first homogeneous asymmetric hydrogenation conducted on an industrial scale.

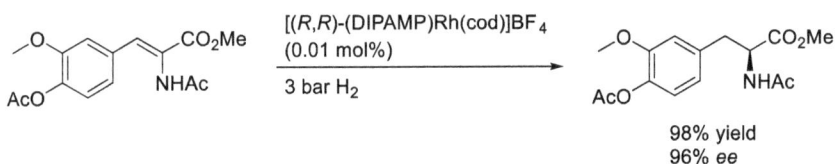

Scheme 3.2: Asymmetric hydrogenation in the Monsanto process for the *L*-DOPA production.

erate enantioselectivity, with up to 15% *ee* in the hydrogenation of a *L*-DOPA precursor. This result showed the feasibility of the concept in general and was the starting point for the development and screening of several chiral ligands for this transformation. A breakthrough was achieved with the introduction of a C_2-symmetric bidentate diphosphine ligand (*R,R*-DIPAMP) or DIOP (2,3-O-isopropylidene-2,3-dihydroxy-1,4-bis(diphenylphosphino)butane) reported by Kagan [13–16]. C_2-Symmetric ligands have since dominated the field of asymmetric catalysis for a long time and combine different attractive properties. The number of undesired competing isomeric transition states that lower the enantioselectivity of the reaction are fewer, compared to reactions with non-symmetric ligand complexes. This feature also reduces the complexity of NMR spectra of C_2-symmetric ligand complexes and facilitates mechanistic studies, when compared to metal complexes with non-symmetric ligands [13, 14]. The enantioselectivity of the asymmetric hydrogenation increased from 28% ee for *PPhMe(C_3H_7) to 50–60% ee for PAMP to 95% ee when using the C_2-symmetric DIPAMP ligand. These are examples of only a few of many chiral ligands tested (Scheme 3.3) [17].

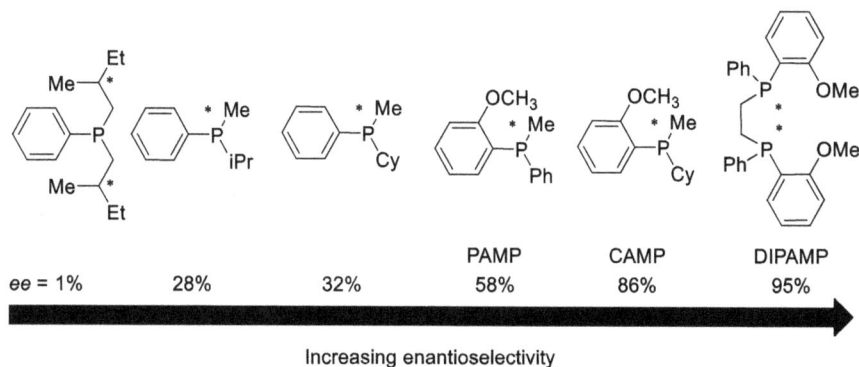

Scheme 3.3: Influence of the ligand structure on the enantioselectivity of the asymmetric hydrogenation in the *L*-DOPA route.

Additionally, the synthesis of C_2-symmetric ligands frequently benefits from a dimerisation step and facilitates the synthesis of the ligand. This strategy was also applied in the synthesis of (R,R-DIPAMP) (Scheme 3.4). In addition to the dimerisation, the formation of diastereomeric phosphinates with (–)-menthol is a key step in the synthesis. This approach is often applied for the separation of enantiomerically enriched compounds from racemates. Usually the other enantiomerically pure counterpart is re-isolated after separation of the diastereoisomers [18].

C_2-Symmetric ligands do not always show superior performance compared to their monodentate counterparts, for example in the case of DICAMP vs. CAMP with DICAMP giving 60–65% ee and CAMP resulting in 88% ee [17].

The elucidation of a reaction mechanism is in most cases complicated and time consuming. The multi-step character of a catalytic reaction, the possibility of intermediates being formed as different geometrical isomers, the differentiation between off-cycle species and species that contribute to the catalytic cycle, and the availability and access to the right equipment to collect data under the applied reaction conditions, especially for pressurised reactions, are substantial problems to overcome.

Nevertheless, the mechanism for the hydrogenation with the *Wilkinson's* catalyst is among the most investigated and best-characterised catalytic cycles (Scheme 3.5) [19]. It

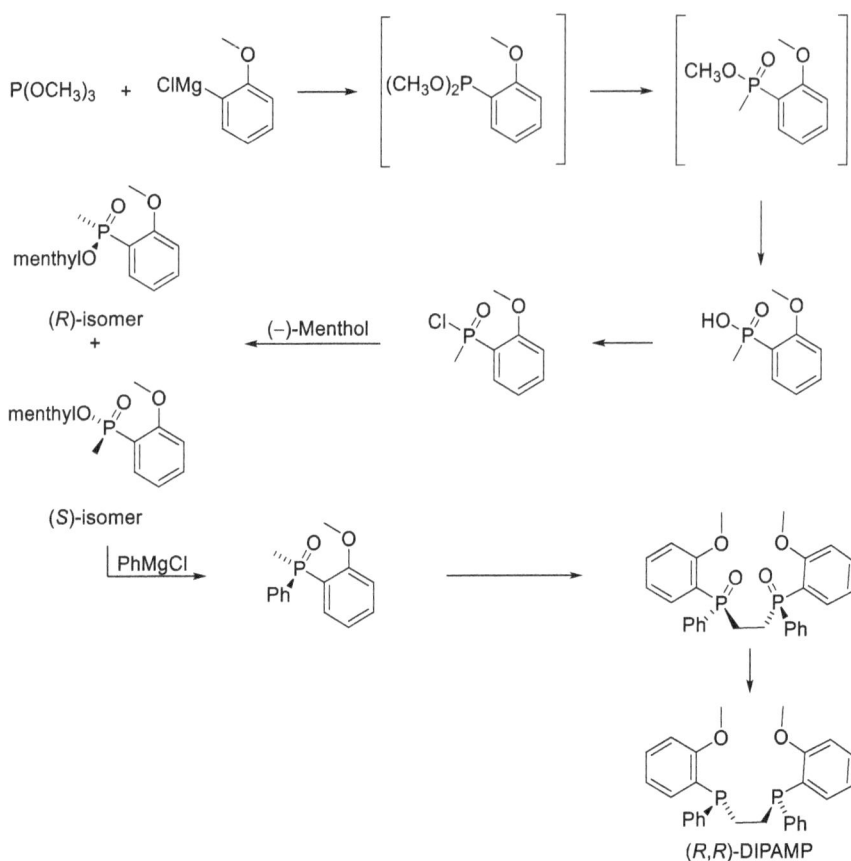

Scheme 3.4: Synthesis of the chiral bidentate ligand (R,R-DIPAMP), applied in the industrial synthesis of L-DOPA.

highlights several key observations that have been considered and the knowledge can be transferred to other reactions. An important feature of this metal complex is the labile nature of the triphenylphosphine ligand which can dissociate from the metal centre and generate a vacant coordination sphere to allow a substrate to coordinate. This is necessary to enter the catalytic cycle after oxidative addition of the *Wilkinson's* catalyst to molecular hydrogen to form a Rh(III) 18 valence electron dihydride complex. The addition of phosphine ligands shifts the equilibrium to off-cycle compounds, and therefore reduces the TOF of the catalyst. After coordination of the double bond to the metal centre the complex undergoes migratory insertion, which is the rate determining step of the catalytic cycle. In an irreversible reductive elimination step, the alkane is generated. Usually functional groups such as C=O, CN, NO$_2$, aryl, and CO$_2$R are not hydrogenated.

The hydrogenation rate of substituted unfunctionalised olefins mirrors the relative binding affinity of the double bond. This observation can be partly explained by

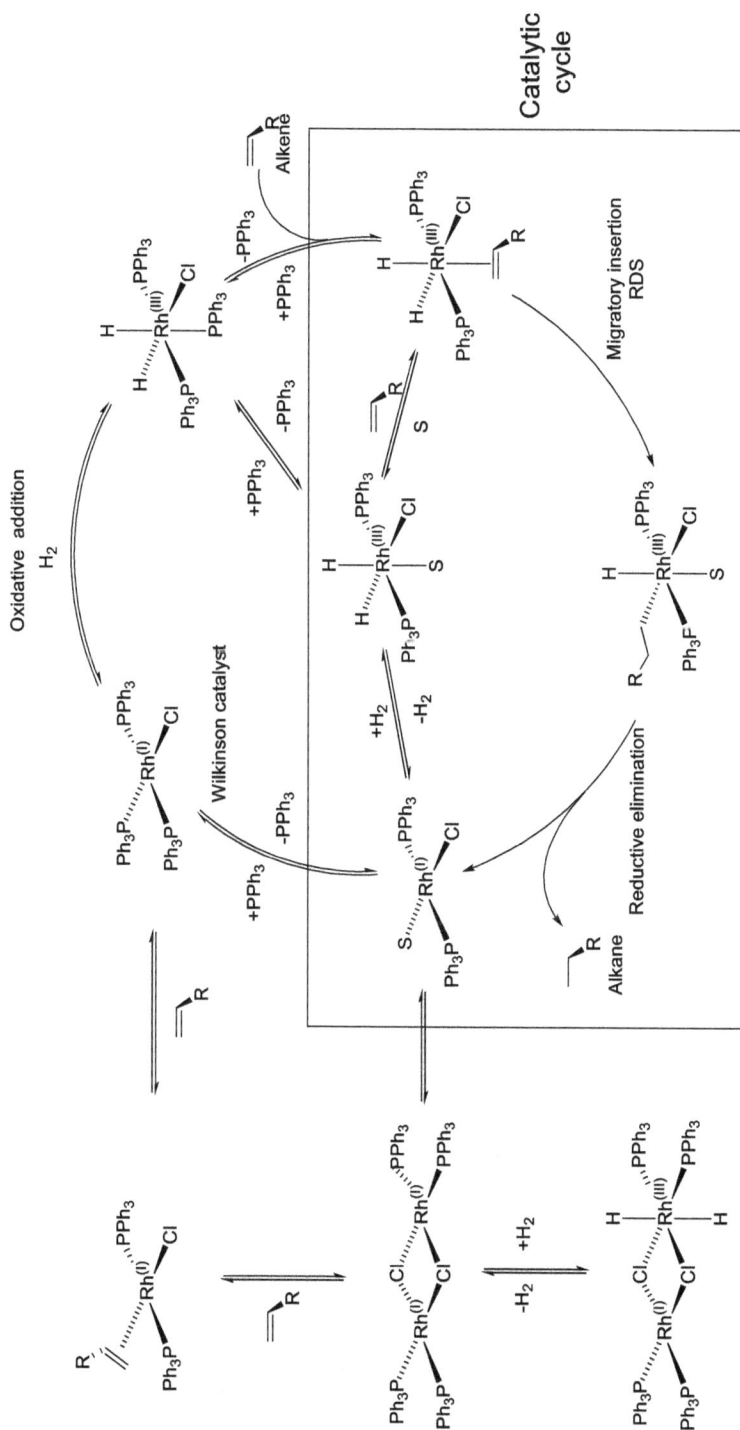

Scheme 3.5: Catalytic cycle of hydrogenations with *Wilkinson's* catalyst.

steric factors (Scheme 3.6). Double bonds next to polar groups are usually hydrogenated faster. Coordinating functional groups adjacent to the double bond also usually
increase the rate of the hydrogenation and are even required for certain types of catalysts to achieve high selectivity and reactivity.

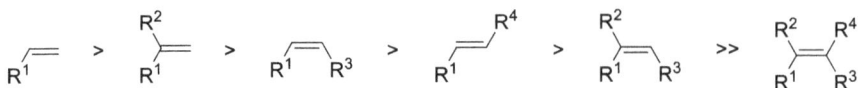

Scheme 3.6: Relative hydrogenation rate of unfunctionalised olefins with the *Wilkinson*'s catalysts.

Another interesting mechanistic aspect for asymmetric reactions can be explained
using the example of the *L*-DOPA hydrogenation. A catalytic reaction usually consists
of more than one step and kinetic considerations play a major impact in the understanding of the overall reaction. The different steps leading to the final product can be
irreversible or in equilibrium. Competing rates determine the concentration of each
intermediate so that the overall rate equation of a catalytic reaction is usually rather
complicated. For asymmetric reactions, the chiral information is fixed as soon as an
irreversible step occurs after the introduction of the stereochemical information. In
the Rh-catalysed asymmetric hydrogenation it was initially assumed that the most stable catalyst-alkene adduct leads to the major enantiomer of the product (Scheme 3.7).
However, kinetic studies conducted by Halpern [20] and Brown [21] showed that of the
two alkene-adducts, the most stable intermediate **II** results in the formation of the
minor enantiomer. Consequently, the less stable alkene adduct **III** reacts faster and results in the formation of the main enantiomer **V**. This is also known as the major-
minor principle; for a more detailed explanation of this concept see the literature [22].

Because neither the dihydride complex nor the alkyl intermediate was observed
in the case study, it is reasonable to assume that the oxidative addition of molecular
hydrogen to the metal centre is the rate-determining step in this reaction. However,
this can change based on the catalyst and the substrate [23].

The *Monsanto Process* started the production of *L*-DOPA in 1974 and is the first
commercialised process with a catalytic asymmetric step employing a chiral metal-
complex as catalyst (Scheme 3.8). The reaction was conducted with a slurry of the prochiral olefin in an alcohol-water mixture and substrate to catalyst ratios of 20,000:1
were reached [4, 17]. Nowadays, the production of *L*-DOPA is conducted by the fermentative process [24] (see Chapter 10: Biocatalysis).

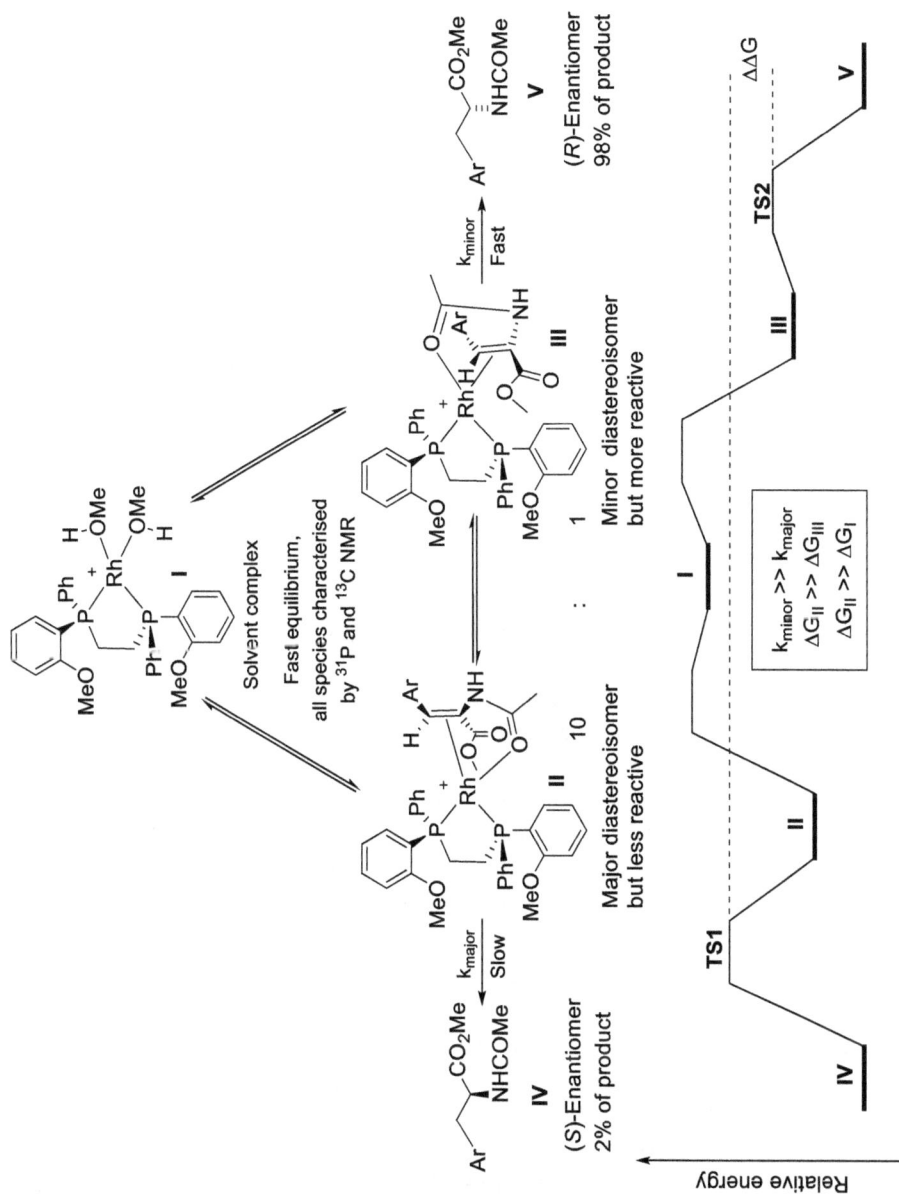

Scheme 3.7: Different equilibriums involved in the enantioselective hydrogenation.

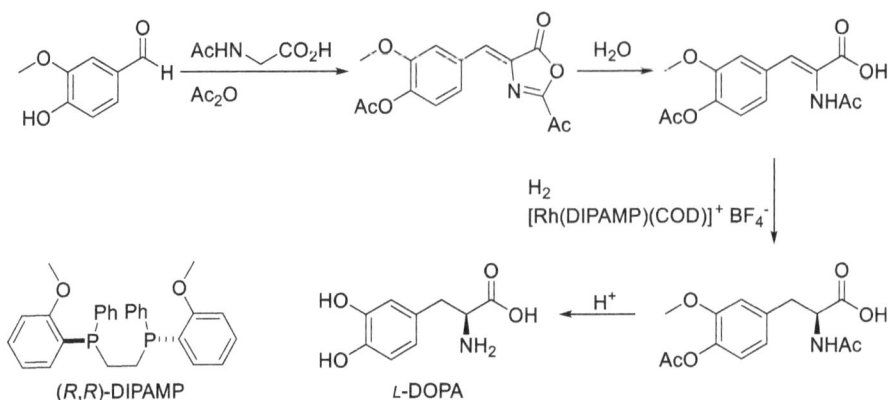

Scheme 3.8: *Monsanto process* for the synthesis of *L*-DOPA.

3.2.2 Menthol (BASF Process)

L-Menthol is one of many compounds which are produced via different routes by different companies, for example BASF, Symrise or Takasago (these routes are discussed in Chapter 2: Heterogeneous hydrogenation and Chapter 7: Rearrangement reactions). It is used in various products due to its disinfectant properties and as a flavour and fragrance component in many products [25].

The BASF menthol process is estimated to run in the range of 3,000 to 5,000 metric tons per year (2010) [26]. The key step in the whole synthesis is the asymmetric hydrogenation of neral, the (Z)-isomer of citral, which is a key building block in the BASF product portfolio and a precursor in the synthesis of vitamins and many flavour and fragrances compounds. The first stereogenic centre in the synthetic sequence can be introduced by asymmetric hydrogenation with perfect chemoselectivity in favour for the α,β-unsaturated-aldehyde over the trisubstituted olefin and the aldehyde function. The stereogenic centre dictates the configuration of the other two stereogenic centres which are introduced in the cyclisation step that follows. The product is then obtained via a heterogeneous hydrogenation of the terminal olefin (Scheme 3.9).

Several important parameters had to be considered to correlate the catalyst and substrate behaviour. First, the enantioselectivity of the hydrogenation strongly depends on the (Z/E)-ratio of citral with neral (the (Z)-isomer) and geranial (the (E)-isomer), yielding the opposite enantiomers of citronellal in the Rh-catalysed hydrogenation. Therefore, an efficient access to one or the other (E/Z)-isomer is required, and a catalyst system which did not cause isomerisation of the conjugated C=C double bond during the reaction had to be found (Scheme 3.10).

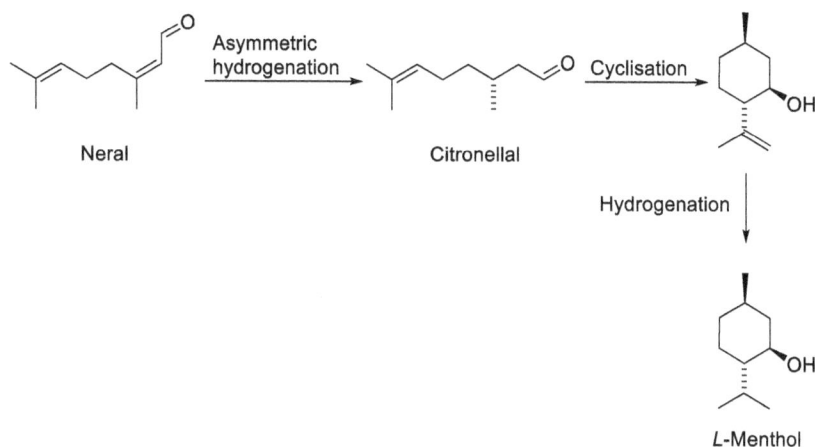

Scheme 3.9: Citral-based synthesis of *L*-menthol by BASF.

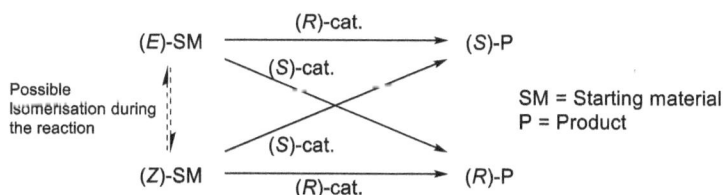

Scheme 3.10: Reaction regime observed for double bond hydrogenation (the configuration of the product depends on the double bond-configuration and the catalyst configuration and can differ based on the substrate and the nature of the ligand).

Prior to the work of BASF, the enantioselective hydrogenation of geranial was reported but gave unsatisfying results with up to 84% *ee* [27]. Additionally, the catalysts loadings under the then-reported reaction conditions were too high for an industrial application.

The key to success was obtained by additional mechanistic insight into the catalyst system (Scheme 3.11). Rh-precursors containing no CO-ligand showed only low reactivity in the reaction. This indicates that a CO-ligand is involved in the catalytic cycle and led to the assumption that additional CO in the reaction mixture might increase the concentration of active catalytic species. Indeed, first experiments using syngas (a mixture of H_2 and CO) resulted in an increase of turnover numbers by a factor of 10. Further mechanistic insights indicated that if the CO pressure is too low, the formation of the dimer (complex 1) is favoured, whereas a higher CO pressure impedes the formation of the catalytically active Rh-monocarbonyl complex. Consequently, careful adjustment of the CO-partial pressure would have a considerable influence on the behaviour of the catalytic system. As already described in this chapter, there are two ways to reduce the cost of a catalyst in an industrial process. Use a catalyst system which can operate in sufficient s/c-ratios that the catalyst can be discarded after use or achieve a sufficient recycling of the catalyst. Catalyst recycling is especially interesting if continuous processing is an option. In this case, different pressures of the H_2/CO gas mixture were applied for the different process steps. Under workup conditions the hydridocarbonyl complex (complex 2) is converted to the more stable bis-ligand-complex (complex 1); complex 2 is then regenerated applying higher pressure of syngas. The hydrogenation itself is run by adding a small amount of CO to the hydrogen gas. This strategy resulted in a setup (Figure 3.2) which allowed for over 100,000 TONs of the catalyst, and the catalyst can be regenerated again [28].

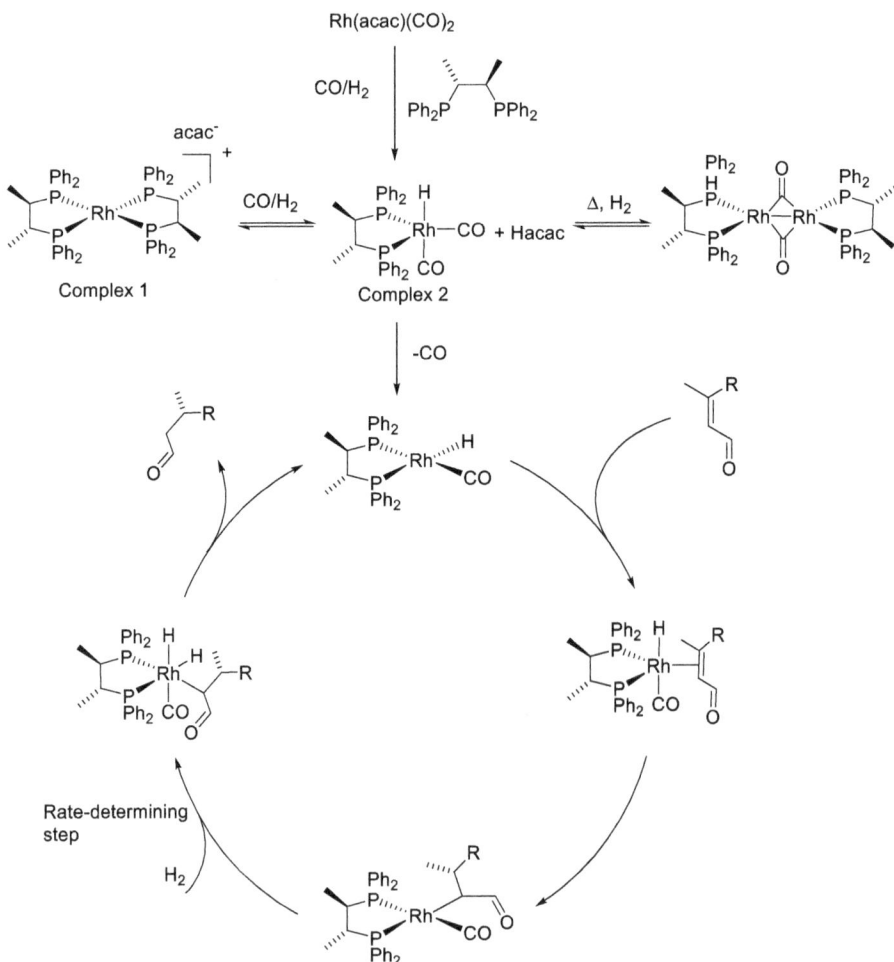

Scheme 3.11: Proposed mechanism of the asymmetric Rh-catalysed hydrogenation of α,β-unsaturated aldehydes.

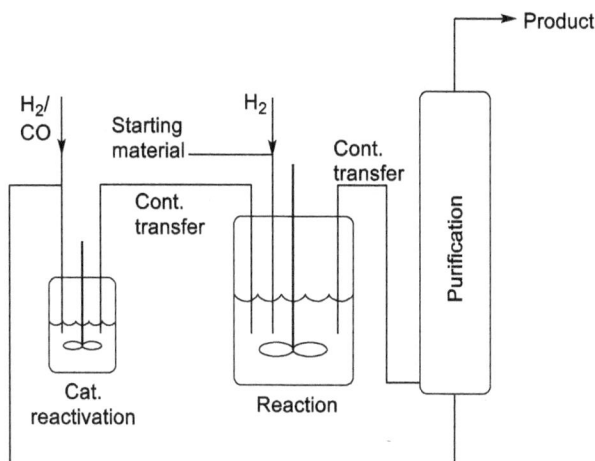

Figure 3.2: Simplified scheme of the hydrogenation process.

3.2.3 Methylene succinamic acid

Another example, which is striking for the power of this methodology, can be found in the pharma sector. The asymmetric hydrogenation of 2-methylene succinamic acid produces an important building block for several biologically active compounds. In this case, the terminal olefin is hydrogenated with an Rh-based catalyst using Et-DuPhos as ligand. Full conversion with 96% *ee* was achieved at an s/c of 100,000:1 and an average turnover frequency of 13,000 h^{-1} was achieved (Scheme 3.12) [29]. In this case, the results highly depend on the substrate quality; even trace amounts of chloride must be avoided since this inhibits the conversion. The *ee* of the product was improved by crystal digestion in isopropanol, and the Rh-content was reduced to less than 1 ppm.

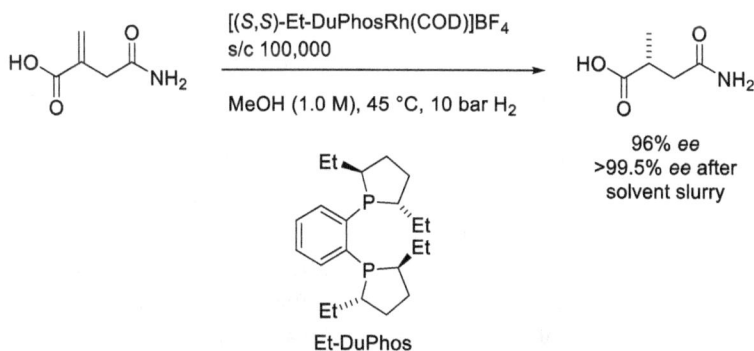

Scheme 3.12: Asymmetric hydrogenation of 2-methylene succinamic acid.

3.2.4 Aliskiren

Different catalytic systems consisting of a metal and a chiral ligand are discussed in this chapter. Whereas the choice of the correct metal is, in most cases, a rather rational decision or can be derived from many examples demonstrated since the first successful enantioselective hydrogenations, the search for the optimal ligand is, in most cases, a rather semi-empirical approach. Given the enormous number of ligands which have been reported in the meanwhile, sometimes, it seems like searching for a needle in a haystack. For sure, many ligands can be ruled out for industrial applications due to their price and availability; however, with the high reactivity reported for some hydrogenations (TONs reaching several millions), even a ligand synthesis of several steps is not an ultimate knockout criteria. Therefore, the ligand is often more expensive than the metal in many catalytic systems for asymmetric transformations.

In this context, the screening of huge ligand libraries has gained importance over the years, and access to high throughput apparatus is standard in each research lab dealing with the development of catalytic methods, these days. One interesting example that demonstrates the power of high throughput screening was reported in the asymmetric hydrogenation of an Aliskiren precursor. Aliskiren is a renin inhibitor that is used for the treatment of high blood pressure and was approved in 2007 in the USA and Europe. The hydrogenation of an α-isopropyl cinnamic acid derivative (Scheme 3.13) is among the most investigated targets for asymmetric hydrogenation and has been tackled by different research groups [30, 31].

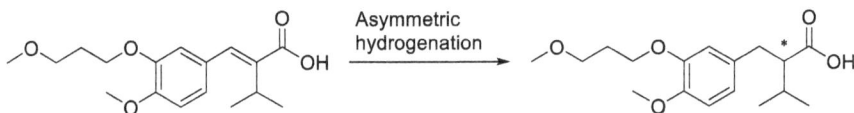

Scheme 3.13: Asymmetric hydrogenation as key step in the synthesis of aliskiren.

The asymmetric hydrogenation of the aliskiren precursor is one where the mixed ligand approach was successfully applied. This concept was independently applied in a cooperation by the Feringa group and DSM and the group of Reetz [32, 33]. Both groups used phosphoramidites that were previously developed for copper-catalysed 1,4-additions to cyclic enones [34], and have in the meanwhile found many applications in organic synthesis [35].

The big advantage is that these ligands can be synthesised in a straightforward manner in one or two steps, starting from a BINOL backbone. This feature allows fast access to a large variety of different ligands combining BINOL (and possible derivatives) with an enormous number of existing amines and makes this ligand class highly modular and versatile. Therefore, it would be highly advantageous if the ligand synthesis also could be conducted in an automated way to improve the throughput of a catalyst screening and avoid the time-consuming ligand synthesis. Such equipment was also used for the hydrogenation of the aliskiren precursor at DSM Pharmaceutical Products (now InnoSyn) [36]. The set-up contains an orbital shaker in which the different reagents for the ligand synthesis were mixed. After the synthesis of the different phosphoramidites, the ligand-containing solutions were filtered and mixed with the catalyst precursors (metal source), the substrates and, optionally, additives. For the hydrogenation, different chiral phosphoramidites were tested in combination with different additives such as phosphines, amines and others [36].

If one assumes that, in the catalytic cycle, two ligands are coordinated to the metal centre, then the addition of a second, different ligand would lead to an equilibrium of three possible species (Scheme 3.14) [35]. At first glance, this looks like a complicated situation with different species involved. However, in the case where one species is more reactive then the others or if one of the complexes is more stable and therefore formed preferentially, this approach can be beneficial. Optimally, both situations occur. This approach significantly increases the number of possible chiral catalysts because chiral monodentate ligands can be mixed with achiral ligands to form a new chiral complex.

$$[Rh(P^{1*})_2]^+ \rightleftharpoons [Rh(P^{1*}P^2)]^+ \rightleftharpoons [Rh(P^2)_2]^+ \quad * = \text{Indicates chirality}$$

Chiral · Chiral · Not chiral

Scheme 3.14: Different ligand complexes induced using two different monodentate ligands.

The initial screening of the aliskiren precursor with the phosphoramidite "instant ligand library" in combination with a rhodium precursor introduced sufficiently high enantioselectivities. However, reaction rates were rather low. A combination of poor σ-donating phosphoramidites and electron rich phosphines, such as triphenylphosphine, resulted in an acceleration of the reaction rate. The accelerated reaction rate and high enantioselectivity are caused by the mixed ligand complexes and consequently, the formation of this mixed-complex should be optimised. The influence of this ratio can be easily seen (Table 3.1).

As mentioned in Chapter 1: Introduction and fundamental aspects, an important economic criterion in industry is the space-time yield. With the best system, at this point, a TON of 1,000 mol mol^{-1}, a TOF of ~400 mol mol^{-1} h^{-1} and ee of 90% were achieved under the most suitable conditions for a future production process. In this case, a compromise between throughput and selectivity was made, running the reaction at 55 °C at 80 bar hydrogen pressure. Nevertheless, the reactivity for the application in an industrial process was rather moderate. Therefore, an additional screening was conducted, retaining the 3,3'-dimethyl-BINOL moiety and testing a set of different amines. It emerged that the catalyst performance was further improved by using cyclohexylamine for the formation of the phosphoramidite and tri-*ortho*-tolyl-phosphine as the phosphine ligand. With these conditions, it was even possible to achieve full conversion at substrate catalyst ratios of >10,000 with a TOF of 1,800 mol mol^{-1} h^{-1} and 90% ee. This is

Table 3.1: Various parameters that affect the outcome of the asymmetric hydrogenation [37].

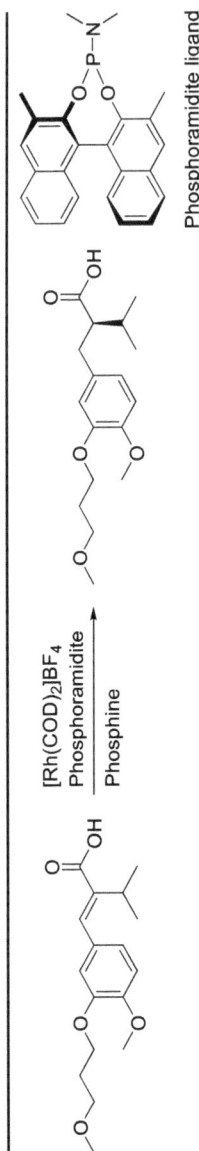

Phosphoramidite ligand

No	PA/Rh	TPP/Rh	Temp (°C)	Time (h)	H_2 (bar)	Vol% H_2O in i-PrOH	Conv. (%)	ee (%)
1	2	1	85	2	20	0	100	80
2	1	1	85	18	20	0	100	66
3	1	2	85	18	20	0	100	76
4	2	1	25	18	20	0	86	85
5	2	1	25	4	20	30	67	94
6	2	1	25	4	20	50	55	92

PA=phosphoramidite, TPP=triphenylphosphine.

a striking example of the synergistic performance of two different ligands in an asymmetric catalytic reaction.

The performance of such a mixed system can be explained by the "Respective Control Concept", which was introduced by Achiwa [38] and was explained using the well-characterised intermediate in the hydrogenation of acetamido-acrylic acid derivatives with bidentate ligands. In this structure, the substrate is coordinated to the metal centre via the C=C double bond and oxygen of the amide function (Figure 3.3). Consequently, the phosphorous atoms of the ligand(s) can adapt a *cis* or *trans*-orientation to the double bond of the substrate. The two P atoms can occupy various positions with respect to the substrate. The coordination residue located in *cis*-position to the substrate is closer to the prochiral double bond and therefore, was postulated to be responsible mainly for enantiocontrol through steric interactions with the substrate. The coordinating residue in *trans*-position is better positioned for electronic interaction with the substrate and hence should mainly affect the reaction rate and coordination behaviour by electronic parameters. Therefore, the steric and electronic properties should be considered individually and varied independently, to obtain an "optimal" ligand for a specific substrate.

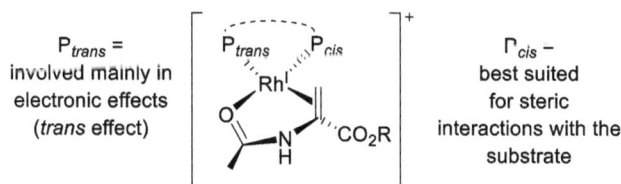

Figure 3.3: The Respective Control Concept by Achiwa explained with a substrate bound to the metal centre.

This behaviour was demonstrated in the asymmetric hydrogenation of dimethyl itaconate. The performance in the asymmetric hydrogenations of ligands with two different phosphine units was tested and compared to that of analogous ligands with two identical phosphine units. The results obtained in the hydrogenation also demonstrated that switching the positions of the *cis*- and *trans*phosphine units has a strong effect on the *ee*-value and conversion (Table 3.2).

Although these results emphasise the potential advantages of tuning the coordinating sites of a bidentate ligand individually, this concept does not guarantee improved catalyst performance. If isomeric complexes are formed, in which the two phosphine units have switched positions, all efforts to optimise the ligand may be useless. However, in the absence of such complications, catalyst performance can be superior to using only C_2-symmetric ligands.

Table 3.2: Asymmetric hydrogenation of dimethyl itaconate using a ligand with different phosphorus residues.

No	P_{cis}	P_{trans}	Catalyst loading (mol%)	Conversion [%]	ee (%)
1	PPh$_2$	PPh$_2$	0.1	36	5
2	P(p-Me$_2$NPh)$_2$	PPh$_2$	0.1	55	68
3	PPh$_2$	P(p-Me$_2$NPh)$_2$	0.1	>99	93

3.2.5 Pregabalin

Pregabalin is a potent anticonvulsant drug used to treat epilepsy and pain [39]. It is marketed under the brand name Lyrica® by Pfizer. In drug usage statistics, it is listed at position 72 with more than 10,000,000 prescriptions in 2017 (United States) [40]. Since the biological activity is derived from the (S)-enantiomer, an asymmetric production process is required. One elegant option is utilising asymmetric hydrogenation as the key step. Different options, for example with DuPhos as catalyst, have been reported [38]; however, the most efficient at that time was the use of a P-chiral phosphine (Trichickenfootphos, TCFP) a ligand developed by Pfizer, in combination with Rh as the active metal [41]. Low catalyst loadings, a higher substrate concentration, an excellent yield and high enantioselectivity are key features of this transformation. Using only 0.0037 mol% catalyst loading, 98% yield in combination with 98% ee was achieved (Scheme 3.15).

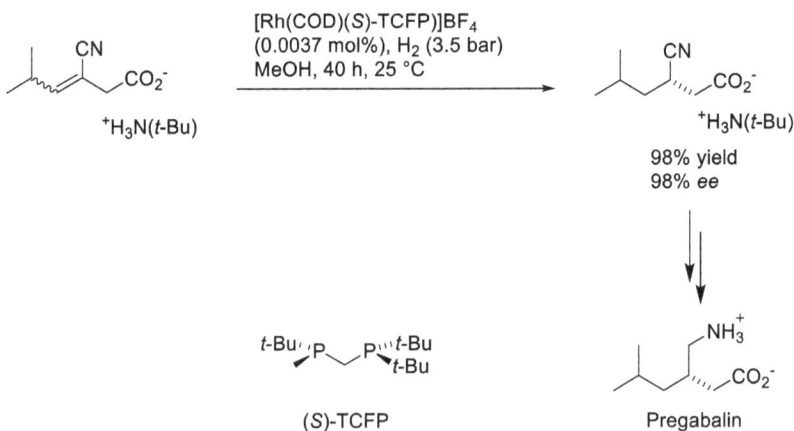

Scheme 3.15: Rh-catalysed asymmetric hydrogenation in the synthesis of pregabalin.

The production process of pregabalin was changed in 2006 to an enzymatic approach, which surpassed the hydrogenation route in terms of cost effectiveness and environmental performance (Chapter 10: Biocatalysis) [42].

3.2.6 (+)−Biotin

(+)−Biotin is a water-soluble vitamin and plays an essential role as a co-factor for carboxylation reactions in biochemical processes [43]. A synthesis of (+)−biotin, which was developed at technical scale by Lonza, uses a diastereoselective hydrogenation approach as the key step in introducing the two-nitrogen substituted stereogenic centres (Scheme 3.16). Starting from diketene, the synthesis of the tetrasubstituted hydrogenation substrate was achieved in a straightforward manner. Optimisation of the reaction conditions (including ligand screening) resulted in a high diastereoselectivity of >99:1, using a ferrocene derived Josiphos ligand.

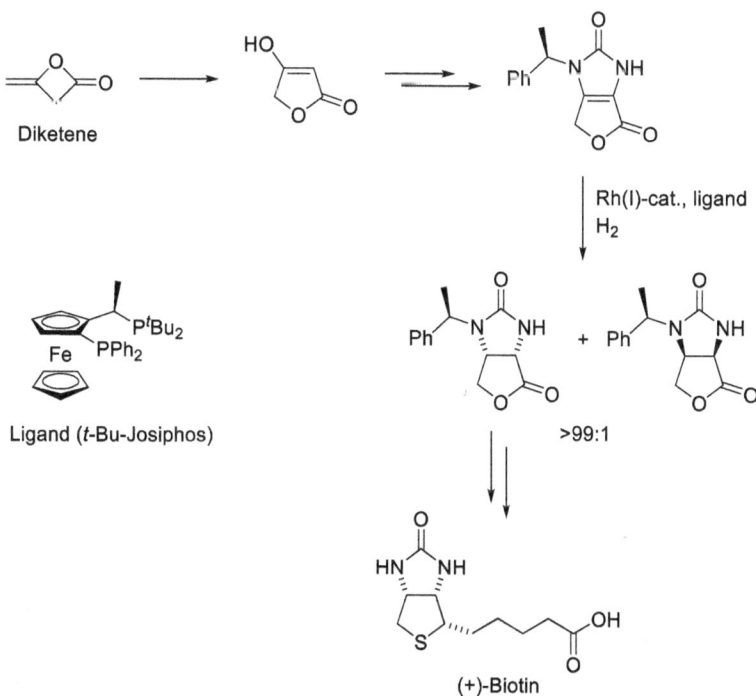

Scheme 3.16: Diastereoselective hydrogenation approach to obtain (+)−biotin.

This process was, in the meanwhile, replaced by cheaper alternatives [44]. The expensive chiral auxiliary, (R)-methylbenzylamine, which was required to obtain the high selectivity in the diastereoselective hydrogenation step, is destroyed upon cleavage. This is one major drawback of the depicted process.

A breakthrough in applying hydrogenation technology in the synthesis of (+)−biotin was achieved in a cooperation between DSM Nutritional Products and the catalysis group of Solvias [45, 46]. The key intermediate in all commercial syntheses of (+)−biotin is the D-lactone (Scheme 3.17). The synthesis of the D-lactone was successfully achieved, starting from fumaric acid, which was converted to the meso-anhydride, as depicted in Scheme 3.17, and the two stereogenic centres were introduced by anhydride desymmetrisation via asymmetric hydrogenation. Both Rh- and Ir-based catalysts can be used in combination with atropisomeric ligands and achieved excellent enantioselectivity in combination with full conversion [47] (for an explanation of atropisomers, see box on page 113).

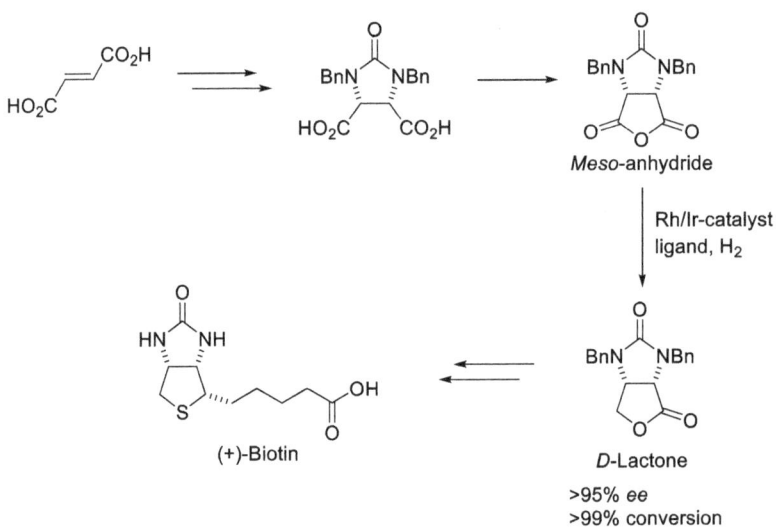

Scheme 3.17: Introduction of two stereogenic centres in the (+)−biotin synthesis by anhydride desymmetrisation *via* asymmetric hydrogenation.

3.2.7 Sitagliptin

Sitagliptin (sold under the brand name Januvia™) is an orally active pharmaceutical ingredient that was discovered by Merck and is used for the treatment of type 2 diabetes mellitus [48]. One key step towards an efficient production process was the discovery of an efficient asymmetric hydrogenation for unprotected enamides (Scheme 3.18) [49]. Earlier, it was assumed that an acetyl-group attached to the nitrogen was required to increase chelation between the substrate and catalyst and hence increase the reactivity and selectivity [19]. However, the introduction and removal of this anchoring group is not trivial and makes this approach a bottleneck for industrial applications [50]. But with the implementation of a *t*-Bu-*Josiphos*-based Rh-catalyst, this limitation was circumvented and excellent selectivity (97% *ee*) with high yield was observed (Scheme 3.18). Additionally, 95% of the precious rhodium was recovered. Further optimisation in terms of pH-value with stoichiometric addition of ammonium chloride led to further improvements [51]. This phenomenon was explained by enamine-imine-tautomerisation, which is affected by the pH-value, playing a crucial role in this reaction. Deuterium labelling experiments indicated that the reaction proceeds via the imine rather than the enamide form [52].

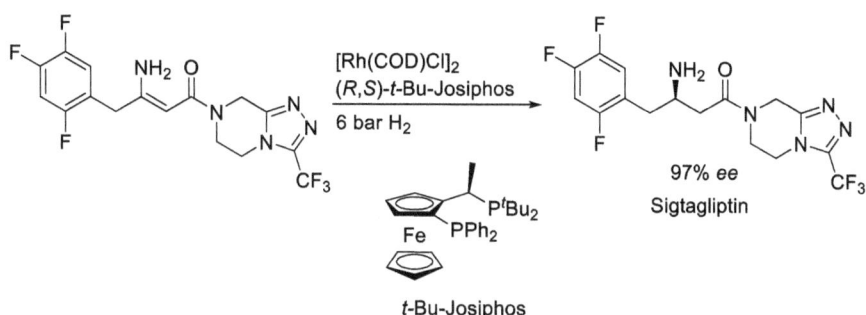

Scheme 3.18: Rh-catalysed asymmetric enamide hydrogenation as the key step in the synthesis of sitagliptin.

This new process with the asymmetric hydrogenation as the key step was awarded the Presidential Green Chemistry award in 2006. The application of the new process resulted in the reduction of 220 kg waste per kg product and increased the overall yield by nearly 50%. Over the product lifetime, Merck expected to eliminate the formation of at least 150 million kg of waste, including nearly 50 million kg of aqueous waste [53]. This is also a great example of how continuous improvement in production processes is required and achieved. This compound won a second Presidential Green Chemistry award in 2010, replacing the previously awarded hydrogenation step with an enzymatic reduction (Chapter 10: Biocatalysis). In this case, an increased productivity of 56%, increased overall yield of 10–13% and waste reduction by 19% was achieved [54].

3.2.8 Pantolactone

Another example which demonstrates the power of Rh-catalysed asymmetric hydrogenation was reported for the synthesis of pantothenic acid (vitamin B_5). (*R*)-Pantothenic acid is essential for the biosynthesis of coenzyme A for other biochemical processes. Pantothenic acid is supplied almost exclusively in the form of the calcium salt with world market volume of ca. 18,000 t/a at prices of 30–60 \$/kg [44] in 2017.

The calcium salt of pantothenic acid is industrially produced via reaction of (R)-pantolactone and calcium β-alaninate (Scheme 3.19). (R)-Pantolactone is obtained by optical resolution from racemic pantolactone [55].

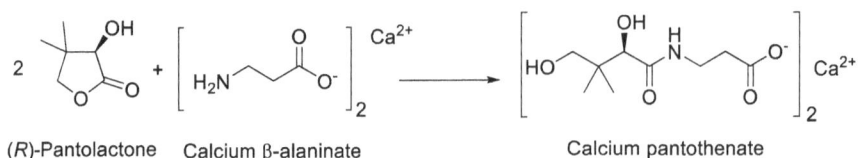

(R)-Pantolactone   Calcium β-alaninate   Calcium pantothenate

Scheme 3.19: Last step in the industrial synthesis of the calcium salt of pantothenic acid.

Alternatively, (R)-pantolactone can be obtained by the enantioselective hydrogenation of 2-oxopantolactone (Scheme 3.20). With a rhodium catalyst containing m-TolPOPPM ligand, a high enantiomeric excess of 91% ee and high turnover numbers (TON) of 200,000 could be reached (Scheme 3.20) [44]. This ligand class can be synthesised from 4-hydroxyproline [56, 57]. Later, rhodium complexes bearing an amidophosphine-phosphinite ligand with cyclohexyl substituents were reported to reach excellent enantioselectivities of >96%. Catalyst loadings of s/c 10,000 and 70,000 were reported [58]. Chiral ruthenium-bisphosphine complexes can be used, but both the activities and selectivities are generally low [59, 60]. Iridium-based catalysts have also been reported for this reaction [61].

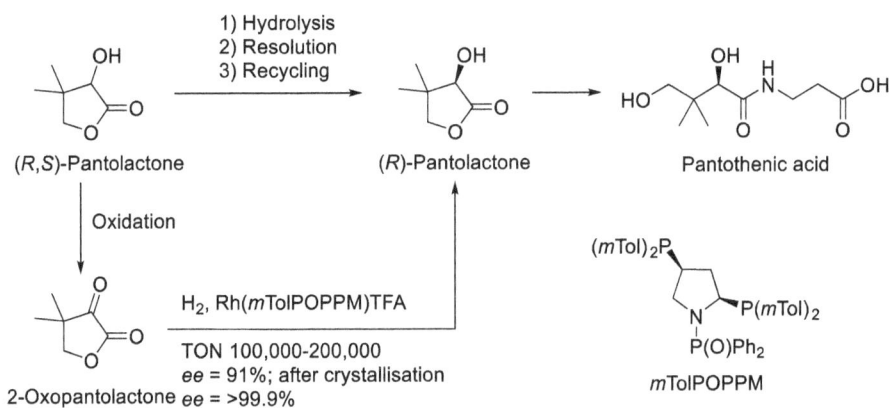

(R,S)-Pantolactone   (R)-Pantolactone   Pantothenic acid

2-Oxopantolactone

H$_2$, Rh(mTolPOPPM)TFA

TON 100,000-200,000
ee = 91%; after crystallisation
ee = >99.9%

mTolPOPPM

Scheme 3.20: Asymmetric hydrogenation as the key technology in the synthesis of the chiral key intermediate (R)-pantolactone for the synthesis of pantothenic acid.

3.3 Ru-catalysed hydrogenations

3.3.1 Hydrogenation of allylic alcohols

Allylic alcohols are among the most suitable substrates for the asymmetric hydrogenation with chiral Ru-based catalysts. A striking example is the hydrogenation of geraniol and its (Z)-isomer nerol to optically active citronellol, which is a compound used in the flavour and fragrance industry (Scheme 3.21). Furthermore, it is used as starting material for various compounds in that industry, for example rose oxide (Chapter 4: Oxidations) [62].

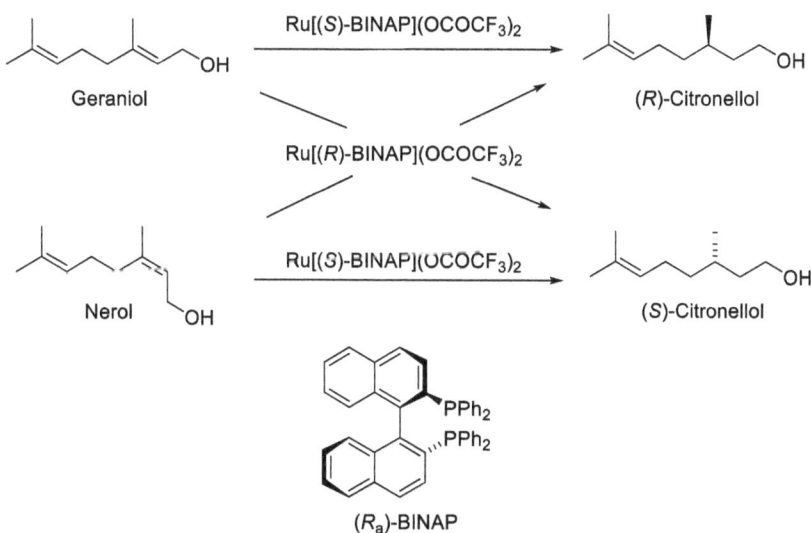

Scheme 3.21: Asymmetric hydrogenation of allylic C=C-bond in geraniol and nerol.

High catalyst turnover numbers of 50,000 and a high substrate concentration of 50% are attractive features of this asymmetric hydrogenation [63]. The unfunctionalised C=C-bond remains untouched. Takasago uses this technology for a 300 t production per year [64]. With the right choice of double bond geometry in the starting material in combination with the correct enantiomer of chiral ligand, both enantiomers of citronellol can be easily accessed in 97% ee. The ligand of choice is BINAP and its derivatives that exhibit axial chirality (see box below).

The hydrogenation of longer isoprenoids such as tetrahydrofarnesol was reported and gave similar selectivities and yields. These reactions were conducted in kg-scale [65, 66]. The desired double-bond isomer can be obtained via distillation, which adds additional efforts to such a synthesis.

Axial Chirality/Atropisomers

Atropisomers are compounds containing a hindered rotation axis such as allenes or ortho-substituted biphenyl-groups with different substituents attached. In the case of hindered rotation along the axis, these compounds behave as enantiomers. The configuration of these compounds can be determined by *Newman* projection looking via the hindered rotation axis (Scheme 3.22). In this case, the substituents of the nearest plane (e.g. nearest aryl ring) are ranked by priority, based on *Cahn–Ingold–Prelog* rule moving from the substituent of highest priority in the first plane along the shortest path to the substituent of highest priority in the plane located behind. The sense of rotation determines the configuration *R* or *S*. Usually, the subscript 'a' is used while referring to axial chirality.

The most commonly used atropisomeric ligands in the synthesis of fine chemicals are those based on a biphenyl and binaphthyl scaffold, for example phosphoramidites (Table 3.1), BINAP, BINOL and derivatives (Schemes 3.21, 3.25, 3.27, 3.29, 3.31) and biphenyl derived ligands such as MeOBIPHEP and SegPhos (Schemes 3.23, 3.24).

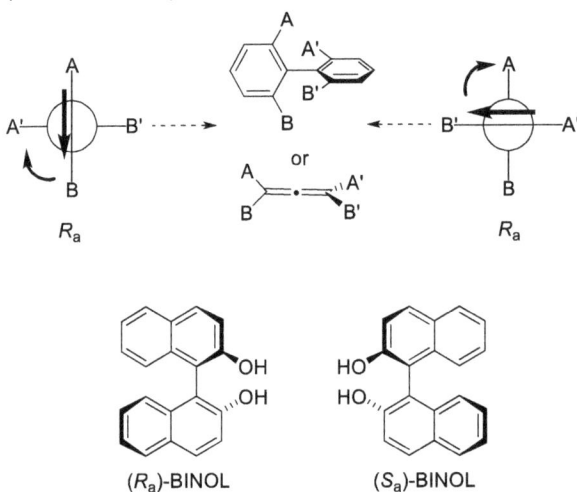

Scheme 3.22: Determination of axial chirality.

3.3.2 Pivotal glutarate intermediate for candoxatril

Another example of how the applied catalyst can change in the course of process optimisation is demonstrated in the synthesis of pivotal glutarate, an intermediate in the synthesis of candoxatril, a cardiovascular drug developed by Pfizer in the mid-1990s [67]. In a first attempt, 0.01 mol% of *Noyori*'s catalyst ([(R)-BINAP(p-cymene)RuCl]Cl) was used in the hydrogenation of an unsaturated sodium carboxylate, and an enantioselectivity of 94% *ee* was achieved. The major drawback of this protocol was double-bond migration which resulted in a 4:1 ratio of the desired product and the corresponding vinylic ether [68].

A further improvement was achieved with a Rh(I)DuPhos catalyst. In this case, an s/c of 3,500 with a TOF = 6,000/h and an *ee* of >99% was achieved, and no isomerisation of the starting material was observed. This method was applied on 12 kg reaction scale [69]. Finally, a production of 2 tons of the API used for clinical trials in phase 3 was achieved, using a Ru-based catalyst. MeOBiphep as ligand, in combination with a ruthenium precursor, resulted in 99.4% *ee* and 84% yield of the crude product (Scheme 3.23). The reaction was run in batch sizes of 231 kg in an autoclave of 4,000 l capacity [70].

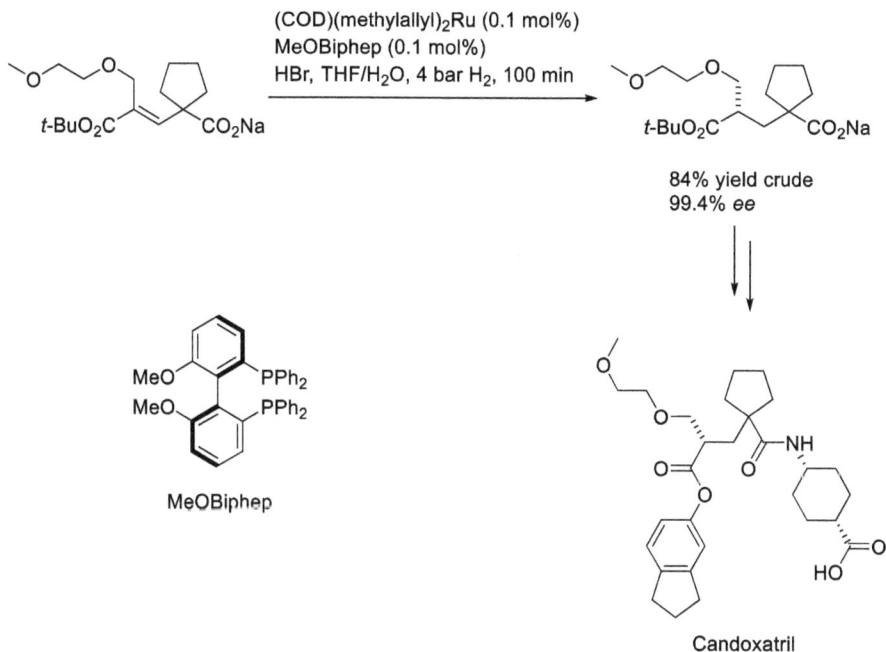

Scheme 3.23: Asymmetric hydrogenation as key step in the synthesis of candoxatril.

3.3.3 Artemisinin

Artemisinin is a naturally occurring compound found in sweet wormwood and is used for the treatment of malaria, resulting in significantly reduced mortality rates. The discovery of this compound and its potential for the treatment of malaria was honoured with the Nobel Prize in Medicine to *Tu Youyou* in 2015 [71].

Its synthesis can be performed via a semisynthetic route starting from artemesinic acid obtained via fermentation (Chapter 10: Biocatalysis). Applying a diastereoselective hydrogenation and photochemical oxidation (Chapter 4: Oxidations) as key conversions resulted in an efficient process, which was the first source of non-plant-derived-artemisinin on an industrial scale. This process was industrialised in Sanofi's production facility in Garessio, Italy, in 2013, with a production volume of up to 60 tons per year, and an average batch size of 370 kg [72]. The initial chemistry route reported by Amyris also uses homogeneous hydrogenation in the first step. In this case, *Wilkinson's* catalyst was used with a catalyst loading of 0.05 mol% in toluene at 47 bar hydrogen pressure. After 19 h at 80 °C, nearly quantitative yields and a diastereomeric ratio of 96:4 were obtained, without any significant over-reduction to tetra-hydroartemisinic acid [73]. These results were further improved by Takasago in a screening which was conducted for Sanofi (Scheme 3.24). Key drivers were the high cost of Rh and the comparable high catalyst loadings in combination with long reaction times. The preferred catalyst system was a Ru-based catalyst with dtbm-Segphos as ligand. In this case, a diastereoisomer ratio of 95:5 with 99% yield was achieved in 5–6 h reaction time. The catalyst loading was reduced to 0.0125 mol% and the hydrogen pressure to 22 bar. Even lower catalyst loadings of 0.00625 mol% resulted in full conversion, but a longer reaction time was required [74].

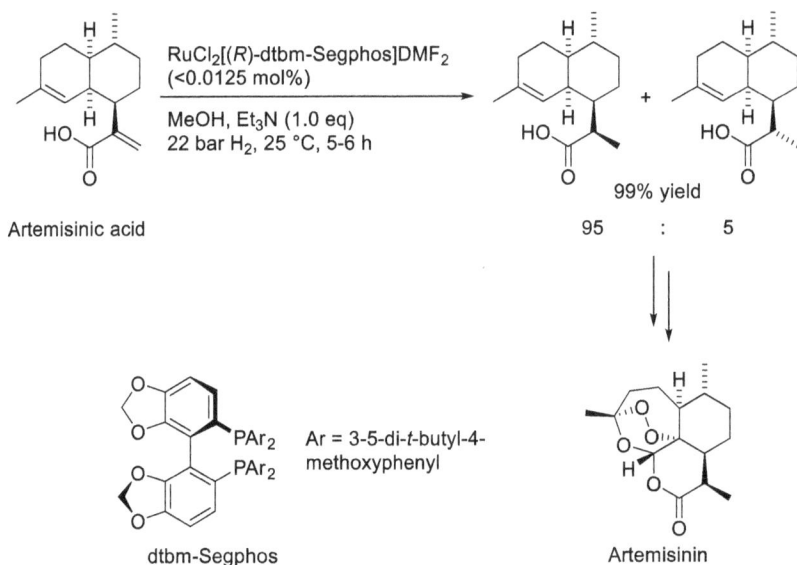

Scheme 3.24: Diastereoselective hydrogenation applied in the synthesis of artemisinin.

3.3.4 Ketone reductions

The asymmetric hydrogenation of ketones is among the most important chemical reactions in achieving the synthesis of enantiomerically enriched secondary alcohols. The first successful report by the research group of *Noyori* at Nagoya university in Japan used RuCl$_2$(BINAP) as catalyst. A highly selective hydrogenation of various β-ketoesters was reported (Scheme 3.25) [63]. One of the advantages of this ligand system is that both enantiomers are commercially available or can be prepared from inexpensive (±)-1,1′-bi-2-naphthol [75] relatively simply.

The mechanism of the ketone hydrogenation is depicted below (Scheme 3.26); it also shows one of the major drawbacks of this catalyst system. Usually, only substrates that bear additional coordinating groups (in this case, the ester function) are

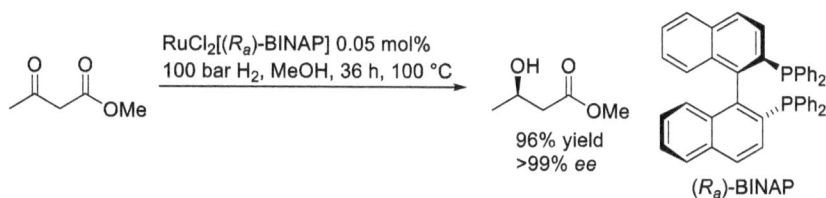

Scheme 3.25: Asymmetric hydrogenation of β-ketoesters reported by the research group of *Noyori*.

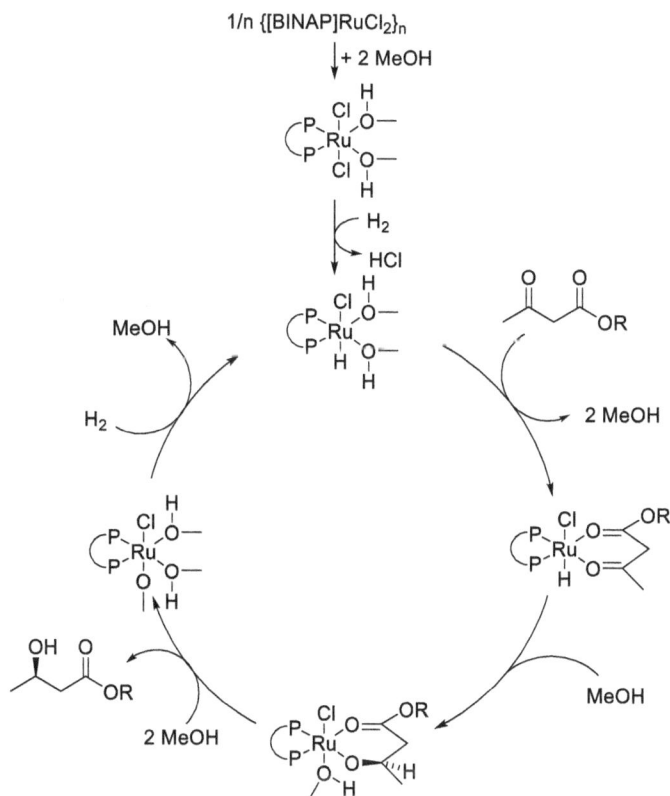

Scheme 3.26: Mechanism of the RuCl$_2$(BINAP)-catalysed reduction of β–ketoesters.

suited for this type of catalyst. In general, excellent results are obtained for β-ketoesters, β-aminoketones and β-hydroxyketones.

One application of this system is in the asymmetric hydrogenation of hydroxy acetone to propane-1,2-diol, an intermediate used for industrial synthesis of the antibacterial levofloxacin (Scheme 3.27) [1]. This reaction is run on 50 t scale per year by Takasago [76], achieving 94% ee with a TON of 2,000. More recently even better catalytic systems have been reported [77].

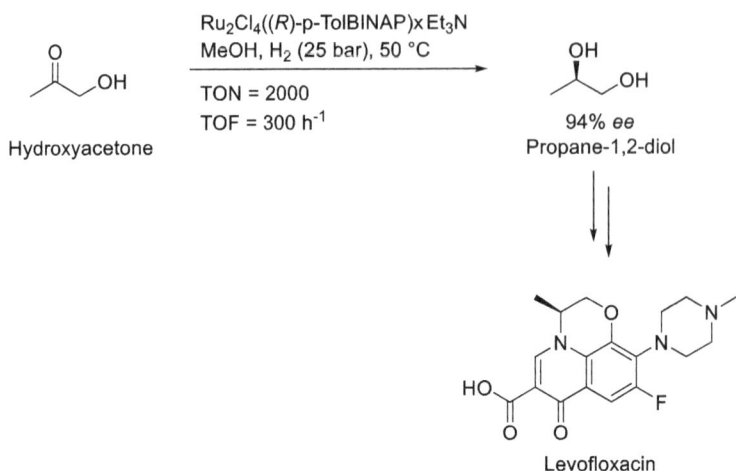

Scheme 3.27: Industrially applied asymmetric hydrogenation hydroxy acetone to propane-1,2-diol.

As described above, the previously mentioned catalyst for ketone hydrogenation requires an adjacent coordinating group. This problem was overcome by a second gener-

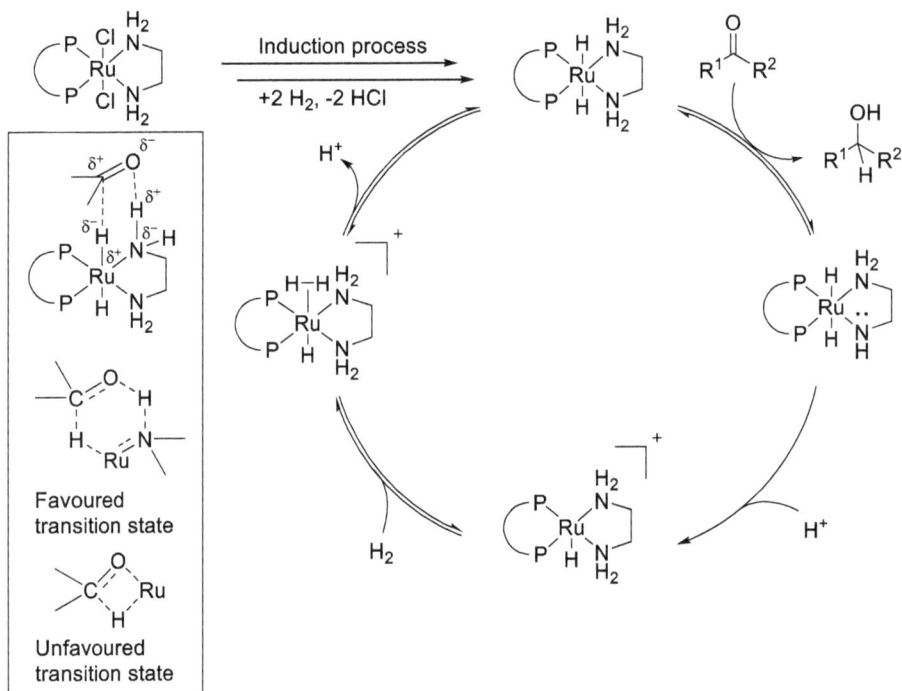

Scheme 3.28: Mechanism of the ketone hydrogenation with Ru(II)-PP-diamine catalysts.

ation of these catalysts, also developed by *Noyori* and his research group. In addition to a chiral diphosphine ligand, these catalysts also contain a chiral diamine ligand. The mechanism of these catalysts is different from the previous one, and it is assumed that there is no coordination of the substrate to the Ru-centre (Scheme 3.28) [78].

With these catalysts, a wide variety of aromatic, heteroaromatic and α,β-unsaturated ketones can be reduced with high enantioselectivity and excellent yields. Also, purely aliphatic ketones can be reduced with these catalysts, with moderate selectivity, though.

An application of these Ru-bisphosphine-diamine catalysts was demonstrated in the synthesis of a potent antipsychotic agent BMS 18110. The step was industrialised resulting in 97% *ee* and 97% yield, using only a catalyst loading of 0.01 mol% (Scheme 3.29). The aromatic fluoride and 2-amino-5-fluoropyrimidine moieties were not touched by the catalyst [1, 79, 80].

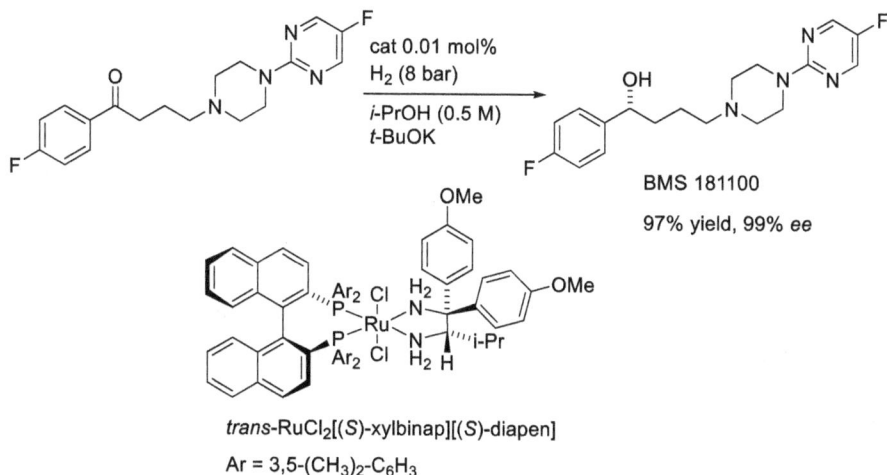

Scheme 3.29: Asymmetric ketone reduction in the synthesis of BMS 181,100.

3.3.5 Dynamic kinetic resolution via asymmetric hydrogenation

Another application of Ru-based catalysts is the dynamic kinetic resolution of racemic β-ketoesters with a configurationally labile α-stereogenic centre. In this case, the reduction can result in only one single stereoisomer out of the four possible stereoisomers with a theoretical yield of 100% (Scheme 3.30) [81]. The basis of this process is that the enolisation of the β-ketoesters must be significantly faster than the hydrogenation of the unwanted enantiomer of the hydrogenation substrate, to achieve good results. In this case, with the right choice of catalyst type and enantiomer, the reaction can be driven towards the selective formation of each stereoisomer. Also, other functional groups or substrates can be converted using this concept and, usually, high enantio- and diastereoselectivities can be obtained [82–84].

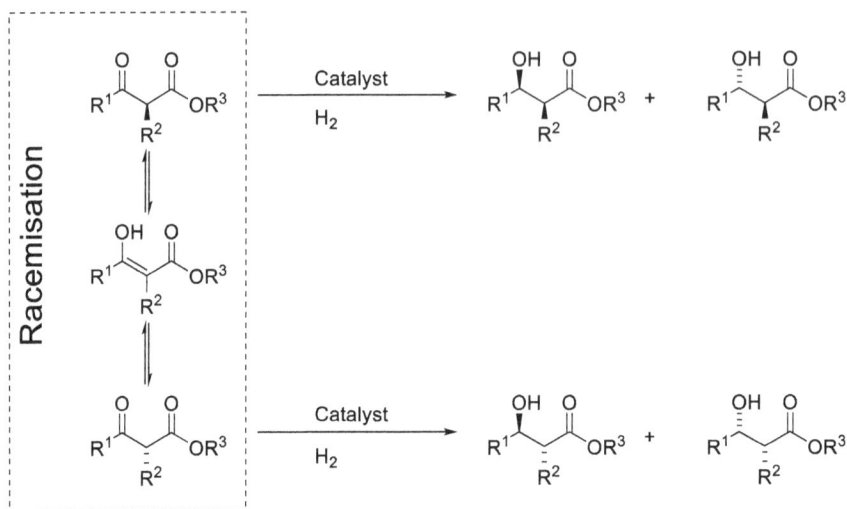

Scheme 3.30: Dynamic kinetic resolution of β-ketoesters via asymmetric hydrogenation.

This technology was applied in the synthesis of various biologically active compounds and was highlighted by the industrialised synthesis of carbapenems by Takasago (Scheme 3.31) [1]. The hydrogenation was conducted in dichloromethane using a (R)-BINAP-based ruthenium catalyst, which gave the (2S,3R)-enantiomer in 99% *ee* and with a *syn:anti* ratio of 94:6. This data indicates that the (S)-enantiomer of the substrate is hydrogenated 15 times faster than the (R)-enantiomer, and the isomerisation of the stereogenic centre in the α-position is 92 times faster than the hydrogenation of the (R)-enantiomer [85–87]. This reaction was conducted in a hydrogenation reactor with a volume of 15 m³ [1].

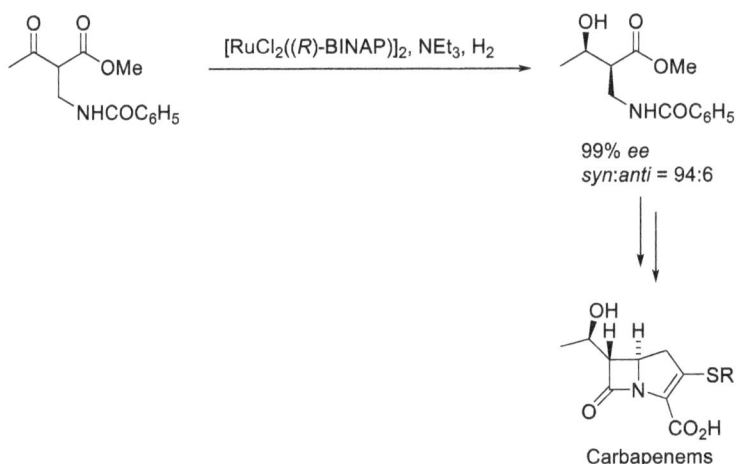

Scheme 3.31: Dynamic kinetic resolution in the industrialised synthesis of carbapenems by Takasago.

3.3.6 Homogeneous ester hydrogenation

One transformation that is often used on small scale in the laboratory that is significantly less common in the manufacture of fine chemicals is the reduction of carboxylic acids and esters to alcohols. The main reason for this is that strong reducing
agents are required, such as lithium aluminium hydride or lithium borohydride. Not
only can the handling of these reagents require extra safety protocols, but the work
up of the reactions can be difficult and a stoichiometric amount of waste (aluminium
or boron respectively) is generated. A much more elegant solution would be hydrogenation. Heterogeneous copper chromate catalysts are known and applied, for example, in the hydrogenation of fatty acids [88]; however, these also have the drawback
of forcing reaction conditions (high temperatures, and pressures) and also the need to
deal with copper and more specifically chromium waste at the end of the reaction.

In recent years there have been significant advances in the use of homogeneous
catalysts for the hydrogenation of esters. In particular, ruthenium complexes have
been developed and commercialised by a number of companies and additionally catalysts based on cheaper metals (such as iron, manganese and cobalt) have been reported by academic laboratories.

The first really successful Ru-based catalysts for ester hydrogenation were developed by Elsevier *et al.* using [Ru(acac)$_3$] and the tridentate ligand "TriPhos" (Figure 3.4)
[89, 90]. However, to our knowledge these catalysts have not been applied on an industrial scale. The second breakthrough came through the use of ligands containing both
nitrogen and phosphorus donors. Milstein *et al.* reported the use a ruthenium catalyst
containing a tridentate PNN-ligand for the hydrogenation of a variety of aromatic and
aliphatic esters [91, 92]. The company Takasago developed and commercialised ruthe-

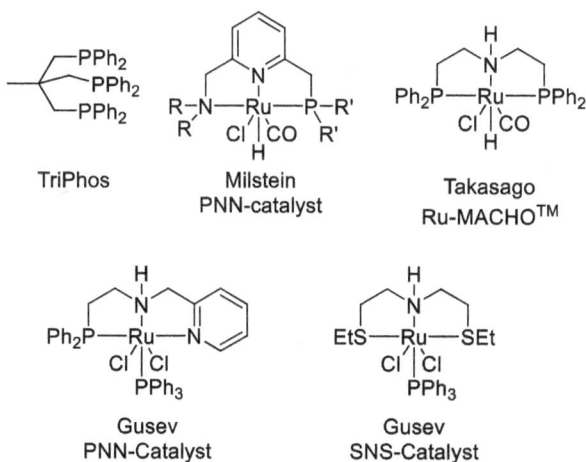

Figure 3.4: Ligands and catalysts applied in the homogeneous ester hydrogenation.

nium catalysts based on a tridentate PNP-ligand and named the catalyst "Ru-MACHO" [93]. Gusev *et al.* developed ligands based on tridentate PNN- and SNS-ligands which have been commercialised by the catalyst company Johnson Matthey [94, 95].

A further development was the application of ruthenium catalysts that use two bidentate PN-ligands, or even one tetradentate PNNP-ligand [96–99]. These ruthenium complexes have been reported for the synthesis of a number of commercial aroma compounds, such as in the synthesis of Ambrox™ from Sclareolide™ (Scheme 3.32) [90]; however, it has not been stated if this process was applied on an industrial scale.

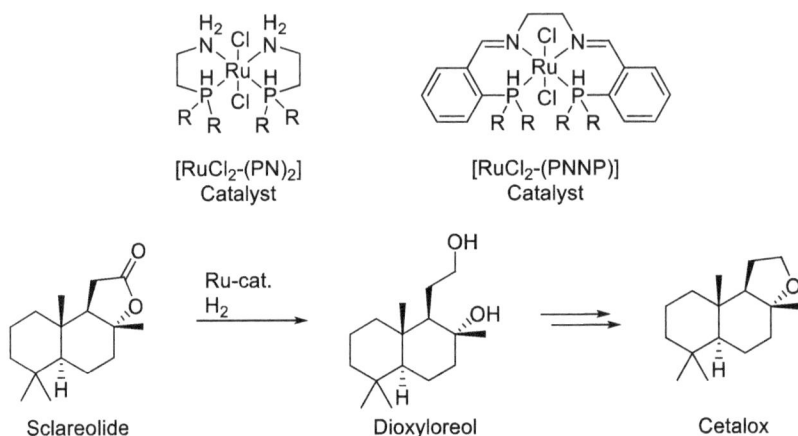

[RuCl$_2$-(PN)$_2$]
Catalyst

[RuCl$_2$-(PNNP)]
Catalyst

Sclareolide

Dioxyloreol

Cetalox

Scheme 3.32: Homogeneous ruthenium catalysed hydrogenation of Sclareolide™.

The one published example of a homogeneous ester hydrogenation being performed on a large scale is the production of (*R*)-1,2-propanediol by Takasago [93]. This diol is a useful chiral building block for pharmaceuticals and has been produced by Ru-catalysed hydrogenation of hydroxyacetone since the 1990s. However, the optical purity of the product was less than 99%, which was problematic for certain applications. Takasago used their Ru-MACHO catalyst for the hydrogenation of (*R*)-methyl lactate, which is readily available in >99% ee. An incredibly efficient process was developed where, per batch, 2.2 t of (*R*)-methyl lactate was hydrogenated with only 6.4 kg of Ru-MACHO in methanol with approximately 40 bar of hydrogen to give 1.5 t of (*R*)-1,2-propanediol in 92% yield with only a tiny reduction of the *ee* (Scheme 3.33). The work was extended to produce a menthol derivative, also applying the Ru-MACHO catalyst for the hydrogenation of an ester [93]. These examples show the efficiency of the homogeneous hydrogenation of esters, and it is likely that many more examples will be reported in the future.

Original Process

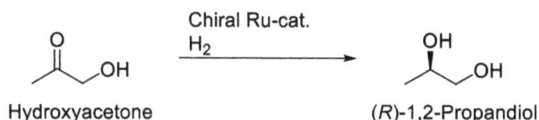

Chiral Ru-cat.
H$_2$

Hydroxyacetone → (R)-1,2-Propanediol

New Process

Ph$_2$P—Ru—PPh$_2$
Cl CO

Ru-MACHO
H$_2$

Ru-MACHO™

(R)-Methyl lactate → (R)-1,2-Propanediol

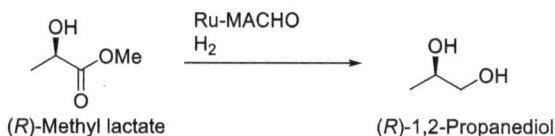

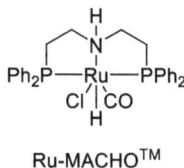

Scheme 3.33: Production of (R)-1,2-propanediol via homogeneous Ru-catalysed hydrogenation.

3.4 Ir-catalysed hydrogenation

3.4.1 Hydrogenation of imines

3.4.1.1 Metolachlor

Metolachlor is the active ingredient in Dual Magnum®, a widely used herbicide, which was originally sold as a racemic mixture of its four possible stereoisomers. Besides the stereogenic centre, the compounds contain a chiral axis introduced by the restricted rotation of the aryl-nitrogen bond (for more information on axial chirality, see box on page 113). However, 95% of the biological activity is induced by the 1'S diastereoisomers [100]. Therefore, the production of an enantiomerically enriched metolachlor seemed a logical consequence.

For the development of an enantioenriched metolachlor, four different strategies were envisioned: Enamide hydrogenation (inspired by L-DOPA), nucleophilic substitution with an asymmetric hydrogenation of methoxy acetone to introduce the stereogenic centre, imine hydrogenation and catalytic alkylation with racemic methoxyisopropanol. It was reported that after first initial experiments, only the imine hydrogenation remained as a realistic option for industrial implementation (Scheme 3.34) [101, 102].

Using xyliphos as ligand, the selectivity of the asymmetric hydrogenation applied in production is rather moderate, 79% ee. Screening of different ligands showed higher enantioselectivities, but, in each case, a loss of catalyst reactivity. The xyliphos ligand is simply the best compromise found so far between catalyst reactivity and stability, and enantioselectivity that can be reached. A key to the success was the combination of acetic acid and TBAI (tetrabutyl ammonium iodide) as additives. None of these additives on their own significantly improved catalyst reactivity; however, the combination of both resulted in a TOF exceeding 1,800,000 mol/mol/h, with substrate to catalyst ratio of 2,000,000. Full conversion is reached within only 4 h applying 80 bar H$_2$-pressure, and

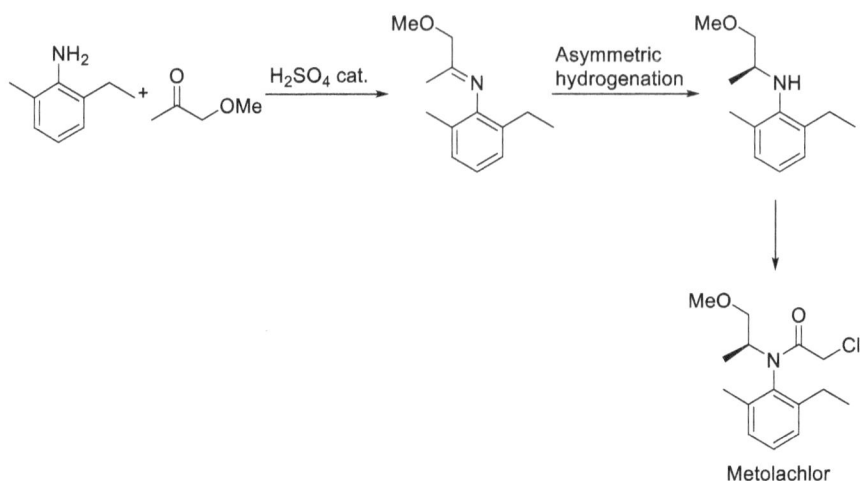

Scheme 3.34: Industrial synthesis of (S)-metolachlor in >10,000 (t/year) by *Syngenta*.

as already mentioned, an *ee*-value of 79% was obtained (Scheme 3.35). The first production of the new asymmetric hydrogen process started in November 1996. Different mechanistic studies have been conducted to investigate the mechanism of this reaction and explain the high reactivity of the catalytic system. The formation of a dinuclear iodo-bridged Ir^{III} iodo complex might play a crucial role. However, the nature of the high reactivity is, so far, not completely understood [103, 104].

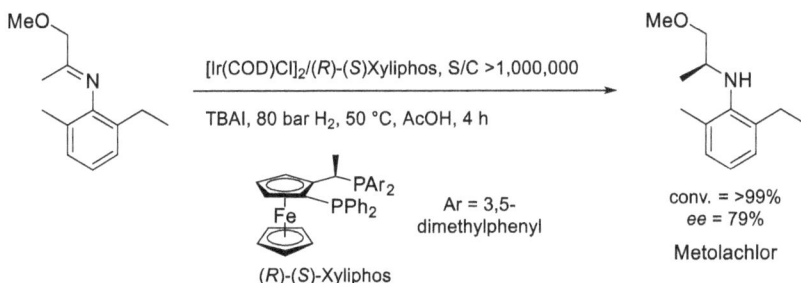

Scheme 3.35: Asymmetric hydrogenation as the key step in the industrial production of (S)-metolachlor.

3.4.1.2 Dextromethorphan

A further example of an efficient imine hydrogenation and how the specific tuning of the ligand residues affects the outcome of the reaction is found in the synthesis of dextromethorphan, an active pharmaceutical ingredient that has an impact on the central nervous system [105]. In a route reported by Lonza [106], the stereogenic centre was introduced using an Ir-based catalyst containing the versatile ferrocenyl-

Table 3.3: Ligand influence in the asymmetric hydrogenation towards dextromethorphan.

R^1	R^2	S/C	ee [%]	Relative activity
Et	t-Bu	200	13.7 (R)	1
t-Bu	Ph	200	14.9 (S)	1.7
Ph	i-Pr	200	35.5 (S)	2.3
i-Pr	t-Bu	200	54.6 (R)	4.0
Ph	t-Bu	1500	85.5 (S)	125
Ar	t-Bu	1500	89.2 (S)	200

Ar=4-MeO-3,5-dimethylphenyl

based ligands developed by *Togni* and *Spindler* [107]. A screening of various derivatives of the ligand family highlights the huge impact the applied ligands can have on the catalyst reactivity and the enantioselectivity (Table 3.3) [108]. As substrates, the H_3PO_4 or the H_2SO_4 iminium salts can be used, which allows the use of intermediates of the at that time established route. These starting materials were also reported by Hoffmann La Roche [109]. However, a rather moderate enantioselectivity (up to 89% *ee*) in combination with a substrate-to-catalyst ratio of only up to 1,500 was achieved. The enantiomeric purity is enriched in the following synthetic steps towards dextromethorphan; nevertheless, moderate chemoselectivity and overreduction of the C=C-bond leaves room for improvement.

A broad screening of various chiral diphosphine ligands bearing different backbones in combination with various metals was carried out and resulted in significantly improved results. The best results were obtained combining the iminium H_2SO_4-salt with an iridium or rhodium-based catalyst containing a sterically demanding (S)-3,5 *t*-butyl-MeOBIPHEP ligand. Further optimisation of the reaction conditions resulted in s/c of 10,000 for the Rh-based catalyst achieving 94% *ee* and 97% yield (Table 3.4). With the iridium-based catalyst the s/c was further reduced to 20,000; however, the enantioselectivity was lower (84%), requiring an additional step to increase the *ee* [110]. This hydrogenation was scaled up at to a several hundred-gram scale [111].

Table 3.4: Comparison of the asymmetric imine hydrogenation towards dextromethorphan with Rh- and Ir-based catalysts.

Parameter[a)	[RhL*(COD)]Cl	[IrL*(COD)]Cl
S/C	10,000	20,000
H_2 (bar)	35	35
Solvent	MeOH (20 wt%)	MeOH (20 wt%)
Temperature (°C)	40	80
ee (%)	94	85
Conversion (%)	>99.9 (15 h)	>99 (10 h)
Yield (%)	97	97
Base	Et$_3$N (10 mol%)	*i*-Pr$_2$NEt (1 mol%)

a) Hydrogenation of the H_2SO_4 salt of the compound depicted in Table 3.3;
L* = (S)-3,5-*tert*-butyl-MeOBIPHEP.

3.4.2 Hydrogenation of ketones with iridium catalysts

As previously shown, iridium-based catalysts can be highly efficient in the hydrogenation of imines. More recently, the hydrogenation of simple ketones was demonstrated, using tridentate spiro-based P-N-N-ligands. This additional coordination site on the ligand was introduced to inhibit catalyst deactivation. High turnover numbers can be easily achieved for ketones (Scheme 3.36, a); even TONs of up to 4,550,000 were reported. However, to achieve these extraordinary results, long reaction times of 15 days and a hydrogen pres-

a)

[Ir(COD)Cl$_2$] / L*
10 bar H$_2$, EtOH, KOt-Bu, RT

up to:
S/C = 5,000,000
TON = 4,550,000
TOF = 12.600/h

up to 99% yield
up to 99.9% ee

b)

[Ir(COD)Cl$_2$] / L*
8 bar H$_2$, EtOH, KOt-Bu, RT

up to:
TON = 1,230,000

up to 98% yield
up to 99.8% ee

c)

[Ir(H$_2$)ClL*], 10-70 bar H$_2$
EtOH, KOt-Bu, RT

up to:
TON = 1,000,000
TOF = 1388/h

up to 97% yield
up to 99.9% ee

L*=

—PAr$_2$

Ar = 3,5-(t-Bu)$_2$C$_6$H$_3$

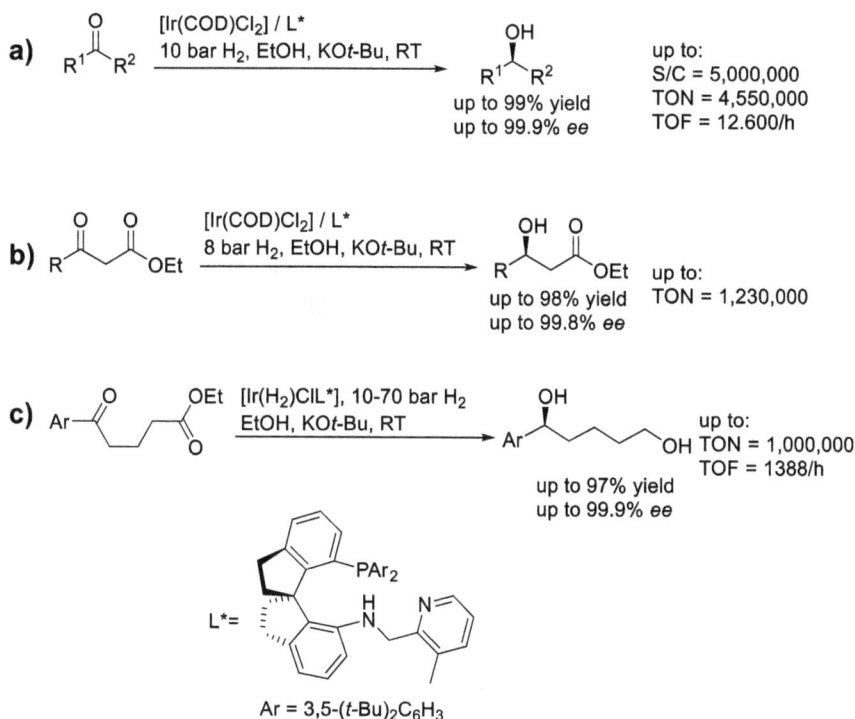

Scheme 3.36: Application of P-N-N-based iridium catalysts in the hydrogenation of various substrates.

sure of 60–100 bar were required [112]. Besides ketones, the asymmetric hydrogenation of β-aryl β-ketoesters [113] (Scheme 3.36, b) and δ-ketoesters [114] (Scheme 3.36, c) were also reported with excellent enantioselectivities, in combination with low catalyst load-ings. In the last example, even the carboxylic ester group is reduced to generate a 1,5-diol. A drawback of these systems is the labour-intensive synthesis of the ligand.

3.4.3 Hydrogenation of unfunctionalised doubled bonds

All catalytic systems discussed so far in this chapter that were applied in the hydro-genation of tri- or tetra-substituted C=C double bonds require certain functional groups adjacent to the prochiral C=C double bond. These catalytic systems usually show only very low reactivity and selectivity in the hydrogenation of "unfunctional-ised" C=C double bonds. For the hydrogenation of these substrates, iridium-based cat-alyst bearing hetero-bidentate ligands have emerged as the most efficient option, over the last few years. Two major findings lead to significantly increased performance of these chiral versions of the *Crabtree* catalyst. The enantioselectivity of the catalysts was significantly increased by the introduction of heterobidentate ligands. The first catalysts contained N,P-ligands which were introduced in 1993 by three different research groups independently (*Helmchen, Pfaltz, Williams*) for allylic substitution reactions [115–117].

Meanwhile, catalysts containing N, C, S and O heteroatoms directly coordinated to the metal centre also showed high enantioselectivities for a broad range of unfunctionalised substrates [118]. This finding is a further development of the "Respective Control Concept" introduced in Section 3.2.4.

The second finding is closely connected to the counter ion of the catalyst. The turnover numbers of the catalysts were significantly improved by replacement of the original PF_6 counterion with the weakly coordinating BAr_F (tetrakis[3,5-bis(trifluoromethyl)phenyl]borate) counterion (Scheme 3.37). Kinetic studies conducted with iridium complexes bearing different counterions provided a plausible explanation to this observation [119]. Catalysts containing stronger coordination counterions such as PF_6 showed first order rate dependence on the olefin concentration, whereas for catalysts containing weaker coordinating counterions such as BAr_F, the rate dependence was close to zero order.

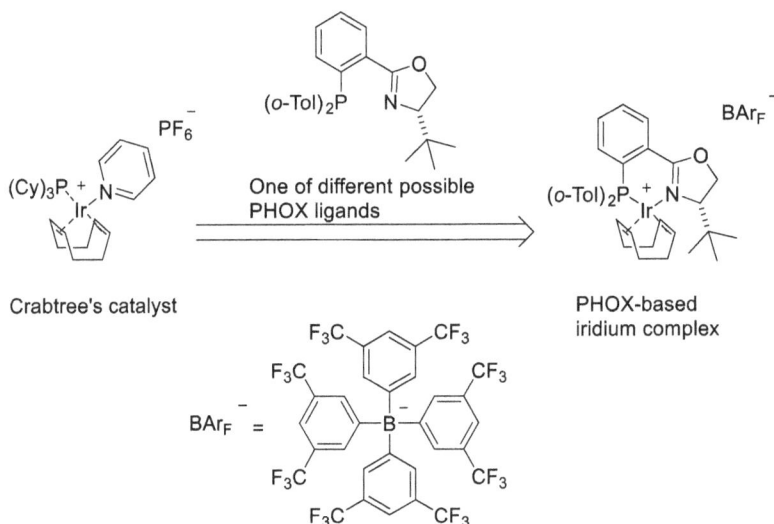

Scheme 3.37: Ir-N,P-metal complexes that were successfully applied in the hydrogenation of unfunctionalised olefins.

This rate difference may be explained by the stronger coordination ability of the PF_6 counterion, hampering coordination of the olefin to the Ir-centre. Therefore, the reaction with the olefin becomes rate-determining. The BAr_F ion does not impede olefin coordination, and the catalyst remains saturated with substrate. Therefore, migratory insertion is much faster than the deactivation reaction in case of the BAr_F salt, and the catalyst has a much longer lifetime. The PF_6 analogue undergoes migratory insertion and deactivation at similar rates and forms an inactive trinuclear iridium hy-dride cluster [120]. With this replacement, the catalyst loading can be reduced >50 fold, still obtaining full conversion and high selectivity (Scheme 3.38). Additionally, BAr$_F$-based iridium complexes reveal higher stability against moisture and oxygen and can be handled in air. Iridium-BAr$_F$ complexes are even purified by column chro-matography on silica gel, if necessary, which facilitates purification of the metal com-plexes significantly.

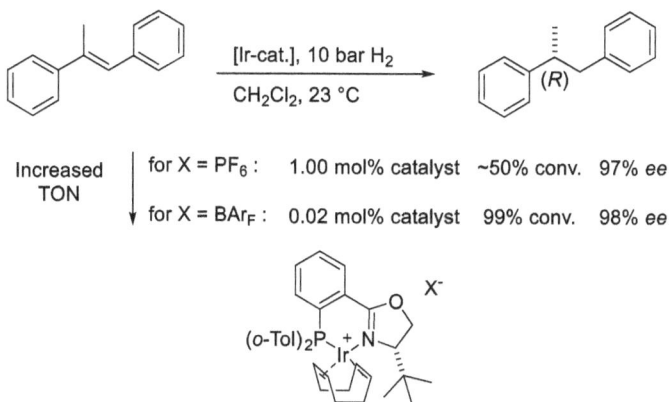

Scheme 3.38: The influence of the counterion in the asymmetric hydrogenation of *trans*-methyl stilbene.

Since the initial report, different substrate classes, for example α,β-unsaturated car-boxylic esters, furans, vinyl boronates and allylic alcohols were successfully hydroge-nated in high selectivity, although not commercially applied [121]. It is likely that for most homogeneous catalytic systems there is no single catalyst which shows high enantioselectivity and reactivity for all substrate classes, and the best catalyst is hard to predict, without a screening.

The most striking examples were reported to introduce the stereogenic centres to the vitamin E side-chain and different side-chain precursors (Scheme 3.39) [122–125]. In all cases, three double bonds were hydrogenated in one step, and two new stereo-genic centres were introduced. In this case, catalysts containing pyridyl phosphinites as ligand emerged as the catalysts of choice. Excellent selectivity was obtained yield-ing the desired (R,R)-stereoisomer in up to 96% selectivity. Further improvement was achieved by introducing a more sterically demanding substituent on the pyridine moi-ety to obtain the (R,R)-enantiomer in up to 98.5% yield [126].

A drawback of these catalysts is that coordinating groups compete with the coordi-nation of the double bond to the iridium centre and therefore reduce the turnover numbers of the catalyst. Therefore, usually weakly coordinating organic solvents, pre-dominately dichloromethane, were used because stronger coordinating solvents com-pete with the substrate for the free coordinating side of the catalyst. This competition

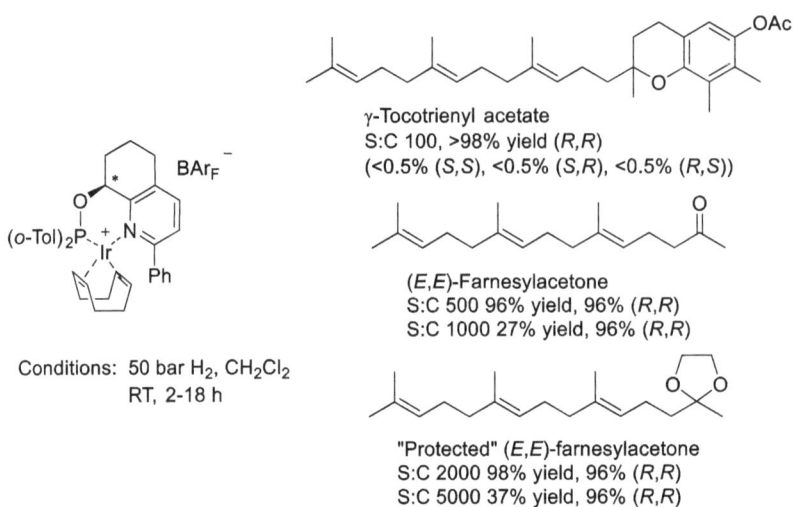

γ-Tocotrienyl acetate
S:C 100, >98% yield (R,R)
(<0.5% (S,S), <0.5% (S,R), <0.5% (R,S))

(E,E)-Farnesylacetone
S:C 500 96% yield, 96% (R,R)
S:C 1000 27% yield, 96% (R,R)

"Protected" (E,E)-farnesylacetone
S:C 2000 98% yield, 96% (R,R)
S:C 5000 37% yield, 96% (R,R)

(o-Tol)$_2$P

BArF

Ph

Conditions: 50 bar H$_2$, CH$_2$Cl$_2$
RT, 2–18 h

Scheme 3.39: Asymmetric hydrogenation of vitamin E and different side-chain precursors.

can even result in complete inhibition for strong coordinating solvents such as acetonitrile [127, 128]. These insights led to further increase of the turnover numbers of the catalyst in the hydrogenation of (E,E)-farnesylacetone. By "protecting" the keto-group of farnesylacetone as a ketal, the substrate-to-catalyst ratio was significantly increased from 500 to 2,000 achieving full conversion, excellent yields and retaining the high selectivity of the reaction (Scheme 3.39) [122].

Another example for the application of these types of iridium catalysts was recently reported for the asymmetric hydrogenation used in the production of Inpyrfluxam (Scheme 3.40) [129–132], a succinate dehydrogenase inhibitor fungicide [133]. This hydrogenation is an example on how efficient cooperation between academia and industry can push ideas towards an application. By careful tuning of the ligand structure, including the backbone, the phosphorus moiety, the aryl group and the counterion, a significant breakthrough in catalyst performance was obtained. The largest influence was observed by tuning the aryl residues as first observed by A. *Pfaltz* and coworkers [134]. The rather exotic hexafluoroisopropanol (HFIP) was selected as solvent of choice since it significantly outperformed alternative solvents in terms of catalytic activity. The high costs of this solvent make it essential to almost completely recovery and reuse it, to have an economically viable and environmentally sustainable process. Furthermore, the hydrogenation reactor had a large effect on the performance, by using a BUSS advanced loop reactor (Figure 2.3) it was possible to reduce the substrate:catalyst ratio to 25,000:1 and still achieve 99% conversion and 97% *ee* within 3 h under optimised conditions. At the end of the reaction, >90% of the iridium metal can be scavenged utilising an active charcoal treatment.

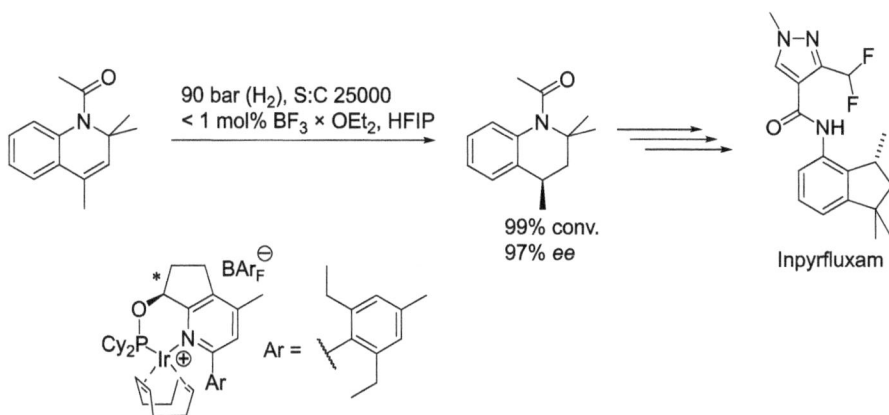

Scheme 3.40: Asymmetric hydrogenation used in the industrial production of the fungicide Inpyrfluxam.

3.5 Asymmetric transfer hydrogenation of C=O and C=N

Asymmetric transfer hydrogenation reactions are well established on a laboratory scale [135]. Compared to a "normal" hydrogenation in which gaseous hydrogen is used, transfer hydrogenation processes use other hydrogen donors. One of the most prominent examples of this technique is coal liquification with tetralin as hydrogen donor. However, this is not a catalytic process [136]. Transfer hydrogenations do not require pressurised equipment, and most of the applied hydrogen donors are comparably inexpensive. However, running chemical reactions under increased pressure is usually not an issue in the chemical industry, and gaseous hydrogen is cheaper, compared to the common hydrogen donors. Therefore, the impact of transfer hydrogenations is significantly lower, compared to hydrogenations applying gaseous hydrogen. The list of possible hydrogen donors is long. Most frequently, formic acid in combination with Et$_3$N, i-PrOH or sodium formate in water are reported. However, many more are known and reported; among these hydrogen donors that can serve as a source of "renewable hydrogen" such as glucose or glycerol are also reported [137, 138]. For an overview of the different prices of the hydrogen donors, see Table 2.1, Chapter 2: Heterogeneous hydrogenation.

From an industrial point of view, HCO$_2$H/Et$_3$N is the preferred choice, because the reactions can be conducted at high concentrations. The removal of the formed CO$_2$, which drives the equilibrium to the product side, can be even accelerated by applying a constant exchange of reactor-head space. In contrast, the dilution required for reactions with i-PrOH to drive the reaction to completion and the requirement to remove the formed acetone mean that use of i-PrOH is less favoured. For applications with sodium formate, solubility in water is usually the limiting factor.

Most frequently, ruthenium-based catalysts are applied in asymmetric transfer hydrogenations. However, iridium- and rhodium-based catalysts have also been reported in applications on scale. Other metals are rather exceptions [139]. Although many different ligands have now been reported [126], among the most commonly applied are *N*-sulfonylated 1,2-diamines [140]. The hydrogenation of these systems follows a similar mechanism as previously described for the Ru(II)-PP-diamine-systems by molecular hydrogen (Scheme 3.28). The main difference is the addition of the hydrogen to the metal-centre, which is illustrated here with *i*-PrOH, and acetone is formed (Scheme 3.41) [141].

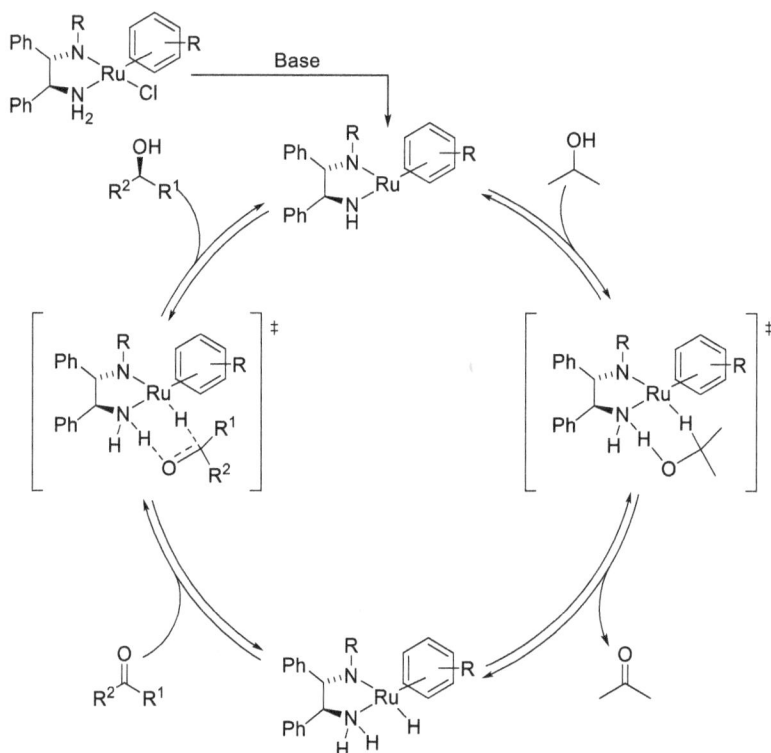

Scheme 3.41: Mechanism of the asymmetric transfer hydrogenation of a ketone with *i*-PrOH using a (sulfonyl-diamine)RuCl(arene)-catalyst.

Among various possible substrates, examples of the homogeneous transfer hydrogenation of C=O and C=N are reported in hundreds of kilograms or a higher scale, and a selection is shown below.

3.5.1 Dorzolamide

Dorzolamide is a carbonic anhydrase inhibitor applied to reduce intracellular pressure such as intraocular pressure [142]. Several approaches were conducted to achieve the diastereoselective hydrogenation of the sulfone depicted in Scheme 3.42, which is the key intermediate in the synthesis [143]. The major problem for this hydrogenation is the methyl group next to the sulfone group, which prevents the nucleophilic attack of a hydride for most reducing methods from the required face to result in the formation of the *trans* diastereoisomer. Additionally, the stereogenic centre adjacent to the sulfone group is prone to isomerisation via a retro-*Michael* mechanism. Therefore, classical *Noyori* catalysts for the asymmetric ketone hydrogenation cannot be applied. This challenging reduction was accomplished with a TsDPEN RuCl (*p*-cymene)-catalyst using formic acid and triethylamine as reagents. With 0.17 mol% catalyst loading, the synthesis of the secondary alcohol was achieved with 86% yield, with selectivities of 98% *de* and *ee*-value of 99.9%. This process was developed by ZaCh System-Zambon Chemicals [144] and was conducted in several batches of 100 kg each [145].

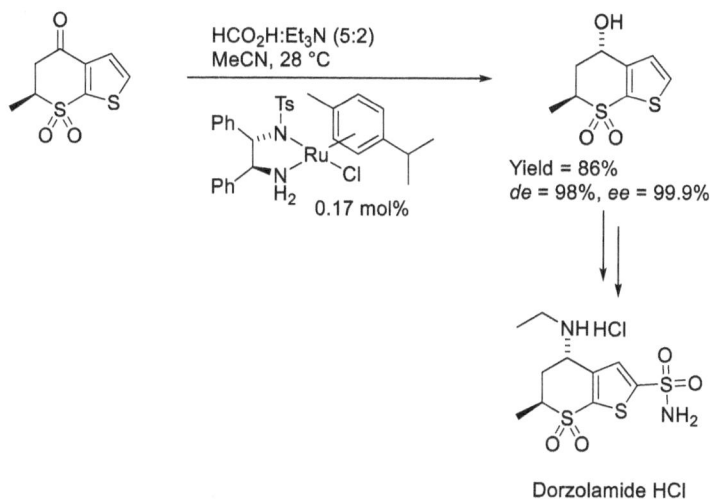

Scheme 3.42: Asymmetric transfer hydrogenation applied in the synthesis of dorzolamide.

3.5.2 Styrene oxide

Another example that uses asymmetric transfer hydrogenation in several 100 kg scale is applied in the manufacture of styrene oxide [146]. In this case, the application of RuCl[TsDPEN](*p*-cymene) as catalyst results 95% *ee*, but only limited reactivity was obtained with 60% conversion and an s/c of 100 (after 16 h). Better results were achieved using a water-soluble *ortho*-sulfonated 1,2-diphenylethlenediamine as ligand. This resulted in 94% *ee* with 87% yield using an s/c of 100 and allowed the recovery of the catalyst [147]. These results were outperformed by Rh- or Ir-based catalysts [146, 148]. Using Cp*IrCl(*R,R,R*)-CsDPEN] catalyst resulted in full conversion and >99% *ee* with an s/c of 1000 (Scheme 3.43).

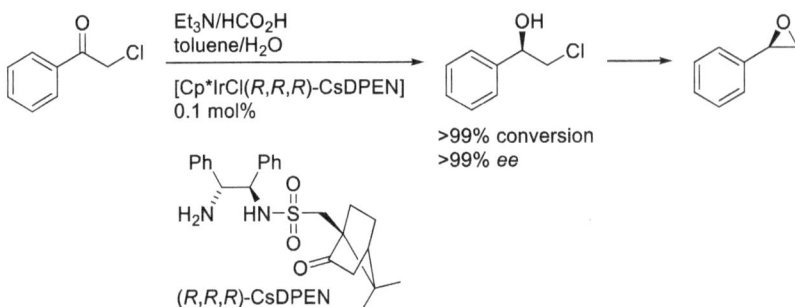

Scheme 3.43: Industrial styrene oxide production with an asymmetric transfer hydrogenation as key.

3.5.3 (*R,R*)-*trans*-Actinol for zeaxanthin

An example of the successful ketone transfer hydrogenation using *i*-PrOH as the hydrogen donor is the preparation of (*R,R*)-*trans*-actinol, which is an intermediate in the synthesis of the carotenoid zeaxanthin. Zeaxanthin is a yellow-orange coloured xanthophyll that is naturally occurring in plants and is also found in high concentrations in the macula of the human eye. It is an antioxidant and is used in nutritional supplements, often with the related carotenoid lutein, for eye-health and as a possible treatment for age-related macular degeneration. It is sold as a single isomer [149].

The production process for (3*R*,3'*R*)-zeaxanthin starts from ketoisophorone, which undergoes a biocatalytic reduction to (*R*)-levodione to set up the initial stereochemistry Scheme 3.44. This is followed by a diastereoselective reduction of the ketone to produce the alcohol-stereogenic centre that appears in the final product; the initially generated stereogenic centre is lost in the subsequent steps. Various catalysts were tested for the ketone hydrogenation (including nickel-alloy and Ru-BINAP catalysts), but none of them could produce a high enough chemo- and diastero-selectivity for the (*R,R*)-isomer [150]. Applying the chiral [(*S,S*)-TsDPEN RuCl(*p*-cymene)] transfer hydrogenation catalyst using isopropanol as the hydrogen source, >94% of the desired (*R,R*)-*trans*-actinol could be obtained. Interestingly, in the screening of a range of ligands, it was found that the existing stereogenic centre in the (*R*)-levodione had the greatest influence on the stereochemical outcome of the reaction; using a racemic TsDPEN ligand resulted in only a small reduction in diasteroselectivity. In fact, excellent results were obtained using the achiral mono-tosyl ethylenediamine ligand [151]. This process was scaled up and optimised further to give a final process that was operated at s/c > 1,000. The reaction is performed at below ambient pressure to remove the produced acetone (from the *i*-PrOH) and drive the reaction to completion.

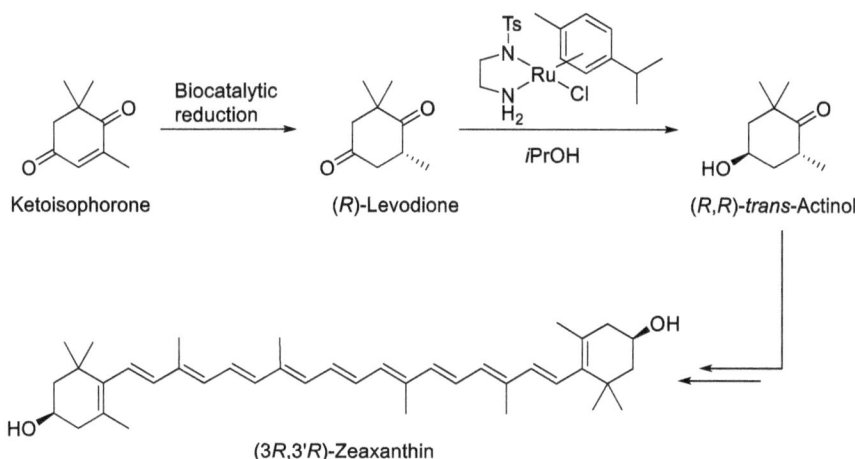

Scheme 3.44: Diastereoselective transfer hydrogenation of (*R*)-levodione to (*R*,*R*)-*trans*-actinol in the synthesis of zeaxanthin.

3.5.4 Almorexant

Almorexant is a dual orexin receptor antagonist developed by *Actelion* and *GlaxoS-mithKline* that affects the sleep-wake cycle [152]. The development was stopped in 2011 in clinical phase II due to undisclosed issues in the safety profile [153]. In the development towards an industrially applicable synthesis, an asymmetric transfer hydrogenation of a prochiral imine was applied to produce several hundred kilograms of product. During the reaction, a by-product, the *N*-formyl derivative, was obtained (2%) due the reaction of the product with formic acid. The formation of this by-product depends highly on the applied reaction time. The combination of two strategies was the key to success in this case. Using a 1:1 ratio of HCO_2H/Et_3N resulted in an increased reaction rate, whereas the use of the MsOH-imine salt as substrate without prior formation of the free-base reduced the reaction with HCO_2H [154]. With this strategy, the amount of this by-product was reduced to ~1%. Overall, the asymmetric transfer hydrogenation resulted in an *ee*-value of 90% using *Noyori's* *N*-sulfonylated 1,2-diamines catalyst with 0.067 mol% catalyst loading. After workup, the desired product was obtained in 87% yield and 99.7% *ee* (Scheme 3.45). Usually, HCO_2H/Et_3N is used as hydrogen donor in the hydrogenation of imines, because *i*-PrOH cannot serve as a hydrogen source in combination with these catalysts, as demonstrated by Noyori [155].

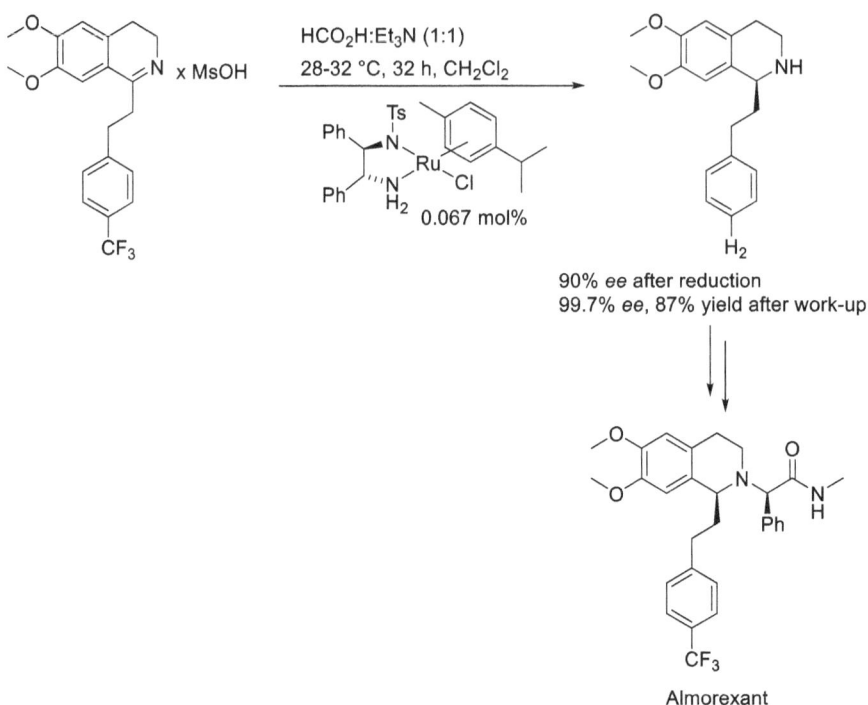

Scheme 3.45: Asymmetric C=N transfer hydrogenation in the pilot campaign to obtain almorexant.

References

[1] Noyori R. Asymmetric catalysis: Science and opportunities (Nobel Lecture). Angew Chem Int Ed 2002, 41, 2008–2022.

[2] Knowles W S. Asymmetric hydrogenations (Nobel Lecture). Angew Chem Int Ed 2002, 41, 1998–2007.

[3] Sharpless K. Searching for new reactivity (Nobel Lecture). Angew Chem Int Ed 2002, 41, 2024–2032.

[4] Knowles W S. Asymmetric hydrogenation. Acc Chem Res 1983, 16, 106–112.

[5] Etayo P, Vidal-Ferran A. Rhodium-catalysed asymmetric hydrogenation as a valuable synthetic tool for the preparation of chiral drugs. Chem Soc Rev 2013, 42, 728–754.

[6] Carlsson A, Lindqvist M, Magnusson T O R. 3,4-Dihydroxyphenylalanine and 5-hydroxytryptophan as reserpine antagonists. Nature 1957, 180, 1200–1200.

[7] Osborn J A, Wilkinson G. Tris(triphenylphosphine)halorhodium(I). Inorg Synth 1967, 67–71.

[8] Korpiun O, Mislow K. New route to the preparation and configurational correlation of optically active phosphine oxides. J Am Chem Soc 1967, 89, 4784–4786.

[9] Horner L, Balzer W D, Peterson D J. Phosphororganische Verbindungen 53 Konfigurationsstabilität von optisch aktivem, -metalliertem methyl-n-propyl-phenyl-phosphin. Tetrahedron Lett 1966, 7, 3315–3319.

[10] Mislov K. Pyramidal inversion barriers of phosphines and arsines. T New Yor Acad Sci 1973, 35, 227–242.

[11]   Horner L, Siegel H, Büthe H. asymmetric catalytic hydrogenation with an optically active phosphinerhodium complex in homogeneous solution. Angew Chem Int Ed Engl 1968, 7, 942.

[12]   Knowles W S, Sabacky M J. Catalytic asymmetric hydrogenation employing a soluble, optically active, rhodium complex. Chem Commun 1968, 1445–1446.

[13]   Whitesell J K. C2 symmetry and asymmetric induction. Chem Rev 1989, 89, 1581–1590.

[14]   Kagan H B, Dang P-T Asymmetric catalytic reduction with transition metal complexes. I. Catalytic system of rhodium(I) with (-)-2,3–0-isopropylidene-2,3-dihydroxy-1,4-bis(diphenylphosphino)butane, a new chiral diphosphine. J Am Chem Soc 1972, 94, 6429–6433.

[15]   Dang T P, Kagan H B. The asymmetric synthesis of hydratropic acid and amino-acids by homogeneous catalytic hydrogenation. Chem Commun 1971, 481.

[16]   Dang T P, Kagan H B. Asymmetric catalytic reduction with transition metal complexes I. A catalytic system of rhodium(I) with (-)-2,3-(9-Isopropylidene-2,3-dihydroxy-1,4-bis(diphenylphosphino) butane, a new chiral diphosphine. J Am Chem Soc 1972, 94, 6429–6433.

[17]   Knowles W S. Asymmetric hydrogenations (Nobel Lecture 2001). Adv Synth Catal 2003, 345, 3–13.

[18]   Vineyard B D, Knowles W S, Sabacky M J, Bachman G L, Weinkauff D J. Asymmetric hydrogenation. Rhodium chiral bisphosphine catalyst. J Am Chem Soc 1977, 99, 5946–5952.

[19]   Halpern J Mechanistic aspects of homogeneous catalytic hydrogenation and related processes. Inorg Chim Acta 1981, 50, 11–19.

[20]   Halpern J. Mechanism and stereoselectivity of asymmetric hydrogenation. Science 1982, 217, 401–407.

[21]   Brown J M, Chaloner P A. The mechanism of asymmetric homogeneous hydrogenation. Rhodium(I) complexes of dehydroamino acids containing asymmetric ligands related to bis(1,2-diphenylphosphino)ethane. J Am Chem Soc 1980, 102, 3040–3048.

[22]   Schmidt T, Dai Z, Drexler H-J, Hapke M, Preetz A, Heller D. The major/minor concept: Dependence of the selectivity of homogeneously catalyzed reactions on reactivity ratio and concentration ratio of the intermediates. Chem Asian J 2008, 3, 1170–1180.

[23]   Gridnev I D, Higashi N, Asakura K, Imamoto T. Mechanism of asymmetric hydrogenation catalyzed by a rhodium complex of (S,S)-1,2-Bis(tert-butylmethylphosphino)ethane. Dihydride mechanism of asymmetric hydrogenation. J Am Chem So 2000, 122, 7183–7194.

[24]   Min K, Park K, Park D H, Yoo Y J. Overview on the biotechnological production of L-DOPA. Appl Microbiol Biotechnol 2015, 99, 575–584.

[25]   Panten J, Surburg H Flavors and Fragrances, 2. Aliphatic Compounds. In: Elvers. ed. Ullmann's Encyclopedia of Industrial Chemistry, Online ed. Weinheim, Germany, Wiley-VCH Verlag GmbH & Co, 2015.

[26]   McCoy M. Hot market for a cool chemical. Chem Eng News 2010, 88, 15–16.

[27]   Chapuis C, Barthe M, de Saint Laumer J Y. Synthesis of Citronellal by RhI-Catalysed Asymmetric Isomerization of N,N-Diethyl-Substituted Geranyl- and Nerylamines or Geraniol and Nerol in the presence of chiral diphosphino ligands, under homogeneous and supported conditions. Helv Chim Acta 2001, 84, 230–242.

[28]   Née Taylor S C J, Jaekel C. Enantioselective hydrogenation of enones with a hydroformylation catalyst. Adv Synth Catal 2008, 350, 2708–2714.

[29]   Cobley C J, Lennon I C, Praquin C, Zanotti-Gerosa A Highly efficient asymmetric hydrogenation of 2-methylenesuccinamic acid using a Rh-DuPHOS catalyst. Org Process Res Dev 2003, 7, 407–411.

[30]   Chen W, McCormack P J, Mohammed K, Mbafor W, Roberts S M, Whittal J. Stereoselective synthesis of Ferrocene-Based C2-Symmetric diphosphine ligands: Application to the highly enantioselective hydrogenation of α-substituted cinnamic acids. Angew Chem Int Ed 2007, 46, 4141–4144.

[31]   Weissensteiner E, Sturm T, Spindler F. Ferrocenyl Diphosphines and their Use, Solvias Ag, WO02/02578, 10 January, 2002.

[32] Pena D, Minnaard A J, Boogers J A F, de Vries A H M, de Vries J G, Feringa B L. Improving conversion and enantioselectivity in hydrogenation by combining different monodentate phosphoramidites; a new combinatorial approach in asymmetric catalysis. Org Biomol Chem 2003, 1, 1087–1089.

[33] Reetz M T, Sell T, Meiswinkel A, Mehler G A New principle in combinatorial asymmetric transition-metal catalysis: Mixtures of chiral monodentate p ligands. Angew Chem Int Ed 2003, 42, 790–793.

[34] Feringa B L, Pineschi M, Arnold L A, Imbos R, de Vries A H M. Highly enantioselective catalytic conjugate addition and tandem conjugate addition–aldol reactions of organozinc reagents. Angew Chem Int Ed 1997, 36, 2620–2623.

[35] Teichert J F, Feringa B L Phosphoramidites: Privileged ligands in asymmetric catalysis. Angew Chem Int Ed 2010, 49, 2486–2528.

[36] de Vries J G, Lefort L. The combinatorial approach to asymmetric hydrogenation: Phosphoramidite libraries, ruthenacycles, and artificial enzymes. Chem Eur J 2006, 12, 4722–4734.

[37] Felfer U, Kotthaus M, Lefort L, Steinbauer G, de Vries A H M, de Vries J G A. Mixed-ligand approach enables the asymmetric hydrogenation of an α-isopropyl cinnamic acid en route to the renin inhibitor aliskiren. Org Process Res Dev 2007, 11, 585–591.

[38] Inoguchi K, Sakuraba S, Achiwa K. Bisphosphine ligands in rhodium-catalyzed asymmetric hydrogenations. Synlett 1992, 169–178.

[39] Burk M J, de Koning P D, Grote T M, Hoekstra M S, Hoge G, Jennings R A, Kissel W S, Le T V, Lennon I C, Mulhern T A, Ramsden J A, Wade R A. An enantioselective synthesis of (S)-(+)-3-Aminomethyl-5-methylhexanoic acid via asymmetric hydrogenation. J Org Chem 2003, 68, 5731–5734.

[40] https://clincalc.com/DrugStats/Drugs/Pregabalin (25.03.2020)

[41] Hoge G, Wu H-P, Kissel W S, Pflum D A, Greene D J, Bao J. Highly selective asymmetric hydrogenation using a three hindered quadrant bisphosphine rhodium catalyst. J Am Chem Soc 2004, 126, 5966–5967.

[42] Development of a chemoenzymatic manufacturing process for pregabalin. Org Process Res Dev 2008, 12, 392–398.

[43] de Clerq P J. Biotin: A timeless challenge for total synthesis. Chem Rev 1997, 97, 1755–1792.

[44] Müller M A, Medlock J, Prágai Z, Warnke I, Litta G, Kleefeldt A, Kaiser K, Potzolli B. Vitamin B5. In: Elvers, ed. Ullmann's Encyclopedia of Industrial Chemistry, Online ed., Weinheim, Germany, Wiley-VCH Verlag GmbH & Co, 2019.

[45] Bonrath W, Karge R, Netscher T, Roessler F, Spindler F. Biotin – The chiral challenge. Chimia 2009, 63, 265–269.

[46] Bonrath W, Karge R, Netscher T, Roessler F, Spindler F. Chiral Lactones by Asymmetric Hydrogenation – A Step Forward in (+)–biotin Synthesis. In: Blaser H-U, Federsel H-J, ed. Asymmetric Catalysis on Industrial Scale: Challenges, Approaches, and Solutions, 2nd ed. Weinheim, Germany, Wiley-VCH, 2010, 27–39.

[47] Bonrath W, Karge R, Roessler F. Manufacture of Lactones. WO 2006108562. DSM IP Assets BV, 2006.

[48] Kim D, Wang L, Beconi M, Eiermann G J, Fisher M H, He H, Hickey G J, Kowalchick J E, Leiting B, Lyons K, Marsilio F, McCann M E, Patel R A, Petrov A, Scapin G, Patel S B, Roy R S, Wu J K, Wyvratt M J, Zhang B B, Zhu L, Thornberry N A, Weber A E. (2R)-4-Oxo-4-[3-(Trifluoromethyl)-5,6-dihydro[1,2,4]triazolo[4,3-a]pyrazin- 7(8H)-yl]-1-(2,4,5-trifluorophenyl)butan-2-amine: A potent, orally active dipeptidyl peptidase IV inhibitor for the treatment of type 2 diabetes. J Med Chem 2005, 48, 141–151.

[49] Hsiao Y, Rivera N R, Rosner T, Krska S W, Njolito E, Wang F, Sun Y, Armstrong J D, Grabowski E J J, Tillyer R D, Spindler F, Malan C. Highly efficient synthesis of β-Amino acid derivatives via asymmetric hydrogenation of unprotected enamines. J Am Chem Soc 2004, 126, 9918–9919.

[50] Burk M J, Casy G, Johnson N B. A three-step procedure for asymmetric catalytic reductive amidation of ketones. J Org Chem 1998, 63, 6084–6085.

[51] Clausen A M, Dziadul B, Cappuccio K L, Kaba M, Starbuck C, Hsiao Y, Dowling T M. A three-step procedure for asymmetric catalytic reductive amidation of ketones. Org Process Res Dev 2006, 10, 723–726.

[52] Hansen K B, Hsiao Y, Xu F, Rivera N, Clausen A, Kubryk M, Krska S, Rosner T, Simmons B, Balsells J, Ikemoto N, Sun Y, Spindler F, Malan C, Grabowski E J J, Armstrong J D. Highly efficient asymmetric synthesis of sitagliptin. J Am Chem Soc 2009, 131, 8798–8804.

[53] https://www.epa.gov/greenchemistry/presidential-green-chemistry-challenge-2006-greener-synthetic-pathways-award (10.05.2019).

[54] https://www.epa.gov/greenchemistry/presidential-green-chemistry-challenge-2010-greener-reaction-conditions-award (10.05.2019).

[55] Bonrath W, Medlock J, Schütz J, Wüstenberg B, Netscher T. Hydrogenation in the vitamins and fine chemicals industry – An overview in hydrogenation. InTech, 2012, 69–90.

[56] Takahashi H, Morimoto T, Achiwa K. Highly effective catalytic asymmetric hydrogenations of α-Keto Esters and an α-Keto acetal with new neutral chiral pyrrolidine bisphosphine-rhodium complexes. Chem Lett 1987, 16, 855–858.

[57] Broger E A, Crameri Y. Chiral-rhodium-diphosphine Complexes for Asymmetric Hydrogenations. EP 158 875. Hoffmann-La Roche. 1984.

[58] Roucoux A, Thieffry L, Carpentier J-F, Devocelle M, Méliet C, Agbossou F, Montreux A. Amidophosphine–Phosphinites: Synthesis and use in rhodium-based asymmetric hydrogenation of activated keto compounds. Crystal structure of Bis[(μ-chloro)((S)-2-((diphenylphosphino)oxy)-2-phenyl-N-(diphenylphosphino)-N-methyl acetamide)rhodium(I)]. Organometallics 1996, 15, 2440–2449.

[59] Genet J P, Pinel C, Mallart S, Juge S, Cailhol N, Laffitte J A. (R,R)-DiPAMP-Ruthenium (II) (2-methylallyl)2: Synthesis and Selected use in Asymmetric Hydrogenation. Tetrahedron Lett 1992, 33, 5343–5346.

[60] Hapiot F, Agbossou F, Mortreux A. Synthesis of new chiral arene Ruthenium(II) aminophosphinephosphinite complexes and use in asymmetric hydrogenation of an activated keto compound. Tetrahedron Asymmetry 1994, 5, 515–518.

[61] Pugin B, Feng X, Spindler F. Dipsosphine Ligands. WO 2008/000815. Solvias AG, June 30, 2006.

[62] Alsters P L, Jary W, Nardello-Rataj V, Aubry J-M. "Dark" singlet oxygenation of β-citronellol: A key step in the manufacture of rose oxide. Org Process Res Dev 2010, 14, 259–262.

[63] Noyori R, Okhuma T, Kitamura M, Takaya H, Sayo N, Kumobayashi H, Akuragawa S. Asymmetric hydrogenation of β-Keto carboxylic esters. A practical, purely chemical access to β-hydroxy esters in high enantiomeric purity. J Am Chem Soc 1987, 109, 5856–5858.

[64] Blaser H U, Spindler F, Studer M. Enantioselective catalysis in fine chemicals production. Appl Catal A-Gen 2001, 221, 119–143.

[65] Akutagawa S. Asymmetric synthesis by metal BINAP catalysts. Appl Catal A-Gen 1995, 128, 171–207.

[66] Netscher T, Scalone M, Schmid R. Enantioselective Hydrogenation: Towards a Large-Scale Total Synthesis of (R,R,R)-α-Tocopherol in Asymmetric Catalysis on an Industrial Scale. Blaser H-U, Schmidt E, eds. Wiley, 2003, 71–89.

[67] Danilewicz J C, Williams M T. Enantiomeric glutaramide diuretic agents. EP 0342850. Pfizer. 1989.

[68] Challenger S, Derrick A, Masson C P, Silk T V. Stereoselective synthesis of a candoxatril intermediate via asymmetric hydrogenation. Tetrahedron Lett 1999, 40, 2187.

[69] Burk M J, Bienewald F, Challenger S, Derrick A, Ramsden J A. Me-DuPHOS-Rh-catalyzed asymmetric synthesis of the pivotal glutarate intermediate for candoxatril. J Org Chem 1999, 64, 3290–3298.

[70] Bulliard M, Laboue B, Lastennet J, Roussiasse S. Large-scale candoxatril asymmetric hydrogenation. Org Process Res Dev 2001, 5, 438–441.

[71] https://www.nobelprize.org/prizes/medicine/2015/press-release/ (26.03.2020).

[72] https://www.sanofi.com/en/media-room/press-releases/2014/2014-08-12-13-00-00 accessed (26.03.2020).

[73] Paddon C J, Westfall P J, Pitera D J, Benjamin K, Fisher K, McPhee D, Leavell M D, Tai A, Main A, Eng D, Polichuk D R, Teoh K H, Reed D W, Treynor T, Lenihan J, Fleck M, Bajad S, Dang G, Dengrove D, Diola D, Dorin G, Ellens K W, Fickes S, Galazzo J, Gaucher S P, Geistlinger T, Henry R, Hepp M, Horning T, Iqbal T, Jiang H, Kizer L, Lieu B, Melis D, Moss N, Regentin R, Secrest S, Tsuruta H, Vazquez R, Westblade L F, Xu L, Yu M, Zhang Y, Zhao L, Lievense J, Covello P S, Keasling J D, Reiling K K, Renninger N S, Newman J D. High-level semi-synthetic production of the potent antimalarial artemisinin. Nature 2013, 496, 528–532.

[74] Turconi J, Griolet F, Guevel R, Oddon G, Villa R, Geatti A, Hvala M, Rossen K, Göller R, Burgard A. Semisynthetic artemisinin, the chemical path to industrial production. Org Process Res Dev 2014, 18, 417–422.

[75] Takaya H, Akutagawa S, Noyori R. (R)-(+)- AND (S)-(−)-2,2′-Bis(Diphenylphosphino)-1,1′-Binaphthyl (BINAP). Org Synth 1989, 20–32.

[76] Kumobayashi H Industrial application of asymmetric reactions catalyzed by BINAP-metal complexes. Recl Trav Chim Pays-Bas 1996, 115, 201–210.

[77] Saito T, Yokozawa T, Ishizaki T, Moroi T, Sayo N, Miura T, Kumobayashi H. New chiral diphosphine ligands de-signed to have a narrow dihedral angle in the biaryl backbone. Adv Synth Catal 2001, 343, 264–267.

[78] Dub P A, Gordon J C The role of the metal- bound N–H functionality in Noyori- type molecular catalysts. Nat Rev Chem 2018, 2, 396–408.

[79] Ohkuma T, Ishii D, Takeno H, Noyori R. Asymmetric hydrogenation of amino ketones using chiral RuCl2(diphophine)(1,2-diamine) Complexes. J Am Chem Soc 2000, 122, 6510–6511.

[80] Noyori R, Koizumi M, Ishii D, Ohkuma T. Asymmetric hydrogenation via architectural and functional molecular engineering. Pure Appl Chem 2001, 73, 227–232.

[81] Noyori R, Tokunago M, Kitamura M. Stereoselective organic synthesis via dynamic kinetic resolution. Bull Chem Soc Jpn 1995, 68, 36–56.

[82] Huerta F F, Minidis A B E, Bäckvall J-E. Racemisation in asymmetric synthesis. Dynamic kinetic resolution and related processes in enzyme and metal catalysis. Chem Soc Rev 2001, 30, 321–331.

[83] Bhat V, Welin E R, Guo X, Stoltz B M. Advances in stereoconvergent catalysis from 2005 to 2015: Transition-metal-mediated stereoablative reactions, dynamic kinetic resolutions, and dynamic kinetic asymmetric transformations. Chem Rev 2017, 117, 4528–4561.

[84] Pellissier H Recent developments in dynamic kinetic resolution. Tetrahedron 2011, 3769–3802.

[85] Noyori R, Ikeda T, Ohkuma T, Widhalm M, Kitamura M, Takaya H, Akutagawa S, Sayo N, Saito T, Taketomi T, Kumobayashi H. Stereoselective hydrogenation via dynamic kinetic resolution. J Am Chem Soc 1989, 111, 9134–9135.

[86] Kitamura M, Ohkuma T, Tokunaga M, Noyori R Dynamic kinetic resolution in BINAP – Ruthenium(II) catalyzed hydrogenation of 2-substituted 3-oxo carboxylic esters. Tetrahedron Asym 1990, 1, 1–4.

[87] Kitamura M, Tokunaga M, Noyori R. Quantitative expression of dynamic kinetic resolution of chirally labile enantiomers: Stereoselective hydrogenation of 2-substituted 3-oxo carboxylic esters catalyzed by BINAP-Ruthenium(II) complexes. J Am Chem Soc 1993, 115, 144–152.

[88] Kreutzer U R. Manufacture of fatty alcohols based on natural fats and oils. J Am Oil Chemist Soc 1984, 61, 343–348.

[89] Teunissen H T, Elsevier, C J. Homogeneous ruthenium catalyzed hydrogenation of esters to alcohols. Chem Commun 1998, 1367–1368.

[90] Saudan L A. Hydrogenation of Esters. In: Cornils B, Herrmann W A, Beller M, Paciello R, eds. Applied Homogeneous Catalysis with Organometallic Compounds, 3rd ed. Weinheim, Germany, Wiley-VCH, 2018, 645–690.

[91] Zhang J, Leitus G, Ben-David Y, Milstein D. Efficient homogeneous catalytic hydrogenation of esters to alcohols. Angew Chem Int Ed 2006, 45, 1113–1115.

[92] Milstein D. Discovery of environmentally benign catalytic reactions of alcohols catalyzed by pyridine-based Pincer Ru complexes, based on metal–ligand cooperation. Top Catal 2010, 53, 915–923.

[93] Kuriyama W, Matsumoto T, Ogata O, Ino Y, Aoki K, Tanaka S, Ishida K, Kobayashi T, Sayo N, Saito T. Catalytic hydrogenation of esters. development of an efficient catalyst and processes for synthesising (R)-1,2-propanediol and 2-(l-Menthoxy)ethanol. Org Process Res Dev 2012, 16, 166–171.

[94] Spasyuk D, Smith S, Gusev D G. From esters to alcohols and back with ruthenium and osmium catalysts. Angew Chem Int Ed 2012, 51, 2772–2775.

[95] Spasyuk D, Smith S, Gusev D G. Replacing phosphorus with sulfur for the efficient hydrogenation of esters. Angew Chem Int Ed 2013, 52, 2538–2542.

[96] Saudan L A. 85 years of catalysis at Firmenich. Chimia 2019, 73, 684–697.

[97] Saudan L A, Saudan C M, Debieux C, Wyss P. Dihydrogen reduction of carboxylic esters to alcohols under the catalysis of homogeneous ruthenium complexes: High efficiency and unprecedented chemoselectivity. Angew Chem Int Ed 2007, 46, 7473–7476.

[98] Saudan L A, Dupau P, Riedhauser J-J, Wyss P. Hydrogenation of Esters with Ru/Bidentate Ligands Complexes. WO 2006/106483, Firmenich SA, 2006.

[99] Saudan L A, Dupau P, Riedhauser J-J, Wyss P. Hydrogenation of Esters with Ru/tetradentate Ligands Complexes. WO 2006/106484, Firmenich SA, 2006.

[100] Shaner D L, Brunk G, Belles D, Westra P, Nissen S. Soil dissipation and biological activity of metolachlor and S-metolachlor in five soils. Pest Mang Sci 2006, 62, 617–623.

[101] Blaser H-U. The chiral switch of (S)-Metolachlor: A personal account of an industrial odyssey in asymmetric catalysis. Adv Synth Catal 2002, 344, 17–31.

[102] Blaser H-U, Buser H-P, Coers K, Hanreich R, Jalett H-P, Jelsch E, P B, Schneider H-D, Spindler F, Wegmann A. The chiral switch of metolachlor: The Development of a large-scale enantioselective catalytic process. Chimia 1999, 53, 275–280.

[103] Dorta R, Broggini D, Kissner R, Togni A. Iridium–imine and –amine complexes relevant to the (S)-metolachlor process: Structures, exchange kinetics, and C-H activation by IrI causing racemization. Chem Eur J 2004, 10, 4546–4555.

[104] Dorta R, Broggini D, Stoop R, Rüegger H, Spindler F, Togni A. Chiral xyliphos complexes for the catalytic imine hydrogenation leading to the metolachlor herbicide: Isolation of catalyst–substrate adducts. Chem Eur J 2004, 10, 267–278.

[105] Nguyen L, Thomas K L, Lucke-Wold B P, Cavendish J Z, Crowe M S, Matsumoto R R. Dextromethorphan. An update on its utility for neurological and neuropsychiatric disorders. Pharmacol Therapy 2016, 159, 1–22.

[106] Werbitzky O. Process for Preparing Optically Active 1(p-Methoxybenzyl).-1,2,3,4,5,6,7,8-Octahydroisoquinoline. WO9703052. Lonza AG, 1997.

[107] Togni A, Breutel C, Schnyder A, Spindler F, Landert H, Tijani A. A novel easily accessible chiral ferrocenyldiphosphine for highly enantioselective hydrogenation, allylic alkylation, and hydroboration reactions. J Am Chem Soc 1994, 116, 4062–4066.

[108] Imwinkelried R Catalytic asymmetric hydrogenation in the manufacture of d-Biotin and dextromethorphan. Chimia 1997, 51, 300–302.

[109] Broger E A, Scalone M, Wehrli C. Process for the Preparation of Optically Active (R or S)-1-(4-methoxy-benzyl)-1,2,3,4,5,6,7,8-octahydro-isoquinolines. EP 0850931, Hoffmann La Roche. 1998.

[110] Püntener K, Scalone M, Wang S Process for Asymmetric Hydrogenation of Hexahydroquinoline Salts WO 03078399 Roche Vitamins AG, 2003.

[111] Püntener K, Scalone M. Enantioselective Hydrogenation: Applications in Process R&D of Pharmaceutical: In Asymmetric Catalysis on Industrial Scale: Challenges, Approaches and Solutions. Blaser H-U, Federsel Wiley-VCH, 2nd ed. 2011.

[112] Xie J-H, Liu X-Y, Xie J-B, Wang L-X, Zhou Q-L. An additional coordination group leads to extremely efficient chiral iridium catalysts for asymmetric hydrogenation of ketones. Angew Chem Int Ed 2011, 50, 7329–7332.

[113] Xie J-H, Liu X-Y, Yang X-H, Xie J-B, Wang L-X, Zhou Q-L. Chiral iridium catalysts bearing spiro pyridine-aminophosphine ligands enable highly efficient asymmetric hydrogenation of β-Aryl β-ketoesters. Angew Chem Int Ed 2012, 51, 201–203.

[114] Yang X-H, Xie J-H, Liu W-P, Zhou Q-L. Catalytic asymmetric hydrogenation of δ-Ketoesters: Highly efficient approach to chiral 1,5-diols. Angew Chem Int Ed 2013, 52, 7833–7836.

[115] Sprinz J, Helmchen G Phosphinoaryl- and phosphinoalkyloxazolines as new chiral ligands for enantioselective catalysis: Very high enantioselectivity in palladium catalyzed allylic substitutions. Tetrahedron Lett 1993, 34, 1769–1772.

[116] von Matt P, Pfaltz A. Chiral phosphinoaryldihydrooxazoles as ligands in asymmetric catalysis: Pd-catalyzed allylic substitution. Angew Chem Int Ed 1993, 32, 566–568.

[117] Dawson G J, Frost C G, Williams J M J, Coote S J. Asymmetric palladium catalysed allylic substitution using phosphorus containing oxazoline ligands. Tetrahedron Lett 1993, 34, 1769–17772.

[118] Müller M A, Pfatz A. Ligands for Iridium-catalyzed asymmetric hydrogenation of challenging substrates in ligand design in metal chemistry reactivity and catalysis. Wiley&Sons, 2016, 46–65.

[119] Smidt S P, Zimmermann N, Studer M, Pfaltz A A. Enantioselective hydrogenation of alkenes with iridium-PHOX catalysts: A kinetic study of anion effects. Chem Eur J 2004, 10, 4685–4693.

[120] Smidt S P, Pfaltz A, Martínez-Viviente E, Pregosin P S, Albinati A. X-ray and NOE Studies on Trinuclear Iridium Hydride Phosphino Oxazoline (PHOX) complexes. Organometallics 2003, 22, 1000–1009.

[121] Verendel J J, Pàmies O, Diéguez M, Andersson P G Asymmetric hydrogenation of olefins using chiral crabtree-type catalysts: Scope and limitations. Chem Rev 2014, 114, 2130–2169.

[122] Bell S, Wüstenberg B, Kaiser S, Menges F, Netscher T, Pfaltz A. Asymmetric hydrogenation of unfunctionalized, purely alkyl-substituted olefins. Science 2006, 311, 642–644.

[123] Bonrath W, Menges F, Netscher T, Pfaltz A, Wüstenberg B. Asymmetric Hydrogenation of Alkennes Using Chiral Iridium Complexe. WO 2006/066863, DSM IP assets BV, 2006.

[124] Mueller M A, Pfaltz A, Medlock J Hydrogenation of Ketones Having at Least a Carbon-carbon Double Bond in the Gamma, delta-Position. EP 2720996, DSM IP assets BV, 2014.

[125] Bonrath W, Medlock J, Netscher T, Verzijl G, de Vries A. Process of Asymmetric Hydrogenation of Ketals and Acetals. WO 2014/096096 DSM IP assets BV, 2014.

[126] Woodmansee D, Müller M-A, Pfaltz A. Chiral pyridyl phosphinites with large aryl substituents as efficient ligands for the asymmetric iridium-catalyzed hydrogenation of difficult substrates. Chem Sci 2010, 1, 72–78.

[127] Blackmond D G, Lightfoot A, Pfaltz A, Rosner T, Schnider P, Zimmermann N. Enantioselective hydrogenation of olefins with phosphinooxazoline-iridium catalysts. Chirality 2000, 22, 1000–10009.

[128] Müller M-A, Pfaltz A. Asymmetric hydrogenation of α,β-unsaturated nitriles with base-activated iridium N,P ligand complexes. Angew Chem Int Ed 2014, 53, 8668–8671.

[129] Jones M, Harris D, Struble J, Hayes M, Koeller K, Özgun K C, Schirmer H, Heinrich J, Bächle F, Goudedranche S, Schotes C. Development of a practical process for the large-scale preparation of the chiral pyridyl-backbone for the crabtree/pfaltz-type iridium complex used in the industrial production of the novel fungicide inpyrfluxam. Org Process Res Dev 2022, 26, 2407–2414.

[130] Schneekönig J, Liu W, Leischner T, Junge K, Schotes C, Beier C, Beller M. Application of crabtree/pfaltz-type iridium complexes for the catalyzed asymmetric hydrogenation of an agrochemical building block. Org Process Res Dev 2020, 24, 443–447.

[131] Schotes C, Beller M, Junge K, Liu W, Leischner T, Schneekönig J. Process comprising the use of new iridium catalysts for the enantioselective hydrogenation of 4-substituted 1,2-dihydroquinolines. WO 2021/058458, Bayer Aktiengesellschaft, September 22, 2020.

[132] Schotes C, Müler S. On the importance of collaboration in the development of sustainable catalytic processes: The case of inpyrfluxam. ACS Sus Chem Eng 2022, 10, 13244–13253.

[133] Kiguchi S, Inoue T, Matsuzaki Y, Iwahashi F, Sakaguchi H. Discovery and Biological Profile of Inpyrfluxam: A New Broad-Spectrum Succinate Dehydrogenase Inhibitor Fungicide. In: Maienfisch M, eds. Recent Highlights in the Discovery and Optimization of Crop Protection Products. Amsterdam, The Netherlands, Academic Press, 2021, 381–389.

[134] Woodmansee D H, Müller M-A, Neuburger M, Pfaltz A. Chiral pyridyl phosphinites with large aryl substituents as efficient ligands for the asymmetric iridium-catalyzed hydrogenation of difficult substrates. Chem Sci 2010, 1, 72–78.

[135] Wang D, Astruc D. The golden age of transfer hydrogenation. Chem Rev 2015, 115, 6621–6686.

[136] Ren X K, Fang D Y, Jin J L, Gao J S. New proceed achieved in the direct coal liquefaction. Chem Ind Eng Prog 2010, 29, 198–204.

[137] Yoshida M, Hirahata R, Inoue T, Shimbayashi T, Fujita K-I. Iridium-catalyzed transfer hydrogenation of ketones and aldehydes using glucose as a sustainable hydrogen donor. Catalysts 2019, 9, 503–513.

[138] Verma A D, Pal S, Verma P, Srivastava V, Manda R K, Sinha I. Ag-Cu bimetallic nanocatalysts for p-nitrophenol reduction using a green hydrogen source. J of Environ Chem Engin 2017, 5, 6148–6155.

[139] Gladiali S, Taras R. Reduction of Carbonyl Compounds by Hydrogen Transfer in Modern Reduction Methods, Andersson P G, Munslow I J, eds. Weinheim, Wiley-VCH, 2008, 135–157.

[140] Noyori R, Hashiguchi S. Asymmetric transfer hydrogenation catalyzed by chiral ruthenium complexes. Acc Chem Res 1997, 30, 97–102.

[141] Noyori R, Yamakawa M, Hashiguchi S. Metal–ligand bifunctional catalysis: A nonclassical mechanism for asymmetric hydrogen transfer between alcohols and carbonyl compounds. J Org Chem 2001, 66, 7931–7944.

[142] Balfour J A, Wilde M I. Dorzolamide. A review of its pharmacology and therapeutic potential in the management of glaucoma and ocular hypertension. Drugs Aging 1997, 10, 384–403.

[143] Blacklock T J, Sohar P, Butcher J W, Lamanec T, Grabowski E J J. An enantioselective synthesis of the topically-active carbonic anhydrase inhibitor MK-0507: 5,6-dihydro-(S)-4-(ethylamino)-(S)-6-methyl-4H-thieno[2,3-b]thiopyran-2-sulfonamide 7,7-dioxide hydrochloride. J Org Chem 1993, 58, 1672–1679.

[144] Volpicelli R, Andretto M, Cotarca L, Nardi A, Verzini M. Asymmetric reduction process. WO 2012120086. ZaCh System SpA. 2012.

[145] Cotarca L, Verzini M, Volpicelli R Catalytic asymmetric transfer hydrogenation: An industrial perspective. Chim Oggi 2014, 32, 36–41.

[146] Komiya S, Shimizu H, Nagasaki I. Industrial Application of the Asymmetric Reduction of C=O and C=N bonds, including enamides and enamines in comprehensive chirality Elsevier. 2012, 83–103.

[147] Ma Y, Liu H, Chen L, Cui X, Zhu J, Deng J. Asymmetric transfer hydrogenation of prochiral ketones in aqueous media with new water-soluble chiral vicinal diamine as ligand. Org Lett 2003, 5, 2103–2106.

[148] Hamada T, Torii T, Izawa K, Noyori R, Ikariya T Practical synthesis of optically active styrene oxides via reductive transformation of 2-Chloroacetophenones with chiral rhodium catalysts. Org Lett 2002, 4, 4373–4376.

[149] Johnson E J, Hammond B R, Yeum K-Y, Qin J, Wang X D, Castaneda C, Snodderly D M, Russell R M. Relation among serum and tissue concentrations of lutein and zeaxanthin and macular pigment density. Am J Clin Nutr 2000, 71, 1555–1562.

[150] Püntener K, Scalone M. Enantioselective Hydrogenation: Applications in Process R&D of Pharmaceuticals. In: Blaser H-U, Federsel H-J, eds. Asymmetric Catalysis on Industrial Scale: Challenges, Approaches, and Solutions, 2nd ed. Weinheim, Germany, Wiley-VCH, 2010, 13–25.

[151] Crameri Y, Puentener K, Scalone M. Verfahren zur Herstellung von trans-(R,R)-Actinol. EP0915076, F. Hoffmann-La Roche, 1999.

[152] Weller T, Koberstein R, Aissaoui H, Clozel M, Fischli W. Substituted 1,2,3,4-tetrahydroisoquinoline Derivatives, WO2005118548, Actelion Pharmaceuticals Ltd December 2005.

[153] https://web.archive.org/web/20110704194943/http://www.gsk.com/media/pressreleases/2011/2011_pressrelease_10019.htm (06. 04.20200).

[154] Verzijl G K M, de Vries H M, de Vries J G, Kapitan P, Dax T, Helms M, Nazir Z, Skranc W, Imboden C, Stichler J, Ward R A, Abele S A, Lefort L. Catalytic asymmetric reduction of a 3,4-Dihydroisoquinoline for the large-scale production of almorexant: Hydrogenation or transfer hydrogenation? Org Process Res Dev 2013, 17, 1531–1539.

[155] Uematsu N, Fujii A, Hashiguchi S, Ikariya T, Noyori R. Asymmetric transfer hydrogenation of imines. J Am Chem Soc 1996, 118, 4916–4917.

4 Oxidations

4.1 Introduction

Catalytic oxidation is a major field in industrial production, research and development of chemicals. Oxidation processes play an important role in the manufacture of
bulk chemicals (several billion USD per year in the world market) and can be divided
into two groups, gas-phase or liquid-phase oxidations. Liquid-phase oxidations are
often three-phase reaction systems. The oxidant is often oxygen or air (gaseous), the
substrates and products are liquids, whereas the catalyst is in the solid phase. A fixed-
bed liquid–gas-phase reactor is generally required in such reactions (Figure 4.1).

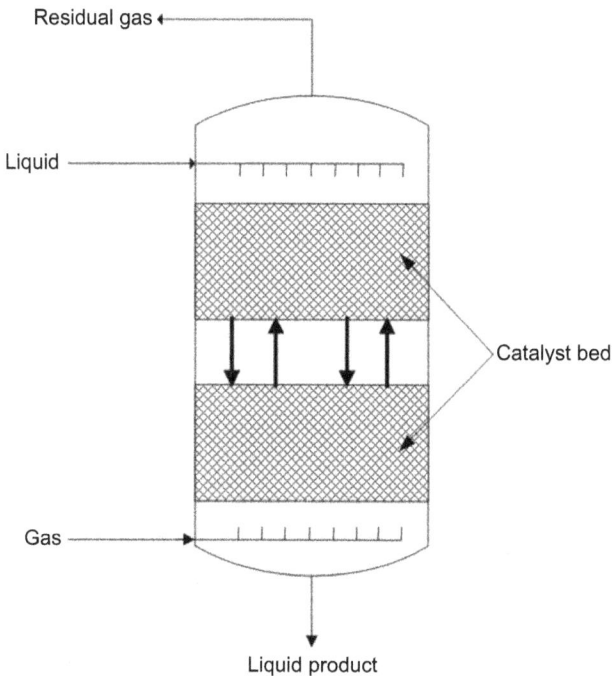

Figure 4.1: Fixed-bed liquid–gas reactor.

Major catalytic oxidation processes are the methanol oxidation to formaldehyde, the
maleic anhydride production from *n*-butane or benzene and phthalic anhydride from
o-xylene or naphthalene. All these processes use air, a heterogeneous catalyst and are
performed at high conversions (mainly >95%). The selectivity is dependent on the process, usually between 75% and 95%.

   Also the oxidation of ammonia to NO and further transformation to NO_2 and nitric acid with O_2 at atmospheric pressure (*Ostwald* process), at >850 °C applying a

https://doi.org/10.1515/9783111102672-004

Pt–Rh catalyst (3% Rh), is a very efficient catalytic oxidation reaction and performed on industrial scale in several million t/a. Although all these processes are highly efficient and well known, they are not further discussed due to their high production volumes.

From a mechanistic point of view, it should be pointed out that oxidation reactions with a heterogeneous catalyst generally follow the *Mars–van Krevelen* theory (see Chapter 1: Introduction and fundamental aspects) [1].

For the fine chemical industry, selective catalytic oxidation methods, which avoid the often-applied stoichiometric reagents, are of importance. However, with the increasing complexity of starting materials, selectivity becomes a critical parameter for these types of reactions. The oxidising agents like air, molecular oxygen or hydrogen peroxide are becoming more popular because of their availability and low price. Working in and with pure oxygen requires special safety installations. For all these processes, new catalysts are needed, which perform with high selectivity and conversion. Furthermore, oxidation agents such as ozone, bleach, hypervalent iodine compounds or potassium hydrogen peroxosulphate are finding an increased number of applications.

Recent trends in industrial selective catalytic oxidations are the use of new raw materials (e.g. from renewable sources), new catalytic oxidation processes (such as oxidative dehydrogenation), and gas-phase oxidation. Especially the development of new reactor and engineering concepts, the better understanding of oxidation reactions on the surface of the catalyst and the search for new (heterogeneous) catalysts are challenges for innovation.

4.2 Oxidation of alcohols

From an industrial point of view, the selective dehydrogenation (a special type of oxidation, involving the removal of hydrogen from an organic compound) of alcohols to their corresponding carbonyl compounds is of high importance. The dehydrogenation of alcohols (the opposite of the hydrogenation of carbonyl compounds (see Chapter 2 Heterogeneous hydrogenations and Chapter 3: Homogeneous hydrogenations)) is a clean reaction, and is mainly endothermic, 84 kJ/mol. The reaction itself is usually catalysed by Cu- or Zn-based catalysts, such as ZnO, CuO, Cu–Cr_2O_3 and follows first-order reaction kinetics. Typical examples from the bulk chemical industry are the manufacture of formaldehyde from methanol in presence of a Fe_2O_3–MO_3 catalyst with conversion of 95–99% and selectivity of 91–94%, ethanal (Cu–Cr_2O_3, X = 30–50%, S = 90%), acetone, 2-butanone (700 kt/a, ZnO or Cu-based catalysts, S = 90%, X = 98%) or cyclohexanone (ZnO or Cu–Cr_2O_3, 400–450 °C, X = 90%, S = 95%) from their corresponding alcohols [2–5].

The selective oxidation of primary alcohols to the corresponding aldehydes is a challenge for the fine chemical industry. Further aldehyde oxidation to the carboxylic acid or CO_2 should be avoided. This can be achieved by the use of new catalysts and/ or modified reactor concepts and oxidation conditions. From an industrial point of view, citral synthesis via oxidation is an interesting example that fulfils all criteria of modern catalysis. Starting from formaldehyde and isobutene, isoprenol is manufactured by a *Prins* reaction (Scheme 4.1) (see Chapter 8: Acid–base-catalysed reactions). The oxidation of isoprenol to isoprenal is carried out with oxygen with a very low residence time (10^{-3} s). The continuous reaction is performed at >320 °C in the gas phase in the presence of an Ag/SiO_2 catalyst under isothermal conditions. The advantage of this process (carried out in 40,000 t/a scale) is the high selectivity and the avoidance of toxic reagents [6–8].

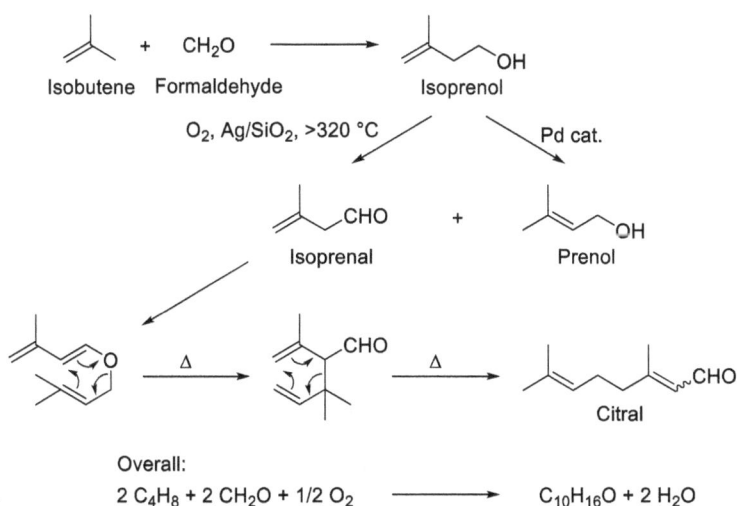

Scheme 4.1: From isobutene to citral–isoprenol oxidation.

4.3 Oxidation of phenols and aromatic compounds

α-Tocopherol is the most important compound of the vitamin E group (for a detailed discussion on Tocopherol, see [9]; for the alkylation reaction, see Chapter 8: Acid–base-catalysed reactions). In its manufacture, trimethylhydroquinone (TMHQ) is condensed with isophytol [9]. Trimethyl-1,4-benzoquinone (TMQ) is obtained by the oxidation of 2,3,6-trimethylphenol (TMP) or 2,3,5-TMP using oxygen or peroxides. TMHQ is finally obtained by the catalytic hydrogenation of trimethyl-1,4-benzoquinone (Scheme 4.2) (see Chapter 2: Heterogeneous hydrogenation). Inorganic acids or salts as oxidants as well as molecular oxygen, air or H_2O_2 combined with homogeneous or heterogeneous catalysts are applied for this transformation. In processes where inorganic acids and salts are used, TMP can be efficiently oxidised to TMHQ, but in catalytic protocols, TMP is more easily converted to TMQ than TMHQ. Several by-products can be produced, for example in the oxidation of the related compound 2,6-dimethylphenol, the prevailing reaction is the oxidative coupling to form the corresponding diphenoquinone. In the oxidation of TMP, TMQ is the main product with 2,2′,3,3′,6,6′-hexamethylbiphenyl-4,4′-diol as the main by-product because the corresponding hexamethyl diphenoquinone is not stable [5, 10].

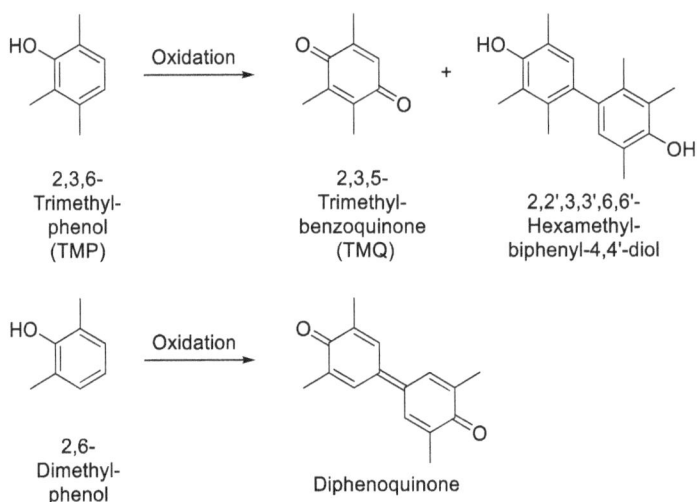

| 2,3,6-Trimethyl-phenol (TMP) | 2,3,5-Trimethyl-benzoquinone (TMQ) | 2,2′,3,3′,6,6′-Hexamethyl-biphenyl-4,4′-diol |

2,6-Dimethyl-phenol   Diphenoquinone

Scheme 4.2: Synthesis of TMQ via 2,3,6-trimethylphenol oxidation and oxidative coupling of 2,6-dimethylphenol.

The oxidation of TMP can also be carried out on an industrial scale using stoichiometric amounts of inorganic salts as oxidants, generating stoichiometric amounts of inorganic waste. For example, oxidation with MnO_2 after the first conversion of 2,3,6-TMP to 4-sulphonyl-2,3,6-TMP, which yields TMQ [11, 12].

However, industrially preferred processes (due to the lower formation of inorganic waste) use oxygen or air or H_2O_2 as oxidants. These methods are also of high importance for the efficient production of economically important compounds in the life science industry [13].

Homogeneous Co catalysts, for example the Co-*Schiff* base complex, salcomine [14, 15] are known as oxidation catalysts. However, one drawback is the fact that it is difficult to recover and recycle the relatively expensive *Schiff* base. Copper catalysts, for example $CuCl_2$ and $CuBr_2$ are widely used for the oxidation of TMP in the presence of molecular oxygen at a large scale [16, 17]. Furthermore, other transition metal halides (such as Cr, Fe, Mn, Ni and Zn) have been reported to catalyse the oxidation of TMP [18]. The combination of copper halides and earth metal halides like magnesium chloride are also applied in the oxidation of phenols [19].

High catalyst loadings are necessary to achieve acceptable conversions especially when Cu-based catalysts are applied; the reason is the catalyst-deactivation which limits this approach. Lowering the catalyst amount while maintaining high conversion of TMP and high selectivity to TMQ is possible by the addition of stabilisers, for example nitrogen-based complexing agents, such as hydroxylamines, oximes or amines [20, 21]. Alternatively stabilising agents such as ionic liquids are also effective [22].

Applying oxygen as an oxidant on a large or industrial scale means that measures must be taken to avoid the risk of explosion of easily flammable organic solvents. One option is to utilise biphasic solvent mixtures of a longer chain alcohol and/or an aromatic solvent with water to suppress the hazardous reaction conditions [23]. Alternatively, a biphasic reaction mixture of water with a carboxylic acid containing a chain of 8–11 carbon atoms can be used, which reduces the possibility of safety incidents [24].

The application of heteropoly acids, for example $H_{3+n}PMo_{12-n}V_nO_{40}$, or $H_{3+n}PMo_{12-n}V_nO_{40}$ on charcoal has been applied for the oxidation of 2,3,6-TMP with molecular oxygen or H_2O_2 [25, 26]. The catalyst can be readily recycled by employing a biphasic system containing a water–acetic acid phase and an organic solvent that is immiscible with water [27]. Several alternative methods for TMP oxidation in the presence of H_2O_2 were developed and are described in the literature, but they are industrially not applied on a large scale [28].

Another approach to TMHQ starts from phenol. Methylation in the gas phase with methanol results in the formation of 2,6-dimethylphenol (see Chapter 5: Gas-phase reactions). The oxidation in the presence of a copper catalyst, for example $CuCl_2$, and oxygen followed by hydrogenation reaction with Pd/carrier, such as Pd/C or Pd/SiO_2 results in 2,6-dimethylhydroquinone with ca. 80% yield (Scheme 4.3). The introduction of the methyl group in *meta*-position was successfully achieved in a *Mannich*-type reaction. Hydrolytic splitting results in TMHQ with 94% yield (see Chapter 2: Heterogeneous hydrogenation) [29].

The transformation of aromatic precursors such as ψ-cumene to TMQ (see Chapter 8: Acid–base-catalysed reactions) is an alternative method to the oxidation of alkyl phenols discussed above. A highly selective catalyst for this oxidation is methyltrioxorhenium (MTO) in combination with *Lewis* base additives, working as ligands. The oxidising agent is hydrogen peroxide (H_2O_2) as a 30% or 50% aqueous solution (Scheme 4.4). This catalyst system can also be applied in TMP oxidation. The oxidation reactions of alkyl-aromatic compounds such as ψ-cumene or xylene or phenols perform well in diluted aqueous solution of H_2O_2 in the presence of methanol. Besides water or methanol, the oxidation can be performed in polar aprotic solvents. The performance of MTO in the oxidation of ψ-cumene is strongly influenced by the solvent. By switching from AcOH and Ac_2O to nitromethane, chloroform or dimethyl carbonate, non-hydroxylated arenes can be oxidised in s/c of 50. Polar aprotic solvents are preferred. Furthermore, the addition of ligands like salen-type compounds allows, based on higher activity, the use of 30% aqueous H_2O_2. In addition, the use of (*N*-salicylidene)aniline-derived *Schiff* bases increases the selectivity to 85% [30]. An alternative catalytic oxidation system is based on *N*-heterocyclic carbene

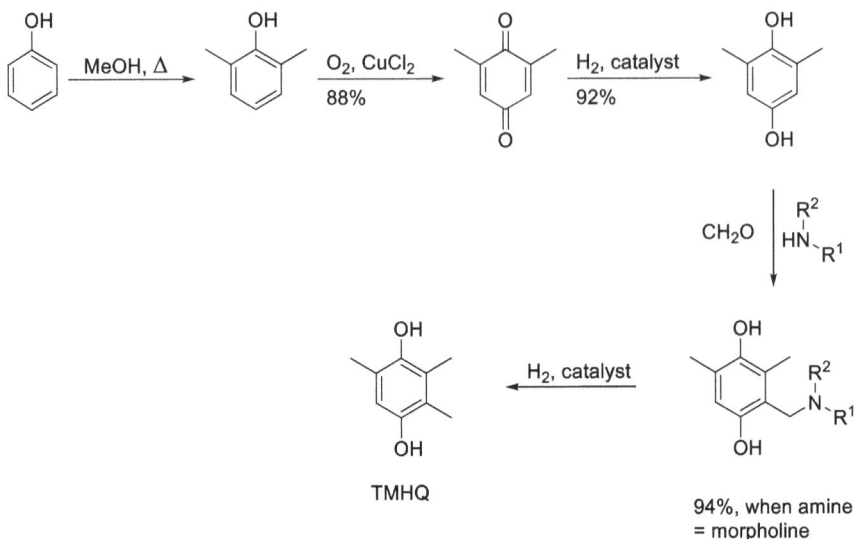

Scheme 4.3: Synthesis of TMHQ, applying the oxidation of dimethylphenol to 2,6-dimethylquinone.

complexes of iron(II), such as [Fe(II)(NCCN)(CH$_3$CN)$_2$] with NCCN = bis(o-imidazol-2-ylidenepyridine)methane. When H$_2$O$_2$ is used as an oxidant, the presence of amphiphiles is beneficial at a reaction temperature <0 °C; with s/c 100, a conversion of around 85% is observed [31].

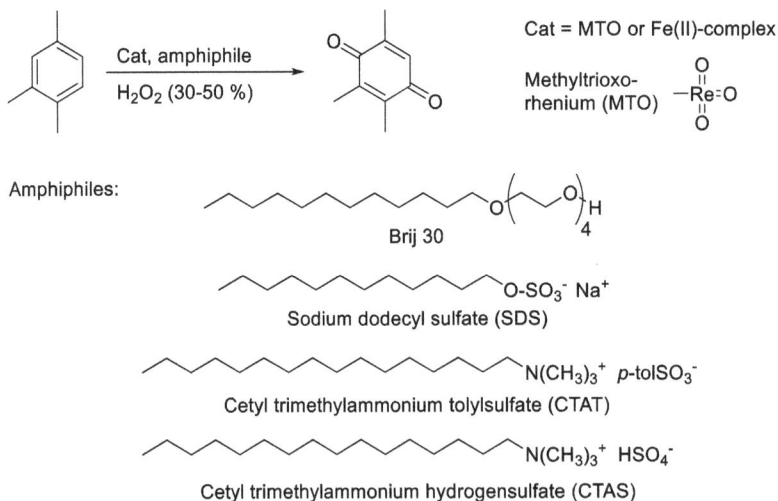

Scheme 4.4: MTO-catalysed oxidation of ψ-cumene.

In human and animal nutrition, vitamin K plays an important role as co-factor in the γ-carboxylation of specific glutamic acid residues in the precursor of the protein prothrombin, converting the protein into the biologically active form. Vitamin K_1 (phylloquinone, Scheme 4.5) and water-soluble forms of menadione (sodium hydrogensulphite) (Scheme 4.6) are used in pharmaceutical applications and as feed additives for poultry, swine and other animals [32]. The manufacture of the most important representative of this group, vitamin K_1, is based on the use of monoacylated starting materials, for example from menadiol (Scheme 4.5) [33, 34]. The monobenzoate derived from menadiol (menaquinol) reacts with isophytol in good yield to the crystalline dihydro-vitamin K_1 derivative in a BF_3 etherate-catalysed reaction [35]. After recrystallisation for the enrichment of the (E)-isomer, the (E)-isomer is saponified to the corresponding hydroquinone followed by an oxidation with oxygen or H_2O_2 to vitamin K_1 (for the alkylation reaction, see Chapter 8: Acid–base-catalysed reactions).

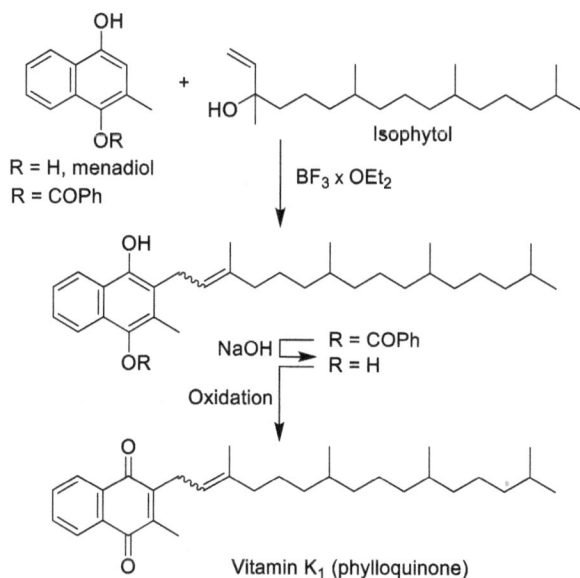

Scheme 4.5: Synthesis of vitamin K_1.

The key building block for vitamin K production is menadiol, which is obtained by catalytic hydrogenation of menadione (vitamin K_3, Scheme 4.6). The processes for the production of the monobenzoate are based on the treatment of 2-methyl naphthalene with (over-)stoichiometric amounts of strong oxidants such as CrO_3 in sulphuric or acetic acid or hydrogen peroxide in acetic acid, or iron(III) chloride/H_2O_2, or nitric acid [27, 36, 37].

Approaches to a more selective menadione synthesis by oxidation have been described using an excess of 30% aqueous hydrogen peroxide in acetic acid at 100 °C

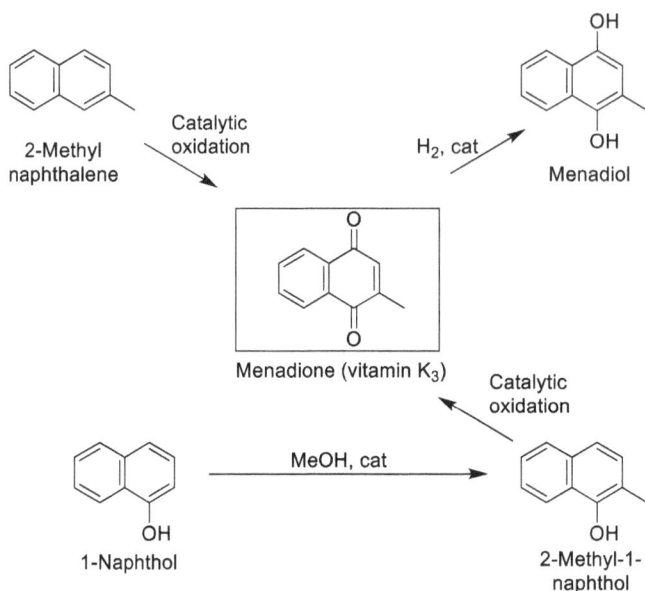

Scheme 4.6: Preparation and use of menadione.

resulting in a selectivity of greater than 90%. In this case, heterogeneous catalysts such as supported Au-nanoparticles, Fe-phthalocyanine-complexes on silica or TiSi-2 can be applied, thus avoiding mineral acids and heavy metals like chromium [38]. It is reported that the oxidation can be performed by applying hydrogen peroxide, *tert*-butylhydroperoxide and oxygen when using the mentioned various solid-supported catalysts, but also by the non-catalysed reaction with molecular oxygen, with even higher selectivity (Scheme 4.6) [39].

The MTO catalysed oxidation to vitamin K_3 (menadione) with high regioselectivity (1,4- vs. 5,8-quinone, 85:15 ratio) using 85% H_2O_2 as oxidant, in mixtures of acetic acid and acetic anhydride, with a s/c of 100 is an efficient alternative procedure [40].

An alternative approach to vitamin K_3 is the Ru-catalysed (terpyridine-derived ruthenium complexes) oxidation of 2-methyl naphthalene. Catalyst loadings of >100 s/c, in the presence of a phase-transfer catalyst (an ammonium, phosphonium salt) as cocatalyst in a two-phasic solvent system (aqueous systems or methanol), applying 30% H_2O_2 as oxidation agent, results in 60% yield (selectivity > 99%) [41]. Iron(III) salts, for example $FeCl_3$ hexahydrate and pyridine-2,6-dicarboxylic acid as catalysts, and H_2O_2 as oxidant were also applied in the oxidation of 2-methyl naphthalene. A yield of 55% can be obtained [42].

The heterogeneous-catalysed oxidation of 2-methyl naphthalene to vitamin K_3 with 30% H_2O_2 produced menadione in 60% yield with a selectivity of around 90% in the presence of Ti-MCM-41 [43]. An alternative approach is the electrochemical oxidation of 2-methyl naphthalene to menadione using the Ce(III)–Ce(IV) redox system which was reported for batch experiments in electrochemical cells [44].

1-Naphtol can also be used as the starting material for the preparation of menadione (Scheme 4.6). The 2-methyl group is introduced by catalytic alkylation with methanol yielding 2-methyl-1-naphthol and subsequent oxidation gives menadione. Polyoxymetallates (*Keggin*-type heteropoly compounds) containing phosphorus, molybdenum, tungsten, vanadium and oxygen were used as oxidants, as well as aqueous hydrogen peroxide in the presence of a niobium-based heterogeneous catalyst system [45].

4.4 Oxidation of various C–C bonds

Linalool, an important intermediate in the synthesis of tocopherol and a key compound in the flavour industry (see Chapter 7: Rearrangement reactions) can be obtained from pinene extracts by hydrogenation, subsequent oxidation to the respective hydroperoxides, hydrogenation to the respective alcohols and pyrolysis to yield linalool (Scheme 4.7). Pinene extracts contain mostly mixtures of α- and β-pinene. After hydrogenation, a mixture of *cis*- and *trans*-pinane is obtained.

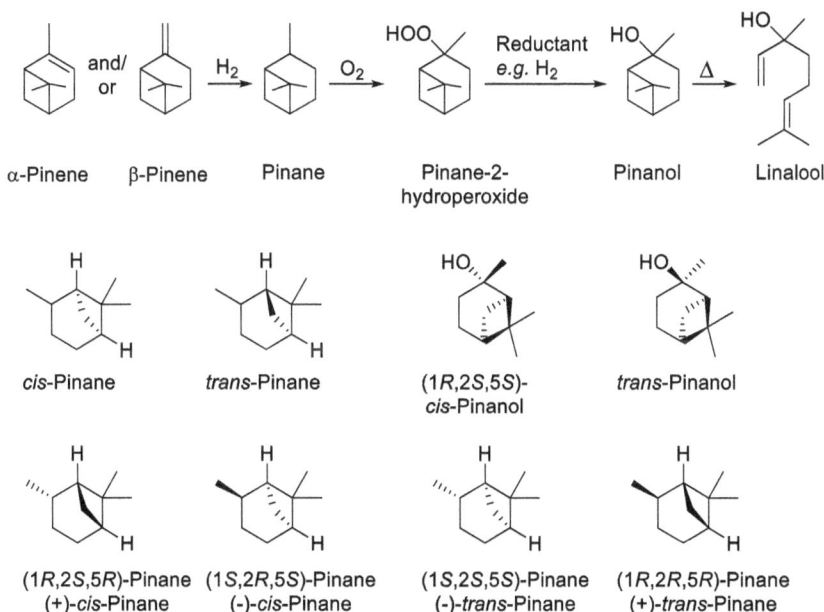

Scheme 4.7: Preparation of linalool from pinenes.

The oxidation of pinanes is carried out in the presence of a catalyst such as Co(OAc)$_2$/Mn(OAc)$_2$/NH$_4$Br (12.8/1.5/5.0 mol%) by direct oxidation with air or oxygen [46]. In both cases, *cis*-pinane (Scheme 4.7) is more reactive than *trans*-pinane because of the steric effect of the *gem*-dimethyl group which hinders the free radical attack on the tertiary C–H bond in 2-position. A selectivity of 54% to *cis*-pinanol and 17% to *trans*-pinanol was obtained at low conversions (Scheme 4.7). The difference in *cis/trans* selectivity is likely to be due to the interaction of the transition metal with the 2-pinanyl radical. When Co(OAc)$_2$ or Mn(OAc)$_2$ were used alone as catalysts and under oxygen-limiting conditions (e.g. air instead of pure oxygen or using a solvent with a lower oxygen solubility), the reaction stopped at the intermediate pinane-2-hydroperoxide [41, 47–49].

The non-catalysed oxidation of pinane is an autoxidation. Oxidation of *cis*-pinane at 100 °C resulted in 15% pinane-2-hydroperoxide which was converted to a mixture of *cis*-pinanol (67%), *trans*-pinanol (17%) and other by-products. Oxidation of *trans*-pinane at 100 °C and subsequent reduction resulted in 17% *cis*-pinanol, 5% *trans*-pinanol and various by-products [49].

Free radicals are generated either by interaction between pinane and O$_2$ or by monomolecular decomposition of pinane-2-hydroperoxide. The recombination of these free radicals does not result in the formation of side products. The variation of the oxygen pressure did not affect the selectivity and thus the reaction steps in the oxidation of pinane involving oxygen are not rate-determining.

In the flavour and fragrance industry, there is a need for new compounds and new efficient processes, especially catalytic processes, to replace processes using stoichiometric reagents. The monoterpene (–)-*cis*-rose oxide is a pyran compound, found in rose oil and fruits and has an appreciated green, floral aroma [50]. Bulgarian roses are a major source of natural rose oxide in the form of fragrant oil. Bulk flavour and perfumery applications of rose oxide as a (–)-*cis*, (+)-*cis*, (–)-*trans* and (+)-*trans* stereoisomeric mixture require a cost-efficient manufacturing process. In the industrial production of rose oxide, citronellol is produced by the enantioselective hydrogenation of nerol or geraniol (see Chapter 3: Homogeneous hydrogenations), followed by an ene-type allylic oxidation with singlet oxygen. The allylhydroperoxide intermediates were transformed into the allylic alcohol and acid-catalysed ring-closure reaction resulted in the formation of a stereoisomeric mixture of rose oxide (Scheme 4.8) [51].

A modern approach to generate singlet oxygen (^{1}O$_2$) from triplet oxygen (^{3}O$_2$) to avoid the dangerous combination of oxygen, organic compounds and light follows the "dark" singlet oxygen generation via catalytic disproportionation of hydrogen peroxide into water and singlet oxygen. It can be carried out in conventional stirred tank reactors by adding hydrogen peroxide to a solution containing the disproportionation catalyst and the substrate [52].

In the presence of Na$_2$MoO$_4$ as a catalyst (4 mol%), a high conversion of citronellol (>95%) could be achieved with only a 1.5-fold excess of 50% aqueous hydrogen peroxide [52]. The critical parameter is the addition rate of hydrogen peroxide, which in turn has

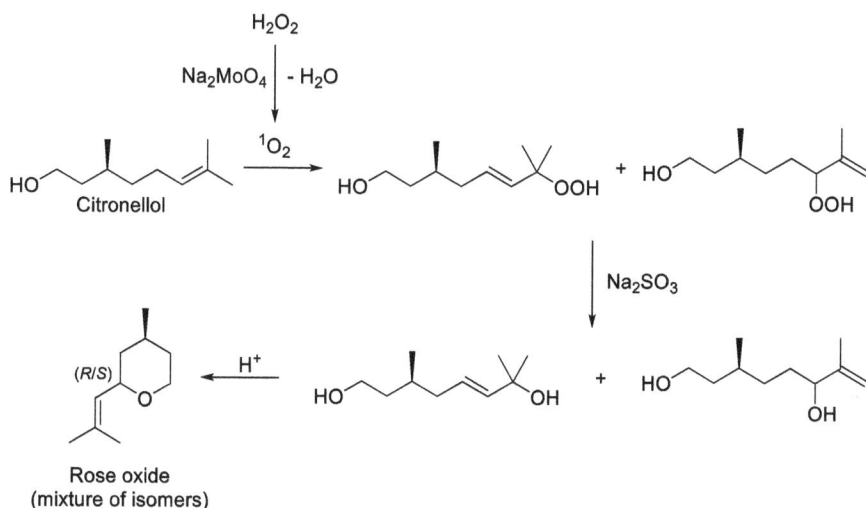

Scheme 4.8: Synthesis of rose oxide via "dark" singlet oxygenation of β-citronellol.

to be adjusted to the reaction temperature. The addition of hydrogen peroxide has to be conducted at the same rate as it is consumed in the reaction (steady-state concentration, $dc/dt = 0$) to achieve sufficient yields. The reaction can be performed at 55 °C, and a 50/50 mixture of the secondary and tertiary hydroperoxides is formed. The latter being the required intermediate for the preparation of rose oxide (Scheme 4.8).

Lauryl lactone is a key intermediate for the production of lauryl lactam (Scheme 4.9) and Nylon 12. Starting from butadiene trimerisation in the presence of a Ti–Al catalyst or bis(π-allyl)Ni catalyst, mainly (E,E,Z)-cyclododecatriene is produced. Using an Al:Ti metal ratio of around 4–5 allows the cyclo-trimerisation to occur in 90% selectivity. The process is based on the fundamental work of G. Wilke and is carried out on a several thousand-tons scale worldwide [53, 54]. Cyclododecatriene hydrogenation at 10–15 bar H_2 pressure and 150–200 °C with a Ni-alloy catalyst results in cyclododecane with >95% conversion and >97% selectivity. Oxidation in the presence of boric acid at around 160 °C at normal pressure results in an alcohol/ketone mixture similar to the cyclohexanol/cyclohexanone process. The oxidation is performed at partial conversion (30%) with 80% selectivity. Lauryl lactam (monomer of Nylon 12) production starts from cyclododecanol/cyclododecanone. This mixture is dehydrogenated to cyclododecanone in the presence of a Cu/Cr or Cu/Al$_2$O$_3$ catalyst at 230–250 °C in a liquid phase process applying air as an oxidant. The treatment with hydroxylamine and *Beckmann* rearrangement at 130 °C followed by water washing results in the formation of the product. The conversion of the process (three steps) is >95% and the selectivity is >97% [55].

A second process to lauryl lactam based on the photochemical oxidation of cyclododecane in the presence of NOCl at 70 °C, 70% conversion and >85% selectivity is carried out in 10,000 t/a scale (Scheme 4.9) [56].

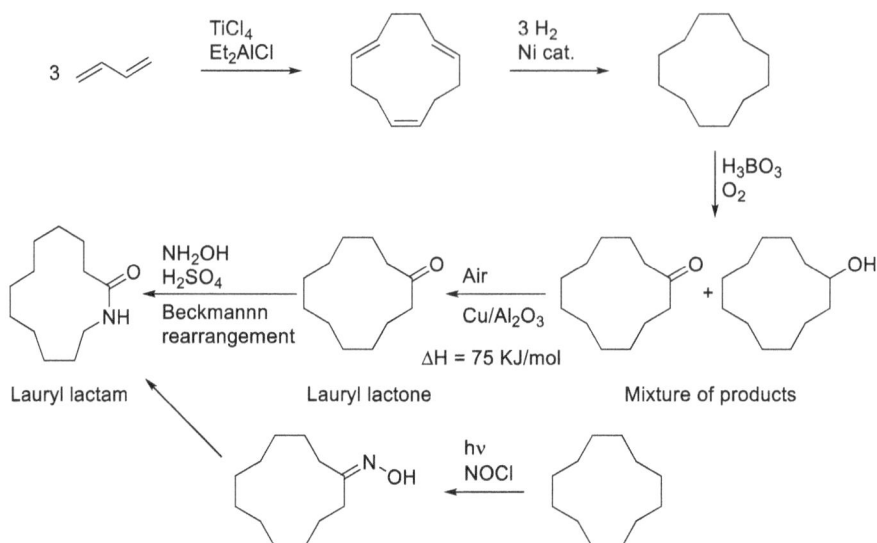

Scheme 4.9: Syntheses of lauryl lactam via *Beckmann* rearrangement.

Furthermore, a process starting from 1,5,9-cyclododecatriene by carbonylation reaction in the presence of a Pd-phosphine catalyst resulting in cyclododecane acid, followed by treatment with nitrosyl sulphuric acid (similar to the SNIA-Viscosa process for ε-caprolactam) is known. The advantage of the process is the green approach by avoiding salt formation [57, 58].

ε-Caprolactam is produced from cyclohexane by photonitrosation reaction (PNC process) using a NOCl/HCl mixture in a photo reactor. The photooxidation is initiated by a mercury lamp (Scheme 4.10, Figure 4.2) (see Chapter 5: Gas-phase reactions and Chapter 7: Rearrangement reactions).

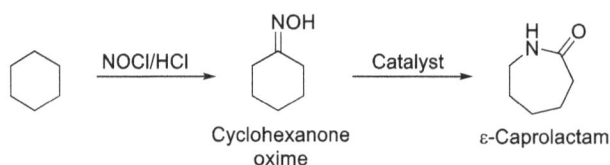

Scheme 4.10: Manufacture of ε-caprolactam from cyclohexane.

β-Picoline is an important intermediate in the field of agrochemicals and produces nicotine amide (see Chapter 5: Gas-phase reactions). The synthesis starts from 2-methylpentane-1,5-diamine, in the presence of a zeolite, such as ZSM-5, and is followed by a dehydration reaction with a supported Pd catalyst (Scheme 4.11) [59, 60]. The transformation of β-picoline into nicotinic amide (vitamin B_3) is performed by ammoxidation resulting in 3-cyanopyridine followed by saponification [61–65].

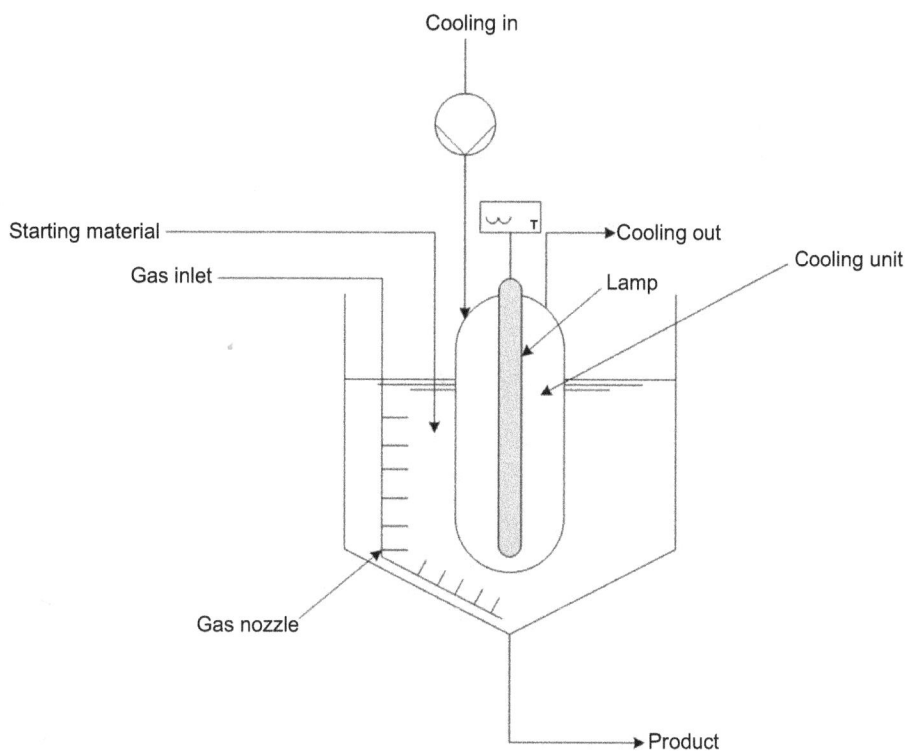

Figure 4.2: Photoreactor for the photonitrosation of cyclohexane (PNC process).

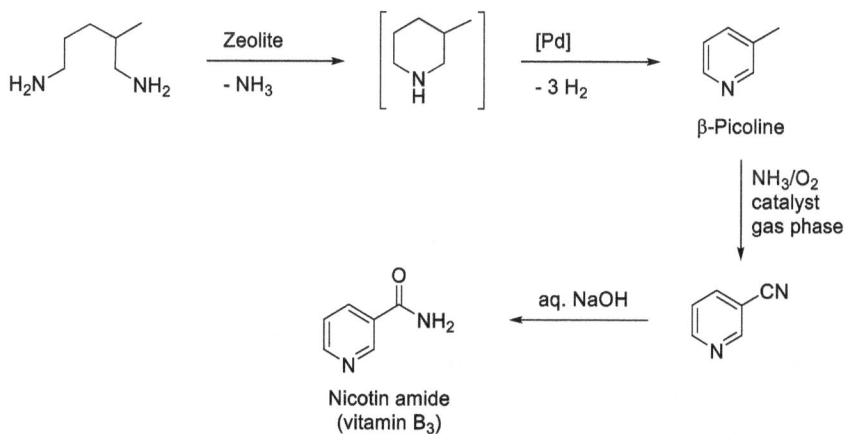

Scheme 4.11: Synthesis and usage of β-picoline.

In the manufacture of the pyrethroide insecticide deltamethrin ((1*R*,3*R*)-[(*S*)-α-cyano-3-phenoxybenzyl-3-(2,2-dibromvinyl)]-2,2-dimethylcyclopropancarboxylate), 3-carene is oxidised with air or oxygen to 3-carene-5-one in the presence of a Cr catalyst (Scheme 4.12). The catalyst is prepared in situ from CrCl$_3$ and pyridine. 3-Carene is obtained mainly from waste streams in paper manufacturing. The TOF of the catalyst applied in the oxidation step is around 850 per min and the reaction follows a radical pathway with a peroxide intermediate [66]. Deltamethrin is used in eliminating a wide variety of household pests, such as spiders, fleas, ticks, carpenter ants, carpenter bees, cockroaches, ant chalks and bed bugs [67].

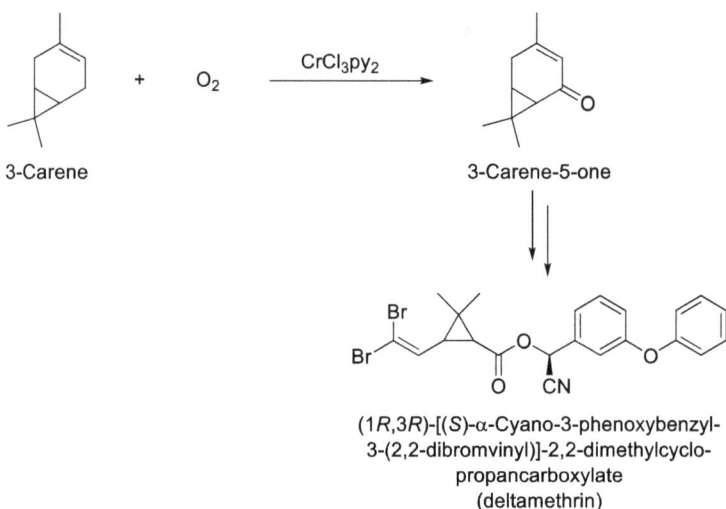

Scheme 4.12: 3-Carene oxidation for a deltamethrin building block.

The application of ozone (O$_3$) in oxidation reactions of alkenes, alkynes and azo compounds is also well described and is performed on an industrial scale (ozonolysis, *Harris* reaction). Ozone treatment of azo compounds results in the corresponding nitrosamine formation, whereas alkenes can be oxidised to aldehydes, ketones, carboxylic acids or alcohols depending on the work up and substitution pattern of the double bond. The ozonolysis of symmetric olefins can even lead to asymmetric products by careful control of the work-up conditions [68].

The reaction mechanism of ozone reactions with alkenes is well investigated (Scheme 4.13). The alkene and ozone form an intermediate molozonide in a 1,3-dipolar cycloaddition, followed by a retro 1,3-dipolar cycloaddition to form the corresponding carbonyl oxide and a ketone or aldehyde, respectively. These two compounds react in another 1,3-dipolar cycloaddition to generate a trioxolane which is relatively stable. This intermediate is converted to the desired products based on the applied work-up conditions [69, 70].

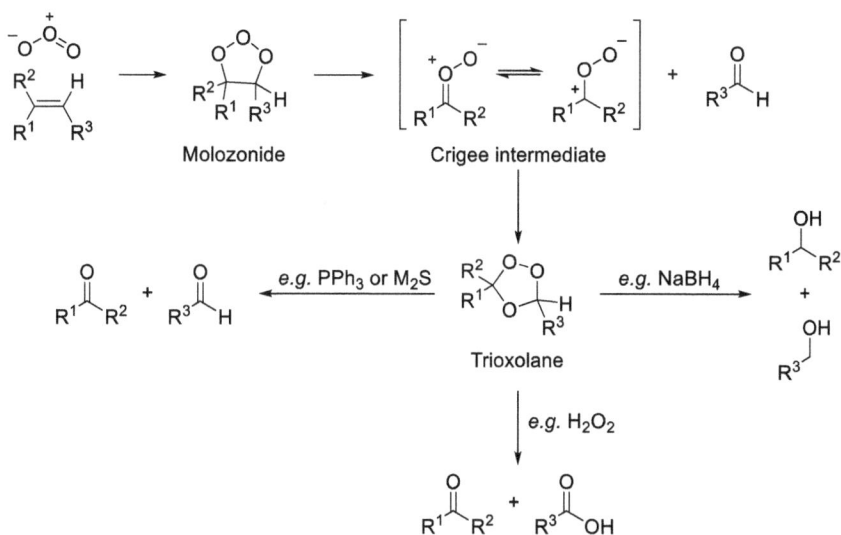

Scheme 4.13: Mechanism of ozonolysis with different work-up options.

Oleic acid ($CH_3(CH_2)_7CH=CH(CH_2)_7COOH$, (9Z)-octadec-9-enoic acid) ozonolysis results in the formation of azelaic acid ($HOOC(CH_2)_7COOH$) and pelargonic acid ($CH_3(CH_2)_7CO_2H$), which are produced on an industrial scale (Scheme 4.14) [71, 72].

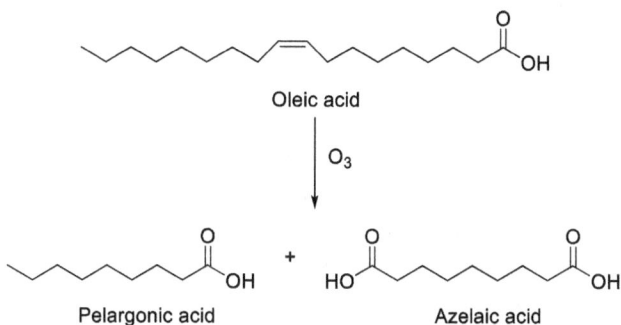

Scheme 4.14: Synthesis of pelargonic and azelaic acid from oleic acid.

The manufacture of glyoxylic acid from malic acid or pelargonic acid (nonanoic acid) or from oleic acid by ozone treatment is of industrial relevance. Nonanoic acid is also produced by hydroformylation of 1-octene followed by oxidation (Oxeno process). It is applied as a plasticizer or herbicide (ammonium salt). Other examples are the ozonolysis of eugenol to the corresponding aldehyde [73]; and the conversion of chrysanthemic acid with ozone being conducted by Lonza in a loop reactor and optimised at 0.5 t/day scale [74]. All of the listed examples are uncatalysed due to the high reactivity of ozone to react with double bonds. Nevertheless, these examples are mentioned to highlight the industrial relevance of ozonolysis as an oxidation method not only in the fine chemical industry.

The Mn salt-catalysed O_3 treatment of alkylated aromatic or heteroaromatic compounds in an organic acid solvent results in the formation of the corresponding aromatic or heteroaromatic acids using a Mn- or Ce-based catalyst (Scheme 4.15) [75]. The reactions are performed at around 15 °C, with a s/c of 100 to 500, in one to three hours reaction time. Conversions >95% and selectivity of 94–98% are obtained.

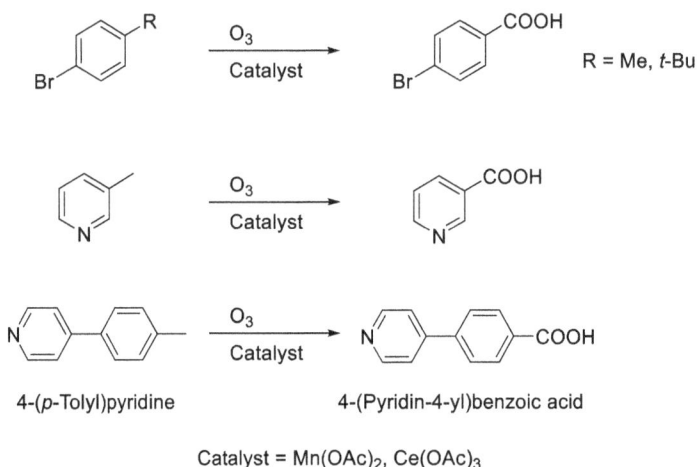

4-(p-Tolyl)pyridine   4-(Pyridin-4-yl)benzoic acid

Catalyst = $Mn(OAc)_2$, $Ce(OAc)_3$

Scheme 4.15: Metal–salt-catalysed carboxylic acid formation by benzylic oxidation in the presence of ozone.

Another benzylic oxidation, that is rather found in the bulk chemical industry, is the air oxidation of mesitylene in the presence of Co or Mn catalysts (Amoco-process). The reaction performs in >95% conversion and >95% selectivity (Scheme 4.16). The resulting trimesic acid (1,3,5-tricarboxybenzene) is applied in the dye industry, polymer industry (e.g. as a softener) and epoxy resin production [76, 77].

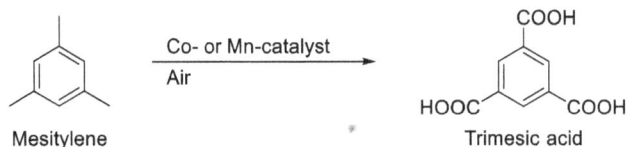

Mesitylene   Trimesic acid

Scheme 4.16: Benzylic oxidation of mesitylene to trimesic acid.

In a similar manner, the catalytic oxidation of pseudocumene results in the formation of trimellitic acid (1,2,4-tricarboxy-benzene) (Scheme 4.17) [76, 78]. The acid is trans-

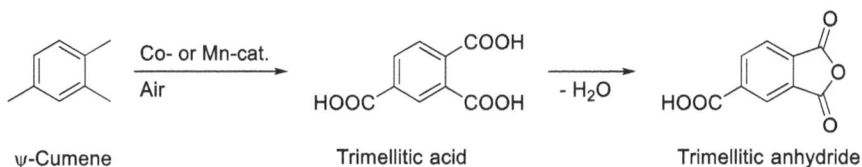

Scheme 4.17: Synthesis of trimellitic anhydride by benzylic oxidation of ψ-cumene.

ferred into the corresponding anhydride, which is used in the polymer industry as a softener and in epoxy resin production [77].

For a better understanding and also in the view of industrial catalytic processes, the palladium-catalysed oxidation of ethylene to acetaldehyde (applied on an industrial scale: *Wacker* process) is fundamental [79]. The process is usually performed under homogeneous conditions. During the reaction, Pd(II) is reduced to Pd(0) and must be re-oxidised applying Cu(II), which is itself reduced to Cu(I). The Cu(I)-intermediate species is itself re-oxidised with oxygen. Overall, only ethene and oxygen are consumed in the production of acetaldehyde (Scheme 4.18). Acetaldehyde is the starting material in the manufacture of acetic acid (by oxidation), acetic anhydride or butadiene (Aldol reaction, *Lebedew* process).

In the one-stage process, carried out at 130 °C and 4 bar, C_2H_2 and O_2 are fed into the reactor with the catalyst mixture ($PdCl_2$ and $CuCl_2$) dissolved in water. The product acetaldehyde is purified by distillation. By-products such as acetic acid, croton aldehyde and chlorinated acetaldehyde are separated. The process is performed in glass-lined or titanium-based equipment to avoid corrosion problems.

An alternative *Wacker* process is carried out in the presence of air at 10 bar and around 110 °C. It is a two-step process, performed in a tubular reactor, where the oxidation and separation steps are isolated from each other. After separation from acetaldehyde, the catalyst is re-oxidised. The crude acetaldehyde is purified by distillation [80].

The yield in the *Wacker* processes is 95%. Chlorinated hydrocarbons, chlorinated acetaldehydes and acetic acid are formed as major by-products [80].

As indicated in Scheme 4.18, the resting state of the Pd(II) catalyst under the most relevant reaction conditions is $[PdCl_4]^{2-}$. The *Wacker* process under industrial conditions is sensitive to the concentration of Cl^- and $CuCl_2$. Furthermore, it is known from kinetic studies that the ethene consumption in the *Wacker* process is first order, a second-order chloride inhibition and a first-order proton inhibition occurs [81]. Additionally, investigations have shown that one key intermediate is $[PdCl_2(C_2H_2)H_2O]$, and water plays an important role in this reaction. The acetaldehyde formation via a *syn* hydroxypalladation is supported by experimental kinetic investigations (under industrial conditions) [81].

The *Wacker* oxidation is also described under heterogeneous conditions. In the liquid phase, catalyst stability is a problem because leaching is often observed. For the application of the *Wacker* chemistry in the gas phase, the catalyst and the oxida-

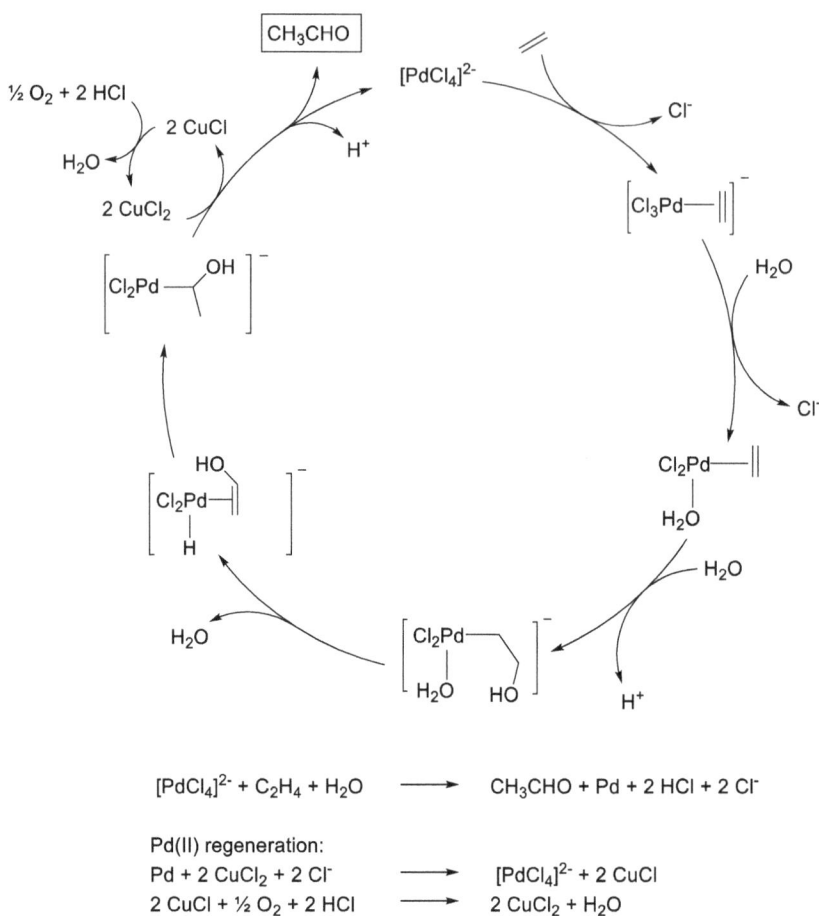

Scheme 4.18: Catalytic cycles of the *Wacker* process.

tion partner must be supported on a carrier. Supports such as zeolites, for example Y-zeolite or heteropoly acids containing vanadium are known for gas-phase *Wacker* oxidations [82, 83]. From a kinetic point of view, a similar reaction behaviour compared to the homogeneous liquid-phase version occurs. The *Wacker* oxidation is first order in ethene and second order in chloride-ion concentration.

A variation of the *Wacker* oxidation performed in DMF with air or oxygen as an oxidant is the *Wacker–Tsuji* oxidation [84]. Here, 1-alkenes are oxidised to 2-ketones. The transformation of 1-decene into 2-decanone in the presence of $PdCl_2/CuCl$ in DMF/ H_2O is performed using oxygen. Higher substituted alkenes react faster than dienes, and the product formation follows "the *Markovnikov*" rule.

The *Wacker* technology is also applied in the synthesis of citral from myrcene (Scheme 4.19), obtained from β-pinene by a gas phase process (see Chapter 5: Gas-phase reactions) [85].

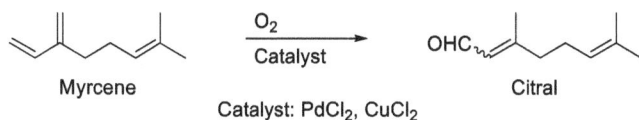

Scheme 4.19: Metal-catalysed myrcene oxidation.

The asymmetric oxidation of organic substrates is a challenging topic, especially to achieve high selectivity and yield. During recent years, modern approaches were developed to bypass problems with by-product formation or lack of selectivity. The (E)-alkene epoxidation with potassium peroxymonosulfate (KHSO$_5$, oxone) in the presence of a fructose derivative as chiral catalyst results in chiral epoxides and is called the Shi epoxidation (Scheme 4.20). In a two-phase solvent system, for example water/hydrocarbon, around 80% yield and 95% *ee* can be achieved. An advantage of this oxidation method is the avoidance of metal salts and problems with their separation and regeneration [86, 87]. The pH value of this reaction is crucial to achieve good results. If the pH value is too low, catalyst decomposition via *Baeyer–Villiger* rearrangement is observed, whereas with a too high pH-value rapid decomposition of the oxidant is detected. Optimal pH values are around 10.5.

The *Shi* epoxidation is also used in the large-scale synthesis of a dipeptide, applied as a protease inhibitor for human immunodeficiency virus. With 30% catalyst loading and stoichiometric oxidant, the epoxide can be synthesised in 65% yield and 88% *ee*. The epoxidation was carried out batch-wise, and the product was purified by crystallisation [88–90].

There are currently only a few examples of photooxidations in industry but they are gaining in importance. Artemisinin can be used for the treatment of malaria and significantly reduces mortality rates. The discovery of this compound and its potential for the treatment of malaria was honoured with the Nobel Prize in Medicine in 2015 to Tu Youyou [91].

Its synthesis can be performed via a semisynthetic route starting from artemisinic acid obtained via fermentation (see Chapter 10: Biocatalysis), followed by diastereoselective hydrogenation (see Chapter 3: Homogeneous hydrogenations). In the final step, photooxidation is applied and results in an efficient process, which was the first source of non-plant-derived artemisinin on an industrial scale (Scheme 4.21) [92]. This process was industrialised in Sanofi's production facility in 2013 in Garessio, Italy with a production volume of up to 60 tons per year with an average batch size of 370 kg [93].

Mechanistically, it was proposed that the last reaction sequence proceeds via a regioselective *Schenck ene* reaction between singlet oxygen and the trisubstituted C=C-bond. The formed peroxide undergoes *Hock* cleavage followed by oxidation with triplet oxygen and cyclisation. Various parameters make the implementation of such a process extremely challenging on an industrial scale. The accumulation of peroxide species is a critical safety issue and must be avoided to ensure a safe process. Also,

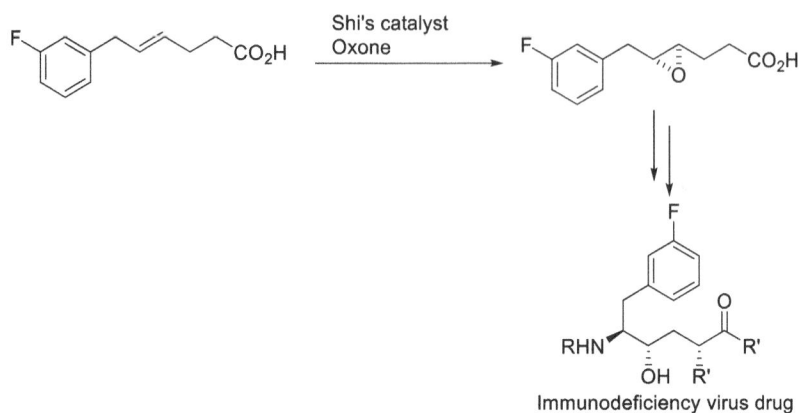

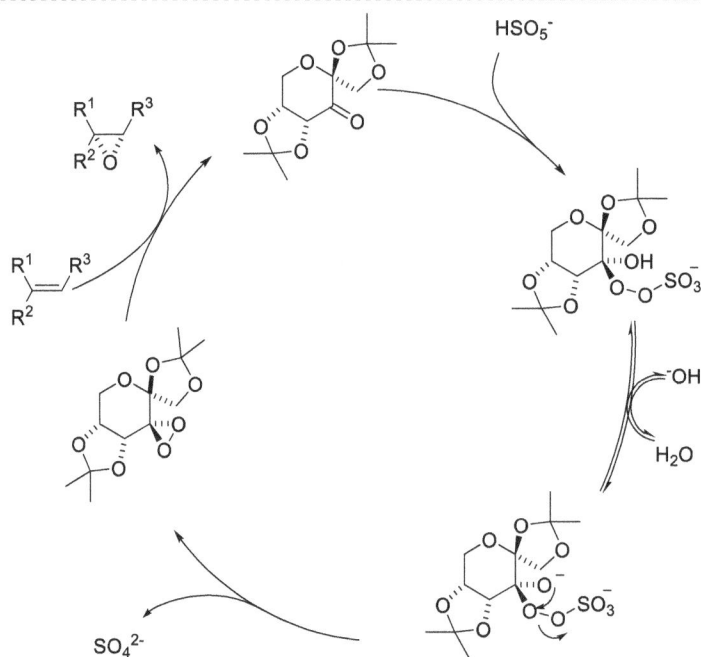

Scheme 4.20: Shi oxidation for the synthesis of an immunodeficiency virus drug and the reaction mechanism.

other aspects such as light transmission, gas transfer and mixing are significantly more complex on the production scale (Figure 4.3). Therefore, also a halogenated solvent was used due to safety concerns, which resulted in a more complicated purification protocol with solvent exchanges. Compared to the first reported synthesis by Amyris which used H_2O_2 and Na_2MoO_4 as catalyst [94], the overall yield from the same starting material was more than 2.5 times higher (55% vs. 19%).

Schenck ene reaction

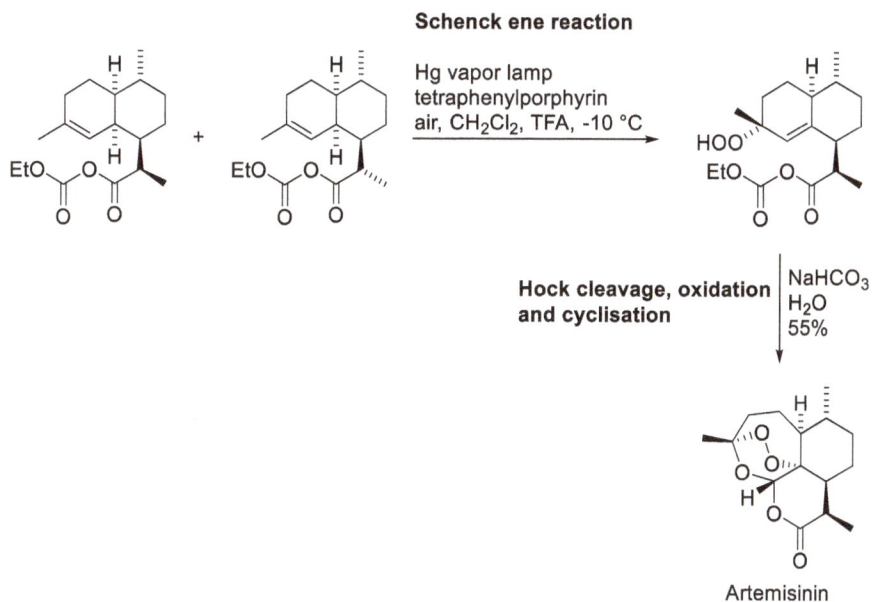

Hg vapor lamp
tetraphenylporphyrin
air, CH$_2$Cl$_2$, TFA, -10 °C

**Hock cleavage, oxidation
and cyclisation**

NaHCO$_3$
H$_2$O
55%

Artemisinin

Scheme 4.21: Photooxidation conducted in the production of artemisinin by Sanofi.

Figure 4.3: Reactor for photooxidation towards artemisinin on production scale (from J. Turconi *et al*, *Org. Process Res. Dev.*, 2014, used with permission).

4.5 Oxidation for silane production

The importance of oxidation reactions is also manifested in the silane and silicone industry. Silicones, also called polysiloxanes, are polymers of siloxanes ($-R_2Si-O-SiR_2-$, with R = organic group). These polymers are colourless compounds and are used as silicone rubber, silicone oil or silicone. Application fields are in personal care, for example cosmetics for skin care; medicine, for example implants and microfluidics; household, for example cooking ware; and the automotive industry, for example lubricants for brakes [95].

The manufacture of polysiloxanes starts from siloxanes, mainly polydimethylsiloxane, obtained by hydrolysis of dimethyl dichlorosilane or the corresponding acetate, with water (Scheme 4.22).

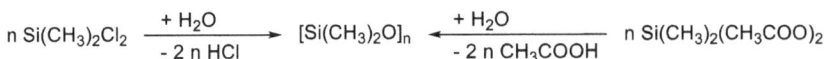

$$n\ Si(CH_3)_2Cl_2 \xrightarrow[- 2\ n\ HCl]{+ H_2O} [Si(CH_3)_2O]_n \xleftarrow[- 2\ n\ CH_3COOH]{+ H_2O} n\ Si(CH_3)_2(CH_3COO)_2$$

Scheme 4.22: Synthesis of polysiloxanes.

The starting material dimethyl dichlorosilane is produced on an industrial scale from silicon (elemental) and methyl chloride in a Cu(I)-catalysed reaction at around 300 °C in a fluidised bed reactor (*Müller–Rochow* process) (Scheme 4.23) [96]. The copper catalyst is generated from copper or copper oxide. Promotors such as zinc or tin are added to the reaction system. The reaction of methyl chloride and silicon results in a mixture of methyl silicon halides, which are separated by distillation (Figure 4.4). The main reaction product dimethyl dichlorosilane is obtained in 70–90% yield, whereas 5–15% methyl trichlorosilane is formed [96]. Based on the boiling points of the silanes, the distillation of the methyl silane mixture, $(CH_3)_2SiCl_2$ (70 °C), CH_3SiCl_3 (66 °C), $(CH_3)_3SiCl$ (57 °C) needs columns with high separating capacities.

$$n\ CH_3Cl + Si \xrightarrow{Cu(I)} ClSi(CH_3)_3,\ (CH_3)_2SiCl_2,\ CH_3SiCl_3,\ SiCl_4,\ ....$$

Scheme 4.23: Synthesis of silanes.

The various alkylchloro silanes obtained in the *Müller–Rochow* process are also starting materials for the manufacture of acyloxysilanes, compounds important for the polysiloxane production [97].

The manufacture of polydimethylsiloxane is performed in continuous reaction mode. Dimethyl dichlorosilane is treated with aqueous hydrochloric acid (22%). After phase separation and water separation followed by neutralisation, the final product is obtained [98].

The siloxane production from dimethyldichloro silane and methanol is also industrialised (Scheme 4.24).

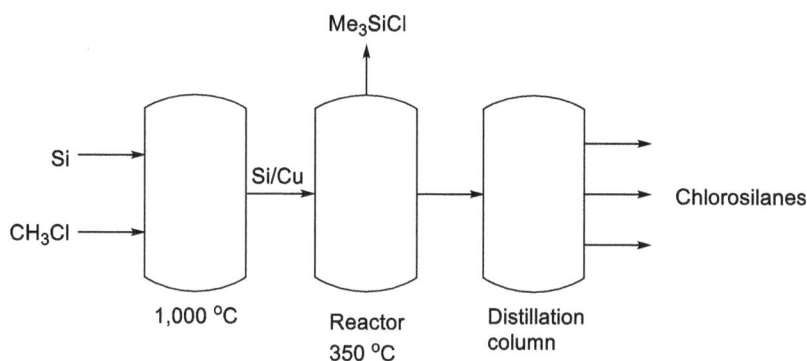

Figure 4.4: Block-flow diagram showing *Müller–Rochow* synthesis of methylchloro silanes.

$$n\,(CH_3)_2SiCl_2 + 2\,n\,MeOH \longrightarrow HO\text{-}[(CH_3)_2SiO]_nH + 2\,n\,CH_3Cl + (n\text{-}1)\,H_2O$$

$$n\,(CH_3)_2SiCl_2 + 2\,n\,MeOH \longrightarrow [(CH_3)_2SiO]_n + 2\,n\,CH_3Cl + n\,H_2O$$

Scheme 4.24: Methanolysis of dimethyl dichloro silane.

An advantage of the process is the reuse of the chlorine as methylchloride, which is utilised in the *Müller–Rochow* synthesis.

References

[1]   Mars P, van Krevelen D W. Oxidations carried out by means of vanadium oxide catalysts. Chem Eng Sci 1954, 3, 41–59.
[2]   Balandin A. Nature of active centers and the kinetics of catalytic dehydrogenation. Adv Catal 1958, 10, 96–129.
[3]   Venkateshwas S, Rao M B. Kinetics of heterogeneous vapor-phase dehydrogenation of 2-ethyl-1-hexanol. Ind J Technol 1982, 20, 150–154.
[4]   Rao U R, Kuloor N R. Dehydrogenation of propyl alcohol in fixed catalyst beds. Ind Chem Eng 1973, 15, 3–11.
[5]   Jover B, Juhasz J, Szabo Z G. Catalytic decomposition of isopropyl alcohol over zinc oxide-magnesium oxide systems with zinc in various surroundings. Z Phys Chem 1978, 111, 239–245.
[6]   Aquila W, Fuchs H, Worz O, Ruppel W, Halbritter K. Process for the Technical Preparation of Unsaturated Aliphatic Aldehydes in a Tube Bundle Reactor. EP 881206, BASF AG, 1998.
[7]   Sauer W, Aquila W, Hoffmann W, Brenner K, Halbritter K. Process for the Continuous Preparation of 2-alkyl-buten-1-als. EP 55354, BASF AG, 1982.
[8]   Hölderich W F, Kollmer F. Oxidation reactions in the synthesis of fine and intermediate chemicals using environmentally benign oxidants and the right reactor system. Pure Appl Chem 2000, 72, 1273–1287.
[9]   Bonrath W, Wyss A, Litta G, Baldenius K-U, von dem Bussche-Hünnefeld L, Hilgemann E, Hoppe P, Stürmer R, Netscher T. Vitamin E (Tocopherols, Tocotrienols). In: Elvers B, ed. Ullmann's Encyclopedia of Industrial Chemistry Online, Weinheim, Germany, Wiley-VCH, 2021.

[10]   Ning Z, Xi Z, Cao G, Zhang X, Hai X. Oxidation of trimethylphenol catalyzed by aqueous soluble oxygen carriers. Oxidation Commun 1999, 22, 527–531.

[11]   Brenner W. Verfahren zur Herstellung von Chinonen. DE 2221624, Hoffmann La Roche, 1972.

[12]   Hirose N K, Hamamura K, Inai Y, Ema K, Banba T, Kijima A. Process for Preparing 2,3,5-trimethylbenzoquinone. EP 0294584, Eisai Co Ltd, 1988.

[13]   Bonrath W, Eggersdorfer M, Netscher T. Catalysis in the industrial preparation of vitamins and nutraceuticals. Catal Today 2007, 121, 45–57.

[14]   Diehl H, Hach C C. Bis(N,N'-disalicylalethylenediamine)-μ-aquodicobalt(II). Inorg Synth III 1950, 196–201.

[15]   Jouffret M. Verfahren zur Herstellung von Trimethyl-p-benzochinon. DE 2450908, Rhone-Poulenc SA, 1975.

[16]   Thoemel F, Hoffmann W. 2,3,5-Trimethyl-p-benzoquinone. DE 3215095, BASF AG, 1983.

[17]   Bartoldus D, Lohri B. Trimethyl Benzoquinone. EP 35635, F. Hoffmann-La Roche AG, 1981.

[18]   Maassen R, Krill S, Jäger B, Huthmacher K. Verfahren zur Herstellung von 2,3,5-Trimethyl-p-benzochinon. EP 1092701, Degussa-Hüls AG, 2001.

[19]   Hsu C Y, Lyons J E. Oxidizing a Phenol to a p-benzoquinone. EP 107427, Suntech Inc, 1984.

[20]   Takehiro K, Orita H, Shimizu M. Method for the Preparation of 2,3,5-Trimethylbenzoquinone. EP 0369823, Agency of Industrial Sciences and Technology, 1990.

[21]   Bodnar Z, Mallat T, Baiker A. Oxidation of 2,3,6-trimethylphenol to trimethyl-1,4-benzoquinone with catalytic amount of $CuCl_2$. J Mol Catal A Chem 1996, 110, 55–63.

[22]   Sun H, Harms K, Sundermeyer J. Aerobic oxidation of 2,3,6-trimethylphenol to trimethyl-1,4-benzoquinone with copper(II) chloride as catalyst in ionic liquid and structure of the active species. J Am Chem Soc 2004, 126, 9550–9551.

[23]   Bockstiegel B, Hoercher U, Laas H. Verfahren zur Herstellung von 2,3,5-Trimethyl-p-benzochinon. DE 4029198, BASF AG, 1992.

[24]   Maassen R, Krill S, Huthmacher K. Verfahren zur Herstellung von 2,3,5-Trimethyl-p-benzochinon. EP 1132367, Degussa-Hüls AG, 2001.

[25]   Kholdeeva O A, Golovin A V, Maksimovskaya R I, Kozhevnikov I V. Oxidation of 2,3,6-trimethylphenol in the presence of molybdovanadophosphoric heteropoly acids. J Mol Catal A: Chem 1992, 75, 235–244.

[26]   Fujibayashi S, Nakayama K, Hamamoto M. An efficient aerobic oxidation of various organic compounds catalyzed by mixed addenda heteropolyoxometalates containing molybdenum and vanadium. J Mol Catal A: Chem 1996, 110, 105–117.

[27]   Vandewalle M-F. Method for Preparing Trimethylbenzoquinone. WO 9818746, Rhone-Poulenc Nutrition Animale SA, 1998.

[28]   Alsters P L, Aubry J-M, Bonrath W, Daguenet C, Hans M, Jary W, Letinois U, Nardello-Rataj V, Netscher T, Parton R, Schütz J, Soolingen J V, Thinge J, Wüstenber B. Selective Oxidation in DSM: Innovative catalysts and technologies. In: Duprez D, Cavani F, eds. Handbook of Advanced Methods and Processes in Oxidation Catalysis, London, Imperial College Press, 2014, Chapter 16, 383–419.

[29]   Bonrath W, Schütz J, Netscher T, Wüstenberg B. Process for Manufacturing TMHQ. WO 2012025587, DSM IP Assets BV, 2012.

[30]   Carril M, Altmann P, Drees M, Bonrath W, Netscher T, Schütz J, Kühn F E. Methyltrioxorhenium-catalyzed oxidation of pseudocumene for vitamin E synthesis: A study of solvent and ligand effects. J Catal 2011, 283, 55–57.

[31]   Lindhorst A C, Schütz J, Netscher T, Bonrath W, Kühn F E. Catalytic oxidation of aromatic hydrocarbons by a molecular iron–NHC complex. Catal Sci Technol 2017, 7, 1902–1911.

[32]   Netscher T, Bonrath W, Bendik I, Zimmermann P-J, Weber F, Rüttimann A. Vitamins, 5. Vitamin K. In: Elvers B, ed. Ullmann's Encyclopedia of Industrial Chemistry, 7th ed. Weinheim, Germany, Wiley-VCH Verlag, 2020, Chapter 5, 71–95.

[33] Hirschmann A, Miller R, Wendler N L. The synthesis of Vitamin K1. J Am Chem Soc 1954, 76, 4592–4594.

[34] Lindlar H. Verfahren zur Herstellung von Kondensationsprodukten. CH 320582, Hoffmann-La Roche AG, 1957.

[35] Isler O, Doebel K. Synthesis of Vitamin K_1 Using Boron Trifluoride Catalysts. US 2683176, Hoffmann-La Roche Inc, 1954.

[36] Puetter H, Bewert W. 2-Methyl-1,4-naphthoquinones. DE 2952709, BASF AG, 1981.

[37] Eremin D V, Petrov L A. Optimization of conditions for preparing vitamin K_3 by oxidation of 2-methylnaphthalene with chromium trioxide in acid solutions. Russ J Appl Chem 2009, 82, 866–870.

[38] Narayanan S, Murthy K V V S B S R, Madhusudan Reddy K, Premchander N. A novel and environmentally benign selective route for Vitamin K_3 synthesis. Appl Catal A: Gen 2002, 228, 161–165.

[39] Kholdeeva O A, Zalomaeva O A, Sorokin A B, Ivanchikova I D, Della P C, Rossi M. New routes to Vitamin K_3. Catal Today 2007, 121, 58–64.

[40] Herrmann W A, Haider J, Fischer R W. Rhenium-catalyzed oxidation of arenes – an improved synthesis of vitamin K_3. J Mol Catal A: Chem 1999, 138, 115–121.

[41] Wienhöfer G, Schröder K, Möller K, Junge K, Beller M. A Novel process for selective ruthenium-catalyzed oxidation of naphthalenes and phenols. Adv Synth Cataly 2010, 352, 1615–1620.

[42] Möller K, Wienhöfer G, Schröder K, Join B, Junge K, Beller M. Selective iron-catalyzed oxidation of phenols and arenes with hydrogen peroxide: synthesis of Vitamin E intermediates and Vitamin K_3. Chem Eur J 2010, 16, 10300–10303.

[43] Anunziata O A, Beltramone A R, Cussa J. Studies of Vitamin K_3 synthesis over Ti-containing mesoporous material. Appl Catal A: Gen 2004, 270, 77–85.

[44] Harrison S, Fiset G, Mahdavi B. Preparation of Quinones by Oxidation of Aromatic Compounds using Electrochemically Prepared Ceric Ion. EP 919533, Hydro-Quebec, 1999.

[45] Monteleone F, Cavani F, Felloni C, Trabace R. A Redox Process for the Production of Menadione and Use of Polyoxometalates as Oxidizing Agents. WO 2004014832, Vanetta SPA, 2004.

[46] Semikolenov V A, Ilyna I I, Simakova I L. Linalool synthesis from α-pinene: kinetic peculiarities of catalytic steps. Appl Catal A: Gen 2001, 211, 91–107.

[47] Sercheli R, Ferreira A L B, Baptistella L H B, Schuchardt U. Transition-metal catalyzed autoxidation of *cis*- and *trans*-Pinane to a mixture of diastereoisomeric pinanols. J Agric Food Chem 1997, 45, 1361–1364.

[48] Yang G-E, Xu L, Huang K Y. Synthesis of pinane hydroperoxide used as an intermediate of linalool. Jingxi Huagong Zhongjianti (Fine Chemical Intermediates) 2004, 34, 39–41.

[49] Fisher G S, Stinson J S, Moore R N. Peroxides from turpentine. Production of technical grade pinane hydroperoxide. Ind Eng Chem 1955, 47, 1368–1373.

[50] Ong P K C, Acree T E. Similarities in the aroma chemistry of gewürztraminer variety wines and lychee (Litchi chinesis Sonn.). Fruit J Agric Food Chem 1999, 47, 665–670.

[51] Ohloff G, Klein E, Schenck G O. Darstellung von Rosenoxiden und anderen Hydropyranderivaten über Photohydroperoxide. Angew Chem 1961, 73, 578.

[52] Alsters P L, Jary W, Nardello-Rataj V, Aubry J-M. "Dark" singlet oxygenation of β-Citronellol: A key step in the manufacture of rose oxide. Org Process Res Dev 2010, 14, 259–262.

[53] Wilke G. Beiträge zur nickelorganischen Chemie. Angew Chem 1988, 100, 189–211.

[54] Keim W. Nickel: Ein Element mit vielfältigen Eigenschaften in der technisch-homogenen Katalyse. Angew Chem 1990, 102, 251–260.

[55] H J-Arpe Industrielle Organische Chemie, 6th ed. WILEY-VCH, Weinheim, 2007, 290–292. für beide; Kugimoto, Junichi; Kawai, Joji; Shimomura, Hideo; Yasumatsu, Ryouta; Ii, Nobuhiro, US 8309714, 2009, Method for production of laurolactam. Ube Industries.

[56] Ollivier J, Drutel D, Elf Atochem S A. Fr, US 6197999, 17 4.2000.

[57]   Taber D F, Straney P J. The synthesis of laurolactam from cyclododecanone via a Beckmann rearrangement. J Chem Edu 2010, 87, 1392–1392.

[58]   Kugimoto J, Kawai J. Process for Producing Laurolactam. US8309714, Ube Industries, 2012.

[59]   Helveling J, Armbruster E, Siegrist W. Process for Preparing 3-methylpiperidine and 3-methylpyridine by Catalytic Cylisation of 2-methyl-1,5-diaminopentane. WO 9422824, Lonza AG, 1994.

[60]   Helveling J. Catalysis at Lonza: From metallic glasses to fine chemicals. Chimia 1996, 50, 114–118.

[61]   Dicosimo R, Burrington J D, Grasselli R K. Ammoxidation of Methyl Substituted Heteroaromatics to Make Heteroaromatic Nitriles. US5028713, Standard Oil Company, 1991.

[62]   Narayana K V, David Raju B, Khaja Masthan S, Venkat Rao V, Kanta Rao P. Reactivity of V_2O_5/MgF_2 catalysts for the selective ammoxidation of 3-Picoline. Catal Lett 2002, 84, 27–30.

[63]   Chuck R. Technology development in nicotinate production. Appl Catal A: Gen 2005, 280, 75–82.

[64]   Bartek J, Robins K, Zigova J. Immobilization of Biocatalyst. WO05040373, Lonza AG, 2005.

[65]   Petersen M, Kiener A. Preparation and functionalization of N-heterocycles. Green Chem 1998, 2, 99–106.

[66]   Rothenberg G. Catalysis: Concepts and Green Applications. Weinheim, Wiley-VCH Verlag GmbH & Co., 2008, 115–116.

[67]   Mestres R, Mestres G. Deltamethrin: Uses and environmental safety. Rev Environ Contam Toxicol 1992, 124, 1–18.

[68]   Claus R E, Schreiber S L. Ozonolytic cleavage of cyclohexene to terminally differentiated products. Org Synth 1986, 64, 150.

[69]   Criegee R. Mechanism of Ozonolysis. Angew Chem Int Ed Engl 1975, 14, 745–752.

[70]   Bailey P S. Ozonation in Organic Chemistry. In: Organic Chemistry, Vol. 39-II, Chapter 2, New York, USA, Academic Press, 1982.

[71]   Cornils B, Lappe P. Dicarboxylic Acids, Aliphatic. In: Ullmann's Encyclopedia of Industrial Chemistry, ed. Elvers. Online ed., Weinheim, Germany, Wiley-VCH Verlag GmbH & Co, 2006.

[72]   David J A, Both S, Christoph R, Fieg G, Steinberner U, Westfechtel A. Fatty Acids. In: Ullmann's Encyclopedia of Industrial Chemistry, ed. Elvers. Online ed., Weinheim, Germany, Wiley-VCH Verlag GmbH & Co, 2006.

[73]   Branan B M, Butcher J T, Olsen L R. Using Ozone in organic chemistry lab: The ozonolysis of eugenol. J Chem Educ 2007, 84, 1979–1981.

[74]   Nobis M, Roberge D M. Mastering ozonolysis: production from laboratory to ton scale in continuous flow. Chemistry Today 2011, 29, 56–58.

[75]   Jary W, Poechlauer P, Ganglberger T. 2003. Process for The Preparation of Aromatic and Heteroaromatic Carboxylic Acids Via the Catalytic Ozonation of The Corresponding Methylated Aromatic and Heteroaromatic Starting Materials. US 7371866, DSM Fine Chemicals Austria NFG GmbH & Co. KG, November 19, 2003.

[76]   Arpe H-J. Industrielle Organische Chemie, 6th ed. Weinheim, Wiley-VCH, 2007, 438–439.

[77]   Szmant H H. Organic Building Blocks of the Chemical Industry. John Wiley & Sons, 1989, 480.

[78]   Fumagalli C, Capitanio L, Stefani G. Process for the preparation of trimellitic acid. US 5250724, Alusuisse Italia Spa, October 10, 1993.

[79]   Jira R. Acetaldehyd aus Ethylen – ein Rückblick auf die Entdeckung des Wacker-Verfahrens. Angew Chem 2009, 121, 9196–9199.

[80]   Eckert M, Fleischmann G, Jira R, Bolt H M, Golka K. Acetaldehyde. In: Ullmann's Encyclopedia of Industrial Chemistry, ed. Elvers. Online ed, Weinheim, Germany, Wiley-VCH Verlag GmbH & Co, 2012.

[81]   Keith J A, Henry P M. The mechanism of the wacker reaction: A tale of two hydroxypalladations. Angew Chem Int Ed Engl 2009, 48, 9038–9049.

[82] Espeel P H, de Peuter G, Tielen M C, Jacobs P A. Mechanism of the wacker oxidation of alkenes over Cu-Pd-Exchanged Y Zeolites. J Phys Chem 1994, 98, 11588–11596.

[83] Evnin A B, Rabo J A, Kasai P H. Heterogeneously catalyzed vapor-phase oxidation of ethylene to acetaldehyde. J Catal 1973, 30, 109–117.

[84] Tsuji J, Nagashima H, Nemoto H A. General synthetic method for the preparation of methyl ketones from terminal olefins: 2-Decanone. Org Synth 1984, 62, 9.

[85] Woell J B. Processes for the Conversion of Myrcene to Citral. EP 439368, Union Camp Corp., 1991.

[86] Wang Z-X, Tu Y, Frohn M, Zhang J-R, Shi Y. An efficient catalytic asymmetric epoxidation method. J Am Chem Soc 1997, 119, 11224–11235.

[87] Frohn M, Shi Y. Chiral Ketone-Catalyzed asymmetric epoxidation of olefins. Synthesis 2000, 14, 1979–2000.

[88] Ager D J, Anderson K, Oblinger E, Shi Y, van der Roest J. An epoxidation approach to a chiral lactone: application of the Shi epoxidation. Org Process Res Dev 2007, 11, 44–51.

[89] Huff J R. HIV protease: a novel chemotherapeutic target for AIDS. J Med Chem 1991, 34, 2305–2314.

[90] Greenlee W J. Renin inhibitors. Med Res Rev 1990, 10, 173–236.

[91] https://www.nobelprize.org/prizes/medicine/2015/press-release/, (26.03.2020).

[92] Turconi J, Griolet F, Guevel R, Oddon G, Villa R, Geatti A, Hvala M, Rossen K, Göller R, Burgard A. Semisynthetic artemisinin, the chemical path to industrial production. Org Process Res Dev 2014, 18, 417–422.

[93] https://www.sanofi.com/en/media-room/press-releases/2014/2014-08-12-13-00-00, (26.03.2020).

[94] Paddon C J, Westfall P J, Pitera D J, Benjamin K, Fisher K, McPhee D, Leavell M D, Tai A, Main A, Eng D, Polichuk D R, Teoh K H, Reed D W, Treynor T, Lenihan J, Fleck M, Bajad S, Dang G, Dengrove D, Diola D, Dorin G, Ellens K W, Fickes S, Galazzo J, Gaucher S P, Geistlinger T, Henry R, Hepp M, Horning T, Iqbal T, Jiang H, Kizer L, Lieu B, Melis D, Moss N, Regentin R, Secrest S, Tsuruta H, Vazquez R, Westblade L F, Xu L, Yu M, Zhang Y, Zhao L, Lievense J, Covello P S, Keasling J D, Reiling K K, Renninger N S, Newman J D. High-level semi-synthetic production of the potent antimalarial artemisinin. Nature 2013, 496, 528–532.

[95] Moretto -H-H, Schulze M, Wagner G. Silicones. In: Ullmann's Encyclopedia of Industrial Chemistry, Weinheim, Germany, Wiley-VCH Verlag, 2005.

[96] Büchel K H, Moretto H H, Woditsch P. Industrial Inorganic Chemistry, 2nd ed. Weinheim, Wiley-VCH, 2000, 296–299.

[97] Büchel K H, Moretto H H, Woditsch P. Industrial Inorganic Chemistry, 2nd ed. Weinheim, Wiley-VCH, 2000, 300.

[98] Büchel K H, Moretto H H, Woditsch P. Industrial Inorganic Chemistry, 2nd ed. Weinheim, Wiley-VCH, 2000, 301.

5 Gas-phase reactions

The fundamental difference between gas-phase reactions and reactions in the liquid phase is the states of aggregation of the substrates. The reactants in gas-phase reactions are typically in the gas phase whereas the catalyst is in the solid phase. This set-up results in some advantages and some disadvantages in the reaction outcome, for example mass-transfer limitations between catalyst and reactants/product may significantly influence the rate of reactions (see Chapter 1: Introduction and fundamental aspects). The diffusion of gases is several magnitudes higher than that of liquids and the movement of particles in the gas phase is significantly faster than in the liquid phase. This behaviour results in lower mass-transfer effects in gaseous reactions versus liquid-phase reactions, reducing disadvantageous mass-transfer limitations in gas-phase reactions. Furthermore, higher selectivity in gas-phase reactions is usually observed compared to reactions in the liquid phase. In many cases, reactions in the gas phase, in contrast to most of the liquid phase processes, are run neat; without the use of an additional solvent which can simplify the subsequent reaction work-up and reduce the amount of waste. Usually, lower pressures can be applied, and the pressure drop over a catalyst bed is smaller.

However, one disadvantage can be the higher reaction temperature required to get the substrate into the gas phase, because the higher temperatures may favour side reactions and/or decomposition. Another drawback with gas-phase reactions is the gas volumes to be transported are higher and require reactors with larger catalyst volumes because of the usually lower pressure in gas-phase reactions compared to liquid-phase reactions. Typical applications for gas-phase reactions are dehydrations, eliminations, cyclisations or rearrangements. Various types of gas-phase reactions can be found in refineries and crackers. High-boiling compounds are converted to more valuable low-boiling compounds during thermal and catalytic cracking. Hydrocracking is a further method usually carried out in the gas phase; hydrogen is added to convert compounds in a reducing environment. Additionally, isomerisations (e.g. *n*-alkanes to *iso*-alkanes in the presence of a catalyst), alkylations (transforming very low-boiling compounds to heavier compounds) or reforming reactions (converting substrates into aromatic compounds) are performed in the gas phase in refineries on a bulk scale.

In the polymer industry, there are also gas-phase processes. The monomers used are often low-boiling compounds that can be vaporised with relatively low energy intake. An example is vinyl acetate, which is a precursor for polyvinyl acetate and polyvinyl alcohol. It is produced on a scale of several million tons per year. It can be prepared by the addition of acetic acid to acetylene. Alternatively, ethylene can be converted to vinyl acetate in the presence of acetic acid, oxygen and a solid catalyst [1]. In this gas-phase process (175–200 °C and 5–10 bar), ethylene is oxidised to acetic acid which reacts with another equivalent of ethylene in the presence of oxygen to form vinyl acetate and water.

https://doi.org/10.1515/9783111102672-005

An example of a gas-phase reaction in the synthesis of fine chemicals is the oxida-tion of 3-picoline (Scheme 5.1). The product, nicotinic acid is also, together with nicotin-amide, called niacin or vitamin B$_3$. A deficiency in niacin causes the disease pellagra, which affects the skin, the gastrointestinal tract and the central nervous system. There are several processes to produce nicotinic acid [2]. The standard liquid-phase process starts from 5-ethyl-2-methypyridine. Oxidation with nitric acid at 230–270 °C and 60–80 bar yields nicotinic acid (Scheme 5.2). The gas-phase process starting from 3-picoline uses air as an oxidant instead of a stoichiometric amount of nitric acid (Scheme 5.1). The process is much more sustainable since this oxidant results in lower amounts of by-products and is less corrosive. A fixed-bed catalyst (vanadium oxide) is used – thus no separation of the catalyst is required, the reaction is run at ambient pressure, and is highly selective, generating only low amounts of by-products (Scheme 5.3) [3].

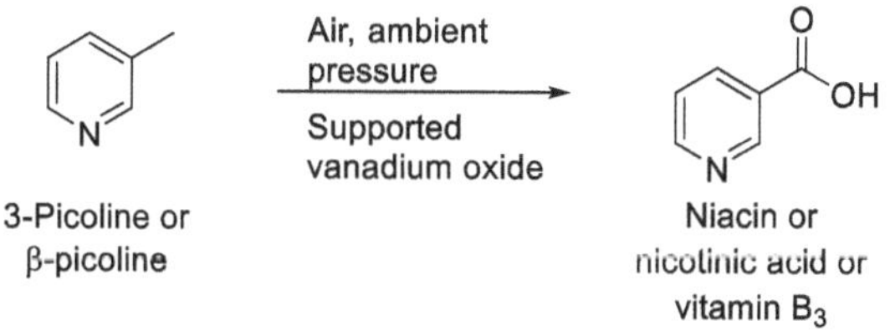

3-Picoline or β-picoline

Air, ambient pressure
Supported vanadium oxide

Niacin or nicotinic acid or vitamin B$_3$

Scheme 5.1: Synthesis of vitamin B$_3$ via gas-phase oxidation of 3-picoline.

5-Ethyl-2-methylpyridine

+ 6 HNO$_3$

230-270 °C
60-80 bar

Niacin or nicotinic acid or vitamin B$_3$

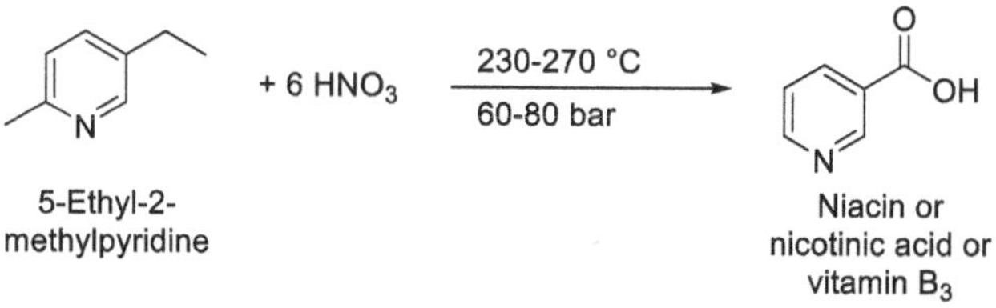

Scheme 5.2: Liquid-phase synthesis of vitamin B$_3$.

Nicotinamide can also be produced from 3-picoline via the corresponding nitrile (Scheme 5.4). In the first step, 3-picoline is converted into 3-cyanopyridine via am-moxidation in the gas phase. A variety of catalysts have been studied for this reac-tion [4–9] – almost all catalysts contain vanadium; vanadium oxide has proven to be the most effective metal oxide for ammoxidation (and oxidation reactions) in the gas phase. Variation of the catalyst is mostly limited to optimisation of the composi-tion, in particular, of the used promoters. These are usually based on alkaline metals and first-row transition metals [10]. 3-Picoline is subsequently hydrolysed to nicotin-amide either chemically under basic conditions or enzymatically applying the bacte-rium *Rhodococcus* [5, 11]. The main advantage of the bioconversion process over the chemical pathway is the higher selectivity. No nicotinic acid is formed when *Rhodo-coccus* is applied (see Chapter 10: Biocatalysis) [12].

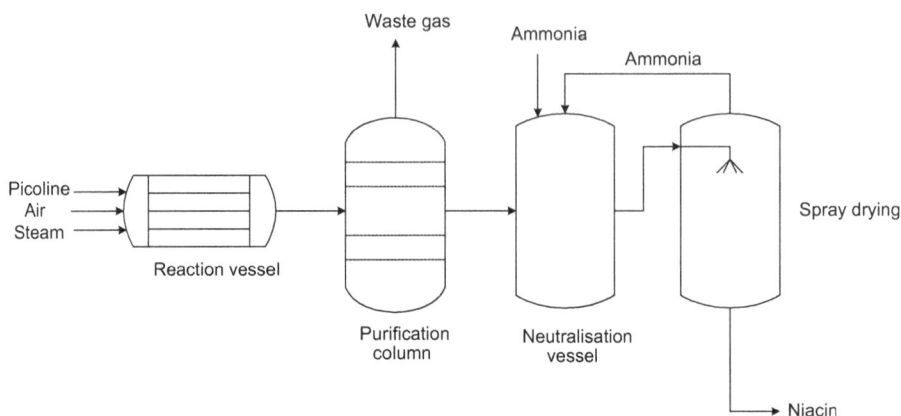

Scheme 5.3: Simplified reaction set-up for the oxidation of 3-picoline.

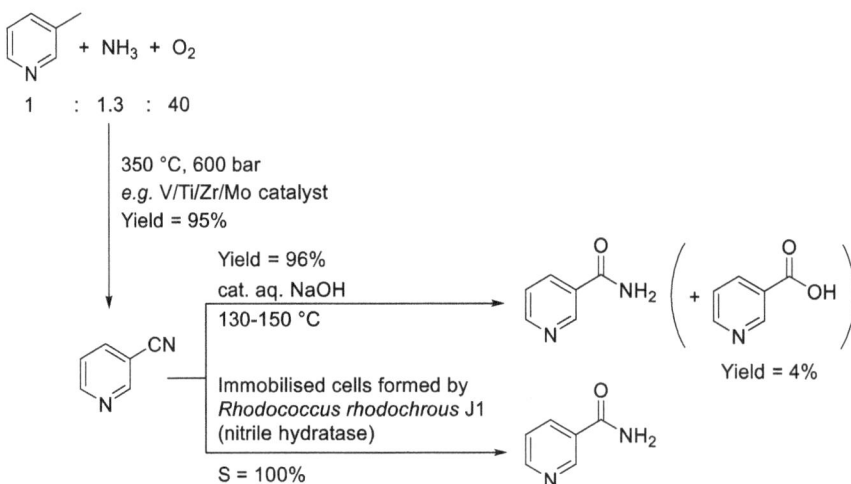

Scheme 5.4: Synthesis of nicotinamide via 3-cyanopyridine.

Besides the use of 3-picoline as the starting material for the synthesis of vitamin B$_3$, it is also used as a starting material for pharmaceuticals and agrochemicals. In most cases, 3-picoline is produced together with pyridine. The synthesis of methyl-pyridines and pyridine is very similar and often a mixture of products are formed which are separated during downstream processing [13–18]; most processes are gas-phase reactions. Mixtures of aldehydes are reacted with ammonia (*Chichibabin* reaction, also written as Tchitchi-babin, Чичибабин) to form pyridines (Scheme 5.5). For example, acetaldehyde, formaldehyde and ammonia react in the gas phase to a mixture of pyridine and 3-picoline. In the first step, acetaldehyde and formaldehyde can react to acrolein. By choosing the appropriate aldehyde and catalyst, the composition of the mixture can be varied, depending on the desired product. The reaction is usually carried out at 350–550 °C and a space velocity of 500–1,000 h^{-1} in the presence of a solid catalyst (e.g. modified SiO_2/Al_2O_3) [19]. Typically, a mixture of 3-picoline and pyridine is obtained. The economically larger product, pyridine, therefore determines the availability and price of 3-picoline.

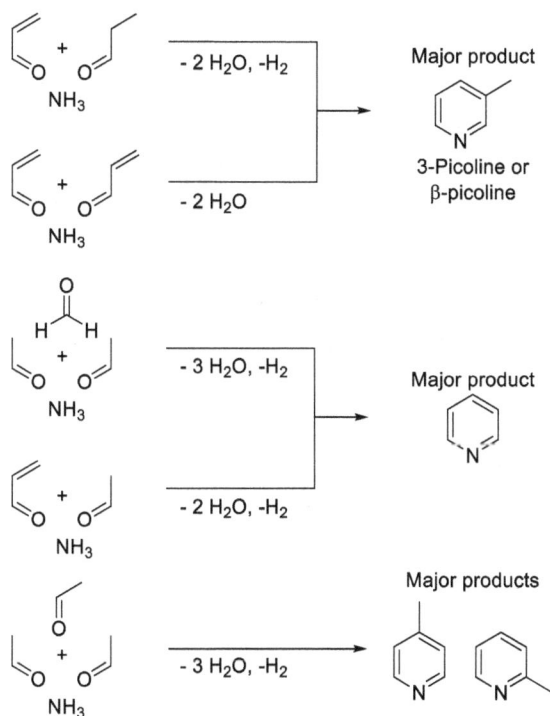

Scheme 5.5: Synthesis of pyridine and methyl-pyridines by a *Chichibabin* pyridine reaction in the gas phase.

In order to separate the 3-picoline from the pyridine market, alternative feedstocks and production processes for picoline have been evaluated. A viable alternative is the cyclisation of 2-methyl-1,5-diaminopentane to 3-methyl piperidine, followed by dehydrogenation to 3-picoline (Scheme 5.6). The cyclisation is carried out in the gas phase at 300–400 °C in the presence of a zeolite catalyst in a fixed-bed reactor [20]. The subsequent gas-phase dehydrogenation is achieved in the presence of a Pd/Al$_2$O$_3$ catalyst at ca. 300 °C [21]. This pathway has the disadvantage of dependency on another product; 2-methyl-1,5-diaminopentane is produced from methylglutaronitrile, a by-product of adiponitrile production (used for nylon 6,6). However, 3-picoline produced via this pathway has a very high isomeric purity and methylglutaronitrile is a readily available and inexpensive starting material.

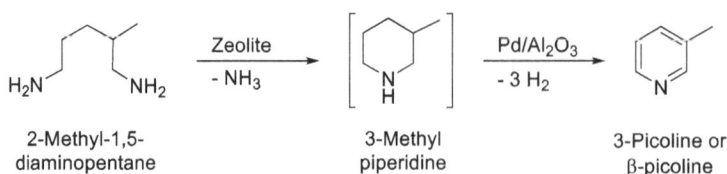

Scheme 5.6: Conversion of 2-methyl-1,5-diaminopentane to 3-picoline via two gas-phase reaction steps.

Another aromatic compound prepared in the gas phase in the fine chemical industry is vanillin (Scheme 5.7). The compound is of great importance in the flavour industry (see Chapter 8: Acid–base-catalysed reactions). The chemical production processes of vanillin start from catechol via 2-methoxyphenol (guaiacol) (see Chapter 4: Oxidations). The gas-phase methylation of catechol with methanol in the presence of a rare-earth orthophosphate catalyst at 500–650 K yields guaiacol. Various rare-earth metals were studied, for example lanthanum, cerium or samarium, optionally doped with an alkali or earth alkali metal [22]. The previous synthesis was a two-step pathway. First, catechol was converted into its corresponding sodium salt. Afterwards, a methylating agent was added (e.g. methyl chloride) in a polar anhydrous solvent. The reaction yields are high for both processes. However, the gas-phase process generates fewer by-products (one equivalent of water vs. one to two equivalents of sodium chloride) [23].

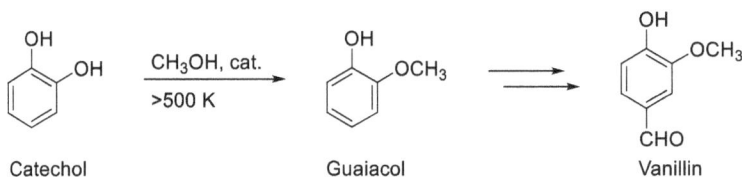

Scheme 5.7: Synthesis of vanillin via gas-phase methylation of catechol.

Methylated phenols, such as dimethylphenols (xylenols) and methylphenols (cresols) are precursors for a variety of products (Figure 5.1). For instance, o-cresol is used as a precursor for herbicides (e.g. 4,6-dinitro-o-cresol, 4-chloro-2-methylphenoxyacetic, γ-(4-chloro-2-methyl-phenoxy)butyric acid), dyes and fragrances (e.g. carvacrol); m-cresol as a precursor of 2,3,6-trimethylphenol (intermediate in the vitamin E synthesis, see Chapter 4: Oxidations) and (−)-menthol (see Chapter 2: Heterogeneous hydrogenations and Chapter 3: Homogeneous hydrogenations), for insecticides (e.g. the intermediate m-phenoxybenzaldehyde), and fragrances (e.g. thymol); p-cresol as a precursor for an anti-oxidant (2,6-di-tert-butyl-p-cresol (BHT, see Chapter 8: Acid–base-catalysed reactions)), fragrances (e.g. anisaldehyde, raspberry ketone), and together with m-cresol in dyes and fire-resistant hydraulic fluids. Xylenols are used as precursors for resins (especially 2,6- and 3,5-xylenol), antioxidants (mixtures of 2,4- and 2,5-xylenol), pesticides and herbicides (mainly 2,6- and 3,5-xylenol) and as a precursor of 2,3,5-trimethylphenol (inter-mediate in the vitamin E synthesis) [24].

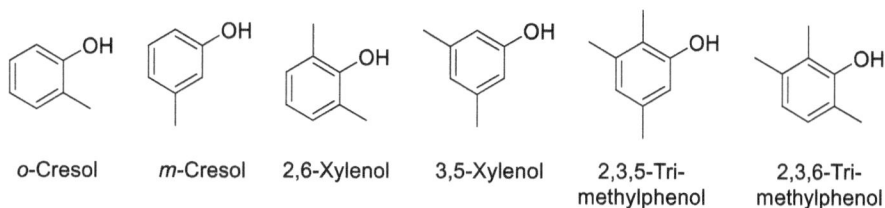

o-Cresol   *m*-Cresol   2,6-Xylenol   3,5-Xylenol   2,3,5-Tri-methylphenol   2,3,6-Tri-methylphenol

Figure 5.1: Methyl-phenol derivatives.

The production of cresols and xylenols, especially o-cresol, and 2,6-xylenol is carried out by gas-phase methylation of phenol with methanol in the presence of a catalyst. At atmospheric or slightly elevated pressure, a mixture of methanol, phenol and water is passed over a catalyst in a multi-tubular fixed-bed reactor at temperatures of 300–500 °C. The reaction temperature required depends on the nature of the catalyst and on the desired composition of the product. The ratio of methanol to phenol also influences the composition of products. Typical catalysts applied are based on magnesium oxide [25–27], iron oxide [28, 29] or γ-alumina [30]. The catalyst itself determines the reaction pathway of the methylation. Phenol and methanol adsorb on basic sites via the hydroxy-groups and thus coordinate in a vertical position to the catalyst. *Ortho* methylation is the favoured reaction pathway with minor amount of *para* methylated products (Figure 5.2). In contrast, the aromatic ring of phenol coordinates to acidic sites in a parallel manner, favouring methylation on all positions, including o-methylation (*meta* methylation is the least likely, due to electronic effects) [31, 32]. The downstream processing of the phenol methylation is relatively costly because of the necessary extensive distillation due to the similar boiling points of various methylated phenols which are formed.

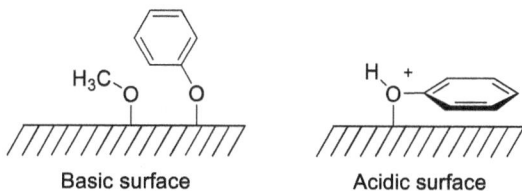

Basic surface   Acidic surface

Figure 5.2: Coordination of phenol on solid catalysts.

The conversion of phenol to aniline was developed by the company Halcon. Aniline is used in a variety of products, the most important one being the building block 4,4,-methylene-di-*para*-phenylene isocyanate, a precursor for polyurethanes. In the production process, nitrobenzene is hydrogenated to aniline (Scheme 5.8) both in the liquid (fixed-bed) and in the gas phase (fluidised bed). In the latter process, supported copper or palladium catalysts are applied. Various modifiers, such as Pb, V, P and Cr, are added to achieve high selectivity. The reaction temperature is about 230–300 °C (4–10 bar) [33].

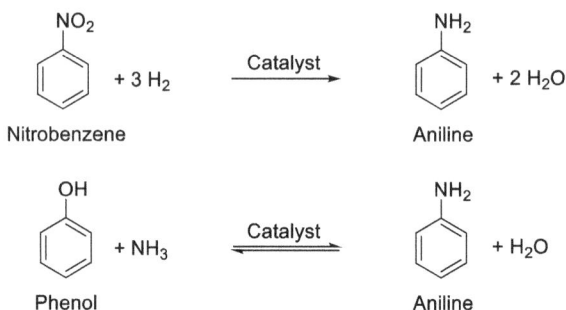

Scheme 5.8: Preparation of aniline.

In an alternative process, phenol is converted into aniline in the gas phase in the presence of ammonia by reductive amination over oxidic catalysts containing Ce, V or Ti on silica–alumina [34–37]. Because the reaction is reversible, a large excess of ammonia is applied to reach high conversions. The reaction is slightly exothermic; thus the reaction temperature (370–425 °C, 200 bar) is kept relatively low in order to shift the equilibrium in the direction of aniline. The yield based on phenol is >98%.

Anthraquinone is the starting compound in the production of a variety of dyes. The production capacity worldwide is around 35,000 t/a [38]. Anthraquinone can be produced by the oxidation of anthracene with chromic acid or by a gas-phase oxidation of anthracene with air in the presence of a catalyst. To circumvent shortages in anthracene supply, a process starting from naphthalene was developed by Kawasaki Kasei Chemicals. In the gas phase oxidation of anthracene with an air/water mixture, a similar catalyst to those used for the oxidation of naphthalene or o-xylene to phthalicanhydride are applied (Scheme 5.9). The catalysts are based on vanadium, mainly iron vanadate or vanadic acid doped with sub-stoichiometric amounts of alkali metal ions [39]. The conversion of anthracene is almost quantitative at temperatures above 300 °C, and the main by-product is phthalic anhydride, which can be easily separated. After a certain time, the activity of the catalyst decreases, which results in a lower yield of anthraquinone and in increased amounts of phthalic anhydride. To prevent the deactivation of the catalyst, a small amount of ammonia is added.

The production process of anthraquinone starting from naphthalene consists of three steps. Naphthalene is oxidised in the gas phase with air in the presence of a vanadium pentoxide catalyst to produce the relatively unstable naphthoquinone; large amounts of phthalic anhydride form at the same time. The products can be sepa-

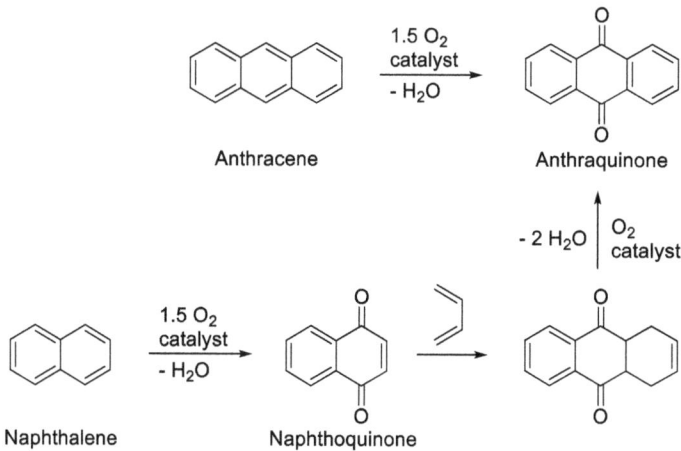

Scheme 5.9: Anthraquinone syntheses.

rated by extraction [40–42]. In the second step, the naphthoquinone is reacted with butadiene in a *Diels–Alder* reaction to give 1,4,4 a,10 a tetrahydroanthraquinone, followed by oxidation with air to anthraquinone.

An important process performed in the gas phase is the manufacture of cyclohexane by benzene hydrogenation. This process is more efficient compared to the cyclohexane distillation from mineral oil fractions. The benzene hydrogenation is carried out in a temperature range of 170 °C to 230 °C, preferably lower than 220 °C at 30 to 50 bar (Scheme 5.10). Reaction temperatures above 220 °C increase the content of benzene and by-products are formed ($\Delta H = -216$ kJ/mol). The formation of by-products, for example methylcyclopentane, starts at around 230 °C. The hydrogenation is performed on industrial scale in the presence of Ni, Pd or Pt catalysts in a fixed-bed reactor system (Scheme 5.11). The sulphur content must be <1 ppm to avoid catalyst deactivation [43].

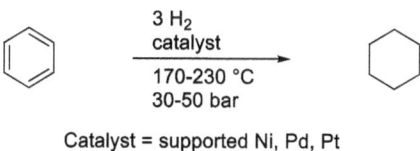

Catalyst = supported Ni, Pd, Pt

Scheme 5.10: Hydrogenation of benzene to cyclohexane.

The worldwide production capacity of cyclohexane was around 5.5×10^6 t/a in 1999 [43]. It is used as a starting material in the manufacture of caprolactam and adipic acid. An alternative liquid-phase benzene hydrogenation process developed by the Institute Francais du Petrole (IFP) is carried out in the presence of a Ni-alloy catalyst at >99% yield. In similar set-ups of liquid-phase hydrogenation reactions, cyclooctane and cyclododecane are produced.

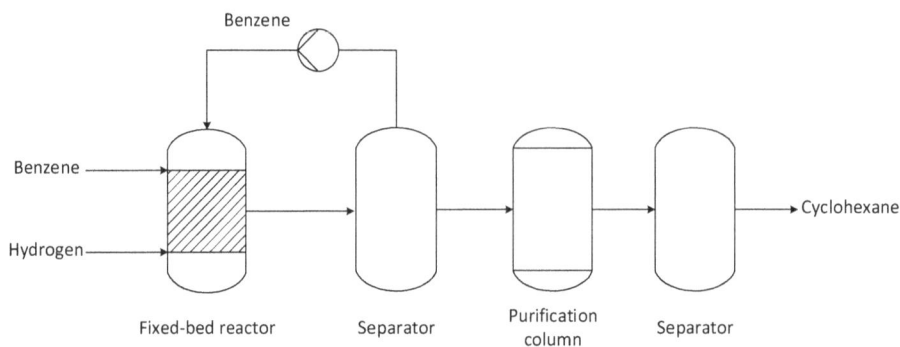

Scheme 5.11: Simplified reaction set-up for benzene hydrogenation.

ε-Caprolactam is an important intermediate in the production of polyamide 6 (Nylon 6). The world production of ε-caprolactam was estimated to be around 3.8 million tons per year in 2006 [44]. It is commonly produced by a liquid-phase *Beckmann* rearrangement (see Chapter 7: Rearrangement reactions and Chapter 4: Oxidations) of cyclohexanone oxime with oleum or sulphuric acid as catalyst (Scheme 5.12). Disadvantages of this process are the formation of large amounts of ammonium sulphate as a side product; depending on the applied process, 1.6–4.4 tons ammonium sulphate are formed per ton of ε-caprolactam. Sumitomo Chemicals Co. Ltd. was the first company that used a heterogeneously catalysed gas-phase process in 2003 on about 60 kt/a scale, applying a zeolite catalyst in a fluidised bed reactor system [44]. Research on the catalyst revealed more attractive catalysts which are lower in cost and can be applied in fixed-bed reactors, which are generally less expensive to install and operate [45]. Niobium-doped zeolites and amorphous NbO_x/SiO_2 catalysts were successfully tested with yields comparable with the catalyst applied by Sumitomo Chemicals Co. Ltd [45–49].

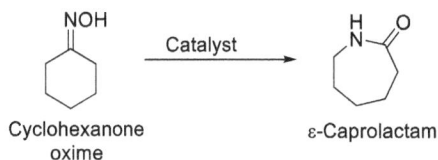

Cyclohexanone
oxime

ε-Caprolactam

Scheme 5.12: *Beckmann* rearrangement to ε-caprolactam.

Several aroma compounds are produced using gas-phase processes, for example citral, a mixture of geranial ((*E*)-3,7-dimethyl-2,6-octadienal) and neral ((*Z*)-3,7-dimethyl-2,6-octadienal), which is an important precursor of *L*-menthol and vitamin A acetate. Furthermore, citral is used as a precursor in the fragrance industry for products in perfumes, soaps and detergents. Citral can be produced from isoprenal and prenol in a hetero *Claisen* rearrangement of an intermediate allyl vinyl ether (Scheme 5.13). Isoprenal and prenol are both manufactured from isoprenol which itself is produced from isobutene and formaldehyde. Isoprenol can be isomerised to prenol or oxidised to iso-prenal (see Chapter 4: Oxidations). The oxidation is carried out in the gas phase at 340–380 °C in the presence of a fixed-bed catalyst (metallic Ag on an inert support). The reaction is run at partial conversion and high selectivity [50, 51].

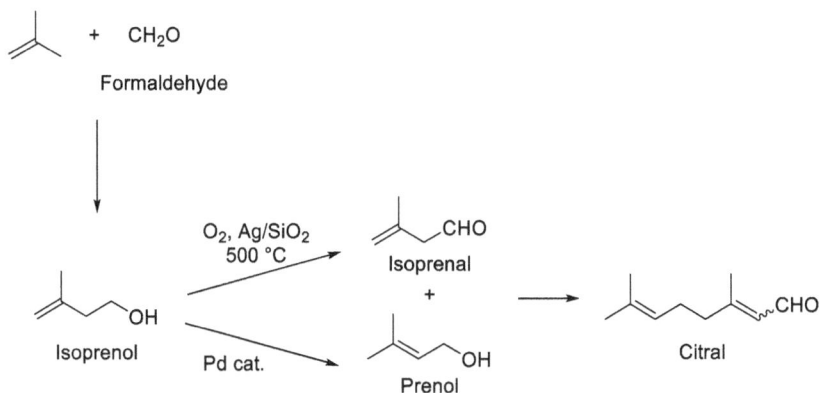

Scheme 5.13: Synthesis of citral, including the gas-phase oxidation of isoprenol.

Linalool ((S)- or (R)-3,7-dimethylocta-1,6-dien-3-ol) can be used as a key intermediate in the synthesis of vitamins K and E [52]. Additionally, it is used as an aroma ingredient itself and also as a starting material in the syntheses of flavour and fragrance com-pounds. α-Pinene is one of the preferred starting materials, as it is a rather inexpensive by-product from the pulp and paper industry [53, 54]. After hydrogenation of α-pinene, *cis*- and *trans*-pinane are obtained (Scheme 5.14), which is followed by oxidation to *cis*-

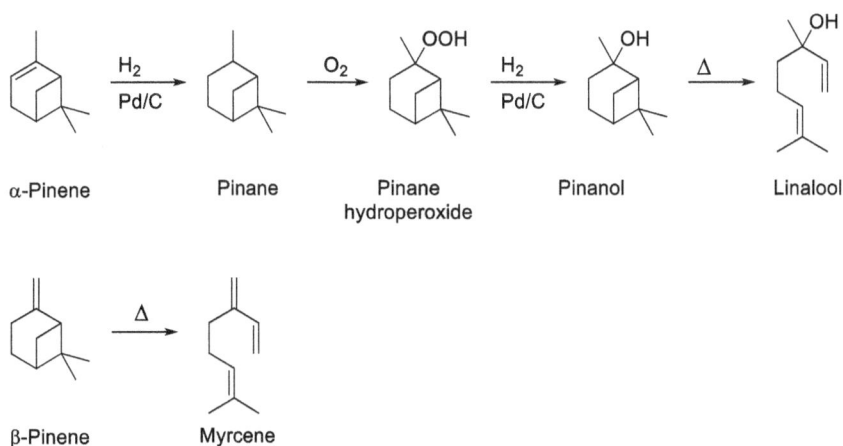

Scheme 5.14: Pinenes as starting materials for linalool and myrcene.

and *trans*-pinane-2-hydroperoxide. Subsequent reduction results in 2-pinanol which is converted to linalool via a gas-phase pyrolysis at 450–600 °C [55–57].

Myrcene is the starting material for a range of industrially important products, for example geraniol, nerol, linalool and isophytol. These compounds are used as reagents in the synthesis of flavour and fragrance compounds and in the synthesis of vitamins A and E and carotenoids [58]. Myrcene is industrially produced by pyrolysis of β-pinene at 400–600 °C, which is obtained with α-pinene from the paper industry (Scheme 5.14) [59, 60]. The reaction is usually carried out over solid catalysts such as copper chromite [61] or amorphous zirconium phosphate [62] or without any catalyst [63].

References

[1]   Schwerdtel W. Vinylacetat auf Basis Äthylen in der Gasphase. Chem Ing Technik 1968, 16, 781–784.
[2]   Blum R. Vitamins, 8. Vitamin B_3 (Niacin). In: Elvers, ed. Ullmann's Encyclopedia of Industrial Chemistry, Online ed. Weinheim, Germany, Wiley-VCH Verlag GmbH & Co, 2020.
[3]   Chuck R. A catalytic green process for the production of niacin. Chimia 2000, 54, 508–513.
[4]   Sembaev D C, Ivanovskaya F A, Guseinov E M, Chuck R J. Catalytic Composition for the Oxidative Ammonolysis of Alkylpyridines. WO9532055, Lonza Ltd., 1995.
[5]   Heveling J, Armbruster E, Utiger L, Rohner M, Dettwiler H-R, Chuck R J. Verfahren zur Herstellung von Nikotonsäureamid. EP770687, Lonza AG, 1997.
[6]   Lukas V H, Neher A, Arntz D. Verfahren zur Herstellung von Cyanopyridinen und dafür geeignete Katalysatoren. EP726092, Degussa AG, 1996.
[7]   Beschke H, Friedrich H, Heilos J. Katalysatoren für die Herstellung von 3-Cyanpyridin. EP0059414, Degussa AG, 1982.
[8]   Dicosimo R, Burrington J D, Grasselli R K. Ammoxidation of Methyl Substituted Heteroaromatics to Make Heteroaromatic Nitriles. US5028713, Standard Oil Company, 1991.
[9]   Narayana K V, David Raju B, Khaja Masthan S, Venkat Rao V, Kanta Rao P. Reactivity of V_2O_5/MgF_2 catalysts for the selective ammoxidation of 3-picoline. Catal Lett 2002, 84, 27–30.
[10] Chuck R. Technology development in nicotinate production. Appl Catal A: Gen 2005, 280, 75–82.
[11] Bartek J, Robins K, Zigova J. Immobilization of Biocatalyst. WO05040373, Lonza AG, 2005.
[12] Petersen M, Kiener A. Preparation and functionalization of N-heterocycles. Green Chem 1998, 2, 99–106.
[13] Watanabe Y, Takenaka S, Koyasu K. α- and γ-Picoline. JP 46039873, Nippon Kayaku Co., Ltd., 1971.
[14] Minato Y, Nishinomiya H, Niwa T. Production of Pyridine Bases and Catalyst Therefor. DE2054773, Koei Chemical Cy., Ltd., 1970.
[15] Minato Y, Yasuda S. A Process for Producing Pyridine Bases. DE 1770870, Koei Chemical Co., Ltd., 1968.
[16] Pyridine Production Catalyst. JP 56026546, Daicel Chemical Industries, Ltd., 1981.
[17] Swift G. Catalytic Gas-Phase Production of Pyridine and 3-methylpyridine from Acrolein and Ammonia. DE 1917037, Imperial Chemical Industries Ltd., 1969.
[18] Beschke H, Friedrich H. Acrolein in the gas phase synthesis of pyridine derivatives. Chem Ztg 1977, 101, 377–384.
[19] Shimizu S, Watanabe N, Kataoka T, Shoji T, Abe N, Morishita S, Ichimura H. Pyridine and Pyridine Derivatives. In: Elvers, ed. Ullmann's Encyclopedia of Industrial Chemistry, Online ed. Weinheim, Germany, Wiley-VCH Verlag GmbH & Co, 2000.

[20] Heveling J, Armbruster E, Siegrist W. Process for Preparing 3-methylpiperidine and 3-methylpyridine by Catalytic Cyclisation of 2-methyl-1,5-diaminopentane. WO 9422824, Lonza AG, 1994.

[21] Pianzola D, Siegrist W. Catalysts for the Preparation of Methylpyridine. WO 11045014, Lonza Ltd, 2011.

[22] Gilbert L, Marcelle J, Le Govic A-M, Tirel P-J. O-Alkylation of Phenolic Compounds via Rare Earth Orthophosphate Catalysts. US 5786520, Rhone-Poulenc Chimie, 1998.

[23] Ratton S. Heterogeneous catalysis in the fine chemicals industry: from dream to reality. Chim Oggi 1998, 16, 33–37.

[24] Fiege H. Cresols and Xylenols. In: Elvers, ed. Ullmann's Encyclopedia of Industrial Chemistry, Online ed. Weinheim, Germany, Wiley-VCH Verlag GmbH & Co, 2012.

[25] Warner G L, Caruso A J. Process for the Alkylation of Hydroxyaromatic Compounds. EP 785180, General Electric Company, 1997.

[26] Warner G L. Method of Ortho-Alkylating Phenol. US 4933509, General Electric Company, 1990.

[27] Lenz -H-H, Roske E, Bär K. Process and Catalyst for Gas Phase Alkylation of Phenol. DE 3524331, BASF AG, 1987.

[28] Sigg R. Process for the ortho-methylation of phenolic compounds. Chemische Werke Hüls AG, DE 3149022, 1983.

[29] Wrzyszcz J, Grabowska H, Kaczmarczyk W. Catalyst for alkylation of phenol with methanol to ortho-cresol or 2,6-xylenol. Instytut Niskich Temperatur I BadaN Strukturalnych Polskiej Akademii Nauk, PL 272062, 1992.

[30] Alscher A, Colling G. A Process for ortho-alkylation of Phenols. DE 2756461, Rütgerswerke AG, 1979.

[31] Ballarini N, Cavani F, Maselli L, Passeri S, Rovinetti S. Mechanistic studies of the role of formaldehyde in the gas-phase methylation of phenol. J Catal 2008, 256, 215–225.

[32] Tabanelli T, Cocchi S, Gumina B, Izzo L, Mella M, Passeri S, Cavani F, Lucarelli C, Schütz J, Bonrath W, Netscher T. Mg/Ga mixed-oxide catalysts for phenol methylation: Outstanding performance in 2,4,6-trimethylphenol synthesis with co-feeding of water. Appl Catal A, Gen 2018, 552, 86–97.

[33] Kahl T, Schröder K-W, Lawrence F R, Marshall W J, Höke H, Jäckh R. Aniline. In: Elvers, ed. Ullmann's Encyclopedia of Industrial Chemistry, Online ed. Weinheim, Germany, Wiley-VCH Verlag GmbH & Co, 2012.

[34] Unnikrishnan P, Srinivas D. Chapter 3 – Heterogeneous Catalysis. In: Joshi S S, Ranade V V, eds. Industrial Catalytic Processes for Fine and Specialty Chemicals, Elsevier, 2016.

[35] Becker M, Khoobiar S. Process for the Production of Organic Amines. US 3860650, Halcon International, 1969.

[36] Russell J L. Process for Preparing Aniline from Phenol. US 3578714, Halcon International, Inc., 1971.

[37] Cornils B Mitsui Petrochem Aniline Process. In: Cornils B, Hermann W A, Wong C-H, Zanthoff H-W, eds. Catalysis from A to Z, Weinheim, Germany, Wiley-VCH Verlag GmbH & Co, 2013.

[38] Vogel A. Anthraquinone. In: Elvers, ed. Ullmann's Encyclopedia of Industrial Chemistry, Online ed. Weinheim, Germany, Wiley-VCH Verlag GmbH & Co, 2012.

[39] Wettstein W. Vanadium Catalyst. DE 1016694, Ciba, 1954.

[40] Matsuura R, Komatsu T, Nomiyama Y, Takahashi H. Substituted tetrahydroanthraquinones. JP 52051356, Kawasaki Kasei Chemicals, 1975.

[41] Process for Purifying Naphthoquinone. GB 2039897, Kawasaki Kasei Chemicals, 1978.

[42] Matsuura R, Kawano K, Matsuzaki K, Narita M. Separation of Naphthoquinone and Phthalic Acid from the Air-Oxidation Products of Naphthalen. JP 54122246, Kawasaki Kasei Chemicals, 1979.

[43] Arpe H-J. Industrielle Organische Chemie, 6th ed. Wiley-VCH, 2007, 384.

[44] Izumi Y, Ichihashi H, Shimazu Y, Sato H. Development and industrialization of the vapor-phase beckmann rearrangement process. Bull Chem Soc Jpn 2007. 80, 7, 1280–1287.

[45] Maronna M M, Kruissink F C, Parton R F, Tinge J T, Hoelderich W F. Preparation of NbOₓ/SiO₂-coated structured packing material and it's catalytic performance in the gas-phase Beckmann rearrangement of cyclohexanone oxime to ε-caprolactame. Cat Commun 2017, 90, 23–26.
[46] Anilkumar A, Hoelderich W. Highly active and selective Nb modified MCM-41 catalysts for Beckmann rearrangement of cyclohexanone oxime to ε-caprolactam. J Catal 2008, 260, 17–29.
[47] Anilkumar A, Hoelderich W. New non-zeolitic Nb-based catalysts for the gas-phase Beckmann rearrangement of cyclohexanone oxime to caprolactam. J Catal 2012, 293, 76–84.
[48] Anilkumar A, Hoelderich W. Gas phase Beckmann rearrangement of cyclohexanone oxime to ε-caprolactam over mesoporous, microporous and amorphous Nb₂O₅/silica catalysts: A comparative study. Catal Today 2012, 289–299.
[49] Anilkumar A, Hölderich W. Production of lactams and carboxylic acid amides by Beckman rearrangement of oximes in the presence of Nb catalysts. WO 2010/063276, June 10, 2010.
[50] Wegner G, Kaibel G, Therre J, Aquila W, Fuchs H. Continuing Method for Producing Citral. WO 08037693, BASF Aktiengesellschaft, 2008.
[51] Aquila W, Fuchs H, Wörz O, Ruppel W, Halbritter K. Verfahren zur kontinuierlichen technischen Herstellung ungesättigter aliphatischer Aldehyde in einem Rohrbündelreaktor. DE 19722567, BASF AG, 1998.
[52] Leiner J, Stolle A, Ondruschka B, Netscher T, Bonrath W. Thermal Behavior of Pinan-2-ol and Linalool. Molecules 2013, 18, 8358–8375.
[53] Swift K A D. Catalytic transformations of the major terpene feedstocks. Top Catal 2004, 27, 143–155.
[54] Erman M B, Kane B J. Chemistry around pinene and pinane: A facile synthesis of cyclobutanes and oxatricyclo-derivative of pinane from cis- and trans-pinanols. Chem Biodivers 2008, 5, 910–919.
[55] Eggersdorfer M. Terpenes. In: Elvers, ed. Ullmann's Encyclopedia of Industrial Chemistry, Online ed. Weinheim, Germany, Wiley-VCH Verlag GmbH & Co, 2012.
[56] Ohloff G, Klein E. Process for the production of linalool. DE 1150974, Studiengesellschaft Kohle mbH, 1961.
[57] Ohloff G, Klein E Die absolute Konfiguration des Linalools durch Verknüpfung mit dem Pinansystem. Tetrahedron 1962, 18, 37–42.
[58] Weiss R. Hydrochlorination of Myrcene. US 2882323, Van Ameringen-Haebler, Inc., 1957.
[59] Mellor J M, Munavalli S. Synthesis of sesquiterpenes. Q Rev Chem Soc 1964, 18, 270–294.
[60] Kolicheski M B, Cocco L C, Mitchell D A, Kamiski M J. Synthesis of myrcene by pyrolysis of β-pinene: Analysis of decomposition reactions. Anal Appl Pyrol 2005, 80, 92–100.
[61] Fuguitt R E, Hawkins J E. The liquid phase thermal isomerization of α-Pinene. J Am Chem Soc 1945, 67, 242–245.
[62] Costa M C C, Johnstone R A W, Whittaker D. Catalysis of gas and liquid phase ionic and radical rearrangements of α- and β-pinene by metal(IV) phosphate polymers. J Mol Catal A: Chem 1996, 104, 251–259.
[63] Stolle A, Bonrath W, Ondruschka B. Kinetic and mechanistic aspects of myrcene production via thermal-induced β-pinene rearrangement. J Anal Appl Pyrol 2008, 83, 23–36.

6 C–C-bond and C–N-bond forming reactions (metal-catalysed)

6.1 Introduction and *Mizoroki–Heck* reactions

Modern trends in the catalytic formation of C–C and C–N bonds, especially in industrial catalysis, are directly connected to platinum group metal-catalysed reactions. During the last two decades, several industrial processes have been implemented utilising metal-catalysed (e.g. Ni, Pd) C–C-bond forming reactions. Scientists R.F. Heck, E.-i. Negishi [1] and A. Suzuki [2] were awarded the Nobel Prize in 2010 for their work in this field. The independently developed Pd-catalysed aryl-olefin coupling reaction allows the synthesis and production of several substance classes, and today it is a workhorse in this field of chemistry [3, 4]. The *Mizoroki–Heck* reaction is often applied in the synthesis of C–C bonds because it avoids the use of strong bases and/or formation of large quantities of waste, which would occur with traditional stoichiometric methods such as the addition of Grignard or lithium species to carbonyl compounds. In addition, the reaction has a high functional group tolerance permitting its use with a diverse range of molecules.

The mechanism of the *Mizoroki–Heck* reaction involves several different organo-palladium-intermediates (Scheme 6.1). The catalytically active Pd(0)-intermediate is often generated from a Pd(II) precursor [5] and is usually formed in the presence of a ligand, for example triphenylphosphine. The Pd(0) species undergoes oxidative addition to an aryl halide followed by the formation of a π-complex with an olefin. Next, the olefin inserts into the Pd–C-bond by *syn* addition. After rotation and β-hydride-elimination, a new palladium–olefin π-complex is formed (not depicted in Scheme 6.1). Reductive elimination in the presence of a base reforms the active Pd(0)-species and the catalytic cycle can begin again. The base, often an inorganic carbonate, is consumed stoichiometrically. If no Pd(0)-source is used directly, the Pd(II) species can be generated by various options including [6, 7] the use of:

1) Phosphines, for example triphenylphosphine in the presence of water. In this case, triphenylphosphine oxide is formed.
2) Amines – Pd(II) can be reduced via hydropalladation followed by reductive elimination.
3) Organometallic compounds such as *n*-BuLi.
4) In the presence of alkenes.

The main driving forces for the development of new processes using metal catalysed C–C and C–N bond forming reactions are the economic and environmental benefits. For the application of noble metals, for example palladium, platinum or gold, a high s/c ratio, catalyst recovery and reuse, and avoiding the contamination of the final product with these metals (<10 ppm) are important to reduce production costs. Fur-

https://doi.org/10.1515/9783111102672-006

thermore, the ligand cost must be minimised or avoided. For the application in the fine chemical industry, an s/c ratio of 4,000 to 20,000 is usually required for implementation. These numbers are in good agreement with the data estimates from literature [8]. The reported catalyst loadings for a range of cross-coupling reactions are constantly decreasing, which makes them increasingly interesting for industrial applications [9].

The aryl-alkene coupling reactions (such as the *Mizoroki–Heck* reaction) were first reported with aryl bromides and iodides, whereas, today, other coupling partners can also be used, including triflates, aryl chlorides, sulfonyl chlorides or aromatic diazonium salts. Usually, *Mizoroki–Heck* reactions are performed in dipolar-aprotic solvents.

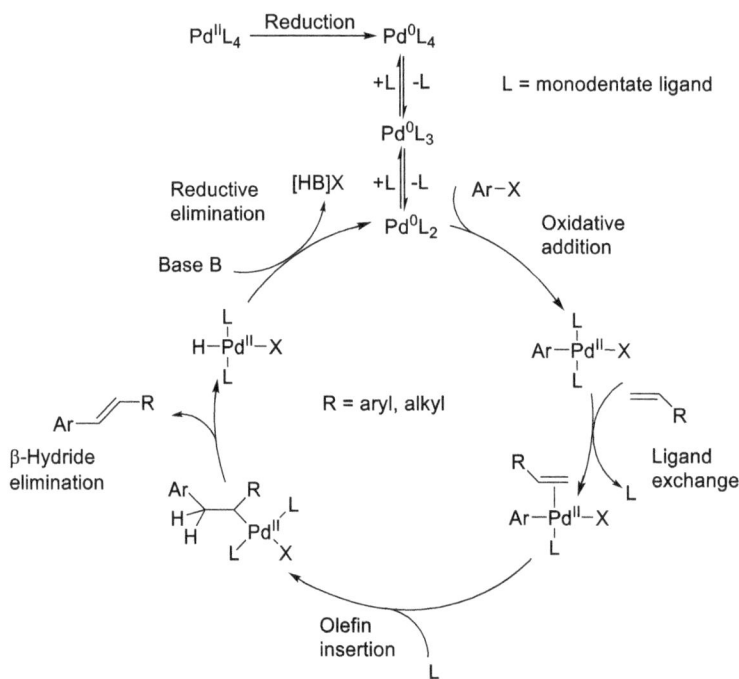

Scheme 6.1: Mechanism of the *Mizoroki–Heck* reaction.

The first example of a *Mizoroki–Heck* reaction implemented on industrial scale was the herbicide Prosulfuron™ (Scheme 6.2) [10]. The synthesis starts from aniline *via* sulfonation, followed by diazotisation to form the substrate for the coupling reaction. The *Mizoroki–Heck* reaction then takes place with trifluoropropene at 15 °C in acetic acid. The Pd catalyst used, $Pd_2(dba)_3$ (dba = dibenzylideneacetone), was adsorbed onto charcoal and then used in the subsequent hydrogenation reaction. The coupling reaction took place with an s/c of 200, and the product was synthesised at an yield of 90% (approximately) per step in several thousand t/a.

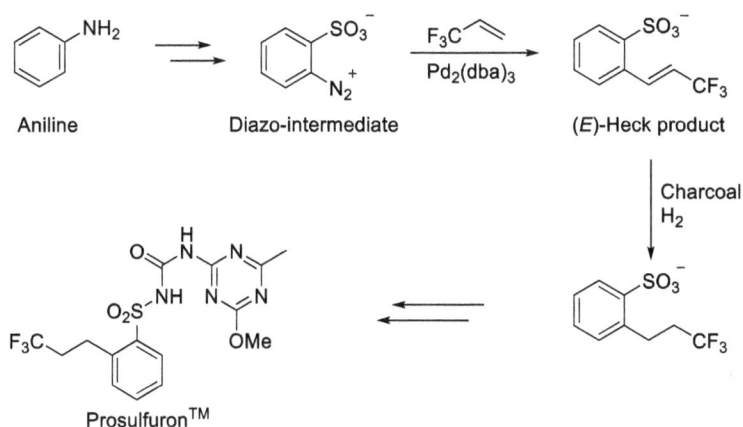

Scheme 6.2: Pd-catalysed coupling to Prosulfuron™.

The pain killer naproxene is also produced by a *Mizoroki–Heck* reaction [11–13] following a route developed by Albermale. The original process started from 2-naphthol and involved multiple steps, including a resolution, since only the (S)-isomer is sold. This process was superseded by a new route with 2-bromo-6-methoxynaphtalene as the key intermediate (Scheme 6.3). The *Mizoroki–Heck* reaction is carried out with ethylene (at 3 kPa pressure) at around 100 °C and s/c >3,000 in THF in the presence of PdCl$_2$/CuCl$_2$ and

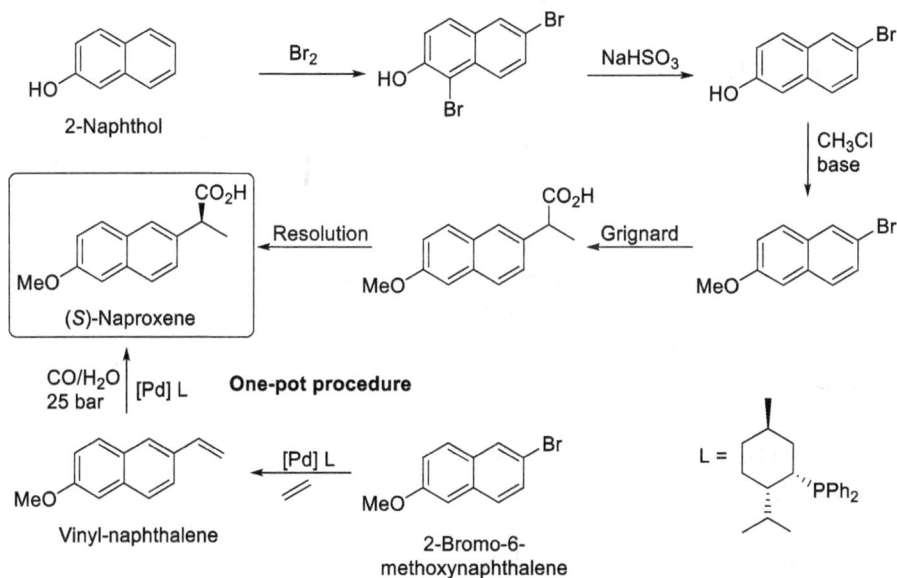

Scheme 6.3: Manufacture of naproxene.

neomenthyl diphenylphosphine as ligand in a two-step one-pot process. The intermediate vinylnaphthalene is transferred by hydro-carbonylation reaction into the final product.

Another application for the *Mizoroki–Heck* reaction is the synthesis of the UV-B filter 2-ethylhexyl-*p*-methoxycinnamate (Eusolex® 2292). This is produced from *p*-bromo anisole and 2-ethylhexylacrylate in the presence of a heterogeneous Pd on carbon catalyst at 190 °C, with Na_2CO_3 as the base (Scheme 6.4). The reaction is performed in NMP and a TON of 6,000, and a TOF of 2,000 h^{-1} can be achieved. Water (15%) has a beneficial effect on the coupling reaction [14]. It is well known from literature that the Pd-species, either Pd(0) or Pd(II), is leached from the support and redeposited on the carrier after the reaction [15–17]. Therefore, it is assumed that the catalytically active species is homogeneous.

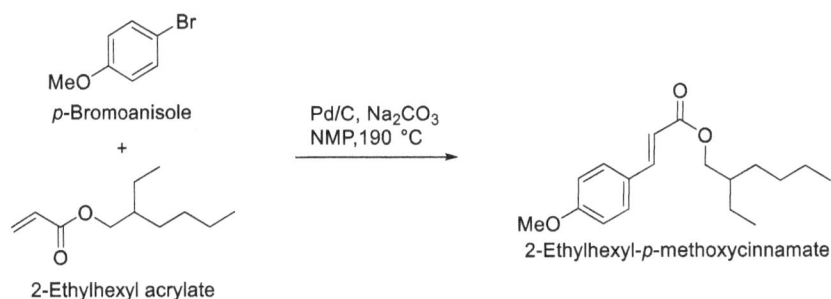

Scheme 6.4: Synthesis of an UV filter by *Mizoroki–Heck* reaction.

Resveratrol (3,4′,5-trihydroxy-(*E*)-stilbene) is a naturally occurring polyphenol found in red wine and is considered to be the origin in the protection of heart and blood vessels from arteriosclerosis [18]. It is produced from 3,5-diacetoxyacetophenone, which is converted to the corresponding styrene by the sequence of hydrogenation, bromination and elimination of HBr (Scheme 6.5) [19] (see Chapter 2: Heterogeneous hydrogenation). The styrene intermediate is then coupled in a *Mizoroki–Heck* reaction with 4-bromophenyl acetate to obtain resveratrol triacetate (mainly the (*E*)-isomer is formed). The catalyst is a palladium complex containing an *N*-ligand (acetophenone-oxime derivative), which mediates the coupling of very electron-rich compounds such as diacetoxy styrene and 4-bromophenyl acetate. Finally, after hydrolysis of the acetoxy groups, resveratrol is obtained.

A cinnamyl nitrile derivative is used as an intermediate in the manufacture of the drug rilpivirine. This is a non-nucleoside reverse transcriptase inhibitor used in the treatment of HIV. The synthesis starts from 4-iodo-2,6-dimethylaniline and uses a *Mizoroki-Heck* reaction catalysed by Pd/C (Scheme 6.6). The reaction takes place in the presence of sodium acetate at a reaction temperature of 140 °C with a catalyst loading of 500–1,000. The product is obtained at a yield of around 80% in a 4:1 ratio of a (*E/Z*)-

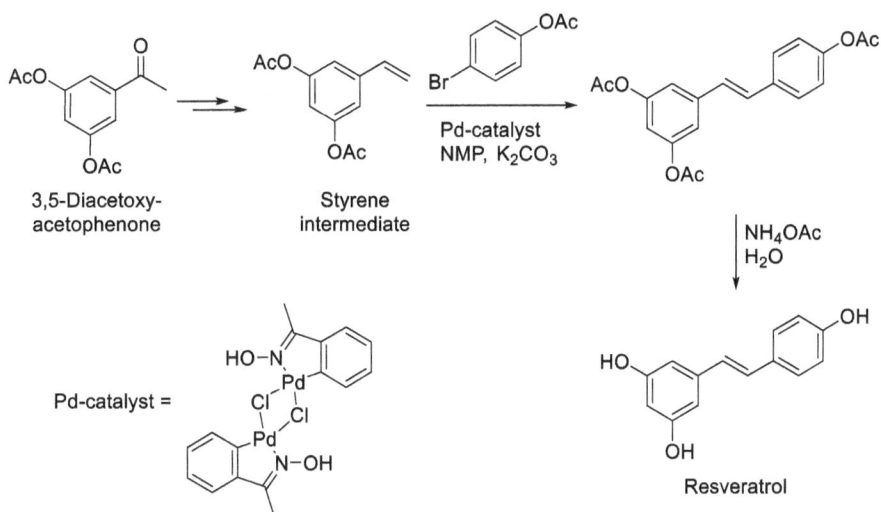

Scheme 6.5: Synthesis of resveratrol.

isomers. The advantage of using a heterogeneous catalyst compared to homogeneous ones is the lower residual metal contamination (<5 ppm) [20–23].

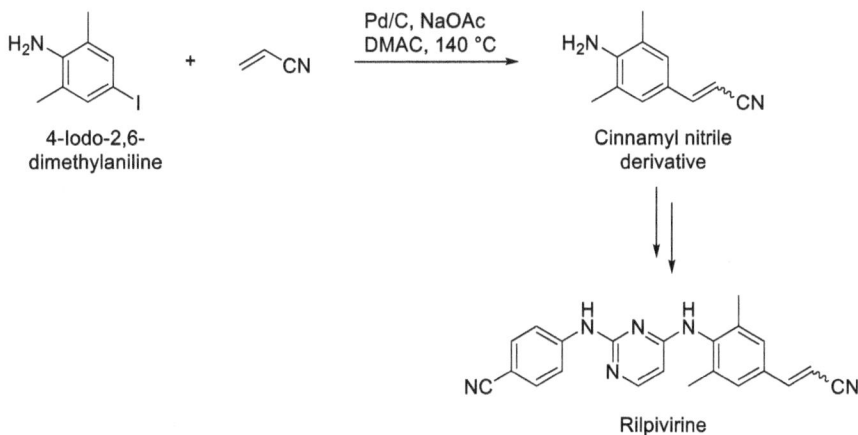

Scheme 6.6: Mizoroki–Heck reaction used in the preparation of rilpivirine.

Mizoroki–Heck reactions can also be carried out with low or ligand-less catalyst loadings and ligand free catalysts, accelerated by microwaves or a combination of microwave and ultrasound irradiation [24, 25]. Pd(OAc)$_2$ or Pd/C can be used as catalysts with an s/c of 1,000–10,000 in the presence of phosphine ligands up to a yield of 98%.

6.2 Other palladium catalysed C–C bond forming reactions

As can been seen above, the *Mizoroki–Heck* reaction can be widely applied for the synthesis of fine chemicals and pharmaceuticals. A range of other related coupling reactions have been developed using palladium and other metals, each of which involves a range of different coupling partners but similar reaction mechanisms. An overview of the most important types is shown in Table 6.1. In the following subsections, we highlight examples of most of these coupling reactions and their application in the synthesis of fine chemicals.

Table 6.1: Summary of important cross-coupling reactions to form C=C bonds.

Reaction name	Metals used	Reactant 1	Hybridisation	Reactant 2[a]	Hybridisation
Mizoroki–Heck	Pd or Ni	R'-C=C	sp^2	Ar–X	sp^2
Sonagashira–Hagihara	Pd & Cu	R'-C≡CH	sp	R–X	sp^3, sp^2
Suzuki–Miyaura	Pd or Ni	R'-B(OR)$_2$	sp^2	R–X	sp^3, sp^2
Kumada–Corriu	Pd, Ni or Fe	R'-MgBr	sp^3, sp^2	Ar–X	sp^2
Hiyama	Pd	R'-SiR$_3$	sp^2	R–X	sp^3, sp^2
Stille	Pd	R'-SnR$_3$	sp^3, sp^2, sp	R–X	sp^3, sp^2
Negishi	Pd or Ni	R'-Zn-X	sp^3, sp^2, sp	R–X	sp^3, sp^2

[a]X= Leaving group, for example halide or triflate.

6.2.1 *Sonogashira–Hagihara* reaction

The Pd-catalysed coupling reaction of aryl halides, aryl triflates or vinyl halides with 1-alkynes was developed in the mid-1970s by K. Sonogashira and N. Hagihara [26]. The mild reaction conditions (e.g. room temperature, presence of a mild base) and sometimes the possibility to work under aqueous conditions allow the synthesis of complex molecules. The reaction mechanism of the *Sonogashira–Hagihara* reaction is not completely understood, but it is believed to be similar to other Pd-catalysed coupling reactions [27]. It is assumed that a Pd-pre-catalyst is activated to form a Pd(0) species that is stabilised by ligands, for example PPh$_3$. The ease of the oxidative addition is increased by increasing the electron-donating ability of the ligand. If the phosphine ligands are bulky, such as P(*o*-Tol)$_3$, it was shown that mono-ligated complexes are formed [28]. An interesting finding is the formation of anionic palladium species, such as [L$_2$Pd(0)Cl]$^-$, which are thought to be the catalytic active species [29]. Such a coordination geometry favours the reductive elimination. Similar to other Pd-catalysed coupling reactions, it can be assumed that in the *Sonogashira–Hagihara* reaction, the aryl halide or triflate or vinyl halide undergoes an oxidative addition to the Pd(0) species (Scheme 6.7). This is believed to be the rate-determining step in the coupling reaction and the actual structure of the Pd(II) species is influenced by the ligand, similar to the

Pd(0) species, as described above. Transmetallation of the oxidative addition product with the Cu-acetylide results in a Pd–aryl/vinyl–alkynyl intermediate and the Cu-complex is regenerated. Reductive elimination forms the coupled product and the Pd(0)-intermediate complex. For the reductive elimination, a *cis*-orientation of the aryl–alkyl substituents on the Pd is necessary; therefore, a *cis-trans* isomerisation reaction is often involved to form the Pd–aryl/vinyl–alkynyl intermediate, which allows the elimination.

The role of the copper co-catalyst in the *Sonogashira–Hagihara* coupling enables the reaction to proceed under mild conditions. The copper catalytic cycle is not well described, and it is thought that in presence of a base, a Cu–π complex is formed. The formation of a Cu–acetylide complex results from the activation and increased acidity of the terminal alkyne proton. The alkynyl group is transferred to the Pd by a transmetallation reaction (Scheme 6.7).

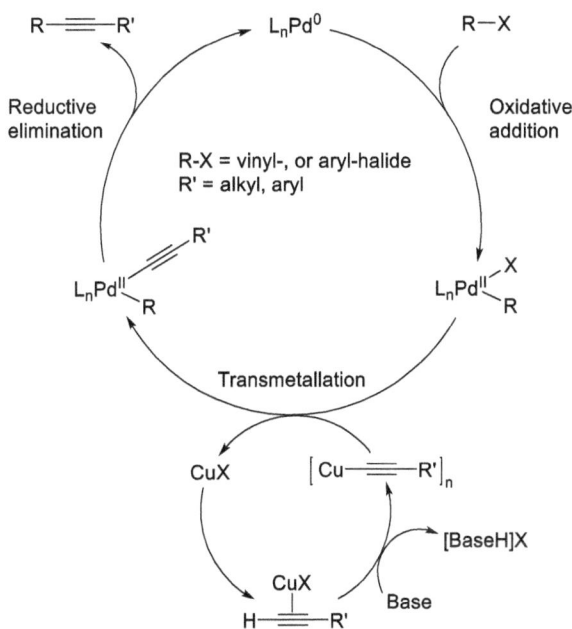

Scheme 6.7: Mechanism of the *Sonogashira–Hagihara* reaction.

In the recent years, there have been investigations of a Cu-free version of the *Sonogashira-Hagihara* coupling with the objective of avoiding several of the drawbacks of this reaction (such as environmentally unfriendly reagents and the formation of alkyne homocoupling products (*Glaser* coupling)). The mechanism of the Cu-free variation of the *Sonogashira–Hagihara* coupling is still under discussion, and it was found that a Pd(0)/Pd(II) species is involved (Scheme 6.8) [30]. The *Sonogashira-Hagihara*-reaction is performed in the presence of a base, for example secondary amines such as morpholine, piperidine or diisopropylamine. The base is added in excess, which often acts as a solvent and reacts with the *trans*-RPdXL$_2$ intermediate [31].

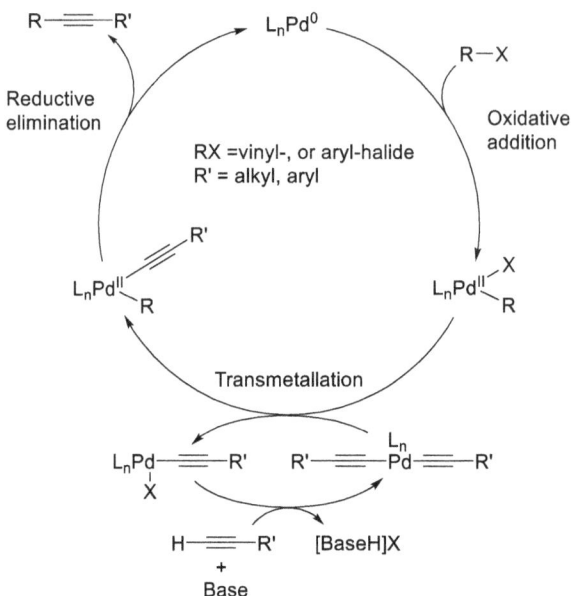

Scheme 6.8: Proposed copper-free mechanism of the *Sonogashira–Hagihara* reaction.

Applications of the *Sonogashira–Hagihara* reaction are found in the synthesis of natural compounds and drugs. The coupling of 3-trifluoromethyl bromobenzene and propargyl alcohol in Et$_2$NH in the presence of a CuI and Pd/C/PPh$_3$ catalytic system leads to an intermediate of cinacalcet (Scheme 6.9), a drug for the treatment of secondary hyperparathyroidism [32].

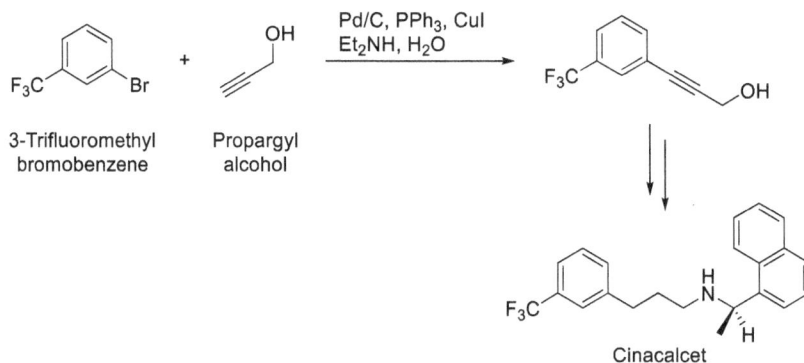

Scheme 6.9: Synthesis of the intermediate of cinacalcet.

A Pd-catalyst is applied in the alkyne–vinyl halide-coupling in the manufacture of terbinafine, which is used as an antimycotic (antifungal) drug (Scheme 6.10). The coupling reaction can be performed in amine solvents such as piperidine using $Pd(PPh_3)_2Cl_2$ or $PdCl_2(CH_3CN)_2$ and CuI as catalysts. The product is formed up to a 96% yield and is produced in scales of multiple tonnes per annum [33]. As an alternative, the coupling can be carried out without Pd in the presence of CuI [34].

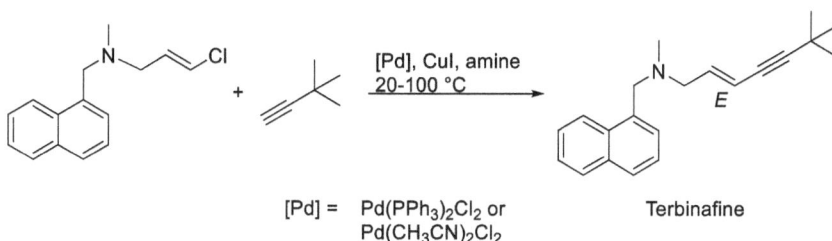

Scheme 6.10: Synthesis of terbinafine (Lamisil™).

The synthesis of benzylisoquinoline or indole alkaloids such as (+)-(S)-laudanosine and (−)-(S)-xylopinine is performed by aryl iodine coupling, with trimethylsilylethyne applying a Pd catalyst (Scheme 6.11). Under the classical *Sonogashira-Hagihara*-reaction conditions, CuI is applied as co-catalyst [35].

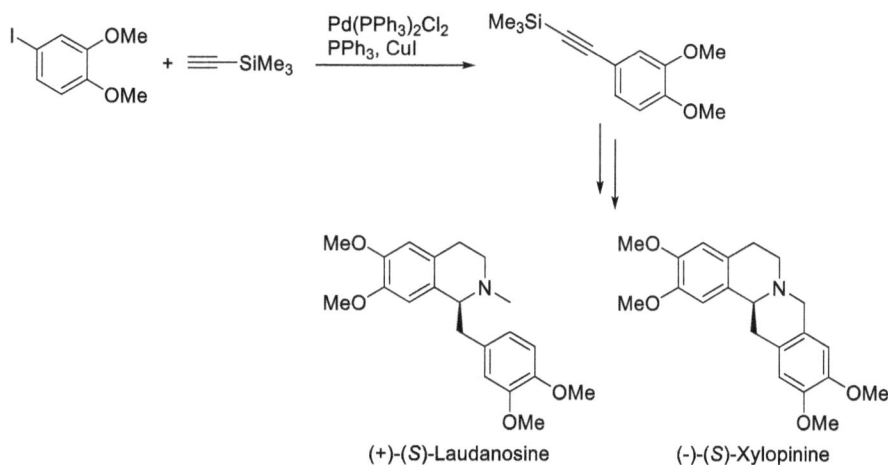

Scheme 6.11: Synthesis of (+)-(S)-laudanosin and (−)-(S)-xylopinine.

The synthesis of altinicline, a nicotinic acetylcholine receptor antagonist and used in the treatment of Alzheimer's disease, starts from methylbutynol. The coupling with a bromo pyridine conducted with Pd/C in the presence of CuI in aqueous dimethoxyethane results in a 92% yield of the coupling product (Scheme 6.12). Further treatment results in the maleate salt of altinicline [36, 37]. An alternative procedure starts from 3-(1-methylpyrolidin-2-yl)pyridine by treatment with BuLi and dichloroethane, followed by I_2 addition and Pd(0) catalysed alkynylation.

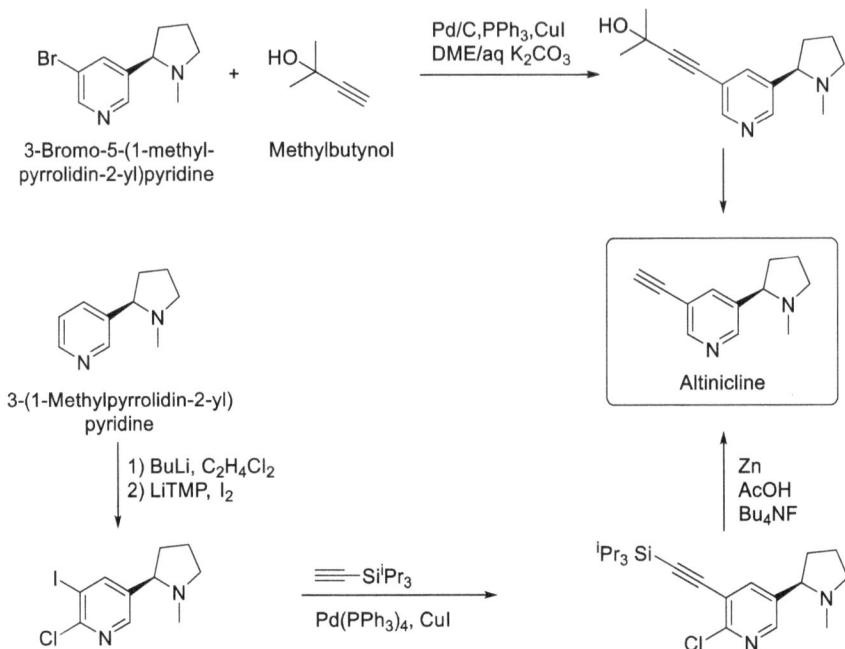

Scheme 6.12: Synthesis of altinicline.

6.2.2 Suzuki–Miyaura-coupling

The Pd-catalysed reaction of a sp^2-hybridised carbon (e.g. an aryl halide) with a boronic acid (or derivative thereof) is the *Suzuki–Miyaura* coupling, and it is used for the synthesis of biphenyls and styrenes. The reaction was first described by A. Suzuki and co-workers in 1979 [38–40]. Suzuki, with Heck and Negishi, was awarded the Nobel Prize in 2010 for his work in developing palladium catalysed cross-coupling reactions. The reaction mechanism of the *Suzuki–Miyaura* reaction is shown in Scheme 6.13. After oxidative addition of the aryl halide to the Pd(0)-complex and exchange of the halogen from the Pd(II)-complex by a base (e.g. an alkoxide), transmetallation with the boronate compound occurs. The boronate is obtained by the treatment of boronic acid with a base. The newly formed Pd-species undergoes reductive elimination, reforming the catalytic active Pd(0) species and forming the product [41–43]. The rate-limiting step in the *Suzuki–Miyaura* reaction is often the oxidative addition if a chloride is used [44]. The main advantages of this coupling reaction are mild reaction conditions, the availability of boronic acids and low toxicity. Furthermore, the reaction also works with triflates as the leaving group. The coupling can also be carried out in aqueous media [45]. Additionally, it was shown that the coupling can be very efficient with catalyst loadings as high as 100,000 [46].

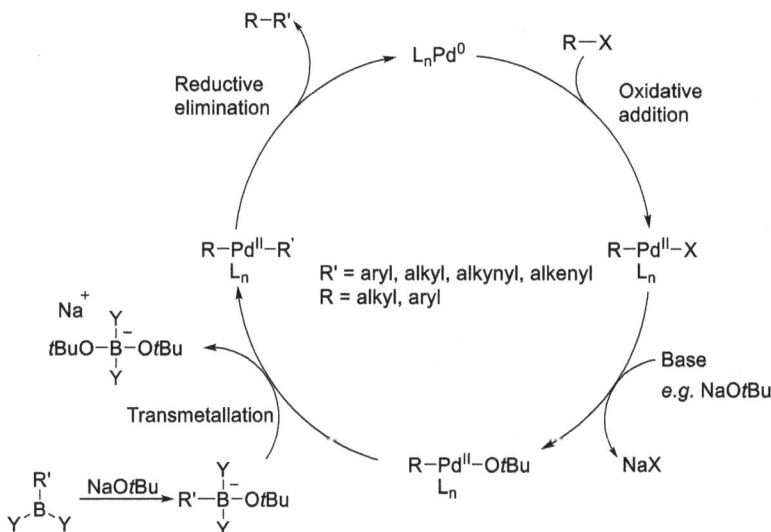

Scheme 6.13: Mechanism of the *Suzuki–Miyaura* reaction.

The role of the base is important for the Pd-catalysed *Suzuki–Miyaura* reactions. The base (usually an alkoxide or hydroxide, HO⁻ is used here) plays three roles. It accelerates the reaction by forming *trans*-[RPd(OH)(PPh₃)₂], a key complex, which, in contrast to *trans*-[RPdX(PPh₃)₂], reacts with R'B(OH)₂. The base (HO⁻) disfavours the reaction by the formation of the unreactive anionic R'B(OH)₃⁻. The overall reactivity is controlled by the base concentration [47]. If trifluoroborates are used in the in the cross-coupling reaction, the base is necessary to achieve the hydrolysis of the trifluoroborate to the corresponding boronic acid, which is the active reagent. Such approaches can be used as a slow release strategy to minimise the accumulation of the labile boronic acids and reduce side reactions [48]. An alternative approach is the use of so-called MIDA-boronates (*N*-methyliminodiacetic acid), which also allow the slow release of the desired boronic acids [49].

Losartan, an AT1 antagonist, is an important drug in the treatment of high blood pressure. It is also used in the treatment of hypertension and diabetic nephropathy and is dispensed as a mono-potassium salt. The *Suzuki–Miyaura* coupling [50] is performed with an *in-situ* prepared catalyst at 60–80 °C in solvents such THF with an s/c 100 in around 95% yield [51, 52] (Scheme 6.14).

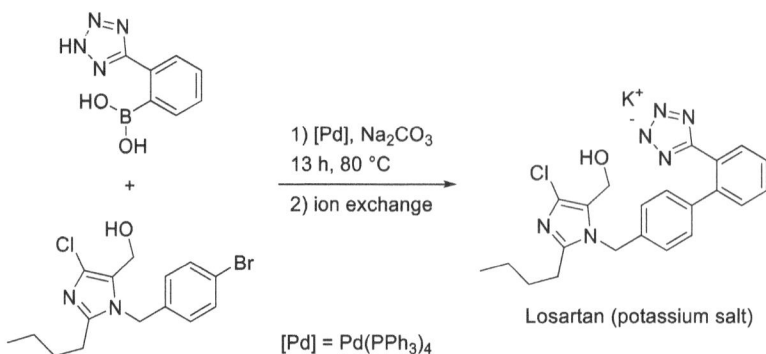

Scheme 6.14: Synthesis of losartan.

The *Suzuki-Miyaura*-coupling is applied at an industrial scale for the manufacture of the fungicide boscalid, on a scale of >1,000 t/a. Boscalid is applied in the wine and fruit farming industry. The coupling of *o*-chloronitrobenzene and chlorobenzene boronic acid can be carried out in the presence of Pd(PPh₃)₄ or Pd/C [53, 54]. The use of Pd/C has the advantage that *o*-iodoaniline can be used to bypass the need to reduce the nitro-group into the corresponding amine (Scheme 6.15).

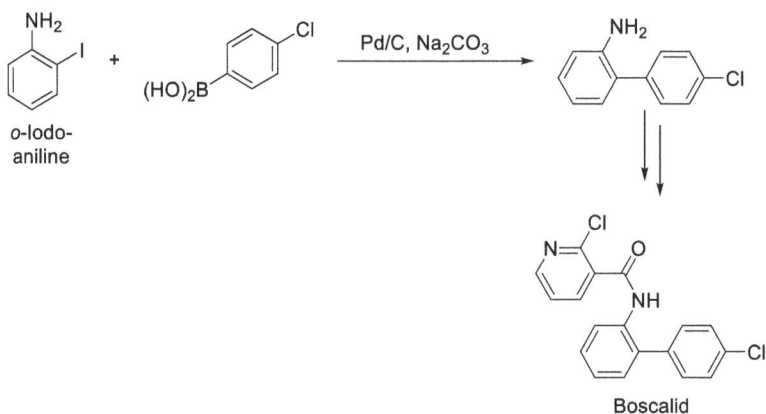

Scheme 6.15: Application of the *Suzuki–Miyaura* coupling in the synthesis of boscalid.

An alternative approach to boscalid is the Pd(PPh₃)₄ catalysed coupling reaction of *o*-chloronitrobenzene and the boronic acid derivative under flow conditions (Scheme 6.16). The formed intermediate is directly transferred to 2-amino-4′-chlorobiphenyl by a hydrogenation reaction in the presence of Pt/C [55]. The coupling reaction is carried out with s/c = 400 in 4:1 ᵗBuOH:H₂O at 160 °C.

The synthesis of the pyrazole carboxamide-based fungicide (applied e.g. for corn, soybeans and potatoes), bixafen, can be carried out with a *Suzuki-Miyaura* reaction of

o-Chloro-
nitrobenzene

Pd(PPh$_3$)$_4$, t-BuOK
t-BuOH/H$_2$O

10% Pt/C, H$_2$

Boscalid

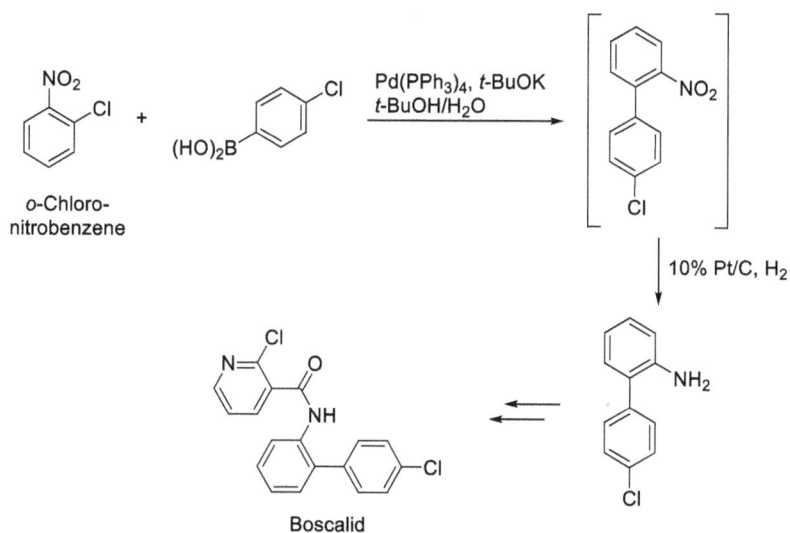

Scheme 6.16: One-pot two-step synthesis of 2-amino-4′-chlorobiphenyl.

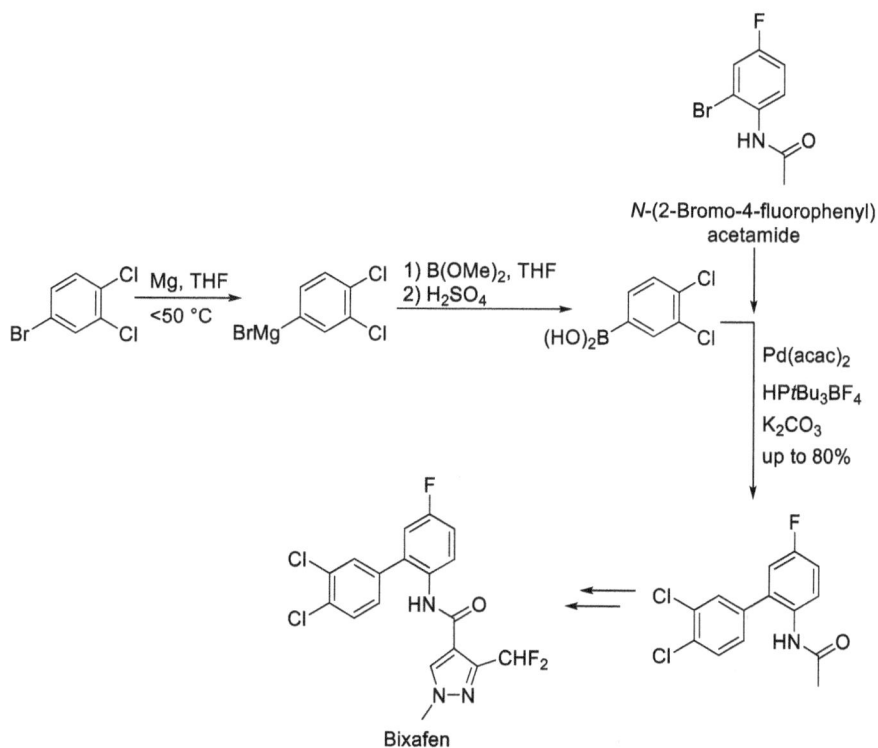

N-(2-Bromo-4-fluorophenyl)
acetamide

Mg, THF
<50 °C

1) B(OMe)$_2$, THF
2) H$_2$SO$_4$

Pd(acac)$_2$
HPtBu$_3$BF$_4$
K$_2$CO$_3$
up to 80%

CHF$_2$

Bixafen

Scheme 6.17: Synthesis of bixafen.

the 3,4-dichlorophenylboronic acid and *N*-(2-bromo-4-fluorophenyl) acetamide in a Pd(acac)$_2$ catalysed reaction, where HP^{t}Bu$_3$BF$_4$ is used as a phase-transfer catalyst in the presence of K$_2$CO$_3$ [56]. After a 13 h reaction time at <50 °C, a yield of around 80% of the coupling product is obtained, which is further transferred under basic conditions into the final product (Scheme 6.17).

Bombykol, a pheromone from silkworm, used by females to attract mates, can be synthesised by a *Suzuki–Miyaura* reaction, starting from (*E*)-(11-hydroxyundec-1-en-1-yl)boronic acid in a Pd(PPh$_3$)$_4$ catalysed reaction (s/c = 20) in the presence of a strong base such as alcoholates in a yield of >80% (Scheme 6.18) [57, 58].

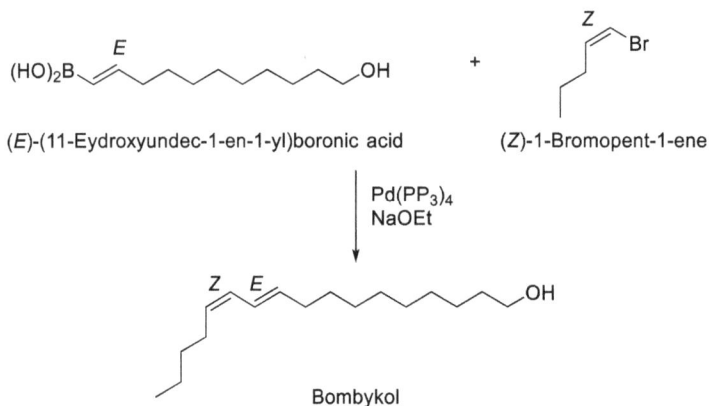

(*E*)-(11-Eydroxyundec-1-en-1-yl)boronic acid   (*Z*)-1-Bromopent-1-ene

Pd(PP$_3$)$_4$
NaOEt

Bombykol

Scheme 6.18: Synthesis of bombykol via Pd-catalysed *Suzuki–Miyaura* reaction.

6.2.3 *Kumada–Corriu* coupling

The alkene–aryl coupling reaction using aryl triflates or aryl halides with a *Grignard* reagent in the presence of Ni or Pd-catalysts was described by M. Kumada and R. Corriu [59]. In this coupling reaction, an organo-magnesium species is the second coupling partner. The main advantage of the *Kumada–Corriu* coupling reaction is the fact that sp^3-hybridised coupling partners can be used, which is not the case with many other coupling reactions. The nickel-catalyst is stabilised with phosphine ligands such as triphenylphosphine, bis(dipenylphophino)ethane (dppe) or similar chelating phosphines. The reductive elimination step is only possible from a *cis*-configuration – which is preferred with chelating phosphine ligands (Scheme 6.19). The oxidative addition of an aryl halide or triflate to the ligand-stabilised Ni(0)-complex (formed from a precursor) results in a Ni(II) species, which is transmetallated with a *Grignard* reagent. An organonickel species and a magnesium halide are then formed. The reductive elimination from a *cis*-configuration results in the active catalyst species and the coupled product. In the case of *trans*-configuration, the reductive elimination occurs after an isomerisation. It was later found that Pd(0) can also replace Ni(0) [60–62].

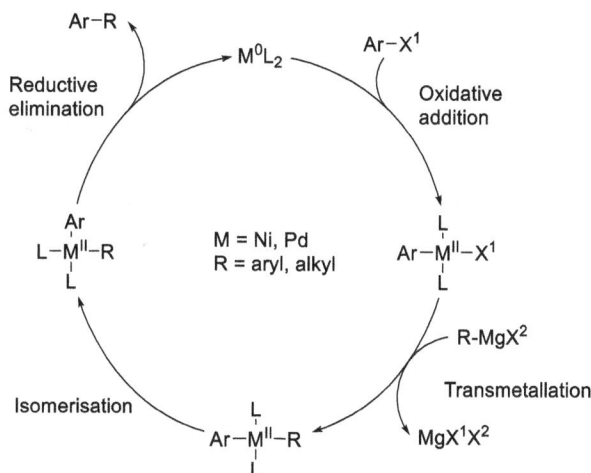

Scheme 6.19: Mechanism of *Kumada–Corriu* coupling.

The *Kumada–Corriu* reaction is applied in several industrial processes; however, the high reactivity and basicity of the *Grignard* reagent can limit the application to relatively simple organic molecules. The commonly used Pd- or Ni-catalysts can be replaced by Fe, Co or Mn-complexes [63–65].

Applications of the *Kumada–Corriu* coupling in large-scale industrial processes include the synthesis of drugs, for example aliskiren (Tekturna™), which is used in the treatment of hypertension [66]. The coupling reaction is carried out in the presence of a Ni-catalyst (s/c = 30) and triethylamine. The Cl-intermediate is synthesised in the

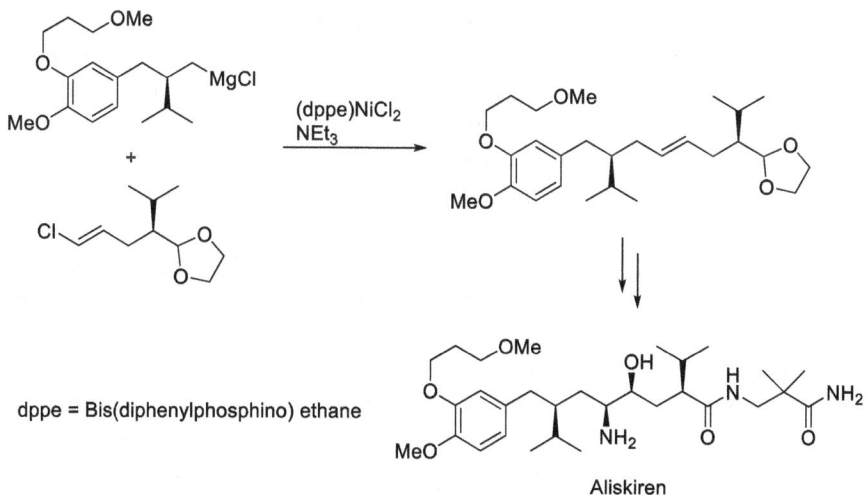

dppe = Bis(diphenylphosphino) ethane

Aliskiren

Scheme 6.20: *Kumada–Corriu* coupling in the synthesis of aliskiren (Tekturna™).

2-(*S*)-configuration by resolution in the presence of pork liver esterase in >99% *ee* and 47% yield from the racemate, which itself is prepared from methyl isovalerate by enol-type chemistry (Scheme 6.20).

Furthermore, the *Kumada–Corriu* coupling is used in the polyene and polymer synthesis such as polyalkylthiophenes [67–69]. The Ni-catalysed coupling takes place in an s/c range of 100–500 in a one-hour reaction time in polar solvents such as THF (Scheme 6.21).

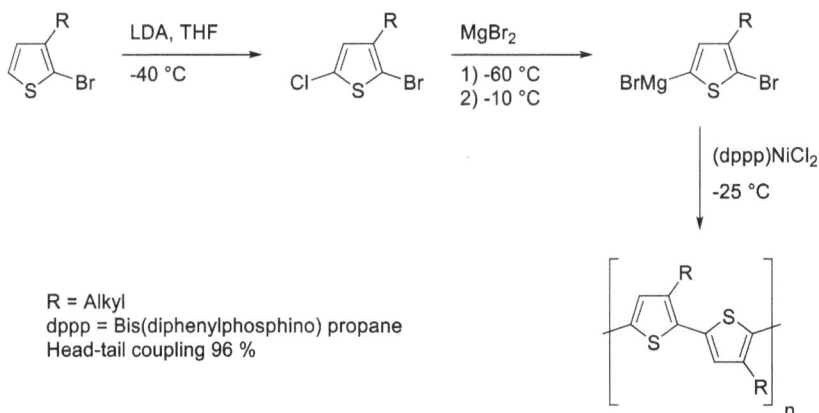

R = Alkyl
dppp = Bis(diphenylphosphino) propane
Head-tail coupling 96 %

Scheme 6.21: Synthesis of polythiophenes via *Kumada–Corriu* coupling.

6.3 Carbonylation reactions

Pd-catalysed carbonylation reactions are another type of the C–C bond forming reaction that has been implemented at an industrial scale. A prominent example is the Hoechst-Celanese process for the anti-inflammatory drug ibuprofen. Ibuprofen is produced in >5,000 t/a in a Pd-catalyst reaction, starting with isobutyl benzene (Scheme 6.22). Acylation reaction of this benzene derivative with acetic anhydride in the presence of HF, followed by a hydrogenation reaction in the presence of Pd/C or a Ni-alloy catalyst results in 1-(4-isobutylphenyl)ethan-1-ol, which is treated in a $PdCl_2$-catalysed carbonylation reaction with CO and aq. HCl using triphenylphosphine as ligand. This reaction sequence replaces the old 6-step *Boots'* process. It is not only cheaper but also generates significantly less waste. The *Boots'* process used $AlCl_3$, a *Darzens*-glycidic ester reaction with chloroacetic acid ester, and an oxime formation with hydroxyl amine (Scheme 6.23). This was followed by dehydration to the corresponding nitrile and finally the hydrolysis to the final product [70], which was commercialised as a racemic mixture of the (*R*)- and (*S*)-enantiomers. For the manufacture of ibuprofen, several modifications to the Boots process are known, all of which are less efficient than the Pd-based procedure [71–74].

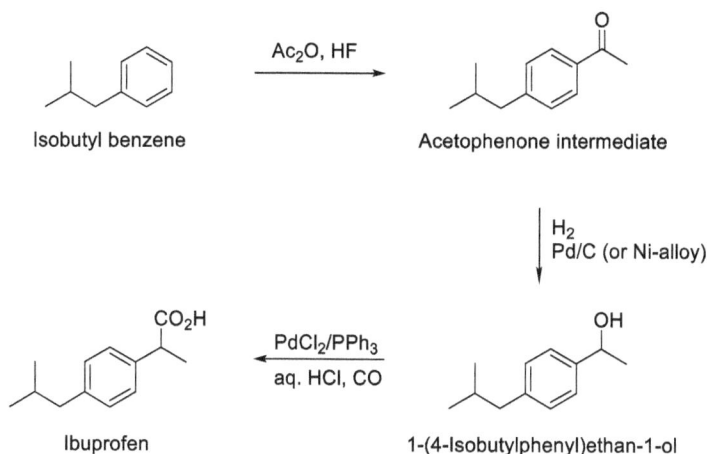

Scheme 6.22: Pd-catalysts in the manufacture of ibuprofen.

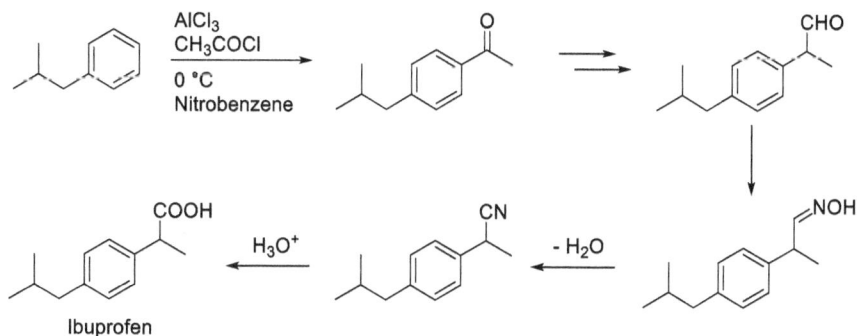

Scheme 6.23: Modified Boots process for the manufacture of ibuprofen.

Similarly, Pd-catalysed amino carbonylation is used in the synthesis of lazabemide (Scheme 6.24). This drug, for the treatment of *Parkinson's* disease, can be synthesised starting with 2,5-dichloropyridine. A catalytic route using a carbonylation was developed that was significantly more efficient than the original eight-step approach [75, 76]. The final product can be produced in only one step in TON = 3,000. The direct amino carbonylation of 2,4-dichloropyridine by a Pd-catalyst has also been described. This compound was also developed as an anti-parkinsonian agent but, to the best to our knowledge, was not commercialised.

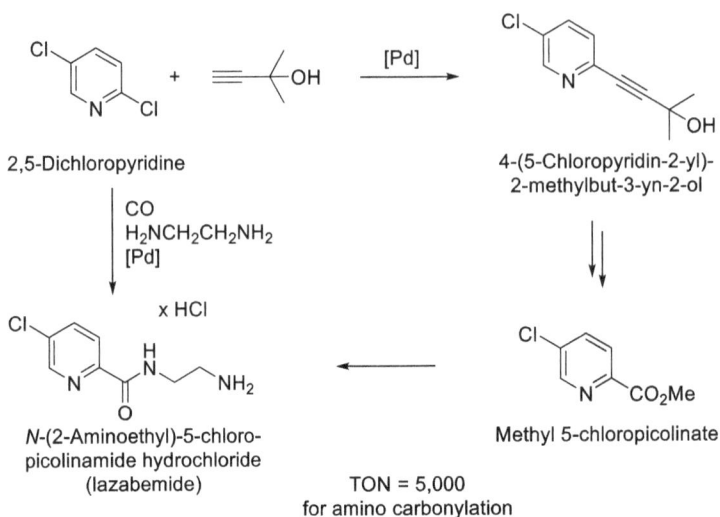

2,5-Dichloropyridine

4-(5-Chloropyridin-2-yl)-2-methylbut-3-yn-2-ol

N-(2-Aminoethyl)-5-chloro-picolinamide hydrochloride (lazabemide)

Methyl 5-chloropicolinate

TON = 5,000
for amino carbonylation

Scheme 6.24: Lazabemide by Pd-catalysed reactions.

6.4 Hydroformylation reactions

The addition of CO and H_2 to olefins in the presence of a catalyst, the so-called hydro-formylation or oxo-synthesis, results in the formation of aldehydes. Hydroformylation reactions are typically catalysed by transition metal catalysts, mainly metal-carbonyl complexes, under homogeneous conditions. The order of reactivity of the catalysts is: Rh > Co > Ir, Ru > Os > Pt > Pd > Fe > Ni. Modified catalysts are more selective, which results in lower energy consumptions compared to non-modified systems. Further-more, water-soluble systems are very selective (n/i) and environmentally friendly. The reactivity of catalytic systems for hydroformylation and ligand influence is [77]: L = PPh_3, $P(OR)_3$ > $P(nC_4H_9)_3$ > NPh_3 > $AsPh_3$.

Industrially applied catalysts are based on $HCo(CO)_4$, $HCo(CO)_3PR_3$ and $HRh(CO)_3PR_3$. In general, hydroformylation reactions are exothermic (30 kcal/mol) and are performed in a temperature range of 90–180 °C at 300 bar pressure and the typical catalyst loadings are at least s/c = 10,000. The hydroformylation reaction was discovered by O. Roelen, and the reaction mechanism was clarified by R. Heck and D. S. Breslow [78]. The products of the oxo-synthesis are around 1,000,000 t/a (ignoring products derived from them).

The various approaches in industrial hydroformylation reactions are summarised in Table 6.2:

Table 6.2: Industrialised hydroformylation processes.

Company or process	Alkene range	Catalyst precursor	Modifying ligand	n/i	T (°C)	p (bar)
RP Ruhrchemie	C_3	[Rh(COD)Cl]$_2$	TPPTS	≥20	up to 130	up to 60
LPO	C_3	HRh(CO)(PPh$_3$)$_3$	TPP	10	up to 120	up to 50
BASF	C_4-C_9	HCo(CO)$_4$	none	4	up to 170	up to 300
Exxon	C_6-C_{12}	HCo(CO)$_4$	–	–	up to 180	up to 300
Shell	C_7-C_{14}	HCo(CO)$_4$	phosphines	7	up to 200	up to 150

RP = Rhône Poulenc, LPO = Low pressure olefin hydroformylation, TPP = triphenylphosphine, TPPTS = sodium tristriphenylphosphine sulphonate.

The co-catalysed reaction mechanism of the oxo-synthesis is shown in Scheme 6.25. After the formation of a HCo(CO)$_3$-species, an olefin binds forming an 18-electron π-complex, which is followed by insertion to form an alkyl-Co-complex. The binding of a molecule of CO and the migratory insertion results in an acyl-Co-complex. Subse-

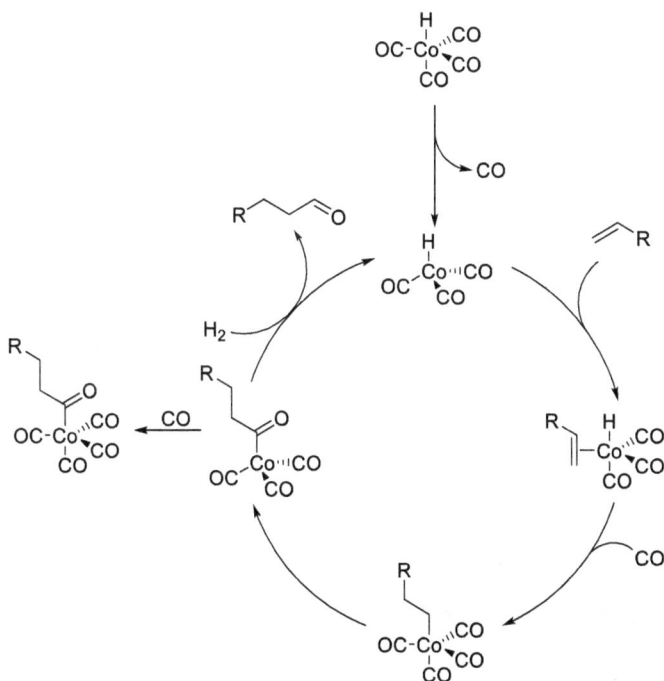

Scheme 6.25: Co-catalysed hydroformylation mechanism.

quent addition of hydrogen releases the formed aldehyde and the active complex $HCo(CO)_3$ is reformed. The respective catalytic cycle with Rh catalysts is similar to the Co catalytic cycle.

The Rh-catalysed olefin hydroformylation in liquid–liquid solvent systems, for example water/organic phase for the manufacture of aldehydes is carried out in the presence of water-soluble ligands such as sodium tristriphenylphosphine trisulphonate (TPPTS). The process is used in the manufacture of a mixture of linear and branched n/i-butyraldehyde in several 100,000 t/a. In the Rh-catalysed hydroformylation of propene, a n/i ratio of 95/5 is achieved. The Rh losses are neglectable and 300,000 tonnes of oxo-products are produced per kg of Rh (Scheme 6.26) [79]. A simplified technical set-up is shown in Scheme 6.27. The advantages of this procedure are low catalyst loading, high n/i-ratio, easy recovery and reuse of the catalyst; TONs of up to 4,000,000 can be achieved.

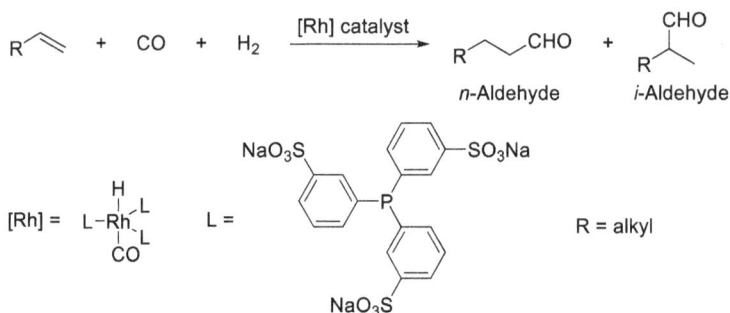

Scheme 6.26: Two-phase solvent system for Rh-catalysed hydroformylation reaction.

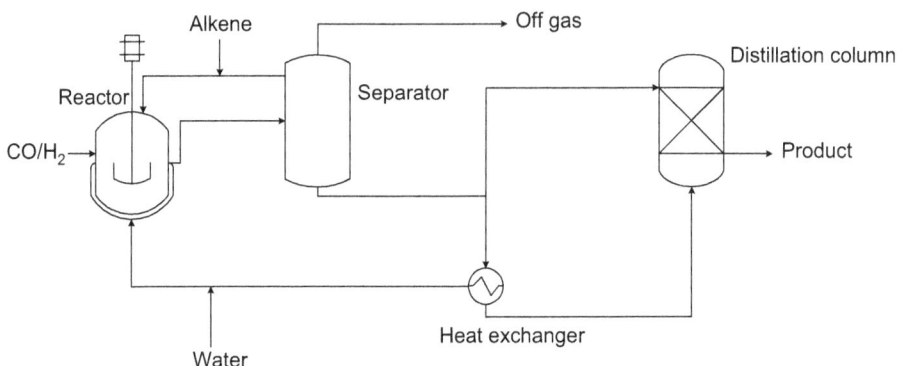

Scheme 6.27: Simplified hydroformylation process of olefins.

The liquid–liquid two-phase hydroformylation of propene and butene is industrially applied in the Rhône–Poulenc–Ruhrchemie process, whereas the ligand synthesis is based on the work of E. G. Kunz. The process is discussed by B. Cornils [80, 81]. More than 600,000 t/a of C_4- and C_5-aldehydes have been produced by this technology. In the last few decades, new types of ligands and technologies for catalyst/ligand recycling such as membranes have been developed [82–85].

In the Rhône–Poulenc–Ruhrchemie process, an Rh catalyst is used; Co-complexes find their application in the hydroformylation of higher olefins. In a hydroformylation process of BASF using higher olefins (C_4–C_9), a Co-hydride complex is used. After the reaction, it is separated by oxidation and acidic treatment. The catalyst is reused and the catalyst losses are replaced by a fresh catalyst. Reaction conditions for the hydroformylation reaction are 150–170 °C and 200–300 bar [86]. Co-catalysts ($HCo(CO)_4$) are also applied in the hydroformylation of C_6–C_{12}-alkenes, the Exxon-process (former *Kuhlmann* or PCUK process). The reaction is performed at 160–180 °C and at around 300 bar. The catalyst is separated from the organic phase by treatment with base extraction, neutralisation and CO treatment [77, 87].

The Shell hydroformylation process of C_7–C_{14} alkenes in the presence of $Co_2(CO)_8$ or $HCo(CO)_4$ and phosphines is carried out at 160–200 °C at 50–150 bar. The reaction products are separated by distillation and directly processed to alcohols by catalytic hydrogenation. The catalyst is recycled from the sump and reused. The products have a high *n*-selectivity and are applied as surfactant precursors [77]. Finally, the application of Rh-catalysts, for example $HRh(CO)(PPh_3)_3$, at 10–50 bar at 60–120 °C in the low pressure oxo process (LPO) is known for propene hydroformylation [77]. The catalyst and product are separated by distillation.

Another important application of the hydroformylation technology in the fine chemical industry is found in the synthesis of aldacet. The 3,4-diacetoxybut-1-ene is converted by a Rh-catalysed hydroformylation to aldacet at around 100 °C (Scheme 6.28) [88, 89]. Aldacet is an important building block in the vitamin A synthesis of BASF and others. At present the world-wide production of vitamin A derivatives is ca. >5,000 t/a [90].

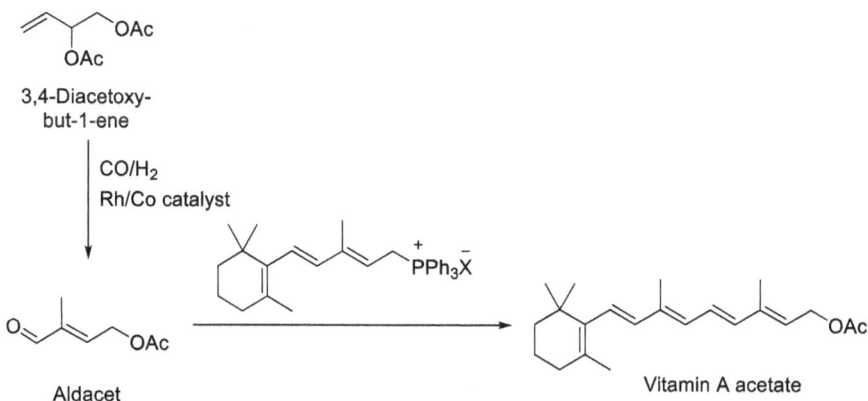

Scheme 6.28: Synthesis of vitamin A acetate.

6.5 SHOP process

The liquid–liquid two-phase catalysis was first established in the SHOP-process (Shell higher olefin process) [91–93]. In this large scale-process (1,000,000 t/a), ethylene is oligomerised to linear C_4–C_{20} α-olefins. The process is carried out at 70–150 bar at 100–150 °C in 1,4-butendiol in the presence of a Ni-catalyst, which is obtained from Ni(cod)$_2$ (cod = cyclooctadiene) and Ph$_2$PCH$_2$COOH. The products of the SHOP process are collected as a separate phase and isolated by decantation. A similar process for butadiene dimerisation into 1,3,7-octatriene in acetonitrile using Pd-phosphine catalysts is also known [94].

The catalytic cycle can involve a Ni–H complex, which is believed to be the active species in the SHOP process. This Ni-H is formed by the dissociation of a cyclooctadiene (Scheme 6.29). Insertion of ethylene in the Ni–H bond results in an alkyl-Ni species that undergoes multiple additional insertions of ethylene to form ethylene-oligomers. This is followed by β-H elimination to yield the α-olefins. The α-olefins are formed in a *Schultz–Flory* distribution, with the amount of C_4–C_{10} α-olefins being greater than >C_{10} α-olefins.

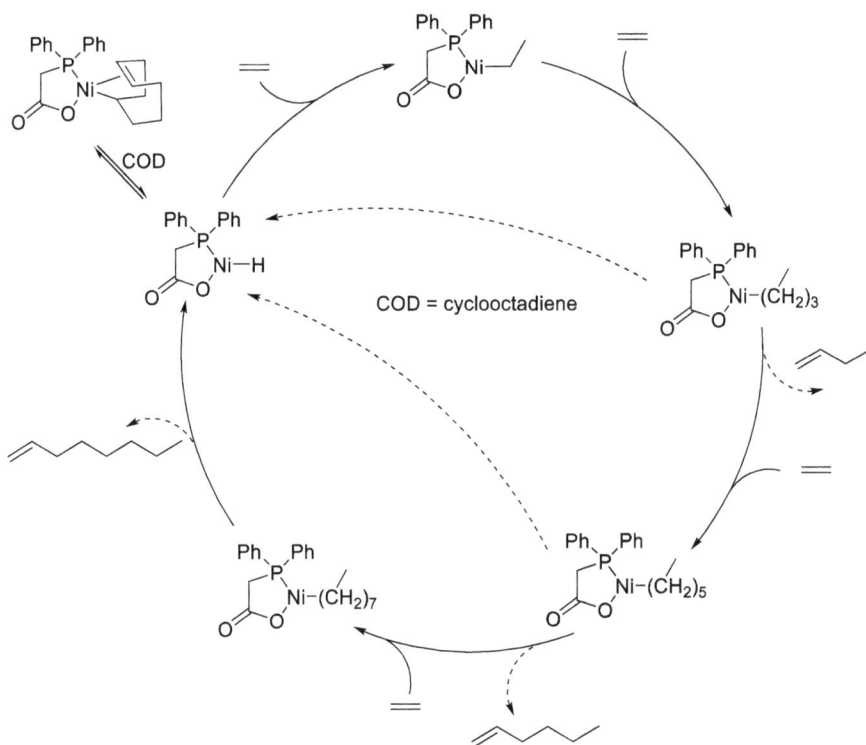

Scheme 6.29: Mechanism of the SHOP process.

The SHOP process is mainly divided into the catalytic steps of ethylene oligomerisation, isomerisation and metathesis reaction. The obtained α-olefins $<C_{10}$ and $>C_{18}$ are isomerised, followed by a metathesis reaction, to obtain C_{11} to C_{14} olefins for hydroformylation into fatty alcohols (Scheme 6.30). These products are useful for plasticisers, lubricants and detergents. A simplified process diagram is displayed in Scheme 6.31.

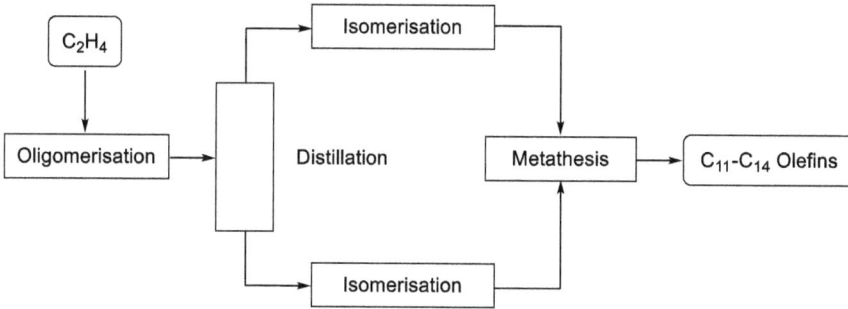

Scheme 6.30: Flow diagram of SHOP process.

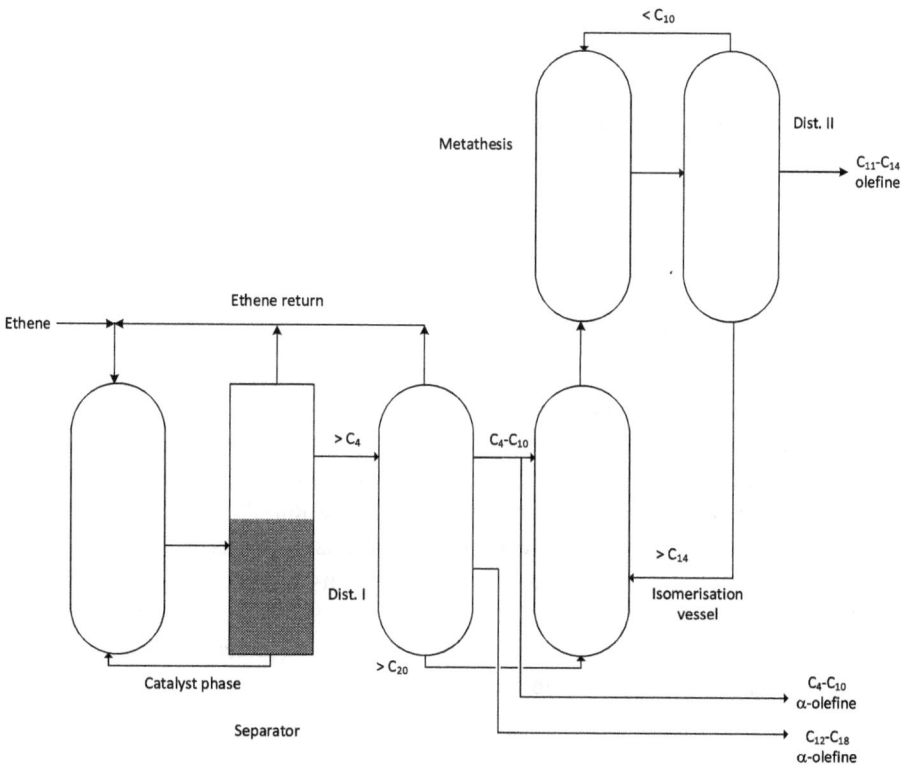

Scheme 6.31: Simplified process diagram of the SHOP process.

The catalyst recycling and reuse in the SHOP process is performed by phase separation of a polar phase (catalyst containing) and a non-polar phase (α-olefin) by a decanter. The polar phase is recharged into the reactor (Scheme 6.32).

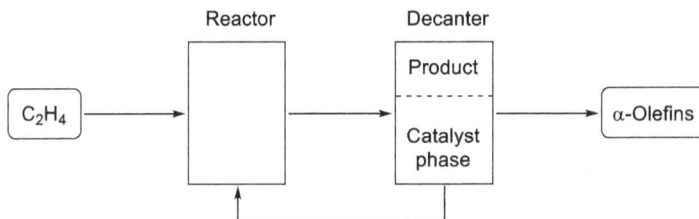

Scheme 6.32: Catalyst recycling and reuse in the SHOP process.

6.6 1,3-Diene functionalisation

The liquid–liquid two-phase catalysis in polar–non-polar phases is also applied in the manufacture of geranylacetone from myrcene in a Rh/TPPTS catalysed process on a scale of several t/a (Scheme 6.33) [95]. Geranylacetone is used in the flavour and fragrance industry and is an important building block in isoprenoid chemistry. The reaction is performed in water or acetonitrile at 20 °C in high regioselectivity (>99%) and yield (81%) at 98% conversion of myrcene.

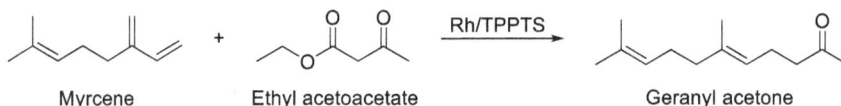

Scheme 6.33: Rh-catalysed C₃-elongation of myrcene in a two-phase solvent system.

The Rh-catalysed β-keto ester elongation reaction of 1,3-dienes is also applied in the manufacture of 6,10,14-trimethylpentadecanone, which is an intermediate in the vitamin E side chain (see Chapter 2: Heterogeneous hydrogenations). β-Farnesene is elongated in the presence of methyl acetoacetate using a Rh/TPPTS catalyst-system at around 90% yield (Scheme 6.34). Hydrogenation results in the formation of the saturated ketone, which is purified by distillation. In the Rh-catalysed elongation reaction, the 1,3-diene is activated by coordination to the metal centre and a π-allyl intermediate is formed (Scheme 6.35). Addition of the β-keto ester results in the formation of the product, and the catalyst is reformed.

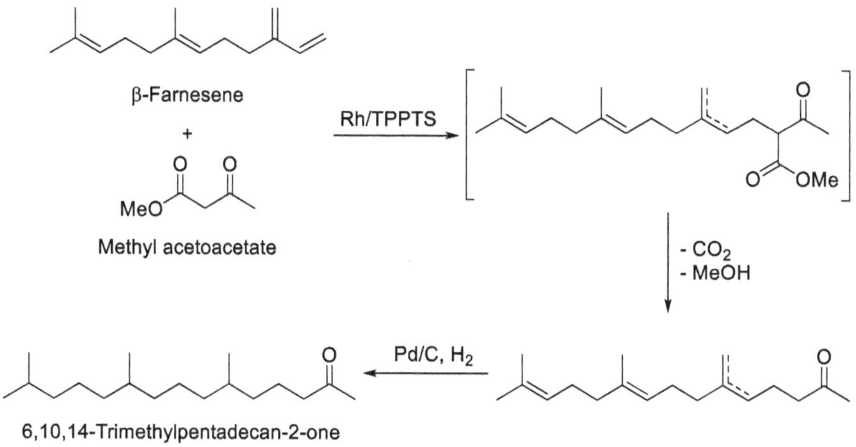
β-Farnesene
+
Methyl acetoacetate
MeO
Rh/TPPTS
OMe
- CO2
- MeOH
Pd/C, H2
6,10,14-Trimethylpentadecan-2-one

Scheme 6.34: Rh-catalysed elongation of β-farnesene.

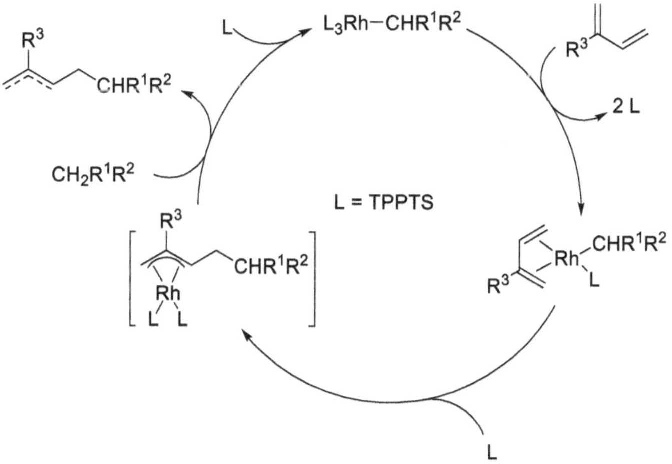
L3Rh−CHR1R2
R3
CHR1R2
CH2R1R2
2 L
L = TPPTS
Rh
L
L

Scheme 6.35: Proposed reaction mechanism of Rh-catalysed 1,3-diene elongation.

6.7 Telomerisation

The linear dimerisation of a 1,3-diene (e.g. butadiene) with the addition of a nucleophile (such as water, alcohol or ammonia) is a telomerisation reaction and is carried out in presence of Pd or Ni catalysts. The reaction can also take place with other dienes such as isoprene or cyclopentadiene. The telomerisation reaction was independently discovered by E. J. Smutny (Shell) and S. Takahashi (Osaka University, Japan) in 1967 [96, 97].

The industrial production of octan-1-ol (Scheme 6.36), starting from butadiene and water, using a Pd-catalyst was established in Japan by Kuraray at a scale of 5,000 t/a. A catalyst containing Pd/triphenylphosphine monosulfonate (TPPMS) is used in a mixture of water/sulfolane and triethylamine, which allows the recycling and reuse of the catalyst. The process is performed under 10–20 bar of carbon dioxide; the role of carbon dioxide has not been entirely understood. It is assumed that after the formation of carbonates from water and carbon dioxide, the carbonate acts as the nucleophile, followed by a hydrolysis of the intermediate telomer carbonate, resulting in octan-1-ol.

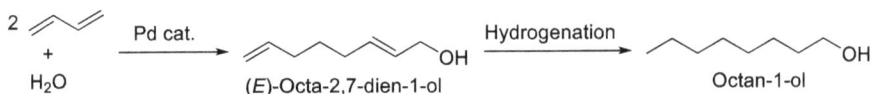

Scheme 6.36: Synthesis of octan-1-ol.

Since 2008, the production of 1-octene from butadiene has been carried out in Tarragona, Spain by Dow Chemicals. In the first step, butadiene undergoes the Pd-catalysed telomerisation with methanol to produce 1-methoxy-2,7-octadiene. This is followed by hydrogenation with a nickel catalyst to produce 1-methoxyoctane (Scheme 6.37). The reaction is carried out with a Pd/TPP (triphenylphosphine) catalyst in a plug-flow reactor system with an s/c = 50,000 [98]. A different telomerisation of butadiene and water using a Pd/TPPTS (sodium triphenylphosphine trisulfonate) catalyst in a loop reactor (to overcome mass transport limitations in the gas–liquid system) has also been described. This reaction set-up allows easy recovery and reuse of the catalyst and a space-time yield of 35 kg m^{-3} h^{-1} at 60 °C without the presence of carbon dioxide [99, 100]. Yields of around 90% over both steps (telomerisation and hydrogenation) have been achieved. Cracking of 1-methoxyoctane results in 1-octene and methanol, which is recycled.

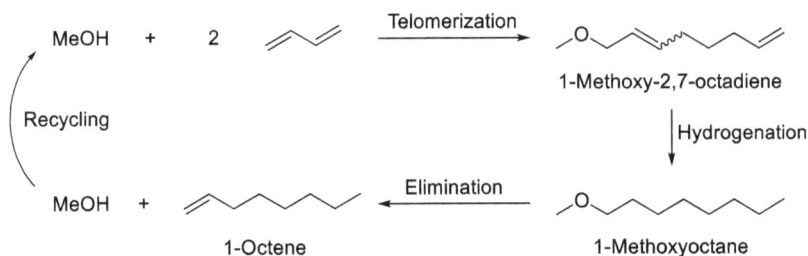

Scheme 6.37: Manufacture of 1-octene by telomerisation.

The mechanism of telomerisation involves a Pd(0) species; the Pd(0) is formed from a Pd(II) precursor by *in-situ* reduction. The Pd(0)-catalyst coordinates two butadiene molecules and then undergoes oxidative addition, forming a bis-allyl-Pd species. Addition of the nucleophile-proton to the η^3-,η^1-octadienyl-complex in the 6-position and direct attack of the nucleophile in the 1- or 3-position of the η^3-octadienyl chain leads to a linear or branched product. Addition of a new butadiene molecule results in the regeneration of the catalyst and decomplexation of the telomer product (see Scheme 6.38) [101]. This mono-palladium complex based mechanism is based on the results of P.W. Jolly and on some intermediates such as η^3-allylpalladium complexes like [Pd(η^3-C$_3$H$_5$)$_2$] and [Pd(η^3,η^3-C$_{12}$H$_{18}$)] synthesised from butadiene and Pd(dba)$_2$, or [Pd(PPh$_3$)η^1,η^3-C$_8$H$_{12}$]. These compounds are well known and described in the literature [102–104]. Their reactivity and role as intermediates in insertion reactions is also well documented.

An alternative second mechanism of telomerisation has been suggested based on the findings of W. Keim and involves a di-palladium bis-allyl species, especially if acetic acid or phenol are used as nucleophiles [105]. In this mechanism, two butadiene molecules coordinate after dissociation of ligands at the bridged-palladium species (Scheme 6.39), followed by a C-C-bond coupling and nucleophilic attack at C-1 or C-3 position, yielding the telomers. If no nucleophilic attack occurs, dimers are formed, for example 1,3,7-octatriene from butadiene.

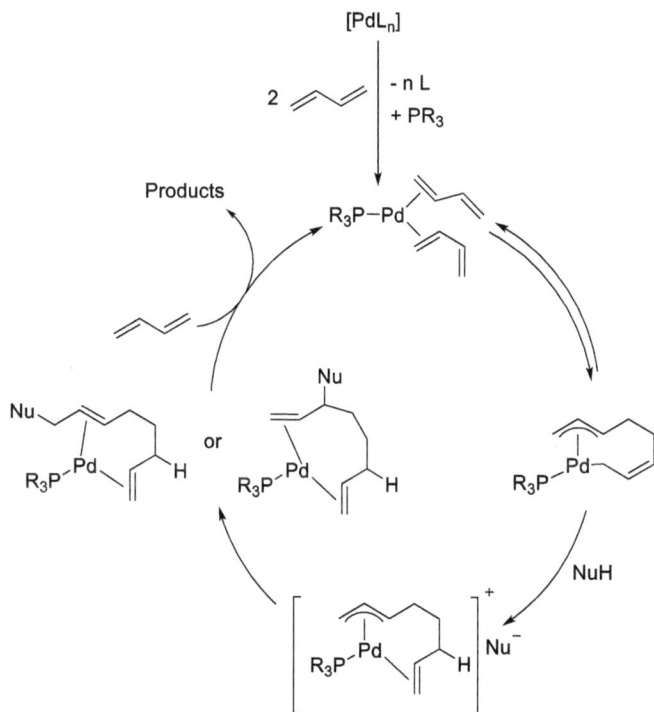

Scheme 6.38: Monopalladium-bisallyl-mechanism of the Pd-catalysed telomerisation reaction of butadiene.

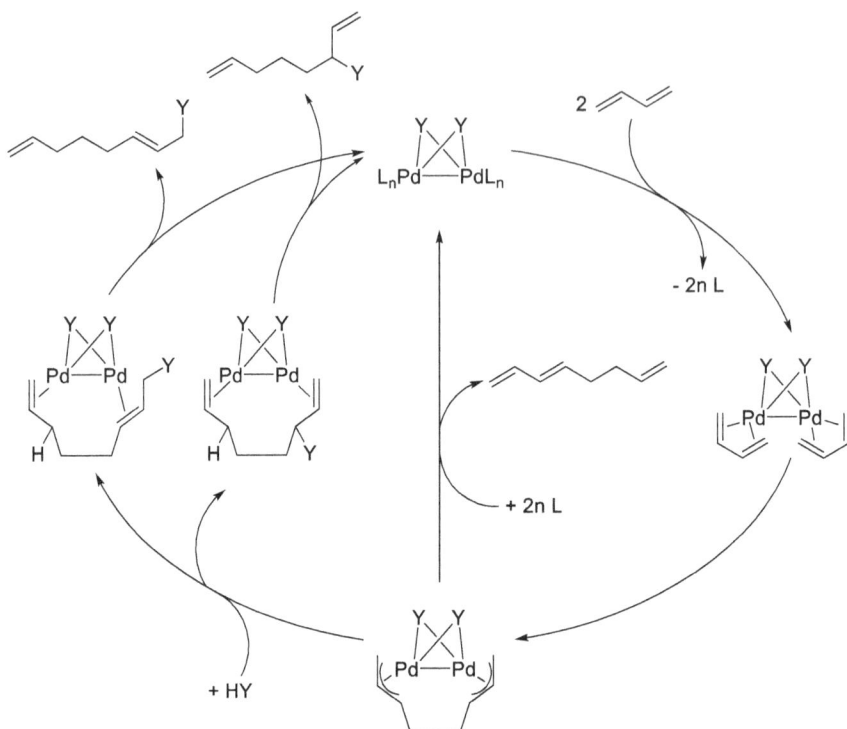

Scheme 6.39: Dipalladium-bisallyl-mechanism of telomerisation.

Nucleophiles that can be used in the butadiene telomerisation vary from polyols to acids and amines. If polyols are used (e.g. ethylene glycol), mono- and di-telomers are formed (Scheme 6.40) [106]. Usually this telomerisation reaction is performed with Pd/TPP or Pd/TPPTS catalysts in mono- or bi-phasic solvent systems at 80 °C. The selectivity of the reaction varies, depending on the reaction conditions; di-telomers are formed under mono-phasic conditions at 18% yield whereas under biphasic conditions, it is <1%. Mono-telomer yields are 74% under biphasic conditions and 53% under mono-phasic conditions. The TON of the reaction is in both cases >1,200.

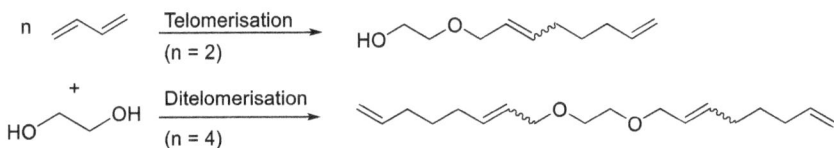

Scheme 6.40: Pd-catalysed telomerisation of butadiene and ethylene glycol.

If the telomerisation of butadiene is carried out with sugars as nucleophile, dienyl ethers are formed (Scheme 6.41). Aldoses like glucose, xylose or higher sugars such as sucrose or poly-carbohydrates such as starch have been converted. The reaction is carried out at >70 °C, with a metal-ligand ratio of 1:2 to 1:3 in DMF or water/triethyl-amine and an s/c >100. The telomerisation of butadiene and starch results in products for surfactants, antifoaming agents or reactive diluents [107–118]. Dienes such as iso-prene and piperylene (1,3-pentadiene) have been used in telomerisation reactions with polyols but industrial applications are currently not known [119, 120].

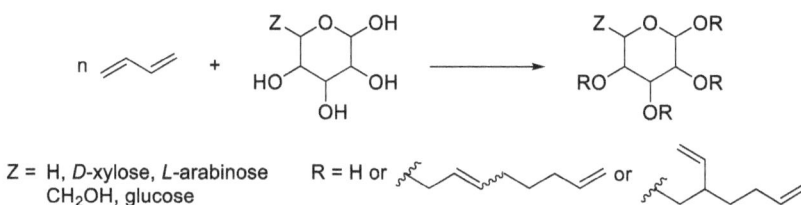

Z = H, D-xylose, L-arabinose R = H or ⌇⌇⌇⌇⌇ or ⌇⌇⌇⌇⌇
CH₂OH, glucose

Scheme 6.41: Butadiene telomerisation with aldoses.

The telomerisation of butadiene with acids is also described in literature. 1,7-Octadiene is formed from butadiene and formic acid with the elimination of carbon dioxide in a 96% yield (Scheme 6.42) [121, 122]. The preferred catalysts for this reaction are Pd–carbene complexes or Pd-phosphine complexes, which allows catalyst loadings of around s/c = 10,000. The reaction is carried out in the presence of a base such as triethylamine in a temperature range of 50–100 °C.

Scheme 6.42: 1, 7-Octadiene from telomerisation reaction.

The telomerisation of butadiene and carbon dioxide in the presence of Pd-based cata-lysts (e.g. $Pd(acac)_2$ and tricyclohexylphosphine (PCy_3)) produces δ-lactone in selectiv-ities of up to 95% after approximately a 15 h reaction time (Scheme 6.43) [123, 124].

The formation of 1-acetoxy-octa-2,7-diene obtained in the telomerisation of buta-diene with acetic anhydride can be explained with the di-palladium-bisallyl mecha-nism (see Scheme 6.39). $Pd/TPP/Ac_2O$ catalyst systems at catalyst loadings of >1,500 at 70 °C in dioxane resulted in a 35% conversion and a selectivity of 85% (Scheme 6.44) [125].

Phthalic acid and butadiene react in a $Pd(OAc)_2/P(OC_6H_4OMe)_3$-catalysed reaction in DMSO to give octyl phthalates, which are used as plasticisers. The catalyst loading s/c is >3,000. The catalysed-ligand system is separated, recycled in the DMSO phase and the product is separated in an isooctane phase (Scheme 6.45) [94].

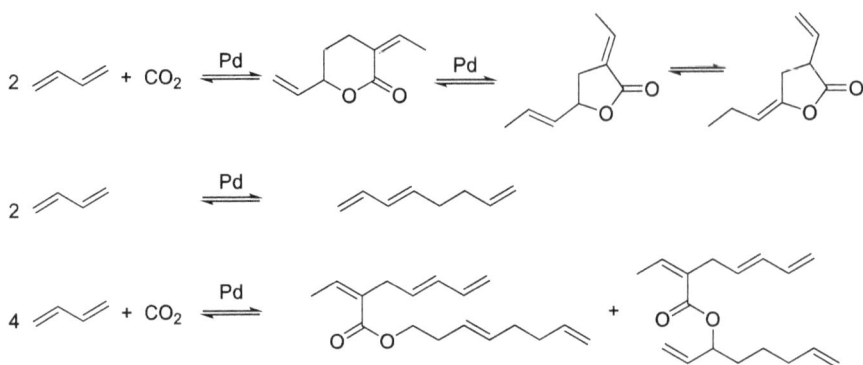

Scheme 6.43: Products obtained of telomerisation of butadiene and carbon dioxide.

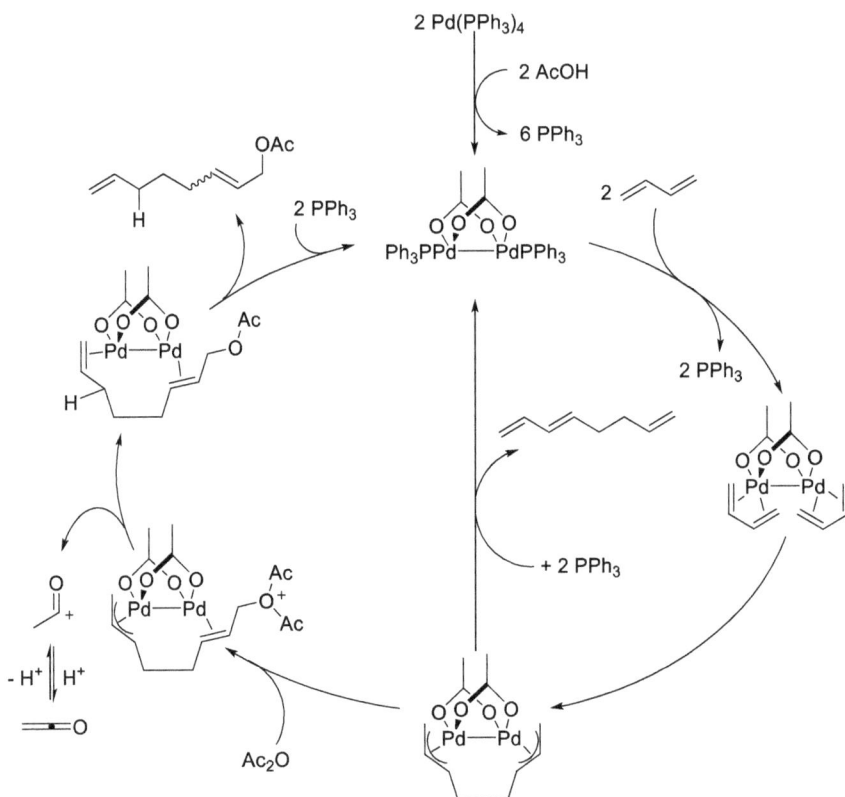

Scheme 6.44: Telomerisation of butadiene with acetic acid.

The telomerisation of butadiene and amines is influenced by the basicity of the amines and the reactivity of the metal, which increases in the series Ni < Pt < Pd. Butadiene reacts with diethylamine using the catalyst system Pd(OAc)$_2$, TPP (triphenyl

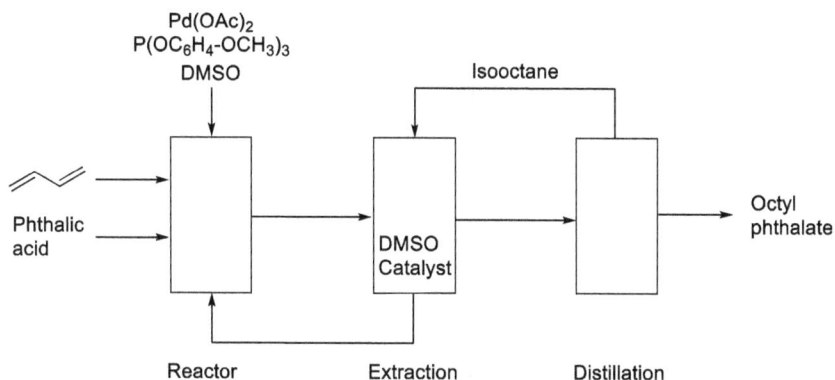

Scheme 6.45: Pd-catalysed octyl phthalate synthesis by telomerisation of butadiene and phthalic acid.

phosphine), triethyl aluminium, with a Pd/P/Al = 1/1/2 to yield mainly (E)-N,N-diethyl-2,7-octadienylamine, a precursor for allylic chlorides and acetates (Scheme 6.46) [126]. The reaction can be carried out in s/c of around 500 in toluene at 40 °C.

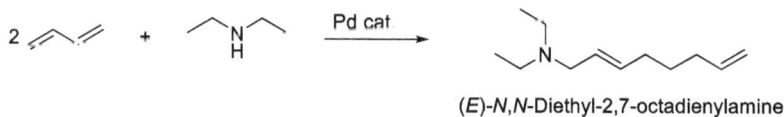

Scheme 6.46: (E)-N,N-diethyl-2,7-octadienylamine by telomerisation.

For the telomerisation of ammonia and butadiene, a two-phase solvent system was chosen to avoid further reaction to secondary or tertiary amines (Scheme 6.47) [127, 128]. The catalyst system Pd/TPPTS in water can easily be recovered and reused. The reactions are carried out at around 100 °C and a mixture of tertiary/secondary and primary amines is obtained in the ratio of 2/26/32.

The palladium-catalysed telomerisation of isoprene and methanol can be carried out in a mono-phasic or biphasic solvent system at s/c of around 1,000 and results in yields of 75% (Scheme 6.48). The product (1-methoxy-2,7-dimethyl-octa-2,7-diene) is an interesting molecule for the flavour and fragrance industry. It can be produced in the thermomorphic system containing water-methanol-isoprene at 60 °C with TPP (single phase) or TPPTS as the ligand. The catalyst-ligand can be separated with the water phase after cooling to room temperature and reused after separation of non-converted isoprene [129].

Dienes from renewable sources such as myrcene (C_{10}) can be used in telomerisation reactions to yield C_{20}-compounds selectively. Nucleophiles such as glycerol or sugar derivatives can be incorporated easily in this synthesis pattern. Therefore, telomerisation is a useful tool to convert renewable building blocks into valuable products in several applications [130]. Similarly, myrcene telomerisation in the presence of amines in ther-

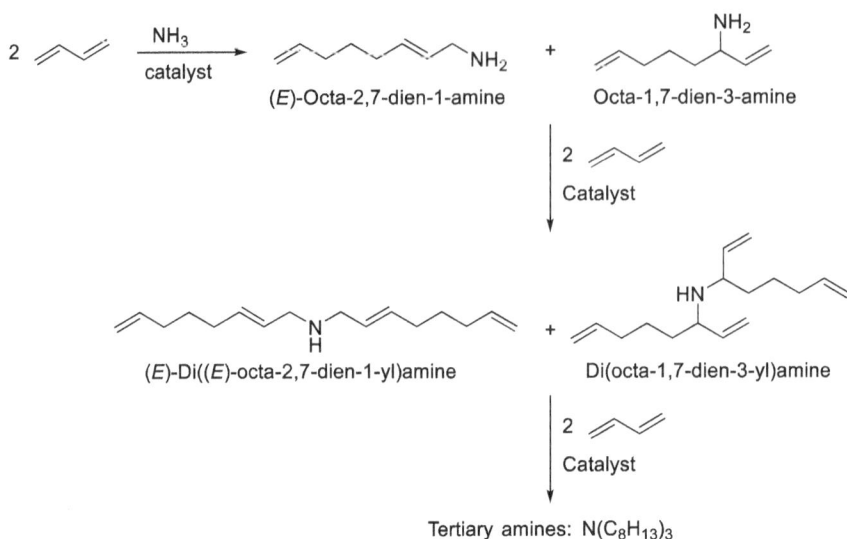

Scheme 6.47: Telomerisation of butadiene and ammonia.

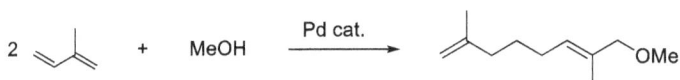

Scheme 6.48: Telomerisation of isoprene and methanol.

momorphic solvent systems results in tertiary amines in high yield; the solvent system can then be recovered and reused [131].

Highly branched industrially relevant C_{21}-esters, useful as non-ionic surfactants, can be synthesised in a Pd-catalysed carboxy-telomerisation of a branched 1,3-diene, preferably β-myrcene, with an alcohol and carbon monoxide (Scheme 6.49). The reaction has an atom economy of 100%. The reaction is suitable for several alcohols, and yields >95% can be achieved. Usually, mono-phosphine ligands are used [132].

The telomerisation of sesquiterpenes (C_{15}-compounds), for example β-farnesene with secondary amines can be performed in a Pd-catalysed reaction. C_{30} compounds are obtained in 94% yield (Scheme 6.50) [133]. The outcome of the reaction depends on the scaffold of the amine, its basicity and activity as nucleophile. The catalyst can be recovered by decantation.

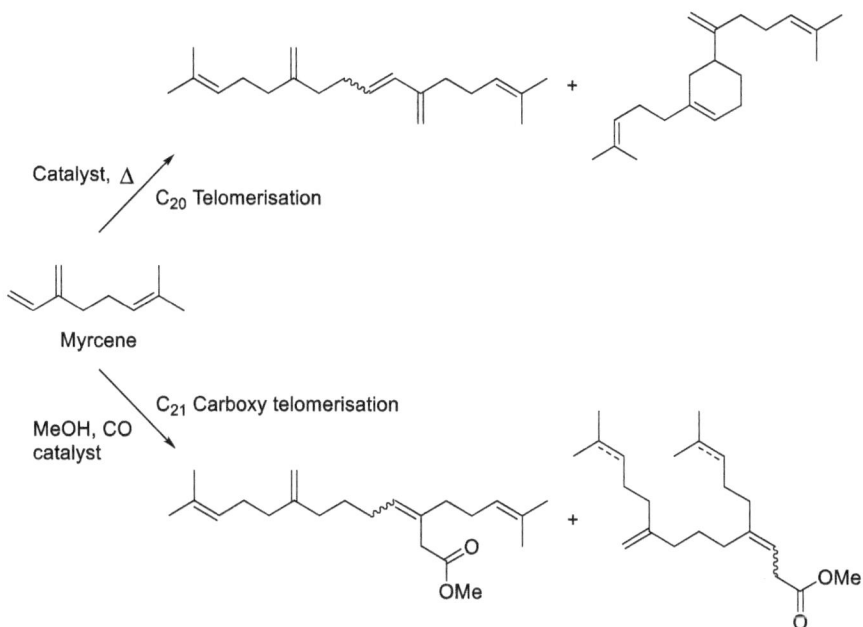

Scheme 6.49: Telomerisation and carboxy-telomerisation of myrcene.

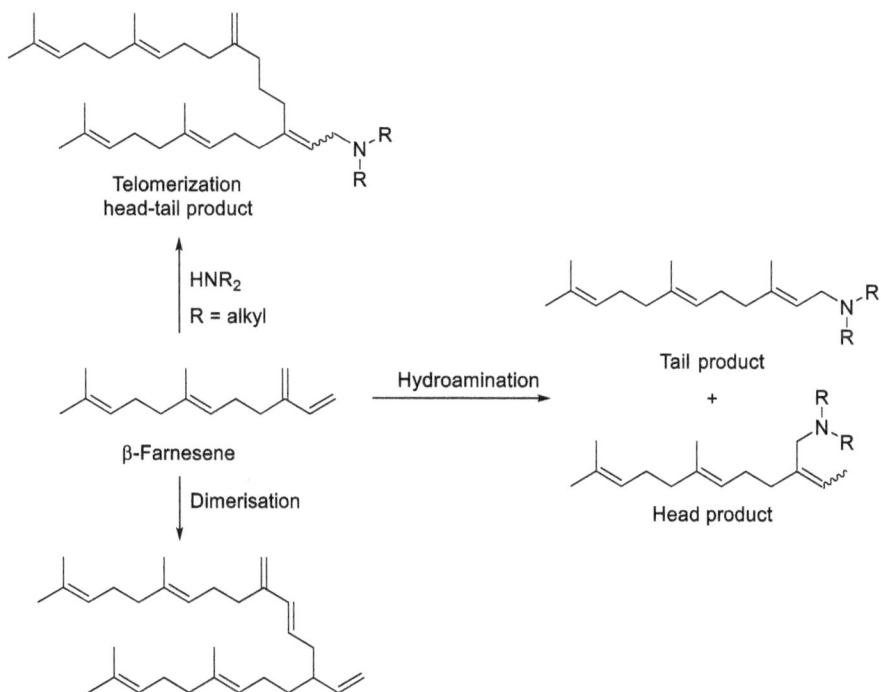

Scheme 6.50: Telomerisation of β-farnesene in presence of amines and competitive reactions.

6.8 Metathesis

In organic chemistry, the phrase "metathesis" describes a reaction for C–C bond formation. In alkene metathesis, C =C bonds are formed, whereas in the alkyne metathesis, C≡C bonds are created. The word metathesis is derived from the Greek language and means "change of position". Over the last few decades, the importance of metathesis increased in academia and industry, especially in organic synthesis and olefin chemistry. Y. Chauvin, R. R. Schrock and R. H. Grubbs were honoured with the Nobel Prize in 2005 for their work on mechanistic investigations of the metathesis reaction and the development of highly reactive catalysts.

First investigations in the field of metathesis reactions were performed by several research groups. K. Ziegler, G. Wilke and co-workers observed the formation of 1-butene from ethylene and were astonished that no higher saturated long-chain hydrocarbons were formed. This was a result of traces of nickel in the reactor which inhibited the propagation in favour of alkene formation, hence stopping the reaction at 1-butene; this was later called the Ni-effect (Scheme 6.51) [134].

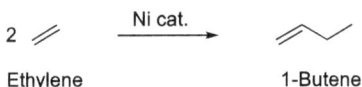

Scheme 6.51: Nickel effect in the ethylene oligomerisation.

R. L. Banks and G. C. Baily found that in the presence of heterogeneous Mo- and W-catalysts, propene reacts into ethylene and butene (Scheme 6.52) [135].

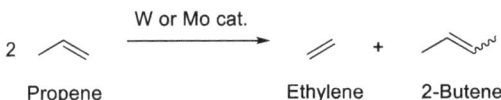

Scheme 6.52: Propene metatheses into ethylene and butene.

The WCl_6 and $EtAlMe_2$ in ethanol/benzene-catalysed reaction of 2-pentene into a 1:1 mixture of 2-butene and 3-hexene was called metathesis (Scheme 6.53) [136]. The reaction is complete in seconds, and an equilibrium mixture of products and starting material is observed.

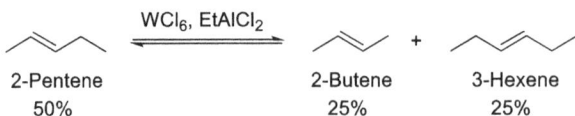

Scheme 6.53: 2-Pentene metathesis reaction.

The reaction of cyclopentene and 2-pentene in the presence of $WOCl_4$ and $SnBu_4$ or $AlEt_2Cl$ results in a 1:2:1 mixture of C_9, C_{10} and C_{11} diene compounds (Scheme 6.54). These findings were explained by Y. Chauvin and form the basis of the discussion about the reaction mechanism [137].

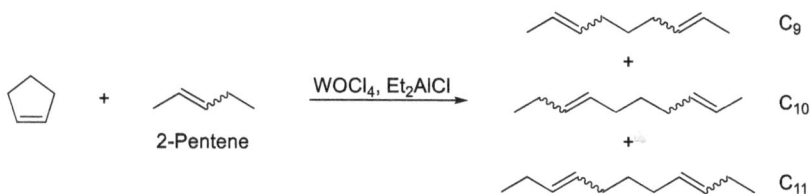

Scheme 6.54: Metathesis reaction of 2-pentene and cyclopentene.

In the reaction mechanism of the metathesis reaction, a four-membered metallocycle intermediate was postulated (Scheme 6.55). Furthermore, a carbene complex, formed by α-hydride elimination was discussed [137].

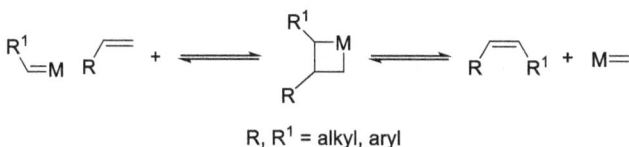

R, R^1 = alkyl, aryl

Scheme 6.55: Simplified reaction mechanism of metathesis reaction.

The postulation of a metallo-cyclobutane intermediate or transition state in the metathesis reaction mechanism is also in agreement with the *Woodward–Hoffmann* rules, because the direct [2+2] cyclo-addition reaction of alkenes is symmetry-forbidden. The major driving force for the metathesis reaction is entropy, because smaller molecules such as ethylene are separated from the reaction system. Especially in ring-closing and cross-metathesis reactions, the entropic effect is favoured because ethylene and propene are removed from the reaction mixture as gases. Starting materials for cross-metathesis (CM) and ring-closure metathesis (RCM) are mainly α-olefins. Furthermore, the formation of five- or six-membered ring compounds is enthalpically favourable in RCM; however, larger macrocycles can also be formed.

New defined metal-carbene complexes developed by Schrock (Mo(VI)-, and W(VI)-complexes with alkoxy and imino ligands) and Grubbs (Ru(II) complexes) have resulted in active and stable catalysts for a broad range of synthetically useful metathesis reactions (Scheme 6.56). Amongst these modified Ru-complexes, the so-called *Hoveyda–Grubbs* catalysts are especially important based on their stability and functional group tolerance [138–140].

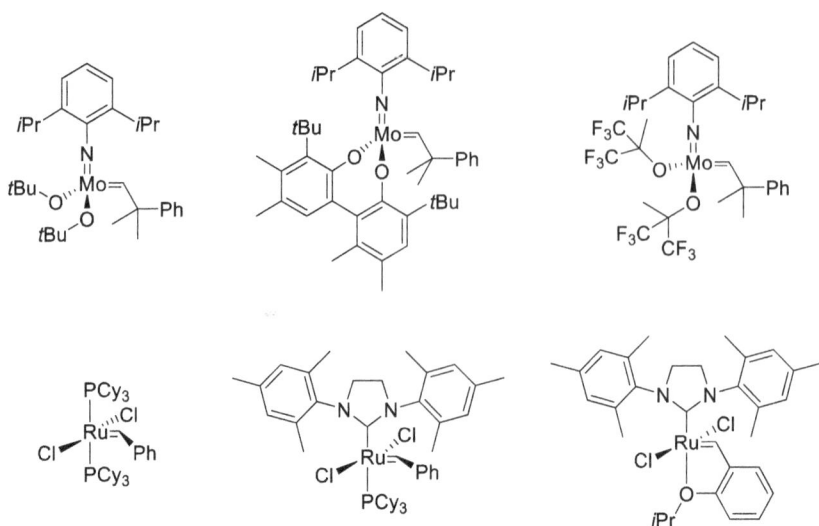

Scheme 6.56: Examples of modern metathesis catalysts.

The application of carbene complexes in a metathesis reaction was first described in 1974 when Casey used a W-carbene-complex in the reaction with isobutene, forming a mixture of products (Scheme 6.57) [141].

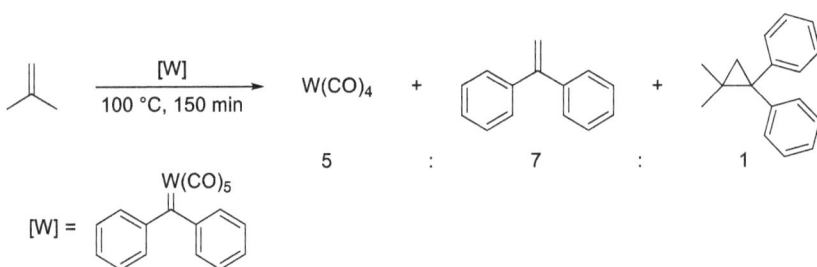

Scheme 6.57: Example of a W-carbene-complex in the metathesis reaction.

The *Schrock*-alkylidene Mo-complexes are highly active but sensitive to air and moisture, resulting in a limited functional group tolerance. The Ru-complexes developed by Grubbs, and their next generation Ru-type catalysts developed by Hoveyda are also highly reactive and less sensitive to most functional groups. In addition, ruthenium catalysts containing an unsaturated *N*-heterocyclic carbene were independently developed by Nolan [142], Grubbs [143] and Herrmann [144] in 1999. These modern types of homogeneous metathesis catalysts work under mild reaction conditions, for example Ru-catalysts are often used at room temperature. Also, the heterogeneous catalyst Re_2O_7/Al_2O_3 works at room temperature, whereas the heterogeneous system WO_3/SiO_2 needs a reaction temperature of around 300 °C.

The application of metathesis reactions as a synthetic tool box is well documented in the literature for cross-metathesis, ring-closing metathesis (RCM), ring-opening metathesis (ROM), ring-opening metathesis polymerisation (ROMP) and enyne metathesis (Scheme 6.58). The acyclic diene metathesis (ADMET) is applied in the synthesis of new polymers [145].

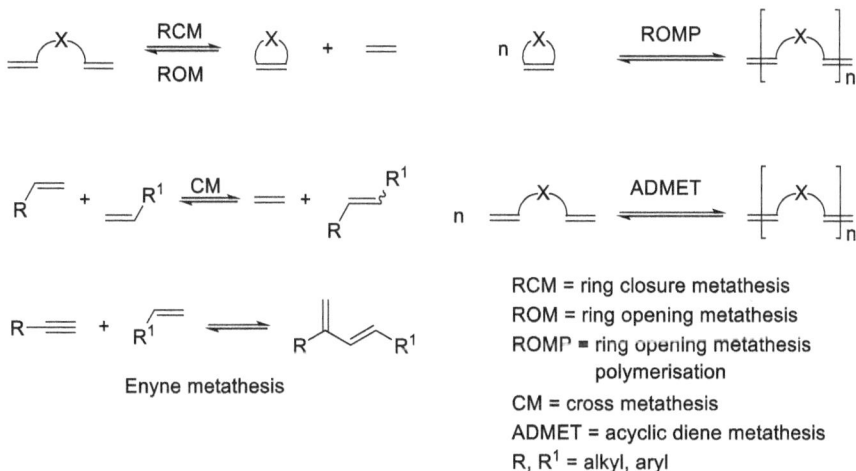

RCM = ring closure metathesis
ROM = ring opening metathesis
ROMP = ring opening metathesis
          polymerisation
CM = cross metathesis
ADMET = acyclic diene metathesis
R, R^1 = alkyl, aryl

Enyne metathesis

Scheme 6.58: Types of metathesis reaction.

Industrial applications of metathesis reactions are the synthesis of 2-butene from ethylene and propene (*Phillips Triolefin* Process), the Shell higher olefin process (see Section 6.5: SHOP process) for the synthesis of α-alkenes, the production of neohexene from isobutene dimers, the 1,5-hexadiene and 1,9-decadiene synthesis from 1,5-cyclooctadiene and cyclooctene and in the synthesis of drugs and natural products.

In the *Phillips Triolefin* Process, originally $Mo(CO)_6$ or $W(CO)_6$ supported on Al_2O_3 were employed as catalysts. It was later found that Re and Mo containing heterogeneous catalysts are more efficient [135, 146]. In another process, a WO_3/SiO_2 catalyst is applied at 350–450 °C. The reverse reaction – conversion of ethylene and 2-butene into propene – is also carried out (Scheme 6.59) [147].

Ethylene   2-Butene   Propene

Scheme 6.59: Manufacture of propene from ethylene and butene (mixture of (*E/Z*)-isomers) by metathesis reaction.

Linear α-alkenes are important intermediate compounds in the manufacture of detergents and fatty alcohols. They are produced by the Shell higher olefin process which was implemented at Royal Dutch Shell and starts from ethylene. The chemical reactions involved are oligomerisation and metathesis reactions (see Scheme 6.30). The production volume is $>10^6$ t/a [148]. The ethylene polymerisation reaction stops growing after around 10 ethylene molecules and ends up in a fraction of C_{12}-C_{18} olefins (40–50% of the product stream). This stream is directly used in downstream processing of fatty alcohols. The remaining material is isomerised using an alkaline on alumina in the liquid phase, that is 1-octene to 4-octene. An olefin metathesis reaction converts a mixture of olefins to 2-tetradecene (C_{14}) [148]. Alternatively, the internal alkenes react in the presence of a Re_2O_7/Al_2O_3 catalyst and ethylene to form a mixture of α-alkenes with an odd and even chain-length. It should be mentioned here that the C_{12}-C_{18} alkenes can be used in a hydroformylation reaction, resulting in aldehydes that are hydrogenated to the corresponding alcohols. These alcohols are used in detergent manufacturing [149].

3,3-Dimethylbut-1-ene (neohexene) is a precursor to synthetic musk aroma and is used in the manufacture of the antifungal drug terbinafine. The synthesis of neohexene is performed from *iso*-octene in a metathesis reaction with ethylene. The production volume of neohexene is around 3,000 t/a. The metathesis reaction is performed with WO_3/SiO_2 and a mixture of ethylene and *iso*-octene at 30 bar and 370 °C, in the presence of MgO (Scheme 6.60) [150].

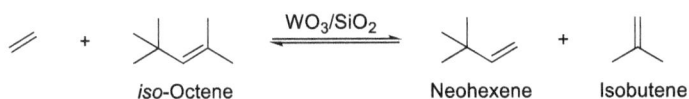

Scheme 6.60: Synthesis of neohexene by metathesis reaction.

The synthetic musk aroma is synthesised from neohexene and *p*-cymene, followed by acetyl chloride treatment (Scheme 6.61) [150]. The resulting naphthalene derivative is known as Amber; it has a woody odour and is used as a fragrance ingredient in laundry products [151].

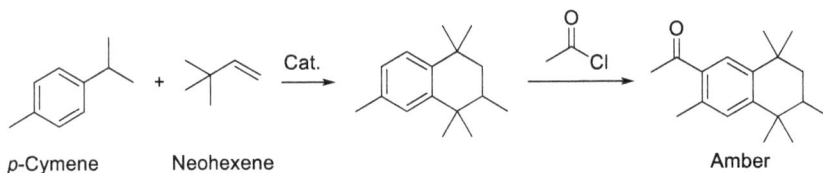

Scheme 6.61: Synthesis of a synthetic musk aroma compound.

1,5-Hexadiene is produced from 1,5-cyclooctadiene (see Chapter 4: Oxidations, butadiene cyclotrimerisation in the syntheses of lauryl lactam via *Beckmann* rearrangement) and ethylene in the presence of Re_2O_7/Al_2O_3 in a fixed-bed reactor set-up (Scheme 6.62) [152].

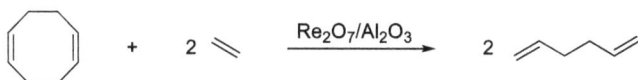

Scheme 6.62: Synthesis of 1,5-hexadiene.

An interesting approach to (*R*)-(-)-muscone via (-)-muscenone starts from (*R*)-(+)-citronellal. A ring-closing metathesis reaction with a homogeneous Ru-complex is the key step of the synthesis; however, until now it has not been commercialised (Scheme 6.63) [153]. The yield of the metathesis ring-closure reaction is 78%, and the reduction of the C =C bond was performed in methanol in presence of Pd/C in 98% yield.

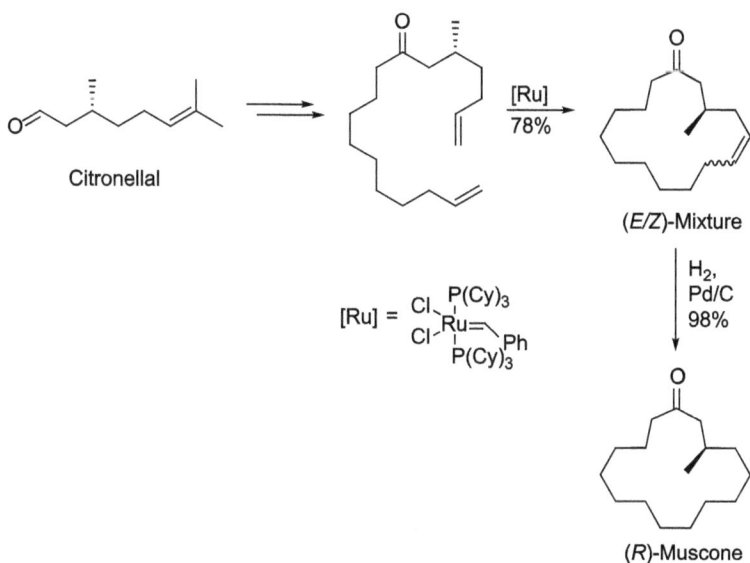

Scheme 6.63: Metathesis reaction approach to (*R*)-(-)-muscone.

6.9 Formation of C–N bonds (*Buchwald–Hartwig* coupling)

The Pd-catalysed coupling reaction of a primary or secondary amine and an aryl-halide or aryl-triflate is the *Buchwald–Hartwig* coupling. The reaction was developed in the 1990s and is carried out in the presence of bases such as ^{t}BuOK or bis(trimethylsilyl)amide (Scheme 6.64) [154–156]. The *Buchwald–Hartwig* reaction is used in the formation of C–N bonds and allows the synthesis of aryl amines. Over the recent few decades, the scope of the reaction has been extended by the establishment of catalyst systems that perform under milder reaction conditions.

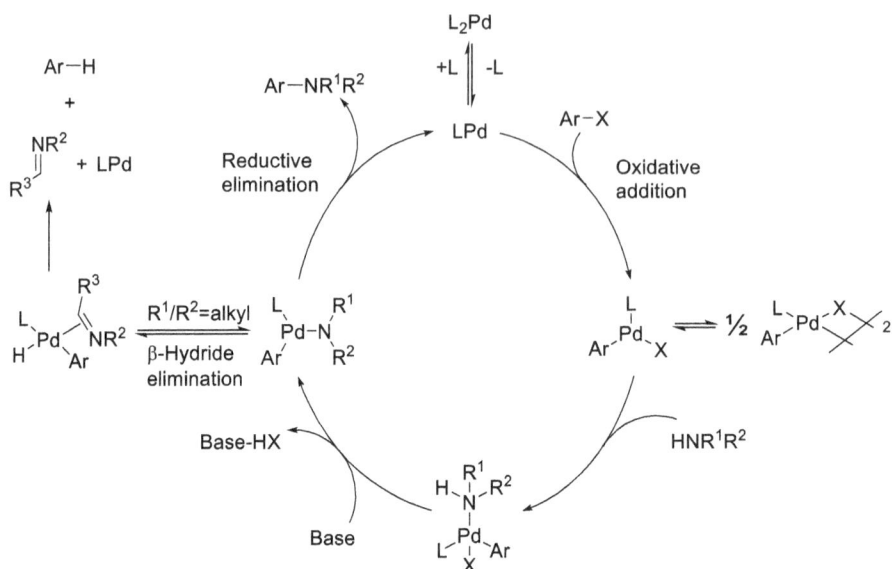

Scheme 6.64: Mechanism of the *Buchwald–Hartwig* amination.

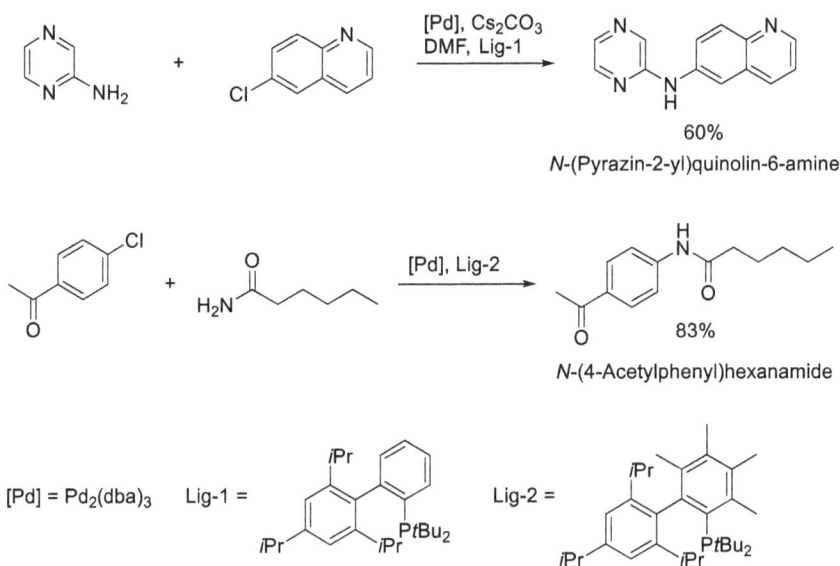

60%

N-(Pyrazin-2-yl)quinolin-6-amine

83%

N-(4-Acetylphenyl)hexanamide

[Pd] = Pd₂(dba)₃   Lig-1 =   Lig-2 =

Scheme 6.65: Coupling reaction of amines with electron-withdrawing groups in presence of sterically hindered ligands.

From a mechanistic point of view, a Pd(0)-species such as a PdL_2-complex, $L = PR_3$, is assumed to be the active catalyst [155]. Similar to the *Mizoroki–Heck* reaction, in the *Buchwald–Hartwig* amination reaction, the oxidative addition of the aryl halide (generally bromide) to the Pd(0) complex is a key step in the catalytic cycle. A potential side reaction, β-hydride elimination, can take place before the reductive elimination if one of the groups on the amine is alkyl, resulting in an imine and a de-halogenated aromatic group. Both electron-rich and electron-poor arenes can be used in the *Buchwald–Hartwig*-reaction; reaction parameters such as reaction time, temperature and catalyst loading influence the progress and outcome of the reaction (Schemes 6.65 and 6.66) [154]. These modifications allow the use of the *Buchwald-Hartwig*-reaction in the industrial synthesis of pharmaceuticals and natural products [157–159].

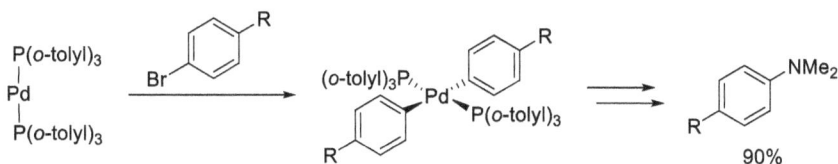

Scheme 6.66: PdL_2 as precursor in the Buchwald–Hartwig amination reaction.

The coupling of aryl iodides or aryl triflates with primary amines in the *Buchwald–Hartwig* amination reaction was achieved by the use of bidentate ligands of the binaphthyl-type, for example bis(diphenylphosphino)binaphthyl (BINAP) or bis(diphenylphosphino)ferrocene (dppf) and yields of up to 95% can be achieved [160–162]. A further development in the *Buchwald–Hartwig* reaction was the use of benzophenone imine which allows the coupling reaction of ammonia synthons and aryl halides to synthesise primary amines (Scheme 6.67) [163, 164]. Alternatively, silylated amines can be used and later deprotected.

A further trend in metal-catalysed coupling reactions, also observed in the *Buchwald-Hartwig*-reaction, is the replacement of expensive Pd-based catalysts by cheaper metal catalysts such as those based on Cu or Ni [165]. More recently, even the nickel-catalysed mono arylation of ammonia has been reported [166]. In this case, Josiphos-nickel complexes were used that improved the stability and activity of the key Ni(0) species [167].

The use of a heterogeneous catalyst such as Pd/C in the presence of phosphine or NHC ligands for the *Buchwald–Hartwig* reaction of aryl halides (iodines, bromines and chlorides) has also been investigated. Here, it must be pointed out that Pd-leaching from the catalyst is mainly responsible for the catalytic activity of these systems [168–171]. An almost full conversion and selectivities of around 95% could be obtained in both protic and non-polar solvents (such as ethanol or toluene) at catalyst loadings of s/c = 1,000.

Bifenazate, a pesticide against spider mites (*tetranychidae*), produced by the Uniroyal Chemical Company, can be synthesised using a *Buchwald–Hartwig* amination. A bromo-biphenyl derivative is coupled to a protected hydrazine with a $Pd(OAc)_2$-BINAP

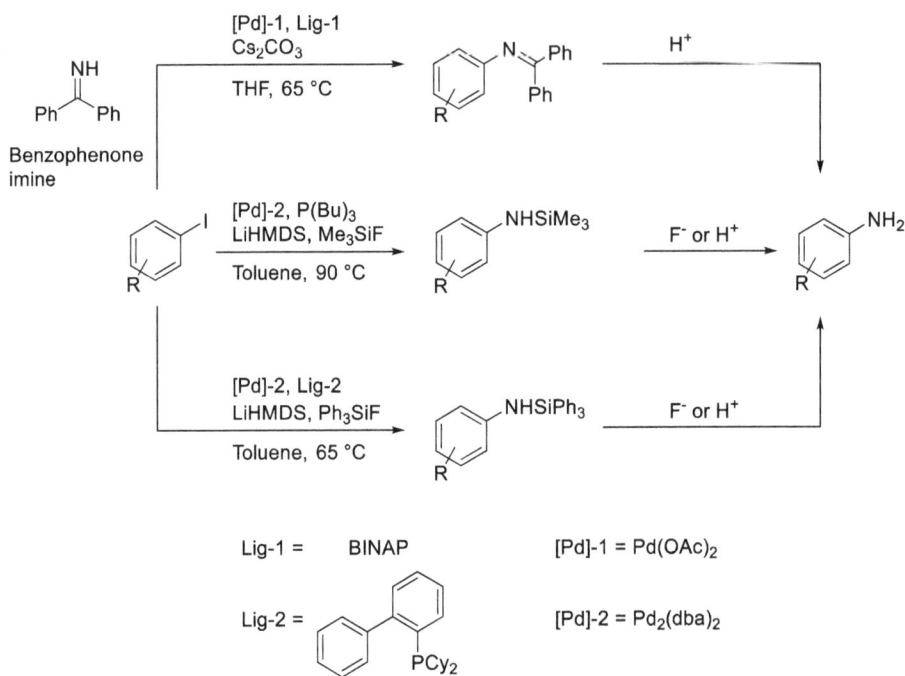

Scheme 6.67: Synthesis of primary amines by *Buchwald–Hartwig* reaction.

catalytic system and a strong base such as sodium *t*-butoxide. Hydrolysis, followed by carbamate formation, results in the active compound (Scheme 6.68) [172].

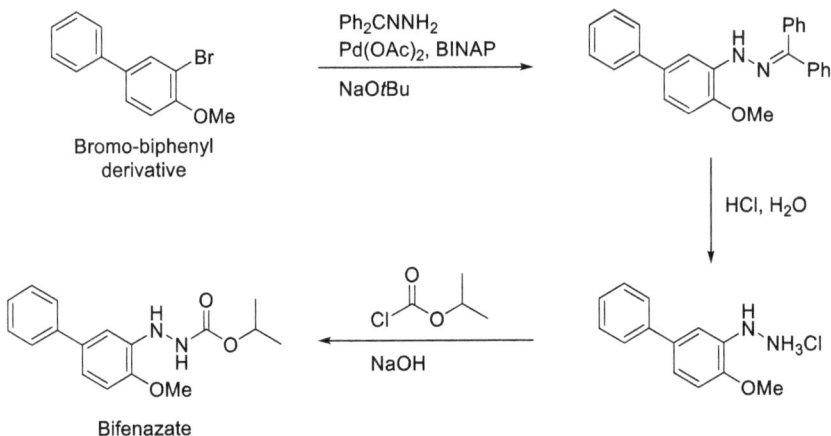

Scheme 6.68: *Buchwald–Hartwig* reaction in the synthesis bifenazate.

Further developments in the formation of C-C and C-N bonds in the field of fine chemical synthesis (which is a rapidly growing area) will be focused on new selective catalysts. A second area of interest will be the replacement of expensive ligands and high-pressure equipment. Therefore, a methodical research and further development of these types of reactions are required.

References

[1]	Negishi E-I. Magical power of transition metals: Past, present, and future (Nobel Lecture). Angew Chem Int Ed 2011, 50, 6738–6764.
[2]	Suzuki A. Cross-coupling reactions of organoboranes: An easy way to construct C=C bonds (Nobel Lecture). Angew Chem Int Ed 2011, 50, 6722–6737.
[3]	Mizoroki T, Mori K, Ozaki A. Arylation of olefins with aryl iodide catalyzed by palladium. Bull Chem Soc Jpn 1971, 44, 581.
[4]	Heck R F, Nolley J P. Palladium-catalyzed vinylic hydrogen substitution reactions with aryl, benzyl, and styryl halides. J Org Chem 1972, 37, 2320–2322.
[5]	Bradshaw M, Zou J, Byrne L, Swaminathan I K, Stewart S G, Raston C L. Pd(II) conjugated chitosan nanofibre mats for application in Heck cross-coupling reactions. Chem Commun 2011, 47, 12292–12294.
[6]	Ozawa F, Kubo A, Hayashi T. Generation of tertiary phosphine-coordinated Pd(0) Species from Pd(OAc)$_2$ In the catalytic heck reaction. Chem Lett 1992, 2177–2180.
[7]	Amatore C, Carre E, Jutand A, M'Barki M A. Rates and mechanism of the formation of zerovalent palladium complexes from mixtures of Pd(OAc)$_2$ and tertiary phosphines and their reactivity in oxidative additions. Organometallics 1995, 14, 1818–1826.
[8]	Torborg C, Beller M. Recent applications of palladium-catalyst coupling reactions in the pharmaceutical, agrochemical, and fine chemical industry. Adv Synth Catal 2009, 351, 3027–3043.
[9]	Roy D, Uozumi Y. Recent advances in palladium-catalyzed cross-coupling reactions at ppm to ppb Molar catalyst loadings. Adv Synth Catal 2018, 360, 602–625.
[10]	Baumeister P, Seifert G, Steiner H. Process for the Preparation of Substituted Benzenes and Benzene Sulfonic Acid and Derivatives Thereof and a Process for the Preparation of N'N-substituted Ureas. EP 584043, Ciba-Geigy AG, 1992.
[11]	Lin R W, Herndon R C J, Allen R H, Chockalingham K C, Focht G D, Roy R K. Preparation of Carboxylic Compounds and Their Derivatives, WO 9830529, Albemarle Corporation, 1998.
[12]	Tse-Chong W. Process for Preparing Aryl-substituted Aliphatic Carboxylic Acids and Their Esters via Regioselective Hydrocarboxylation of 1-arylalkenes in Presence of Palladium-copper-cyclic Phosphine Catalysts. US 5315026, Ethyl Corporation, 1994.
[13]	Tse-Chong W. Process for Preparing Olefins. US 5536870, Albemarle Corporation, 1996.
[14]	Eisenstadt A. Utilization of Heterogeneous Palladium-on-Carbon Catalyzed Heck Reactions in Applied Synthesis. In: Herkes F E, ed. Catalysis of Organic Reactions, New York, Chemical Industries Series, Marcel Dekker, Vol. 75, 1998, 415–427.
[15]	Köhler K, Heidenreich R G, Krauter J G E, Pietsch J. Highly active palladium/activated carbon catalysts for Heck reactions: Correlation of activity, catalyst properties and Pd leaching. Chem Eur J 2002, 8, 622–631.
[16]	Biffis A, Zecca M, Basato M. Metallic palladium in the heck reactions active catalysts or convenient precursor? Eur J Inorg Chem 2001, 1131–1133.
[17]	Shmidt A F, Mametova L V. Main features of catalysis in the styrene phenylation reaction. Kinet Catal 1996, 37, 406–408.

[18] Renaud S, de Lorgeril M. Wine, alcohol, platelets, and the French paradox for coronary heart disease. The Lancet 1992, 339, 1523–1526.

[19] Härter R, Lempke U, Radspieler R. Process for the preparation of stilbene derivatives. WO 2005023740, DSM IP Assets B. V., 2005.

[20] Schils D, Stappers F, Solberghe G, van Heck R, Coppens N, van den Heuvel D, van der Donck P, Callesvaert T, Meeussen F, de Bie E, Eersels K, Schouteden E. Ligandless Heck coupling between a halogenated aniline and acrylonitrile catalyzed by Pd/C: Development and optimization of an industrial-scale Heck process for the production of a pharmaceutical intermediate. Org Process Res Dev 2008, 12, 530–536.

[21] Guillemont J, Pasquier E, Palandjian P, Vernier D, Gaurrand S, Lewi P J, Heeres J, de Jonge M R, Ko-ymans L M H, Daeyaert F F D, Vinkers M H, Arnold F, Das K, Pnuwels R, Andries K, de Béthune M P, Bettens E, Hertogs K, Wigerinck P, Timmerman P, Janssen P A J. Synthesis of novel diarylpyrimidine analogues and their antiviral activity against human immunodeficiency virus type 1. J Med Chem 2005, 48, 2072–2079.

[22] Baert L E C, Lewi P J, Heeres J. Prevention of hiv-infection with tmc278. WO 2006106103, Tibotec Pharmaceuticals Ltd., 2006.

[23] Stoffels P. WO 2005021001, Tibotec Pharmaceuticals Ltd., September 3, 2004.

[24] Palmisano G, Bonrath W, Boffa L, Garella D, Barge A, Cravotto G. Heck reactions with very low ligandless catalyst loads accelerated by microwaves or simultaneous microwaves/ultrasound irradiation. Adv Synth Catal 2007, 349, 2338–2344.

[25] Cravotto G, Beggiato M, Penoni A, Palmisano G, Tollari S, Leveque J-M, Bonrath W. High-intensity ultrasound and microwave, alone or combined, promote Pd/C-catalyzed aryl-aryl couplings. Tetrahedron Lett 2005, 46, 2267–2271.

[26] Sonogashira K, Tohda Y, Hagihara N. A convenient synthesis of acetylenes: Catalytic substitutions of acetylenic hydrogen with bromoalkenes, iodoarenes, and bromopyridines. Tetrahedron Lett 1975, 4467–4470.

[27] Chinchilla R, Nájera C. The sonogashira reaction: A booming methodology in synthetic organic chemistry. Chem Rev 2007, 107, 874–922.

[28] Stambuli J P, Buhl M, Hartwig J F. Synthesis, characterization, and reactivity of monomeric, arylpalladium halide complexes with a hindered phosphine as the only dative ligand. J Am Chem Soc 2002, 124, 9346–9347.

[29] Amatore C, Jutand A. Anionic Pd(0) and Pd(II) intermediates in palladium-catalyzed Heck and cross-coupling reactions. Acc Chem Res 2000, 33, 314–321.

[30] Gazvoda M, Virant M, Pinter B, Košmrlj J. Mechanism of copper-free Sonogashira reaction operates through palladium-palladium transmetallation. Nat Commun 2018, 9, 1–9.

[31] Jutand A, Négri S, Principaud A. Formation of ArPdXL(amine) complexes by substitution of one phosphane ligand by an amine in trans-ArPdX(PPh$_3$)$_2$ complexes. Eur J Inorg Chem 2005, 631–635.

[32] Szekeres T, Repasi J, Szabo A, Mangion B. A Process for Preparing 3-(trifluoromethyl)-benzenepropanal and Its Derivatives as Intermediates for the Synthesis of Cinacalcet. WO 2008035212, Medichem S.A., June 8, 2007.

[33] Alain M, Feri F, Gaslain Y. A two-step synthesis of terbinafine. Tetrahedron Lett 1996, 37, 57–58.

[34] Castaldi G, Barreca G, Rossi R. Stereoselective Alkynylation Process and Cuprous Catalyst for the Preparation of Terbinafine. DE 60223862, Dinamite Dipharma S.p.A., 2002.

[35] Mujahidin D, Doye S. Enantioselective Synthesis of (+)-(S)-Laudanosine and (−)-(S)-Xylopinine. Eur J Org Chem 2005, 13, 2689–2693.

[36] Bleicher L S, Cosford N D P, Herbaut A, McCallum J S, McDonald I A. A practical and efficient synthesis of the selective neuronal acetylcholine-gated ion channel agonist (S)-(−)-5-Ethynyl-3-(1-methyl-2-pyrrolidinyl)pyridine Maleate (SIB-1508Y). J Org Chem 1998, 63, 1109–1118.

[37]   Wang D X, Booth H, Lerner-Marmarosh N, Osdene T S, Abood L G. Structure-activity relationships for nicotine analogs comparing competition for [3H]nicotine binding and psychotropic potency. Drug Dev Res 1998, 45, 10–16.

[38]   Miyaura N, Yamada K, Suzuki A. A new stereospecific cross-coupling by the palladium-catalyzed reaction of 1-alkenylboranes with 1-alkenyl or 1-alkynyl halides. Tetrahedron Lett 1979, 20, 3437–3440.

[39]   Miyaura N, Suzuki A. Stereoselective synthesis of arylated (E)-alkenes by the reaction of alk-1-enylboranes with aryl halides in the presence of palladium catalyst. J Chem Soc Chem Commun 1979, 19, 866–867.

[40]   Miyaura N, Suzuki A. Palladium-Catalyzed cross-coupling reactions of organoboron compounds. Chem Rev 1995, 95, 2457–2483.

[41]   Smith G B, Dezeny G, Hughes D L, King A O, Verhoeven T R. Mechanistic Studies of the Suzuki Cross-Coupling Reaction. J Org Chem 1994, 59, 8151–8156.

[42]   Matos K, Soderquist J A. Alkylboranes in the Suzuki–Miyaura coupling: Stereochemical and mechanistic studies. J Org Chem 1998, 63, 461–470.

[43]   Miyaura N, Yamada K, Suzuki A. A new stereospecific cross-coupling by the palladium-catalyzed reaction of 1-alkenylboranes with 1-alkenyl or 1-alkynyl halides. Tetrahedron Lett 1979, 36, 3437–3440.

[44]   Pagett A B, Lloyd-Jones G C. Suzuki–Miyaura Cross-Coupling. In Org React 2020, doi: 10.1002/0471264180.or100.09.

[45]   Casalnuovo A L, Calabrese J C. Palladium-catalyzed alkylations in aqueous media. J Am Chem Soc 1990, 112, 4324–4330.

[46]   Martin R, Buchwald S L. Palladium-Catalyzed Suzuki–Mlyaura cross-coupling reactions employing dialkylbiaryl phosphine ligands. Acc Chem Res 2008, 41, 1461–1473.

[47]   Amatore C, Jutand A, Le Duc G. Kinetic data for the transmetalation/reductive elimination in palladium-catalyzed Suzuki–Miyaura reactions: Unexpected triple role of hydroxide ions used as base. Chem Eur J 2011, 17, 2492–2503.

[48]   Lennox A J J, Lloyd-Jones G C. Organotrifluoroborate hydrolysis: Boronic acid release mechanism and an acid-base paradox in cross-coupling. J Am Chem Soc 2012, 134, 7431–7744.

[49]   Knapp D M, Gillis E P, Burke M A. General solution for unstable boronic acids: Slow-release cross-coupling from air-stable MIDA boronates. J Am Chem Soc 2009, 131, 6961–6963.

[50]   Wang Y, Li Y, Li Y, Zheng G, Li Y. Method for the Production of Losartan. WO 2006081807, Ratiopharm GmbH, 2006.

[51]   Larsen R D, King A O, Chen C Y, Corley E G, Foster B S, Roberts F E, Yang C, Lieberman D R, Reamer R A, Tschaen D M, Verhoeven T R, Reider P J, Lo Y S, Rossano L T, Brookes A S, Meloni D, Moore J R, Arnett J F. Efficient synthesis of losartan, a nonpeptide angiotensin II receptor antagonist. J Org Chem 1994, 59, 6391–6394.

[52]   Smith G B, Dezeny G C, Hughes D L, King A O, Verhoeven T R. Mechanistic Studies of the Suzuki Cross-Coupling Reaction. J Org Chem 1994, 59, 8151–8156.

[53]   Eicken K, Rack M, Wetterich F, Ammermann E, Lorenz G, Strathmann S. N-biphenylpyrazolecarboxamides as Fungicides. DE 19735224, BASF A.G., 1997.

[54]   Eicken K, Rang H, Harreus A, Goetz N, Ammermann E, Lorenz G, Strathmann S. Bisphenylamide. DE 19531813, BASF A.G., 1997.

[55]   Glasnov T N, Kappe C O. Toward a Continuous-Flow Synthesis of Boscalid®. Adv Synth Catal 2010, 352, 3089–3097.

[56]   Dockner M, Rieck H, Moradi W A, Lui N. Process for Preparing Substituted Biphenylanilides. WO 2009135598, Bayer Cropscience AG, 2009.

[57]   Oppenheimer J, Emonds M V M, Derstine C W, Clouse R C. Process for the Preparation of Methyl 4-amino-3-chloro-6-(4-chloro-2-fluoro-3-methoxyphenyl)pyridine-2-carboxylate. WO 2013102078, Dow AgroSciences LLC, 2012.

[58] Epp J B, Schmitzer P R, Crouse G D. Fifty years of herbicide research: Comparing the discovery of trifluralin and halauxifen-methyl. Post Manag Sci 2018, 74, 9–16.

[59] Tamao K, Sumitani K, Kumada M. Selective carbon-carbon bond formation by cross-coupling of Grignard reagents with organic halides. Catalysis by nickel-phosphine complexes. J Am Chem Soc 1972, 94, 4374–4376.

[60] Yamamura M, Moritani I, Murahashi S-I. The reaction of σ-vinylpalladium complexes with alkyllithiums. Stereospecific syntheses of olefins from vinyl halides and alkyllithiums. J Organomet Chem 1975, 91, C39–C42.

[61] Kumada M, Tamao K, Sumitani K. Phosphine-nickel complex catalyzed cross-coupling of Grignard reagents with aryl and alkenyl halides: 1,2-dibutylbenzene. Org Synth 1978, 58, 127–133.

[62] Tamao K, Sumitani K, Kiso Y, Zembayashi M, Fujioka A, Kodama S-I, Nakajima I, Minato A, Kumada M. Nickel-phosphine complex-catalyzed Grignard coupling. I. Cross-coupling of alkyl, aryl, and alkenyl Grignard reagents with aryl and alkenyl halides: General scope and limitations. Bull Chem Soc Jpn 1976, 49, 1958–1969.

[63] Sherry B D, Fürstner A. The promise and challenge of iron-catalyzed cross coupling. Acc Chem Res 2008, 41, 1500–1511.

[64] Hess W, Treutwein J, Hilt G. Cobalt-catalysed carbon-carbon bond formation reactions. Synthesis 2008, 3537–3562.

[65] Cahiez G, Duplais C, Buendia J. Chemistry of organomanganese(II) compounds. Chem Rev 2009, 109, 1434–1476.

[66] Johnson D S, Lee Jie J. Aliskiren (Tekturna), the First-in-Class Renin Inhibitor for Hypertension. In: J L J, Johnson D S, ed. Modern Drug Synthesis, Hoboken, John Wiley & Sons, Inc, NJ, 2010, 153–154.

[67] Cheng Y-J, Yang S-H, Hsu C-S. Synthesis of conjugated polymers for organic solar cell applications. Chem Rev 2009, 109, 5868–5923.

[68] Loewe R S, Ewbank P C, Liu J, Zhai L, McCullough R D. Regioregular, head-to-tail coupled Poly(3-alkylthiophenes) made easy by the GRIM method: Investigation of the reaction and the origin of regioselectivity. Macromolecules 2001, 34, 4324–4333.

[69] McCullough R D, Lowe R D. Enhanced electrical conductivity in regioselectively synthesized poly(3-alkylthiophenes). J Chem Soc Chem Commun 1992, 1, 70–72.

[70] Stuart N J, Adams S S. Phenyl Propionic Acids. US 3385886, Boots Pure Drug Company, 1962.

[71] Wolber E K A, Rüchardt C. New syntheses of ibuprofen and naproxen. Chem Ber 1991, 124, 1667–1672.

[72] Fiagl F, Schlosser M. A one-pot synthesis of ibuprofen involving three consecutive steps of superbase metalation. Tetrahedron Lett 1991, 32, 3369–3370.

[73] Lin R W, Herndon R C Jr, Atkinson E E Jr. Process for Recycling and Regenerating Carbonylation Catalyst Used in Synthesis of Ibuprofen. US 5055611, Albemarle Corp., 1991.

[74] White D R. Process for Preparing Esters of α-acetyl-α'-methylsuccinic Acid and Esters of α-methyl-α'-acetyl-α'-(5-methyl-3-oxohexyl)succinic Acid. US 4194053, The Upjohn Company, 1980.

[75] Schmid R. Homogeneous Catalysis with Metal Complexes in a "Pharmaceutical and Vitamins" Company: Why, What for, and Where to Go? Chimia 1996, 50, 110–113.

[76] Scalone M, Vogt P. Amidation of Pyridines. EP 385210, F. Hoffmann-La Roche AG, 1990.

[77] Trzeciak A M, Ziolkowski J J. Perspectives of rhodium organometallic catalysis. Fundamental and applied aspects of hydroformylation. Coord Chem Rev 1999, 190–192, 883–900.

[78] Heck R F, Breslow D S. Carboxyalkylation reactions catalyzed by cobalt carbonylate Ion. J Am Chem Soc 1961, 83, 4023–4027.

[79] Bach H, Bahrmann H, Gick W, Konkol W, Wiebus E. Industrial use of water-soluble ligands for hydroformylation catalysts. Chem Ing Tech 1987, 59, 882–883.

[80] Kuntz E G. Addition of Hydrogen Cyanide to Unsaturated Organic Compounds with at Least One Ethylenic Double Bond. DE 2700904, Rhone-Poulenc, 1977.

[81]   Cornils B. Industrial aqueous biphasic catalysis: Status and directions. Org Process Res Dev 1998, 2, 121–127.

[82]   Gärtner R, Cornils B, Springer H, Lappe P. Sulfonated Arylphosphines. DE 3235030, Ruhrchemie, 1984.

[83]   Livingston J R, Mozeleski E J, Sartori G. A Method for Separating A Water-soluble Noble Metal Complex Catalyst from A Hydroformylation Reaction. WO 9304029, Exxon Chemical Patents, 1993.

[84]   Bexten L, Cornils B, Kupies D. Separation and Purification of Salts of diphenyl(m-sulfophenyl) phosphine, phenylbis(m-sulfophenyl)phosphine, and/or tris(m-sulfophenyl)phosphine. DE 3431643, Ruhrchemie, 1986.

[85]   Wiebus E, Cornils B. Industrial oxo synthesis with immobilized catalyst. Chem Ing Tech 1998, 66, 916–923.

[86]   Dümbgen G, Neubauer D. Grosstechnische herstellung von oxo-alkoholen aus propylen in der BASF. Chem Ing Tech 1969, 974–980.

[87]   Vleeschhouwer P H M, Garton R D, Fortuin J M H. Analysis of limited cycles in an industrial oxo reactor. Chem Eng Sci 1992, 47, 2547.

[88]   Himmele W, Auila W. 1,2-Di(acyloxy)-3-formylbutanes. DE 1945479, BASF, 1971.

[89]   Rheude U, Vicari M, Aquila W, Wegner G, Niekerken J. Process for the Production of the C5-acetate Used for the Manufacture of Vitamin A by the Hydroformylation of Distillation-purified 3,4-diacetoxy -1-butene. EP 1247794, BASF, 2002.

[90]   Wüstenberg B, Müller M-A, Schütz J, Wyss A, Schiefer G, Litta G, John M, Hähnlein W. Vitamins, 2. Vitamin A (Retinoids) Ullmann's Encyclopedia of Industrial Chemistry. Weinheim, Wiley-VCH, 2020.

[91]   Keim W, Shryne T M, Bauer R S, Chung H, Glockner P W, van Zwet H. DE 2054009, Shell Int. Research, May 7, 1971.

[92]   Keim W. Pros and cons of homogeneous transition metal catalysis, illustrated for SHOP [Shell higher olefin process]. Chem Ing Techn 1984, 56, 850–853.

[93]   Keim W. Oligomerisierung von Ethen zu α-Olefinen: Erfindung und Entwicklung des Shell-Higher-Olefin-Prozesses (SHOP). Angew Chem 2013, 125, 12722–12726.

[94]   Keim W, Durocher A, Voncken P. Telomerisation von Olefinen in Zwei-Phasen-Systemen. Erdoel und Kohle, Erdgas. Brennstoffchemie 1976, 1, 31.

[95]   Mercier C, Chabardes P. Organometallic chemistry in industrial vitamin A and vitamin E synthesis. Pure Appl Chem 1994, 66, 1509–1518.

[96]   Smutny E J Oligomerization and dimerization of butadiene under homogeneous catalysis. Reaction with nucleophiles and the synthesis of 1,3,7-octatriene. J Am Chem Soc 1967, 89, 6793–6794.

[97]   Takahashi S, Shibano T, Hagihara N. The dimerization of butadiene by palladium complex catalysts. Tetrahedron Lett 1967, 8, 2451–2453.

[98]   Schaart B J, Pelt H L, Jacobsen G B. Continuous Process for the Telomerization of Conjugated Dienes. US 5254782, Dow Chemical Company, 1993.

[99]   Behr A, Dehn D. Dreiphasige Butadien-Telomerisationen im kontinuierlich betriebenen Schlaufenreaktor. Chem Ingenieur Technik 2008, 80, 1509–1517.

[100] Behr A, Dehn D. Dreiphasige Butadien-Telomerisationen im kontinuierlich betriebenen Schlaufenreaktor Teil 2: Bestimmung der Mikrokinetik. Chem Ingenieur Technik 2008, 80, 1775–1783.

[101] Bruijnincx P C A, Jastrzebski R, Hausoul P J C, Klein Gebbink R J M, Weckhuysen B M. Pd-Catalyzed Telomerization of 1,3-Dienes with Multifunctional Renewable Substrates: Versatile Routes for the Valorization of Biomass-Derived Platform Molecules. In: Meier M A R, Weckhuysen B N, Bruijnincx P C A, eds. Top Organomet Chem, Berlin Heidelberg, Organometallics and Renewables, Springer, Vol. 39, 2012, 45–102.

[102] Benn R, Jolly P W, Joswig T, Mynott R, Schick K P. Intermediates in the palladium-catalyzed reactions of 1,3-dienes. Part 5. Butadiene complexes of nickel, palladium and platinum. Z Naturforsch B 1986, 41B, 680–691.

[103] Benn R, Jolly P W, Mynott R, Raspel B, Schenker G, Schick K P, Schroth G. Intermediates in the palladium-catalyzed reactions of 1,3-dienes. 2. Preparation and structure of (η^1,η^3-octadienediyl) palladium complexes. Organometallics 1985, 4, 1945–1953.

[104] Jolly P W. η^3-Allylpalladium-Verbindungen. Angew Chem 1985, 97, 279–291.

[105] Behr A, Ilsemann G V, Keim W, Krüger C, Tsay Y-H. Octadienyl-bridged bimetallic complexes of palladium as intermediates in telomerization reactions of butadiene. Organometallics 1986, 5, 514–518.

[106] Keim W, Kraus A, Huthmacher K, Hahn R. Method and Catalysts for the Preparation of 6,10- and 6,9-diemthyl-5,10-undecadienyl-2-ones by the Telomerization of Isoprene with Alkyl Acetylacetonates. DE 19730546, Degussa A.-G., 1999.

[107] Bessmertnykh A, Henin F, Muzart J. Palladium-catalyzed telomerization of butadiene with aldoses: A convenient route to non-ionic surfactants based on controlled reactions. J Mol Catal A 2005, 238, 199–206.

[108] Estrine B, Bouquillon S, Henin F, Muzart J Telomerization of butadiene with pentoses in water: Selective etherification. Green Chem 2005, 7, 219–223.

[109] Desvergnes-Breuil V, Pinel C, Gallezot P. Green approach to substituted carbohydrates: Telomerisation of butadiene with sucrose. Green Chem 2001, 3, 175–177.

[110] Donze C, Pinel C, Gallezot P, Taylor P L. Palladium-catalyzed telomerization of butadiene with starch. Adv Synth Catal 2002, 344, 906–910.

[111] Fabry B, Gruber B. Octadienyl Ether Sulfates for Use in Surfactants. DE 4020973, Henkel, 1992.

[112] Raths H-C, Gruber B, Ouzounis D. Unsaturated Ethers of Polyoxyalkylenes. DE 4021478, Henkel, 1992.

[113] Wangemann F. Preparation of Glycerol Ether Sulfates by Sulfation with Sulfur Trioxide. DE 4114243, Henkel, 1992.

[114] Gruber B, Eicken U, Fischer H. Unsaturated Oligoesters Containing Alkadienyl Group as Lacquer Binding Material. DE 4129527, Henkel, 1993.

[115] Müller R, Gruber B, Wangemann F, Seidel K, Hollenberg D. Glyceryl Ether Sulfates in Aqueous Detergents with Mildness to Skins. Wo 9316156, Henkel, 1993.

[116] Hill K, Mahler U. Telomers as Antifoaming Agents. WO 9202284, Henkel, 1992.

[117] Höhlein P, Meixner J, Pedain J. Use of Octadienyl Ethers as Reactive Diluents. DE 4141190, Bayer, 1993.

[118] Gruber B, Eicken U, Stork N. Preparation and Application of Fatty Acid Esters of Alkadienyl Ethers of Polyols as Lacquer Thinner, DE 4129528, Henkel, 1993.

[119] Bunte R, Gruber B, Tucker J. Preparation of Unsaturated Ethers as Surfactant Intermediates. DE 4021511, Henkel, 1992.

[120] Patrini R, Marchionna M. Telomerization of Conjugated Alkadienes. EP 0613875, Snamprogetti SpA, 1994.

[121] Pittman C U, Hanes R M, Yang J J. Selective hydrodimerization of 1,3-butadiene to 1,7-octadiene. J Mol Catal 1982, 15, 377–381.

[122] Brehme V, Neumann M, Bauer F, Röttger D. Producing Dienes by Hydrodimerization in the Presence of Catalysts. WO 2008003559, Evonik Degussa GmbH, January 10, 2008.

[123] Behr A, Heite M. Telomerization of carbon dioxide and 1,3-butadiene. Process development via mini-plant technology. Chem Ing Tech 2000, 72, 58–61.

[124] Dinjus E, Leitner W. New insights into the palladium-catalyzed synthesis of δ-lactones from 1,3-dienes and carbon dioxide. Appl Organomet Chem 1995, 9, 43–50.

[125] Behr A, Beckmann T, Schwach P. Multiphase telomerisation of butadiene with acetic acid and acetic anhydride. J Organomet Chem 2008, 693, 3097–3102.

[126] Antonsson T, Langlet A, Moberg C. Stereochemistry of palladium-catalyzed telomerization of butadiene with diethylamine. J Organomet Chem 1989, 363, 237–241.

[127] Prinz T, Keim W, Drießen-Hölscher B. Zweiphasenkatalyse: Eine Strategie zur Vermeidung von Konsekutivreaktionen am Beispiel der Telomerisation von Butadien und Ammoniak. Angew Chem 1996, 1835–1836.

[128] Traenker H J, Jentsch J, Prinz T, Drießen-Hölscher B, Keim W. Two-phase Telomerization Method and Palladium-ligand Catalysts for the Production of 1-amino-2,7-octadiene from Butadiene and Ammonia. DE 19808260, Bayer, 1999.

[129] Behr A, Fischer T, Grote M, Schnitzmeier D. Process concepts for transition metal-catalyzed telomerization of isoprene with methanol. Chem Ing Tech 2002, 74, 1586–1591.

[130] Fassbach T A, Behr A, Vorholt A J. Telomerisation of Renewables. In: Cole-Hamilton D J, van Leeuwen P W N M, eds. Homogeneous Catalysis with Renewables, Series Catalysis by Metal Complexes, Springer, 39, 2017, 81–91

[131] Faerber T, Behr A, Vorholt A J. Hydroamination and Telomerisation of β-Myrcene. In: Cole-Hamilton D J, van Leeuwen P W N M, eds. Homogeneous Catalysis with Renewables, Series Catalysis by Metal Complexes, Springer, 39, 2017, 177–189.

[132] Vogelsang D, Dittmar M, Seidensticker T, Vorholt A J. Palladium-catalyzed carboxytelomerization of β-myrcene to highly branched C21-esters. Catal Sci Technol 2018, 8, 4332–4337.

[133] Vogelsang D, Fassbach T, Kossmann P, Vorholt J A. Terpene-derived highly branched C30-Amines via palladium-catalysed telomerisation of β-farnesene. Adv Syn Catal 2018, 360, 1984–1991.

[134] Wilke G. Beiträge zur nickelorganischen Chemie. Angew Chem 1988, 100, 189–211.

[135] Banks R L, Bailey G C. Olefin disproportionation. A new catalytic process. Ind Eng Chem Prod Res Dev 1964, 3, 170–173.

[136] Calderon N, Chen H Y, Scott K W. Olefin metathesis – A novel reaction for skeletal transformations of unsaturated hydrocarbons. Tet Lett 1967, 34, 3327–3329.

[137] Jean-Louis Hérisson P, Chauvin Y. Catalyse de transformation des oléfines par les complexes du tungstène. II. Télomérisation des oléfines cycliques en présence d'oléfines acycliques. Die Makromol Chem 1971, 141, 161–176.

[138] Schrock R R. High-oxidation-state molybdenum and tungsten alkylidene complexes. Acc Chem Res 1986, 19, 342–348.

[139] Schwab P, France M B, Ziller J W, Grubbs R H. A series of well-defined metathesis catalysts–synthesis of [RuCl^2(CHR')(PR$_3$)$_2$] and its reactions. Angew Chem Int Ed Engl 1995, 34, 2039–2041.

[140] Schwab P, Grubbs R H, Ziller J W. Synthesis and Applications of RuCl2(=CHR')(PR3)2: The Influence of the Alkylidene Moiety on Metathesis Activity. J Am Chem Soc 1995, 118, 100–110.

[141] Casey C P, Burkhardt T J. Reactions of (diphenylcarbene)pentacarbonyltungsten(0) with alkenes. Role of metal-carbene complexes in cyclopropanation and olefin metathesis reactions. J Am Chem Soc 1974, 96, 7808–7809.

[142] Huang J, Stevens E D, Nolan S P, Petersen J L. Olefin metathesis-active ruthenium complexes bearing a nucleophilic carbene ligand. J Am Chem Soc 1999, 121, 2674–2678.

[143] Scholl M, Trnka T M, Morgan J P, Grubbs R H. Increased ring closing metathesis activity of ruthenium-based olefin metathesis catalysts coordinated with imidazolin-2-ylidene ligands. Tetrahedron Lett 1999, 40, 2247–2250.

[144] Ackermann L, Fürstner A, Weskamp T, Kohl F J, Herrmann W A. Tetrahedron Lett 1999, 40, 4787–4790.

[145] Lindmar-Hamberg M, Wagener K B. Acyclic metathesis polymerization: The olefin metathesis reaction of 1,5-hexadiene and 1,9-decadiene. Macromolecules 1987, 20, 2949–2951.

[146] Mol J C. Industrial application of olefin metathesis. J Mol Cata A Chem 2004, 213, 39–45.

[147] Ghashghaee M. Heterogeneous catalysts for gas-phase conversion of ethylene to higher olefins. Rev Chem Eng 2018, 34, 595–655.

[148] Keim W. Oligomerization of ethylene to α-Olefins: Discovery and development of the Shell Higher Olefin Process (SHOP). Angew Chem Int Ed 2013, 52, 12492–12496.

[149] Reuben B, Wittcoff H. The SHOP process: An example of industrial creativity. J Chem Educ 1988, 65, 605.

[150] Khosravi E, Szymanska-Buzar T. Ring Opening Metathesis Polymerisation and Related Chemistry State of the Art and Visions for the New Century. Springer Science & Business Media, 2012, 250–251.

[151] Fahlbusch K-G, et al. Flavors and Fragrances. In: Ullmann's Encyclopedia of Industrial Chemistry, 7th ed. Wiley, 2007, 45–46.

[152] Chaumont P, John C S. Olefin disproportionation technology (feast) – A challenge for process development. J Mol Catal 1988, 46, 317–328.

[153] Kamat V P, Hagiwara H, Katsumi T, Hoshi T, Suzuki T, Ando M. Ring closing metathesis directed synthesis of (R)-(-)-muscone from (+)-citronellal. Tetrahedron 2000, 56, 4397–4403.

[154] Guram A S, Buchwald S L. Palladium-catalyzed aromatic aminations with in situ generated aminostannanes. J Am Chem Soc 1994, 116, 7901–7902.

[155] Paul F, Patt J, Hartwig J F. Palladium-catalyzed formation of carbon-nitrogen bonds. Reaction intermediates and catalyst improvements in the hetero cross-coupling of aryl halides and tin amides. J Am Chem Soc 1994, 116, 5969–5970.

[156] Wolfe J P, Buchwald S L. Palladium-catalyzed amination of aryl halides and aryl triflates: N-hexyl-2-methyl-4-methoxy-aniline and N-methyl-N-(4-chlorophenyl)aniline. Org Synth 2002, 78, 23.

[157] Muci A R, Buchwald S L. Practical palladium catalysts for C-N and C-O Bond Formation. Topics in Curr Chem 2002, 219, 131–209.

[158] Wolfe J P, Wagaw S, Marcoux J F, Buchwald S L. Rational development of practical catalysts for aromatic carbon–nitrogen bond formation. Acc Chem Res 1998, 31, 805–818.

[159] Hartwig J F. Approaches to catalyst discovery. New carbon-heteroatom and carbon-carbon bond formation. Pure Appl Chem 1999, 71, 1416–1423.

[160] Driver M S, Hartwig J F. A second-generation catalyst for aryl halide amination: Mixed secondary amines from aryl halides and primary amines catalyzed by (DPPF)PdCl2. J Am Chem Soc 1996, 118, 7217–7218.

[161] Wolfe J P, Wagaw S, Buchwald S L. An improved catalyst system for aromatic carbon–nitrogen bond formation: The possible involvement of Bis(Phosphine) palladium complexes as key intermediates. J Am Chem Soc 1996, 118, 7215–7216.

[162] Louie J, Driver M S, Hamann B C, Hartwig J F. Palladium-Catalyzed amination of aryl triflates and importance of triflate addition rate. J Org Chem 1997, 62, 1268–1273.

[163] Wolfe J P, Ahman J, Sadighi J P, Singer R A, Buchwald S L. An ammonia equivalent for the palladium-catalyzed amination of aryl halides and triflates. Tetrahedron Lett 1997, 38, 6367–6370.

[164] Huang X, Buchwald S L New ammonia equivalents for the Pd-Catalyzed amination of aryl halides. Org Lett 2001, 3, 3417–3419.

[165] Hartwig J F. Transition metal catalyzed synthesis of arylamines and aryl ethers from aryl halides and triflates: Scope and mechanism. Angew Chem Int Ed 1998, 37, 2046–2067.

[166] Borzenko A, Rotta-Loria N L, MacQueen P M, Lavoie C M, McDonald R, Stradiotto M. Nickel-catalyzed monoarylation of ammonia. Angew Chem Int Ed 2015, 54, 3773.

[167] Green R A, Hartwig J F. Nickel-catalyzed amination of aryl chlorides with ammonia or ammonium salts. Angew Chem Int Ed 2015, 54, 3768.

[168] Monguchi Y, Kitamoto K, Ikawa T, Maegawa T, Sajiki H. Evaluation of aromatic amination catalyzed by palladium on carbon: A practical synthesis of triarylamines. Adv Synth Catal 2008, 350, 2767–2777.

[169] Komáromi A, Novák Z. Examination of the aromatic amination catalyzed by palladium on charcoal. Adv Synth Catal 2010, 352, 1523–1532.

[170] Fareghi-Alamdari R, Haqiqi M G, Zekri N. Immobilized Pd(0) nanoparticles on phosphine-functionalized graphene as a highly active catalyst for Heck, Suzuki and N-arylation reactions. New J Chem 2016, 40, 1287–1296.

[171] Al-Amin M, Arai S, Hoshiya N, Honma T, Tamenori Y, Sato T, Yokoyama M, Ishii A, Takeuchi M, Maruko T. Development of second generation gold-supported palladium material with low-leaching and recyclable characteristics in aromatic amination. J Org Chem 2013, 78, 7575–7581.

[172] Park S B, Chee G-L, Dekeyser M A. 4-methoxybiphenyl Hydrazone Derivatives. US 6706895, Chemtura Canada, 2002.

7 Rearrangement reactions

7.1 Introduction

Rearrangement reactions are a broad class of organic reactions in which the reactant undergoes a rearrangement to give a structural isomer of the original molecule via migration of an H atom or a larger molecular fragment. From the view of atom economy and E-factor, rearrangement reactions fulfil the criteria of the modern type of chemistry and green chemistry, with all atoms of the starting material being present in the product structure.

In many rearrangement reactions, the migration occurs directly to a neighbouring position. These rearrangements belong to the class of [1,2]-rearrangements or [1,2]-shifts. These reactions are often sigmatropic rearrangements meaning that a σ-bond migrates during the reaction. The nomenclature of rearrangement reactions is described by numbering the atoms directly attached to the bond that is broken with 1 and 1' (Scheme 7.1). The following atoms in the direction of the rearrangement are labelled 2, 3 and so forth starting from 1 and 2', 3' and so forth starting from 1'. After the rearrangement, the new σ-bond is connected to two atoms which characterise the rearrangement. The numbers are listed in square brackets and the prime is removed from the second number.

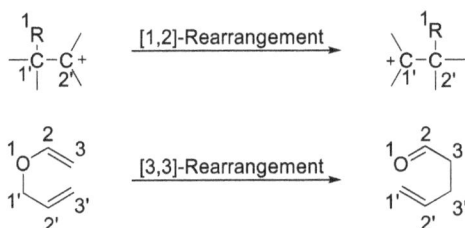

Scheme 7.1: Naming and examples of sigmatropic rearrangements.

Some of the most important industrial rearrangement reactions, such as [3,3]-sigmatropic rearrangements, are used in the synthesis of isoprenoid building blocks such as isophytol, β-ionone or aroma compounds such as methyl heptanone. These are the main focus of this chapter.

7.2 *Wagner–Meerwein* rearrangements

The *Wagner–Meerwein* rearrangement is a predominately acid-catalysed [1,2]-rearrangement in which a hydrogen, alkyl or aryl group migrates from one carbon to a neighbouring carbon to generate a new carbocation. This carbocation reacts with a nucleophile or a proton from a neighbouring atom is eliminated. The driv-

https://doi.org/10.1515/9783111102672-007

ing force of the reaction is that the initially formed carbocation has the tendency to rearrange to a thermodynamically more stable structure. In the terpene chemistry, the *Wagner–Meerwein* rearrangement is of importance in the manufacture of camphene from isoborneol (Scheme 7.2) via alkyl chain migration (C4 → C6) and elimination. Camphene is used in the aroma and flavour industry.

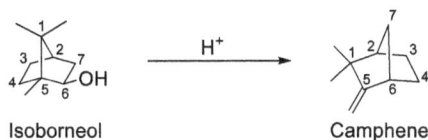

Isoborneol   Camphene

Scheme 7.2: Camphene ex isoborneol.

Furthermore, this reaction is industrially applied in the manufacture of trimethylhydroquinone (TMHQ), an intermediate in the industrial manufacture of vitamin E. Ketoisophorone, obtained from oxidation of β-isophorone, is converted to TMHQ diacetate under acid catalysis in acetic anhydride. Saponification results in the required hydroquinone (Scheme 7.3).

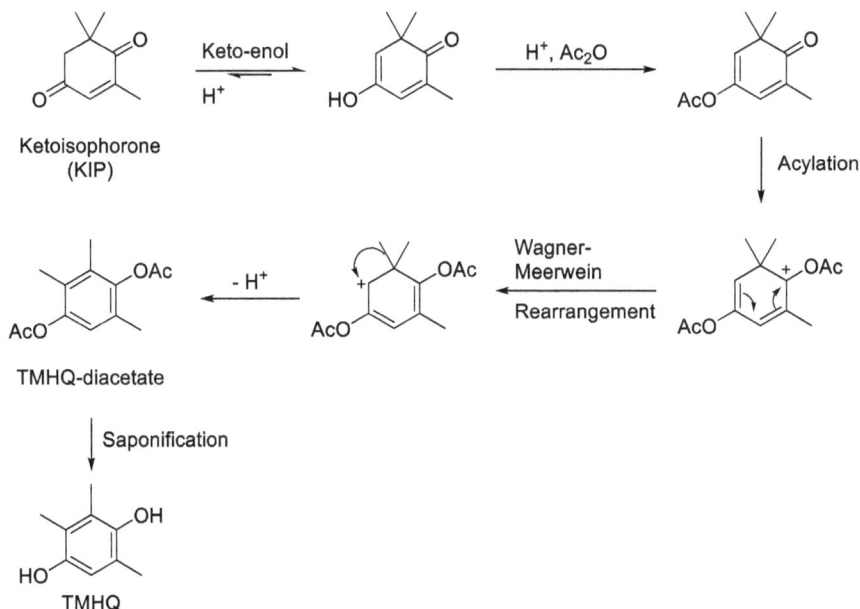

Ketoisophorone
(KIP)

Keto-enol
H⁺

H⁺, Ac₂O

Acylation

TMHQ-diacetate

- H⁺

Wagner-
Meerwein

Rearrangement

Saponification

TMHQ

Scheme 7.3: Manufacture of trimethylhydroquinone (TMHQ) by *Wagner–Meerwein* rearrangement.

7.3 *Beckmann* rearrangement

Another [1,2]-sigmatropic rearrangement used in the production of fine chemicals is the *Beckmann* rearrangement (Scheme 7.4). This reaction is a key step in the synthesis of ε-caprolactam and the painkiller paracetamol – details can be found in Chapter 8: Acid–base-catalysed reactions (for paracetamol) and Chapter 4: Oxidations (for ε-caprolactam) [1].

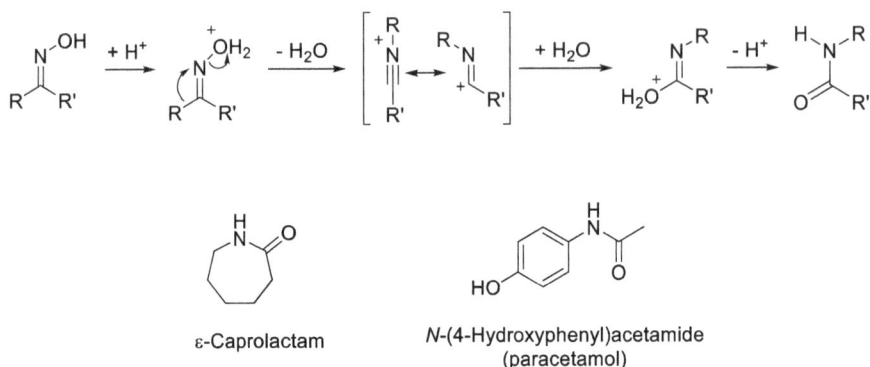

ε-Caprolactam

N-(4-Hydroxyphenyl)acetamide
(paracetamol)

Scheme 7.4: *Beckmann* rearrangement – general reaction scheme.

7.4 [3,3]-Sigmatropic-type rearrangements

Chemically, isoprenoids can be formed by sequential C_2-ethynylation (base-catalysed), for example the transformation of acetone to methylbutynol, followed by C_3-elongation via an acid-catalysed *Saucy–Marbet* or *Carroll* reactions, which are both [3,3]-sigmatropic rearrangements. The corresponding precursors are obtained via isopropenyl methyl ether (IPM) or β-ketoester addition to a tertiary allylic alcohol (Scheme 7.5). An alternative method of C_3-elongation is the Aldol reaction, but a significant amount of by-products can be formed, so the *Saucy–Marbet* or *Carroll* reactions are preferred. The most efficient C_3-elongation process is based on the acid-catalysed *Saucy–Marbet* reaction, using the activated acetone equivalent, IPM as C_3-building block [2].

Modern trends in acid-catalysed reactions, such as these rearrangement reactions, aim to replace *Brønsted* acids such as mineral acids (e.g. hydrochloric acid, sulphuric acid or nitric acid) with solid acids, for example ion-exchange resins or zeolites. This technology allows continuous processing and benefits from reduced amounts of waste generation (see Chapter 8: Acid–base-catalysed reactions).

The *Saucy–Marbet* reaction is divided in three reaction types: propargylic alcohol ester rearrangement (*Saucy–Marbet I*), which is Cu- or Ag-catalysed and not further applied on an industrial scale in the fine chemical industry. This is because the pro-

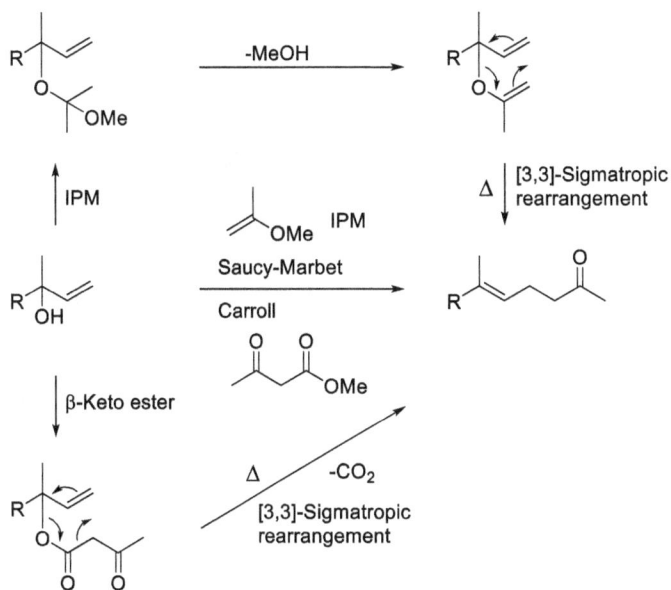

Scheme 7.5: C$_3$-elongation reactions by *Carroll* and *Saucy–Marbet* reaction.

duction of the acetate from the alcohol and acetic anhydride creates acetic acid as a by-product. Additional waste is created based on the limited selectivity and the propargylic alcohol is sensitive to acids. This type of reaction can be applied, for example the synthesis of citral from dehydrolinalool (DLL), using the acid- or base-catalysed rearrangement and saponification of the allene acetate (Scheme 7.6) [3].

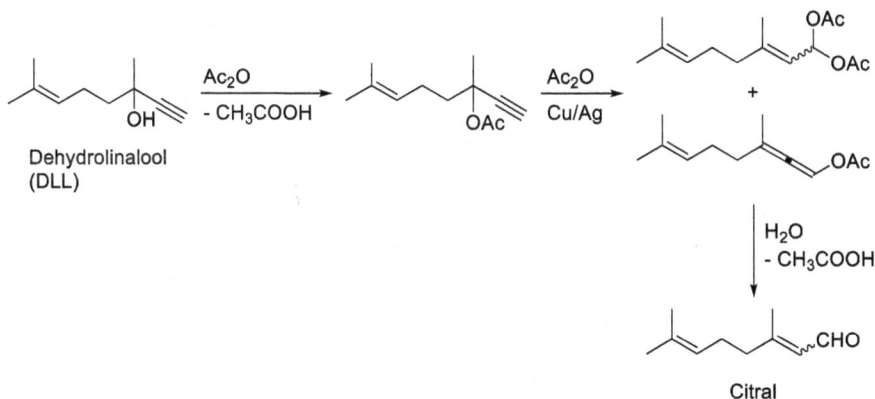

Scheme 7.6: Cu- or Ag-catalysed *Saucy-Marbet* I rearrangement.

The acid-catalysed reaction of tertiary vinyl alcohols and IPM to the corresponding ether derivatives (a *Markovnikov* addition) followed by a [3,3]-sigmatropic rearrangement reaction, results in the formation of γ,δ-unsaturated ketones and is called the *Saucy–Marbet* II reaction. This reaction is of industrial relevance in the production of isoprenoid building blocks such as methyl heptenone (Scheme 7.7). The reaction is usually carried out at 100–160 °C and 10–15 bar pressure in the presence of a *Brønsted* acid, such as phosphites or sulphuric acid. IPM is used in excess because methanol is formed during the reaction which reacts with IPM to give 2,2-dimethoxypropane. This can be separated and reused by conversion back into IPM in a gas-phase reaction [4, 5]. Alternative catalysts for the *Saucy–Marbet* II reaction especially for the synthesis of 6-methyl-5-hepten-2-one, 6-methyl-5-octen-2-one, (5*E*/*Z*)-6,10-dimethyl-5,9-undecadien-2-one, (5*E*/*Z*)-6,10-dimethyl-5-undecen-2-one and 6,10,14-trimethyl-5,9,13-pentadeca-trien-2-one are HOP(O)(OR)$_{2'}$ (R' = Me, Et) or HOP(O)H(OR), HOP(O)H(Ph) or similar types of phosphites [6–8]. The *Saucy–Marbet* II reaction is applied on an industrial scale at several thousand t/a for the production of various compounds.

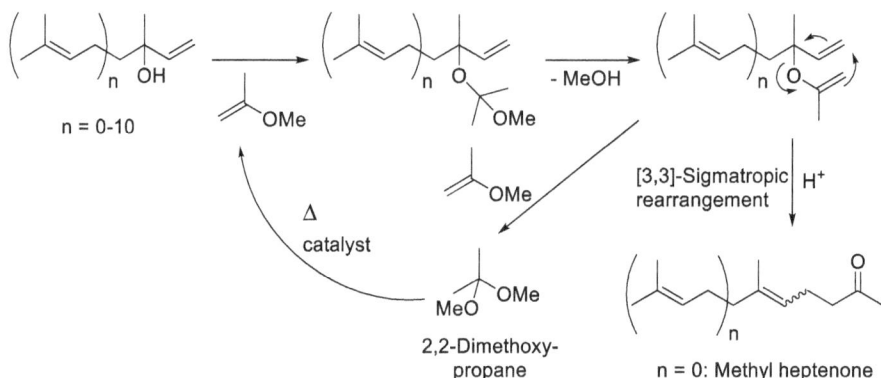

Scheme 7.7: Acid-catalysed *Saucy–Marbet* II reaction.

The *Saucy–Marbet* III reaction is an acid-catalysed reaction of isopropenyl ether with α-alkynols followed by a [3,3]-sigmatropic rearrangement resulting in the formation of β-oxoallenes in high yields. Treatment with base causes the isomerisation of these allenes to the corresponding α,β-unsaturated dienones. This reaction sequence is applied on an industrial scale of several thousand t/a and results in an economic synthesis of ψ-ionones and ψ-irones from DLL and its homologs. The reaction is carried out at a pressure of 10–15 bar, at 80–150 °C in the presence of sulphur-containing *Brønsted* acids such as sulphuric acid or *p*-TsOH (Scheme 7.8) [9]. The *Saucy–Marbet* reactions II and III fulfil the criteria of modern chemical processes since they are efficient, high-yielding, atom-economical and catalysed processes.

Linalool and linalyl acetate are found in a wide range of natural flowers and spice plants. They are of interest to the fragrance industry and are used as perfume compo-

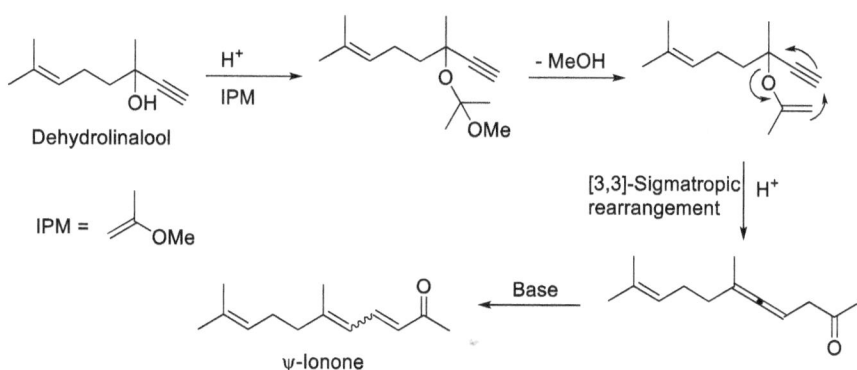

Scheme 7.8: *Saucy–Marbet* III reaction – C_3-elongation and rearrangement of α-alkynols.

nents in soaps, shampoos and lotions. Linalool and linalyl acetate are manufactured in around 50,000 t/a and 10,000 t/a scale by semi-hydrogenation of DLL or dehydrolinalyl acetate by *Lindlar* hydrogenation (see Chapter 2: Heterogeneous hydrogenations) [10, 11]. Linalool is also a key intermediate in the manufacture of isophytol and therefore vitamin E. The reaction sequence from acetone to methylbutynol by catalytic ethynylation, semi-hydrogenation followed by C_3-elongation by *Saucy–Marbet* reaction to methylheptenone and finally catalytic ethynylation and *Lindlar* hydrogenation are industrially applied (Scheme 7.9) [2].

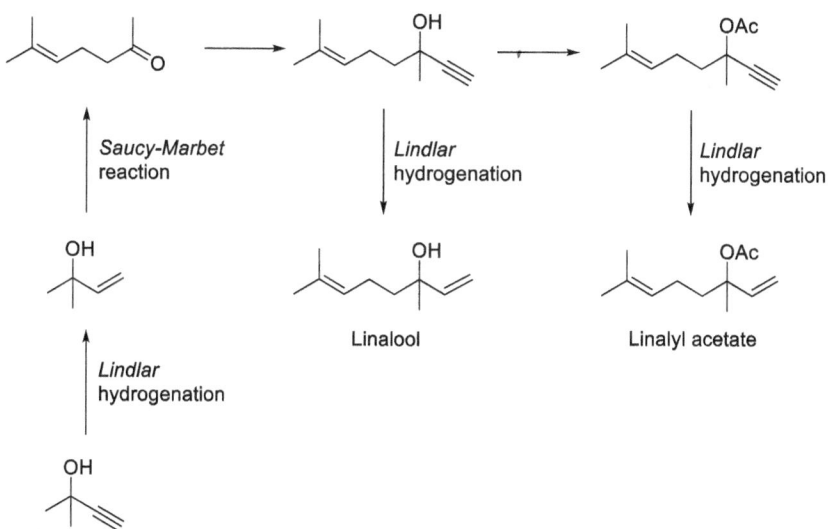

Scheme 7.9: Manufacture of linalool and linalyl acetate.

Structurally related to linalool and linalyl acetate are the compounds geraniol and nerol. They can be extracted from essential oils, synthesised from linalool *via* an allylic rearrangement (Section 7.6), or by the conversion of naturally obtained β-pinene. The first step of the process from β-pinene is a gas-phase process producing myrcene (Chapter 5: Gas-phase reactions). This is followed by treatment with HCl in the presence of a Cu(I)-catalyst in combination with a quartary ammonium salt (PTC conditions) to produce geranyl-, linalyl- and neryl chloride (Scheme 7.10) [12, 13]. Treatment with acetate in the presence of a base such as triethylamine results in geranyl-, linalyl- and neryl acetate. Saponification results in the formation of geraniol and nerol. Today, geraniol/nerol mixtures are mainly produced by the rearrangement of linalool by applying an orthovanadate catalyst [14] or via the BASF process [15]. The (*E*)- and (*Z*)-isomers (geraniol/nerol) can be separated and isolated by fractional distillation.

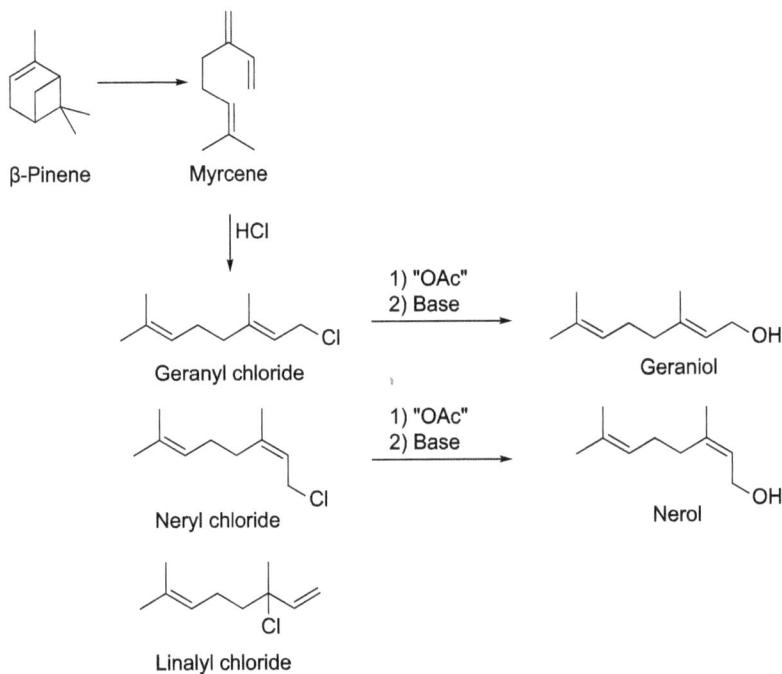

β-Pinene Myrcene

HCl

1) "OAc"
2) Base

Geranyl chloride Geraniol

1) "OAc"
2) Base

Neryl chloride Nerol

Linalyl chloride

Scheme 7.10: Synthesis of nerol and geraniol.

A further aroma compound is nerolidol (3,7,11-trimethyl1,6,10-dodecatrien-3-ol); the synthetic version is an (*E/Z*)-mixture and each of these isomers exists as a pair of enontiomers. (+)-(*E*)-Nerolidol is found in cabreuva oil, and (−)-(*Z*)-nerolidol in *Dalbergia parviflora* wood oil. Starting from linalool and isopropenyl methyl ether, a *Saucy–Marbet* rearrangement produces geranyl acetone (>90% yield) (Chapter 7: Rearrangement reactions). Alternative approaches to geranyl acteone are the C_3-elongation reaction via a *Carroll* reaction or the *Kimmel Sax* reaction of diketene and linalool. C_2-elongation of geranyl acetone using a base-catalysed addition of acetylene (>95% yield) followed by *Lindlar* hydrogenation (95% yield) results in a mixture of nerolidol stereoisomers (Scheme 7.11) [16–18]. Synthetic nerolidol is produced on a scale of several thousand t/a, and is applied in the manufacture of flavor molecules or as an intermediate for vitamins E and K, via farnesyl acetone [19].

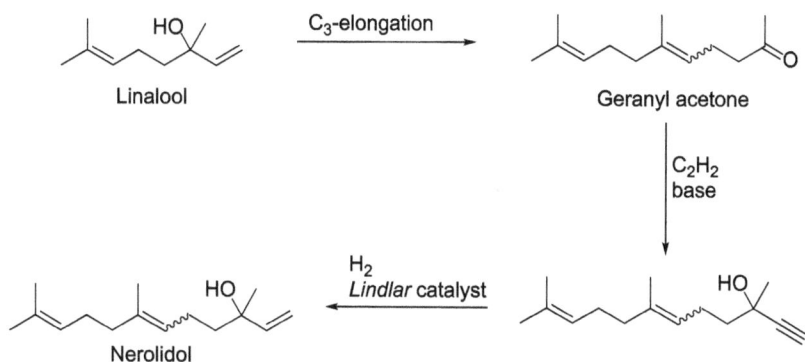

Scheme 7.11: Manufacture of nerolidol.

The *Brønsted* acid-catalysed *Saucy–Marbet* reaction is also applied in the manufacture of isonalines (methyl-substituted ionones), important compounds for the aroma industry. The C_4-elongation of DLL with butenyl methyl ether (BME) is performed in several thousand t/a. BME can be synthesised from 2-butanone in an acid-catalysed addition of methanol followed by the elimination of MeOH (Scheme 7.12). The DLL-BME adduct reacts by elimination of MeOH and rearrangement into a mixture of *n*- and *iso*-methyl-ψ-ionone. The acid-catalysed ring-closure reaction results in the formation of the isonalines, which are separated and purified by distillation. Usually, the reactions are performed with s/c of >10,000 and yields >90% are obtained [9, 20]. An alternative approach for the synthesis of isonalines or their intermediates is the (solid) base-catalysed Aldol reaction of citral and 2-butanone [21–26].

The rearrangement of a β-keto allyl ester followed by decarboxylation, resulting in the formation of γ,δ-unsaturated allyl ketone, is named the *Carroll* reaction (Scheme 7.13) [27]. The *Carroll* reaction is usually carried out at 100–200 °C [28] and is industrially applied in the synthesis of carbonyl compounds, especially in the C_3-elongation of vinyl ethers, for example in the manufacture of geranylacetone from linalool. Production volumes of several thousand t/a demonstrate the industrial relevance of this reaction. Compared to the *Saucy–Marbet* reaction, the atom economy in the *Carroll* reaction is less favourable, because of the loss of one carbon atom during the decarboxylation. Similar to the *Saucy–Marbet* reaction and the *Claisen* reaction, a [3,3]-sigmatropic rearrangement is a key feature of the transformation.

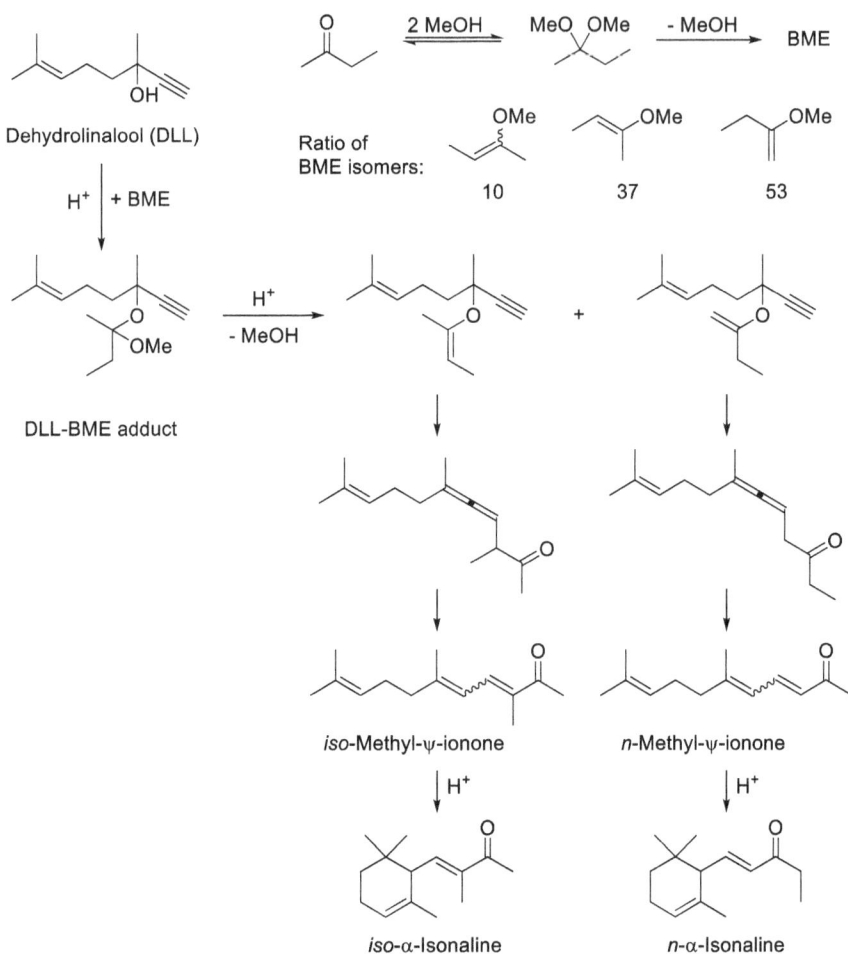

Scheme 7.12: Manufacture of isonalines.

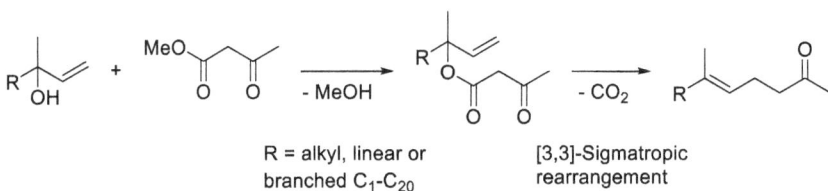

R = alkyl, linear or
branched C$_1$-C$_{20}$

[3,3]-Sigmatropic
rearrangement

Scheme 7.13: C$_3$-Elongation of vinyl alcohols by the *Carroll* reaction.

Two different [3,3]-sigmatropic rearrangements are applied in the industrial manufacture of citral. In the first step, isobutene and formaldehyde are converted by a *Prins*
reaction to isoprenol. Isoprenol on the one hand undergoes an isomerisation to prenol

catalysed by Pd/C and on the other hand an oxidation catalysed by Ag/SiO$_2$ in the gas phase at around 500 °C to isoprenal. Prenol and isoprenal are converted via a multi-step cascade: acetal formation, thermal elimination and finally a *Claisen* followed by a *Cope* rearrangement to citral. The two consecutive [3,3]-sigmatropic rearrangements are all thermally catalysed. This process is performed in >40,000 t/a (Scheme 7.14) [29] and can be seen as benchmark for industrial production, showing a high atom economy and an excellent E-factor.

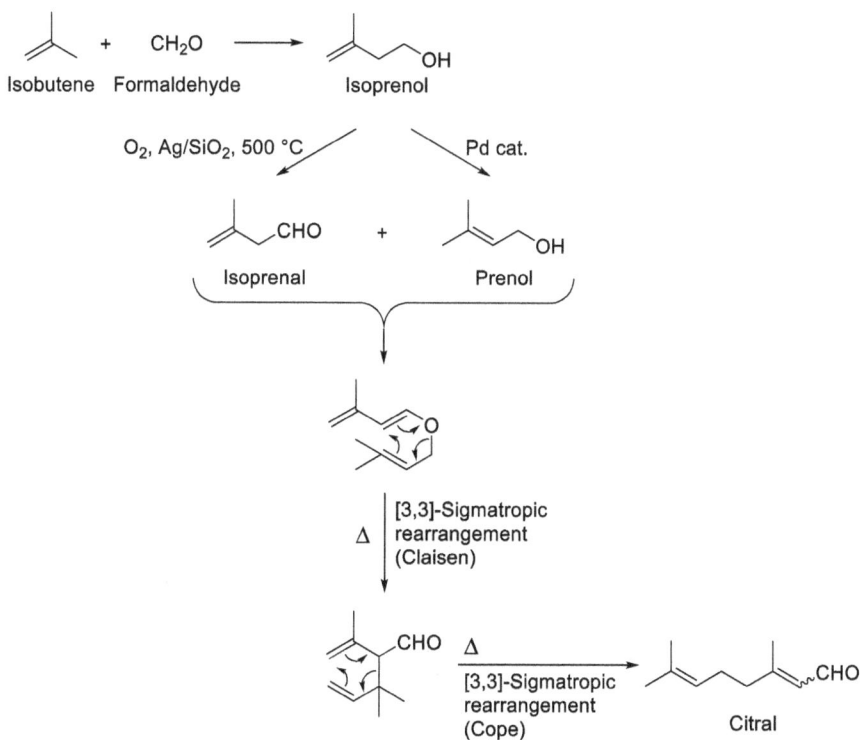

Scheme 7.14: Two different [3,3]-sigmatropic rearrangements applied in the manufacture of citral.

The isomerisation of allene ketones into an (*E/Z*)-mixture of α,β-unsaturated carbonyl compounds is important for the synthesis of ionones and irones as well as flavour compounds. The reaction sequence starting from DLL of C$_3$-elongation by *Saucy–Marbet* reaction and base-catalysed rearrangement of the allene ketone intermediate to ψ-ionone (mixture of (*E/Z*)-isomers) in MeOH at around 0 °C can be performed with NaOH in 95% overall yield (two steps) (Scheme 7.15) [9, 30]. The application of solid base catalysts such as basic ion exchange resins decreases the formation of waste and has the advantage of high selectivity [31].

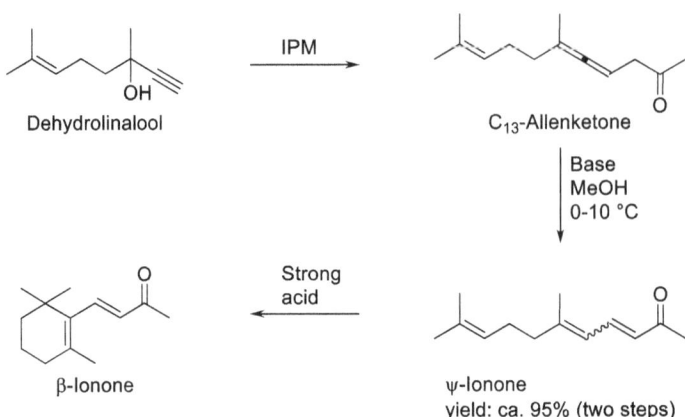

Scheme 7.15: Base-catalysed isomerisation of an allene ketone into ψ-ionone.

Furthermore, the rearrangement of 8-(2,2-dimethylcyclopropyl)-6-methyl-4,5-octadien-2-one into 8-(2,2-dimethylcyclopropyl)-6-methyl-3,5-octadien-2-one followed by an acid-catalysed cyclisation results in the formation of α- and β-irones. The rearrangement reaction is carried out in the presence of NaOH/MeOH, whereas the irone formation is obtained by a BF_3-catalysed ring-closure reaction. The irones are obtained in a mixture of *trans* and *cis* α-, and *cis* β-irones (47:41:10) in 75% yield [32].

In an alternative approach to vitamin A and carotenoid intermediates, *Saucy–Marbet* or *Carroll* reactions have been applied to achieve the synthesis of C_{18}-allene-ketone. Based on the selected reaction conditions various double-bond patterns can be targeted. The polyenes can be synthesised in up to 70% yield as an (*E/Z*)-mixture (Scheme 7.16) [33].

Chrysanthemic acid esters are insecticides of the pyrethrin types (I + II). Pyrethrins are found in the flowers of pyrethrums (*Chrysanthemum cinerariaefolium* and *C. coccineum*) [34]. Pyrethroids constitute the majority of commercial household insecticides. Chrysanthemic acid derivatives are produced industrially as a mixture of *cis*- and *trans* isomers in a cyclopropanation reaction of a diene, followed by hydrolysis of the ester (Scheme 7.17) [35].

The synthesis of chrysanthemic acid esters can also be performed starting from prenol using a *Claisen* rearrangement (*Sagami* approach). The resulting compound undergoes a C_1 elongation with CCl_4 in a Fe-catalysed procedure and is subsequently treated with a base to form the cyclopropane ring (Scheme 7.18) [36]. Several similar procedures following this approach have been described in the literature. Structurally similar compounds such as permethrin are applied as insecticide.

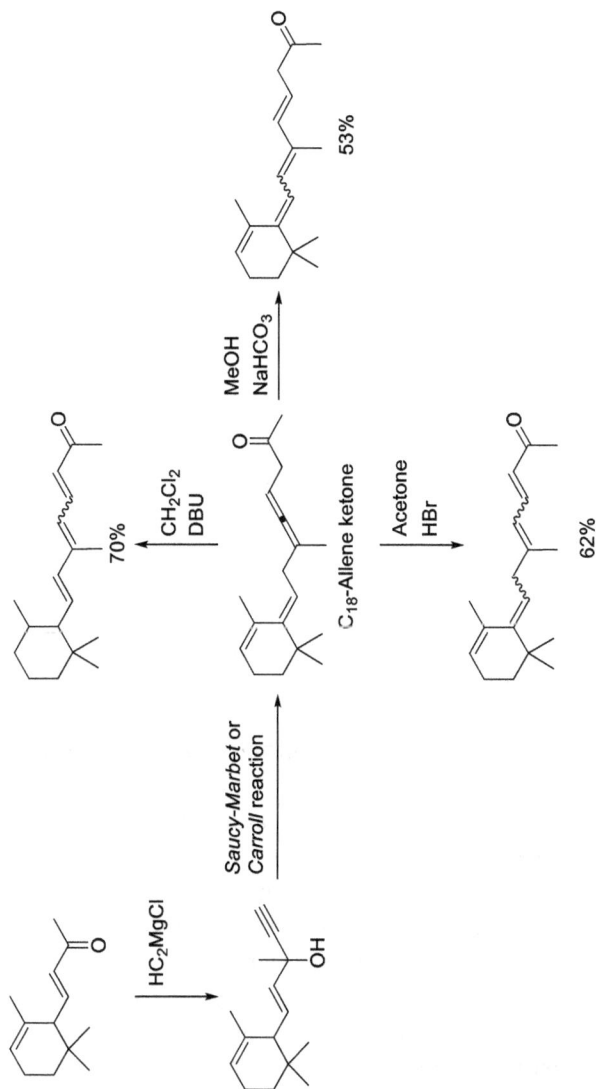

Scheme 7.16: The rearrangement of an allene ketone to form vitamin A and carotenoid intermediates.

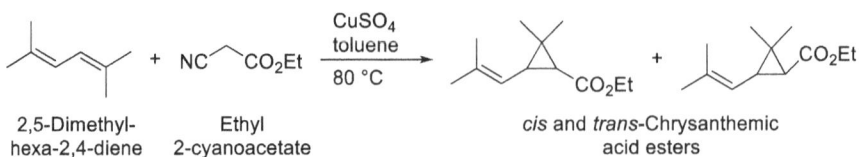

2,5-Dimethyl-   Ethyl   *cis* and *trans*-Chrysanthemic
hexa-2,4-diene   2-cyanoacetate   acid esters

Scheme 7.17: Synthesis of chrysanthemic acid esters.

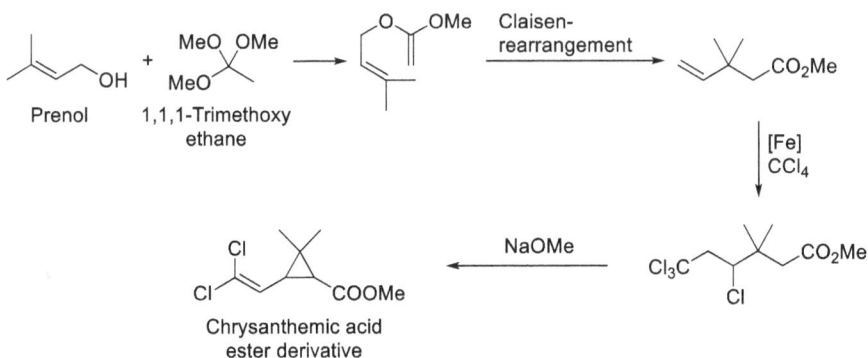

Prenol   1,1,1-Trimethoxy
                    ethane

Chrysanthemic acid
ester derivative

Scheme 7.18: *Sagami* approach via *Claisen* rearrangement to chrysanthemum acid derivatives.

7.5 The *Meyer–Schuster, Rupe–Kambli* and *Pauling* rearrangements

α,β-Unsaturated carbonyl compounds are the key intermediates in the manufacture of carotenoids, fragrance and flavouring compounds and vitamins, products that are produced in several 10,000s t/a [2, 37]. The acid-catalysed transformation of α-alkynols into α, β-unsaturated carbonyl compounds is described as the *Meyer–Schuster* reaction (aldehydes as the formed product) or *Rupe–Kambli* reaction (ketones as product) (Scheme 7.19) [38, 39]. The main disadvantage of the classical approach of the *Meyer–Schuster* and the *Rupe–Kambli* rearrangements is the formation of by-products, such as elimination and polymerisation, since most of the starting materials are acid sensitive.

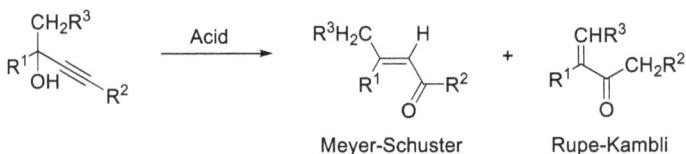

Meyer-Schuster   Rupe-Kambli

Scheme 7.19: Product distribution in the *Meyer–Schuster* and *Rupe–Kambli* rearrangement.

An alternative, efficient and selective method for the rearrangement of α-alkynols into the corresponding α,β-unsaturated aldehydes is the *Pauling* rearrangement, silyl-vanadates are applied as catalysts (Scheme 7.20) [40]. The manufacture of citral from

DLL at 140 °C in the presence of tris(triphenylsilyl)vanadium oxide was performed on an industrial scale on a few thousand t/a (Givaudan); the catalyst could be recycled and re-used. The advantage of this reaction is the high selectivity and simple work-up because the reaction products can be distilled out of the reaction mixture. Yields of 90% with full conversion can be achieved by applying an s/c of 100 (without re-use of catalyst). The α-alkynol reacts with the vanadium catalyst displacing a silanol group. The resulting V-alcohol intermediate undergoes rearrangement to the allene species, which is hydrolysed by the silanol into the allene alcohol, a labile compound that forms directly the carbonyl compound. During the hydrolysis, the V-catalyst is re-formed and can be directly re-used. The products of the *Pauling* rearrangement from V-enolates, which are useful intermediates in the synthesis of carbonyl compounds and their derivatives, are used as starting materials in several reaction types, for example in aldol-type chemistry, *Mannich* reactions or electrophilic aminations [41].

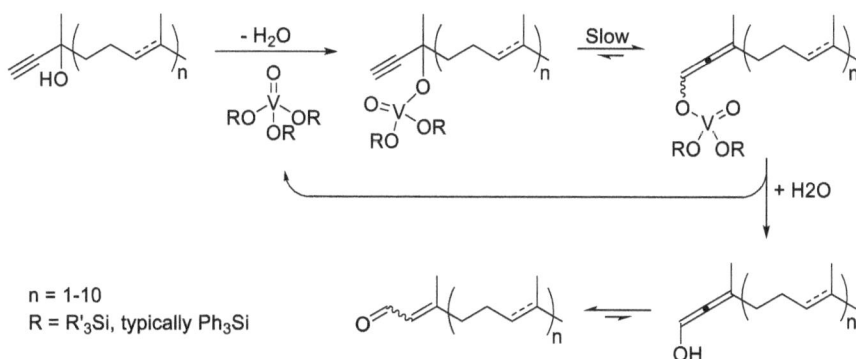

n = 1-10
R = R'₃Si, typically Ph₃Si

Scheme 7.20: Vanadium-catalysed *Pauling* rearrangement of α-alkynols.

α-Alkynols, which are acid- and temperature-sensitive, cannot be reacted to form α,β-unsaturated carbonyl compounds with the above-described silyl-vanadate catalysts. However, with Re complexes, a *Rupe–Kambli* rearrangement is possible [42]. Ethynyl-β-ionol, industrially manufactured by catalytic ethynylation of β-ionone (see Chapter 8: Acid–base-catalysed reactions), reacts in the presence of a perrhenate catalyst to form an (E/Z)-mixture of unsaturated ketones in 60% yield; a by-product is the furan (Scheme 7.21).

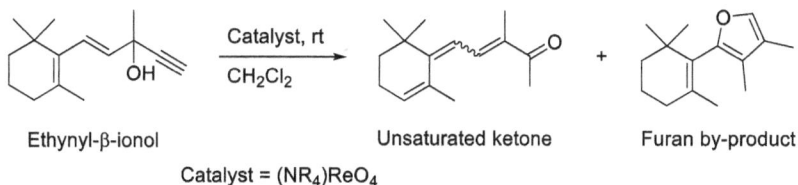

Ethynyl-β-ionol   Unsaturated ketone   Furan by-product

Catalyst = (NR₄)ReO₄

Scheme 7.21: Perrhenate-catalysed *Rupe–Kambli* rearrangement of ethynyl-β-ionol.

Acid- and/or temperature-sensitive α-alkynols can be rearranged into the corresponding α,β-unsaturated carbonyl compounds via a Ru-catalysed *Meyer–Schuster* rearrangement. This approach can be used multiple times on similar substrates to produce intermediates of vitamin A acetate. The sequence followed the (catalytic) addition of acetylene to a ketone followed by the Ru-catalysed rearrangement of the α-alkynol to give an α,β-unsaturated aldehyde. C$_3$-elongation (via an aldol reaction with acetone) and the addition of another equivalent of acetylene results in a similar α-alkynol, which can undergo a second rearrangement to give retinal. Reduction, isomerisation and acylation results in all-(*E*) vitamin A acetate (Scheme 7.22) [43, 44].

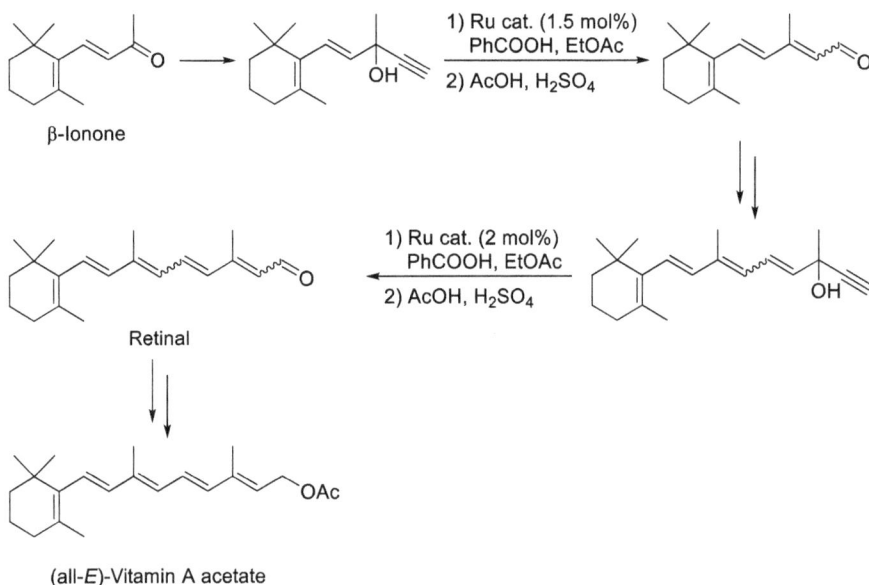

β-Ionone

1) Ru cat. (1.5 mol%)
   PhCOOH, EtOAc
2) AcOH, H$_2$SO$_4$

1) Ru cat. (2 mol%)
   PhCOOH, EtOAc
2) AcOH, H$_2$SO$_4$

Retinal

(all-*E*)-Vitamin A acetate

Scheme 7.22: Synthesis of vitamin A acetate using the Ru-catalysed rearrangement of α-alkynols as key steps.

Another approach to vitamin A aldehyde (retinal) is the isomerisation of polyene carbonyl compounds, such as trienals into conjugated derivatives in the presence of solid base catalysts like sodium aluminium silicate (Scheme 7.23) [45]. A similar rearrangement reaction is the base-catalysed reaction of allene derivatives to vitamin A aldehyde (Scheme 7.24) [46].

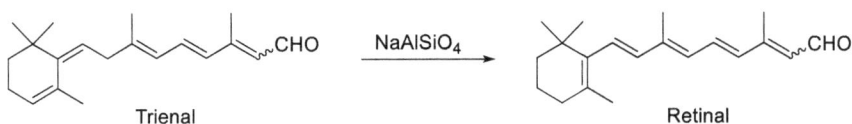

Trienal   NaAlSiO$_4$ Retinal

Scheme 7.23: Synthesis of vitamin A aldehyde by rearrangement of a polyene precursor.

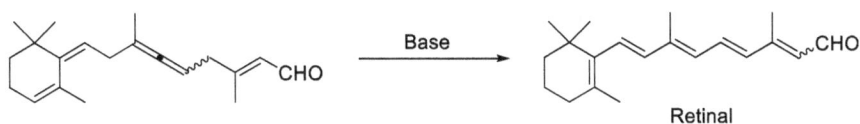

Scheme 7.24: Rearrangement of allene intermediates towards vitamin A aldehyde.

7.6 Allylic rearrangements

An alternative approach to synthetic linalool is the rearrangement of geraniol/nerol obtained by hydrogenation of citral (Scheme 7.25). The rearrangement from the primary into the tertiary allyl alcohol is performed with W(VI)-catalysts such as $WO_2L_nL'_n$, where L and L' are ligands selected from the group of amino alcohols, for example cis-bis[dioxotungsten(VI)] complexes. The reaction is performed at 120–200 °C and a mixture of the primary allyl alcohols geraniol and nerol and the tertiary allyl alcohol linalool is obtained in a ratio of approximately 55:5:35. These types of rearrangement can also be carried out in presence of WO_2Cl_2-based catalysts in presence of amino alcohols such as triethanolamine. The s/c of these reactions is around 1,000 and the metal-to-ligand ratio is 1:1 to 1:3. From a process point of view, the rearrangement reaction can also be carried out in a semi-continuous mode [47–49]. A further advantage of WO_2Cl_2-based catalysts compared to $VO(OR)_3$ or MoO_2-catalysts (which have also been reported for the allyl alcohol rearrangement) is the higher activity and selectivity of the WO_2Cl_2-catalysts [50]. As these reactions are driven by the equilibrium of the starting material and product, this approach can also be used to achieve the reverse reaction.

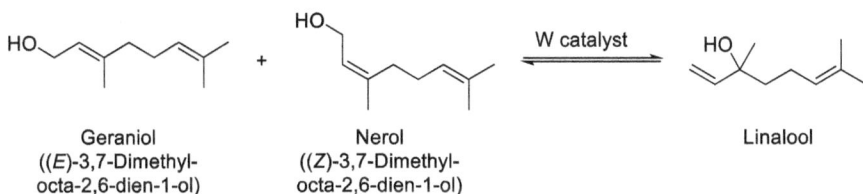

Geraniol	Nerol	Linalool
((E)-3,7-Dimethyl-	((Z)-3,7-Dimethyl-	
octa-2,6-dien-1-ol)	octa-2,6-dien-1-ol)	

Scheme 7.25: [W]-catalysed rearrangement of geraniol/nerol/linalool.

A similar rearrangement of an allylic alcohol, this time from the tertiary to the primary position is performed in the industrial synthesis of vitamin A acetate. 3-Methylpent-1-en-4-yn-3-ol is obtained via ethynylation of methyl vinyl ketone; this then undergoes an acid-catalysed rearrangement to form an (E/Z)-mixture of 3-methylpent-2-en-4-yl-1-ol (Scheme 7.26) – this is described in more detail in Chapter 8: Acid–base-catalysed reactions.

The manufacture of C_5-building block molecules is of industrial importance, especially for the manufacture of isoprenoids like carotenoids or vitamin A. The C_5-compound aldacet ((E)-3-methyl-4-oxobut-2-en-1-yl acetate) is produced from a butadiene and 2-

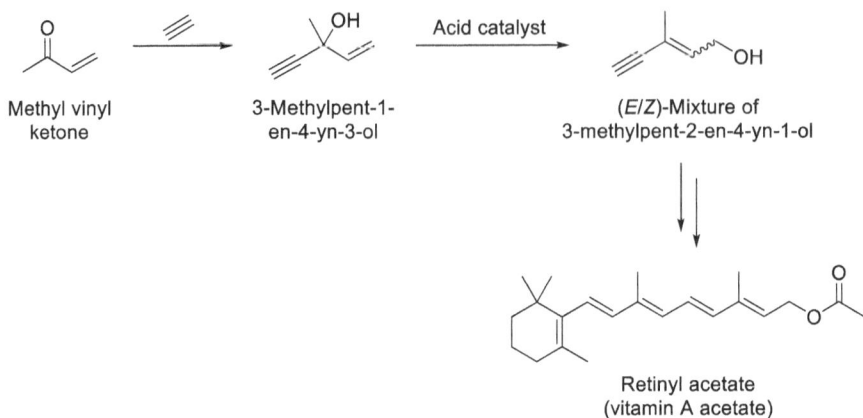

Methyl vinyl ketone → **3-Methylpent-1-en-4-yn-3-ol** → *Acid catalyst* → **(E/Z)-Mixture of 3-methylpent-2-en-4-yn-1-ol**

Retinyl acetate (vitamin A acetate)

Scheme 7.26: Rearrangement of a tertiary alcohol to a primary alcohol in the synthesis of vitamin A acetate.

butene mixture (coming from the C_4-fraction of an oil-cracker) by addition of acetic acid under oxidative conditions. The butene-diacetate undergoes a Cu-catalysed allylic rearrangement to form 3,4-diacetoxybut-1-ene, followed by a Rh-catalysed hydroformylation reaction at 100 °C (Scheme 7.27) [51, 52].

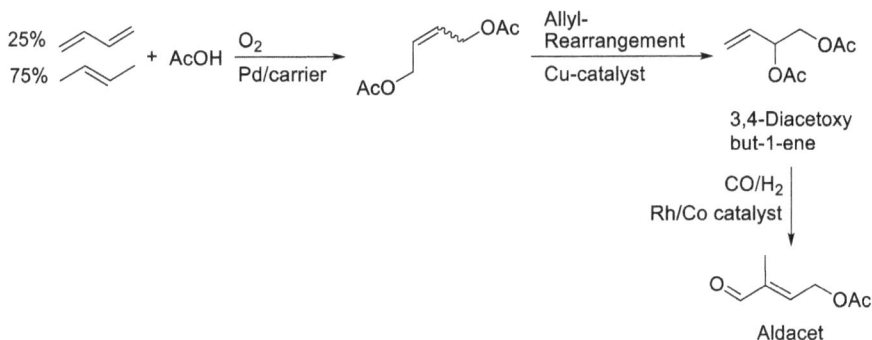

25% / 75% + AcOH $\xrightarrow[\text{Pd/carrier}]{O_2}$ → $\xrightarrow[\text{Cu-catalyst}]{\text{Allyl-Rearrangement}}$ → **3,4-Diacetoxy but-1-ene** $\xrightarrow[\text{Rh/Co catalyst}]{CO/H_2}$ → **Aldacet**

Scheme 7.27: Manufacture of aldacet from butadiene.

Aldacet itself can further be modified by oxidation and chlorination to 4-chloro-3-methyl-4-oxobut-2-enyl acetate (in 93% yield) or hydrogenated using a Pd/C catalyst to afford 4-acetoxy-2-methybutanal (Scheme 7.28) [53, 54]. The hydrogenation reaction is performed with the addition of a base or a basic support to avoid by-product formation; a selectivity of 98% can be achieved.

The synthesis of methyl chrysanthemate, a key compound of this substance class, which can be easily transferred into other esters by a transesterification reaction, starts from 2-methylbut-3-en-2-ol, followed by conversion into prenyl bromide. This is then converted into the sulphinate intermediate which undergoes the key cationic [1,2]-shift

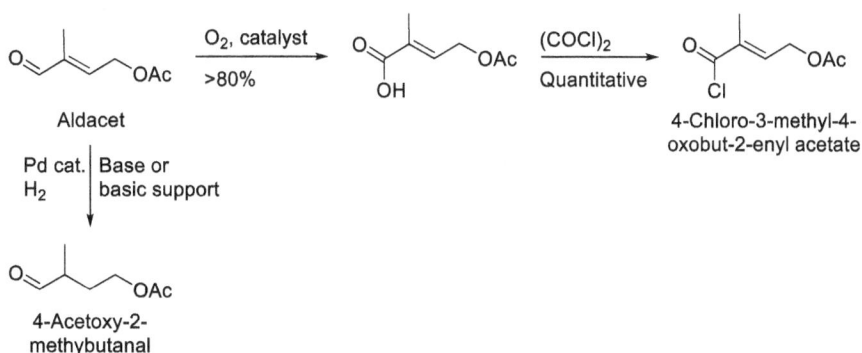

Scheme 7.28: Synthesis of C$_5$-building blocks from aldacet.

to form the corresponding sulphone. *Michael* addition followed by ring closure of the cyclopropane ring results in the chrysanthemic acid derivative (Scheme 7.29) [55, 56].

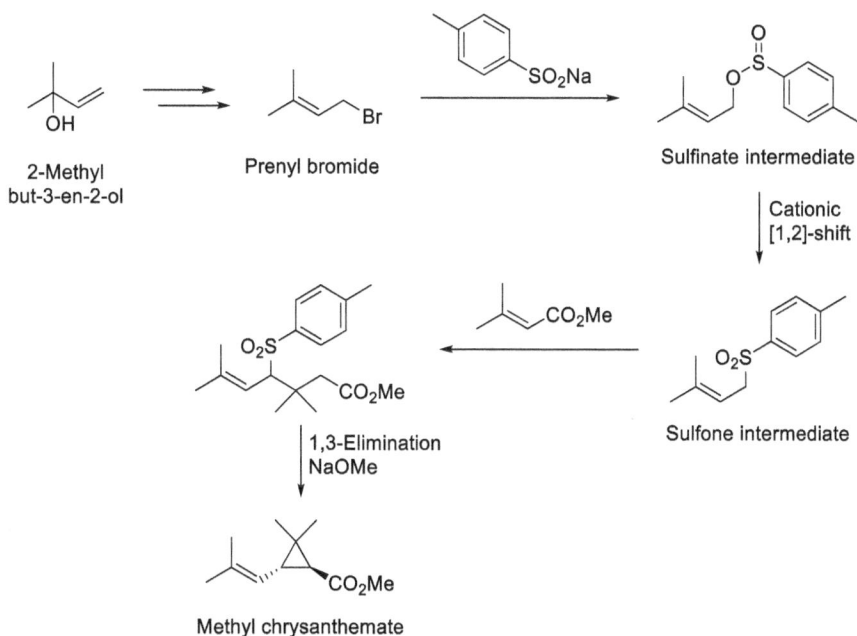

Scheme 7.29: Manufacture of chrysanthemum acid methyl ester from methylbutenol.

From an industrial point of view, another important allyl rearrangement/[1,3]-hydride shift is the enantioselective transformation of an allylamine into a chiral enamine applying a chiral Rh-BINAP complex in the Takasago process for (−)-menthol (Scheme 7.30). Alternative processes to produce (−)-menthol are discussed in Chapters 2 and 3: Heterogeneous and Homogeneous hydrogenations. (−)-Menthol is found in nature in plants such as *Mentha arvensis, Ocimum basilicum* or *Rosmarinus officinalis.* The production volume of (–)-menthol is around 20,000 t/a; the main processes are extraction from natural sources and chemical synthesis [57].

Takasago's menthol process was established in 1983 and contains two separate rearrangement reactions [58]. In the first step, myrcene is treated with diethyl amine/butyllithium. The product *N,N*-diethylgeranyl amine undergoes a [1,3]-hydride shift in the presence of a Rh-(S)-BINAP catalyst to form the corresponding enamine (3 R)-*E*-diethylamino-3,7-dimethyl-1,6-octadiene. This is followed by saponification to give (*R*)-(+)-citronellal in high optical purity (>99% *e.r.*). Citronellal is transformed by a *Lewis* acid-catalysed carbonyl-ene rearrangement in 92% to isopulegol followed by a hydrogenation reaction in the presence of a Ni-alloy catalyst [59, 60] (see Chapter 2: Heterogeneous hydrogenations).

The rearrangement reaction of diethylgeranyl amine is performed with a catalyst loading of around 10,000 on a multi t/a scale. The enamine is separated by distillation, the catalyst is recovered and reused. The original ene-reaction was catalysed by $ZnBr_2$ to afford the product in 92% selectivity. This step was later improved by the change to aluminium tris(2,6-diphenylphenoxide) which gives the product with 99.5% selectivity [61].

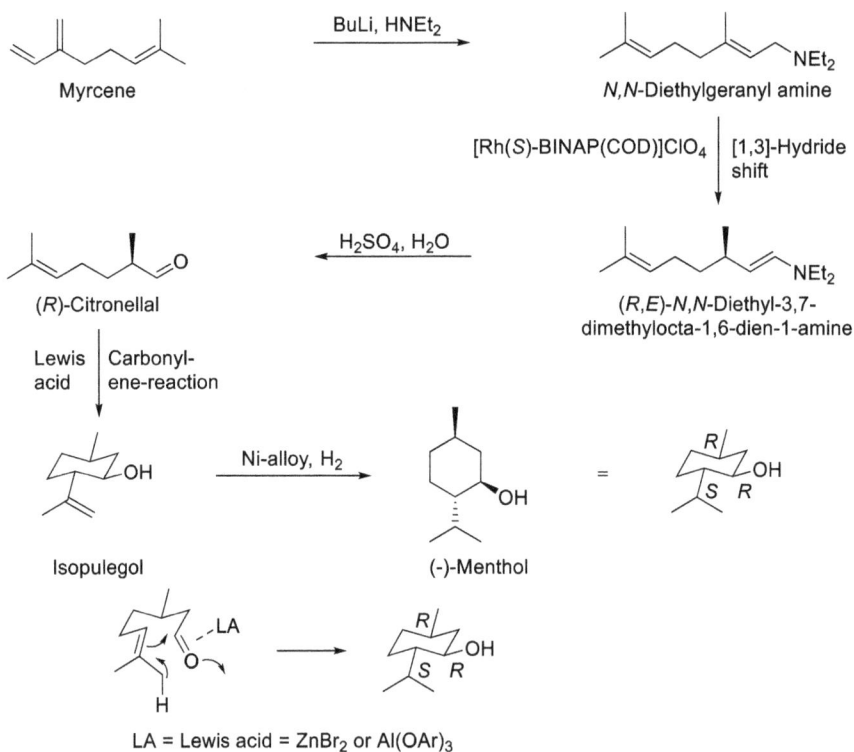

Scheme 7.30: Takasago's (–)-menthol process.

An additional benefit of the Rh-catalysed allylic isomerisation is the flexibility of this methodology. By the application of (R)-BINAP instead of (S)-BINAP as a ligand, the opposite enantiomer can be obtained [62]. This strategy allows access to both (R)- and (S)-citronellal which are the starting points for different flavour and fragrance compounds (Scheme 7.31). This highlights a general approach in the industry, broadening the product portfolio from one cheap key intermediate.

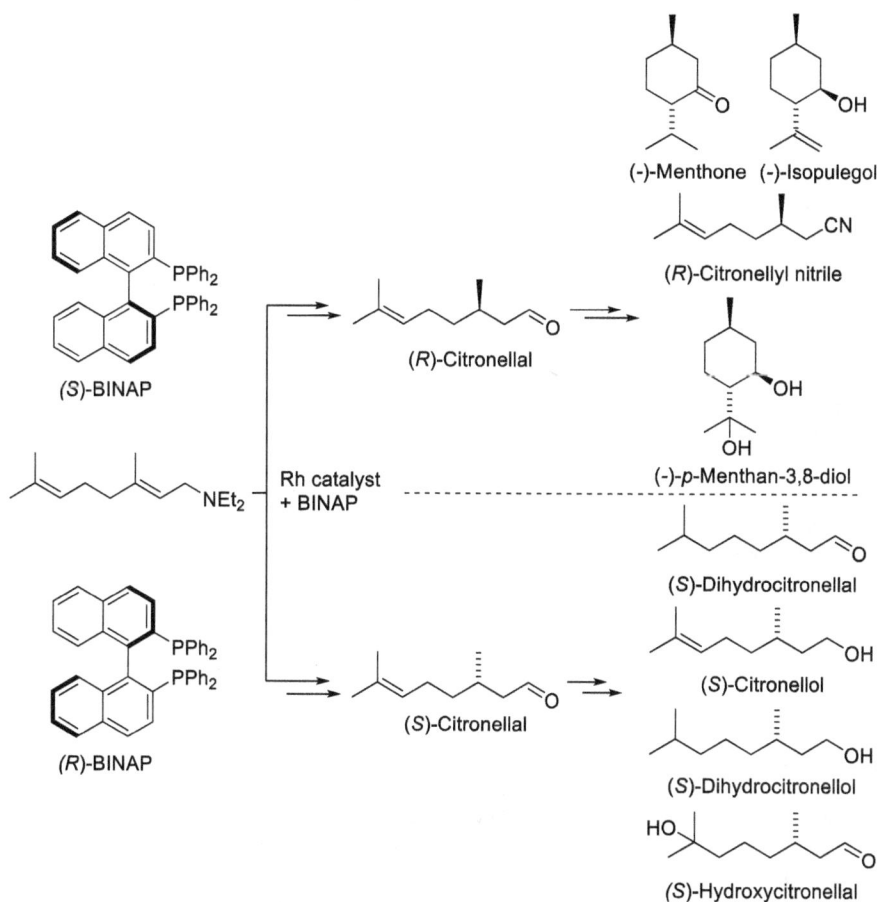

Scheme 7.31: Flexible application of the Rh-catalysed isomerisation for the synthesis of various flavour and fragrance compounds.

References

[1]   Ritz J, Fuchs H, Kieczka H, Moran W C. "Caprolactam". Ullmann's Encyclopedia of Industrial Chemistry. Weinheim, Wiley-VCH, 2011.

[2]   Eggersdorfer M, Laudert D, Létinois U, McClymont T, Medlock J, Netscher T, Bonrath W. One hundred years of vitamins-A success story of the natural sciences. Angew Chem Int Ed 2012, 51, 12960–12990.

[3]   Saucy G, Marbet R, Lindlar H, Isler O. New syntheses of citral and related compounds. Helv Chim Acta 1959, 42, 1945–1955.

[4]   Saucy G, Marbet R. Reaction of tertiary vinylcarbinols with isopropenyl ether. New method for the preparation of γ,δ-unsaturated ketones. Helv Chim Acta 1967, 50, 2091–2094.

[5]   Saucy G, Marbet R. Reaction of tertiary vinylcarbinols with vinyl ethers. New method for the preparation of γ,δ-unsaturated aldehydes. Helv Chim Acta 1967, 50, 2095–2100.

[6]   Aquino F, Bonrath W, Riebel H P. Novel Use of Phenylphosphinic Acid in the Manufacturing of Gamma,delta-unsaturated Ketones. WO 2018091624, DSM, 2018.

[7]   Aquino F, Bonrath W, Riebel H P. Novel Use of Phenylphosphinic Acid in the Manufacturing of Gamma,delta-unsaturated Ketones. WO 2018091623, DSM, May 24, 2018.

[8]   Aquino F, Bonrath W, Riebel H P. Novel Use of Phenylphosphinic Acid in the Manufacturing of Gamma,delta-unsaturated Ketones. WO 2018091622, DSM, 2018.

[9]   Saucy G, Marbet R. New type of β-oxoallene synthesis by reaction of tertiary acetylene carbinols with vinyl ethers. A method for preparation of pseudoionone and related compounds. Helv Chim Acta 1967, 50, 1158–1167.

[10] Bonrath W. New trends in (heterogeneous) catalysis for the fine chemicals industry. Chimia 2014, 68, 485–491.

[11] Bonrath W, Tschumi J, Medlock J. Process for Preparation of 3,7-dimethyl-1-octen-3-ol 2012 WO 2012025559, DSM, 2012.

[12] Mitchell P W, McElligott L T, Sasser D E. Hydrohalogenation of Conjugated Dienes. EP 132544, Union Camp Corp, February 13, 1984.

[13] Webb R L. Nitrogen base catalyzed displacement reaction of allylic terpene halides, US 3031442, April 24, 1962.

[14] Ninagawa Y, Yoshiaki O, Jujita Y, Hosogai T. Verfahren zur Isomerisierung ungesättigter Alkohole, DE 2307468, Kuraray Co, September 27, 1973.

[15] Schaefer B. Natural Products in the Chemical Industry. Berlin, Germany, Springer-Verlag, 2014.

[16] Fahlbusch K-G, Hammerschmidt F-J, Paten J, Pickenhagen W, Schatkowski D, Bauer K, Garbe D, Suburg H. Flavors and Fragrances. In: Elvers, ed. Ullmann's Encyclopedia of Industrial Chemistry, Online ed., Weinheim, Germany, Wiley-VCH Verlag GmbH & Co, 2012.

[17] Aquino F, Bonrath W. Process for the rearrangement of allyl alcohols. EP 2128120, DSM, December 3, 2009.

[18] Bonrath W, Scheer P L, Tschumi J, Zenhaeusern R. Catalytic ethynylation for the production of propargylic alcohols. WO 2004018400, DSM, March 4, 2004.

[19] Bonrath W, Künzi R. Process for the preparation of (5E, E)-farnesylacetone. WO 2009019132, DSM, February 12, 2009.

[20] Bonrath W, Künzi R. Process for Preparation of Allyl and Propargyl Ethers from Tertiary Ethynyl or Vinyl Carbinols and Isopropenyl or Butenyl Methyl Ether. WO 2009138389, DSM, 2009.

[21] Climent M J, Corma A, Iborra S, Velty A. Synthesis of methylpseudoionones by activated hydrotalcites as solid base catalysts. Green Chem 2002, 4, 474–480.

[22] Roelofs J C A A, Van Dillen A J, De Jong K P. Condensation of citral and ketones using activated hydrotalcite catalysts. Catal Lett 2001, 74, 91–94.

[23]   Ma X, Fang W, Xu P, Song J, Li Y, Yin Z, Xu Y, Fu Z, Zhong J. Synthetic Method of α-isomethylionone. CN 108017525, Shandong NHU Pharmaceutical Co., 2018.

[24]   Kawanobe T, Kojo K. Pseudo-isomethylionone. JP 61158945, Hasegawa, T., Co Ltd., Jpn, 1986.

[25]   Horikawa S, Hamada M, Maeda M, Kunishige T, Goto T. Pseudoisomethylionone. JP 48013087, Shiono Koryo Kaisha, Ltd., 1973.

[26]   Pseudoisomethyl- and Isomethylionones. NL 103065, International Flavors and Fragrances I.F.F. (Nederland) N.V., 1962.

[27]   Carroll M F. Addition of α,β-unsaturated alcohols to the active methylene group. Part I. The action of ethyl acetoacetate on linalool and geraniol. J Chem Soc 1940, 704–706.

[28]   Martin Castro A M. Claisen rearrangement over the past nine decades. Chem Rev 2004, 104, 2939–3002.

[29]   Hoelderich W F, Kollmer F. Oxidation reactions in the synthesis of fine and intermediate chemicals using environmentally benign oxidants and the right reactor system. Pure Appl Chem 2000, 72, 1273–1287.

[30]   Bonrath W. Process for Preparing Dienones in High Yield and Purity. WO 2008/092655, DSM, 2008.

[31]   Bonrath W, Karge R, Netscher T, Pressel Y. Isomerization of β-allene Ketones. US 8802898, DSM, 2014.

[32]   Fráter G, Helmlinger D. Process for the Preparation of Irones. EP 0418690, Givaudan, 1991.

[33]   Bienayme H. Intermediates for the Preparation of Vitamin A and Carotenoids and Process for Their Preparation. EP 0647624 A1, Rhone-Poulenc, 1995.

[34]   Rivera S B, Swedlund B D, King G J, Bell R N, Hussey C E, Shattuck-Eidens D M, Wrobel W M, Peiser G D, Poulter C D. Chrysanthemyl diphosphate synthase: Isolation of the gene and characterization of the recombinant non-head-to-tail monoterpene synthase from Chrysanthemum cinerariaefolium. Proc Natl Acad Sci USA 2001, 98, 4373–4378.

[35]   Kelly L F. A synthesis of chrysanthemic ester: An undergraduate experiment. J Chem Educ 1987, 64, 1061.

[36]   Konto K, Matsui M, Negishi A, Takahatake Y. Process for Preparing dihalov-nylcyclopropynecarboxylates. US 4681953, Sagami Chem Res., 1975.

[37]   Hoffmann R W. Industrielle synthesen terpenoider Riechstoffe. Chem Ztg 1973, 97, 23–28.

[38]   Meyer K-H, Schuster K. Umlagerung tertiärer Äthinyl-carbinole in ungesättigte Ketone. Ber Dtsch Chem Ges 1922, 55, 819–823.

[39]   Rupe H, Kambli H E. Ungesättigte Aldehyde aus Acetylen-alkoholen. Helv Chim Acta 1926, 9, 672.

[40]   Pauling H, Andrews D A, Hindley N C. The rearrangement of α-Acetylenc Alcohols to α,β-unsaturated carbonyl compounds by silylvanadate catalysts. Helv Chim Acta 1976, 59, 1233–1243.

[41]   Trost B M, Tracy J S. Catalytically generated vanadium enolates formed via interruption of the Meyer–Schuster rearrangement as useful reactive intermediates. Acc Chem Res 2020, 53, 1568–1579.

[42]   Arnold W, Bonrath W, Pauling H, Pernin R, Thum A. Perrhenate katalysierte Umlagerung von Ethinl-β-ionol. Helv Chim Acta 1996, 79, 646–650.

[43]   Bonrath W, Schuetz J, Netscher T, Wuestenberg B. Process of Producing 7,8-dihydro-C_{15}-aldehyde. WO 2016059150, DSM, 2016.

[44]   Bonrath W, Netscher T, Schuetz J, Wuestenberg B. Method for Producing Specific α,β-unsaturated Aldehydes. WO 2016059152, DSM, 2016.

[45]   Robeson C D, Lindsay J K. Vitamin A Synthesis. US 2676990, Eastman Kodak, 1954.

[46]   Ancell J E, Meilland P. Method for Preparing Vitamin A. WO 00/02854, Rhone-Poulenc, 2000.

[47]   Haese F, Ebel K. Isomerization of Allyl Alcohols with Tungsten-based Catalysts. WO 2003047749, BASF, 2003.

[48]   Ebel K, Stock F. Process and tungstenoxo(VI)-aminoalcohol Complex Catalysts for the Isomerization of Allyl Alcohols. De 19958603, BASF, 2001.

[49] Haese F, Ebel K, Burkart K, Unvericht S, Muenster P. Method for Isomerizing Allyl Alcohols via a Semi Continuous or Continuous Operation. WO 2003048091, BASF, 2003.

[50] Chabardes P, Kuntz E, Varagnat J. Use of oxo-metallic derivatives in isomerization. Reactions of unsaturated alcohols. Tetrahedron 1977, 33, 1775–1783.

[51] Himmele W, Auila W. 1,2-Di(acyloxy)-3-formylbutanes. DE 1945479, BASF, 1971.

[52] Rheude U, Vicari M, Aquila W, Wegner G, Niekerken J. Process for the Production of the C5-acetate Used for the Manufacture of Vitamin A by the Hydroformylation of Distillation-purified 3,4-diacetoxy-1-butene. EP 1247794, BASF, 2002.

[53] Bonrath W, Netscher T, Schütz J, Wüstenberg B. Method for Preparing Acid Chlorides as Intermediates of Synthesis of Vitamin A. Wo 12175395, DSM, 2012.

[54] Bonrath W, Schütz J. Method for Preparation of 4-acetoxy-2-methylbutanal by Catalytic Carbon Carbon Double Bond Hydrogenation. WO 12098067, DSM, 2012.

[55] Matel J J, Huynh C. Synthesis of chrysanthemic acid II. Bull Soc Chim Fr 1967, 985.

[56] Julia M, Guy-Rouault A. Synthesis of cyclopropanes from sulfones. Application to chrysanthemic acid. Bull Soc Chim Fr 1967, 1411.

[57] Schäfer B. Menthol mint versus the Takasago process. Chemie in Unserer Zeit 2013, 47, 174–182.

[58] Emura M, Matsuda H. A green and sustainable approach: Celebrating the 30th anniversary of the asymmetric l-menthol process. Chem Biodivers 2014, 11, 1688–1699.

[59] Nakatani Y, Kawashima K. A highly stereoselective preparation of l-isopulegol. Synthesis 1978, 147–148.

[60] Snider B B. Lewis-acid catalysed ene-reactions Acc. Chem Res 1980, 13, 426–432.

[61] Iwata T, Okeda Y, Hor Y. Method for Producing Isopulegol, JP 2002212121, Takasago, 2002.

[62] Otsuka S, Tani K. Catalytic asymmetric hydrogen migration of allylamines. Synthesis 1991, 9, 665–680.

8 Acid–base-catalysed reactions

8.1 Replacement of *Brønsted* acids

As mentioned in the introduction, modern trends and approaches in industrial catalysis are the development of new types of catalyst which allows new modes of operation, such as continuous processes under heterogeneous conditions, the combination of unit operations and the decrease of by-product formation (minimising the formation of waste). In acid-catalysed reactions, a major source of waste comes from mineral acids and *Lewis* acids, which are difficult to recover and/or recycle.

In the fine chemical industry, only a limited number of catalytic processes, especially heterogeneous processes, were known some years ago. In recent years, the situation has partly changed. In 2005, more than 120 industrial processes were based on solid (heterogeneous) acid catalysis [1, 2]. Additionally, around 13 industrial processes are based on solid base catalysis (heterogeneous).

The main disadvantages of *Brønsted*-acid and Brønsted-base catalysis are:
- Waste formation,
- Corrosion,
- Low selectivity.

The statement of M. Hara *et al.* summarising the situation in the mid-2000s is very impressive [3]:

> Over 15 million tons of sulfuric acid is annually consumed as "unrecyclable catalyst" – which requires costly and inefficient separation of the catalyst from homogeneous reaction mixtures – for the production of industrially important chemicals, thus resulting in a huge waste of energy and large amounts of waste products.

Therefore, there is a need for more selective catalysts, especially due to the ever-increasing production volumes, and a preference for processes using a heterogeneous catalyst.

In recent years, several new processes applying heterogeneous catalyst/solvent systems have been described. Here the replacement of mineral acids by solid acids was the focus. These developments are based on the availability of solid materials and new or modified catalytic concepts such as working in multi-phase systems. Alternatives for mineral acids such as sulphuric acid or phosphoric acid are solid acids like zeolites, heteropoly acids, acidic clays, sulphonated polymers such as acidic ion-exchange resins, or Nafion®-type materials. Nafion® is a perfluoroalkylsulphonic acid resin from perfluoropropionic acid and tetrafluoroethylene, or Nafion® composites, such as Nafion® on silica. An advantage of the application of new types of acid catalysts is the broad range of acidity which is available. For example, solid acids are often strong acids and have the behaviour of super acids. The term "super acid" was

https://doi.org/10.1515/9783111102672-008

introduced by James Bryant Conant in 1927 and is used for an acidity stronger than 100% H_2SO_4 (pKa = −3.0). Later, super acids were used by Georg Olah for stabilisation of carbocations and their chemistry [4, 5]. The main differences between styrene-based and perfluorinated resins are the number of active sites and their acidic strength. The AMBERLYST®-type resins show a *Hammett* acidity of −2.2, whereas per-fluoro modified acids of the Nafion-type have a *Hammett* acidity from −11 to −13 [5–7].

One major class of acid-catalysed reactions is rearrangement reactions. From an atom economy point of view, rearrangement reactions fulfil the criteria of green chemistry because in this type of reaction predominantly all the atoms of the starting material are present in the product, and only a very limited amount of waste is produced. Nevertheless, such types of reactions also require further improvements, such as replacement of the currently used acid and avoiding the use of halogen-containing solvents.

The replacement of mineral acids by strong solid acids such as ion-exchange resins containing sulphonic acid groups is well established in the chemical industry for bulk chemicals such as bisphenol A (BPA). BPA is manufactured from phenol and acetone in several million t/a and applied in plastics industry, for example in the formation of polycarbonates (Scheme 8.1) [8].

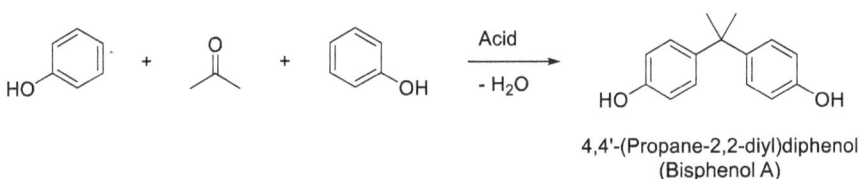

4,4'-(Propane-2,2-diyl)diphenol
(Bisphenol A)

Scheme 8.1: Bisphenol A synthesis.

In the manufacture of vitamin A and derivatives by the Isler route, 3-methylpent-1-en-4-yn-3-ol (produced from methyl vinyl ketone and ethyne) is converted via an allylic rearrangement into the corresponding primary alcohol (Scheme 8.2) [9]. In the past, this rearrangement reaction was performed with sulphuric acid in solvents such as dichloromethane [10]. New processes have been established applying a two-phase solvent system: water and a water-immiscible ether (e.g. diisopropyl ether) in the presence of an acid, such as a *Lewis* or *Brønsted* acid, or solid acids such as acidic ion-exchange resins, for example AMBERLYST® 15, 16, 36 or DOWEX® [11, 12]. The reaction is carried out under mild reaction conditions with s/c >1,000 and has an increased yield and selectivity compared to the sulphuric acid-catalysed process, and it fulfils the principles of green chemistry.

The synthesis of many isoprenoid derivatives such as vitamins and aroma compounds makes use of base-catalysed C_2-elongation and acid-catalysed C_3-elongation reactions [13]. The C_3-elongation reaction is described in Chapter 7: Rearrangement reactions and the C_2-elongation later in this chapter. The key-building block of the C_3-elongation reactions is isopropenyl methyl ether (IPM), which is produced on several

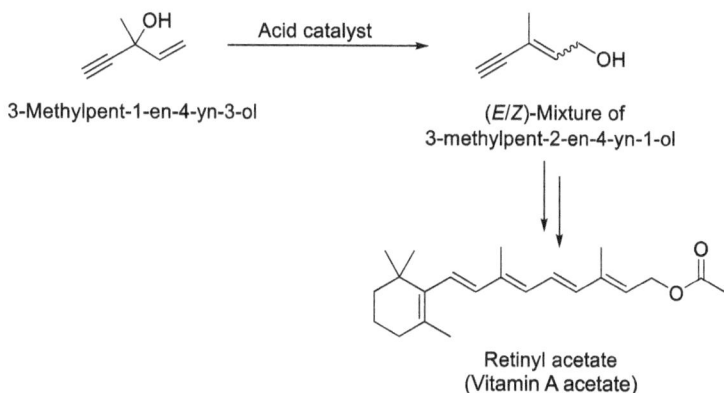

3-Methylpent-1-en-4-yn-3-ol

(*E/Z*)-Mixture of
3-methylpent-2-en-4-yn-1-ol

Retinyl acetate
(Vitamin A acetate)

Scheme 8.2: Solid acid-catalysed allylic rearrangement of 3-methyl-1-en-4-yn-3-ol.

10,000s t/a scale in a gas-phase reaction of 2,2-dimethoxypropane (DMP) on a solid cata-lyst (see Chapter 5: Gas-phase reactions). In the production of 2,2-dimethoxypropane, new technologies have been applied, such as replacement of a homogeneous acid by a solid acid (e.g. an ion-exchange resin) and separation of water with membrane technol-ogy to shift the reaction equilibrium towards the desired product (Scheme 8.3). The new procedure has the advantages of lower waste formation, and increased yield and selec-tivity [14, 15].

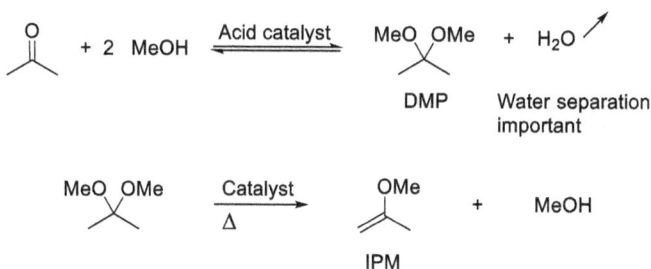

DMP   Water separation
        important

IPM

Scheme 8.3: Synthesis of isopropenyl methyl ether (IPM).

Another type of solid acid which can replace homogeneous acids is a copolymer of tetrafluoroethene and a polyalkylsulphonic acid called Nafion®. Nafion® itself is used in membranes, anode materials or as a strong solid acid. Solid acid-catalysed re-actions by Nafion® are acylation reactions, rearrangement reactions and isomerisa-tion reactions [16, 17].

An interesting application of a Nafion-silica composite is its use as a solid acid cata-lyst for the *Fries*-rearrangement of phenyl acetate to hydroxy-acetophenone in the syn-thesis of paracetamol. The heterogeneous Nafion® composite catalysed reaction results

OH NOH

SOCl$_2$

B

HON

Separation

OH

Paracetamol

C

NH$_2$OH

OH

OH O

A

A: *Fries* rearrangement
B: *Beckmann* rearrangement
C: Hydrolysis and recovery

Phenyl acetate

HF
Ac$_2$O

OH

Phenol

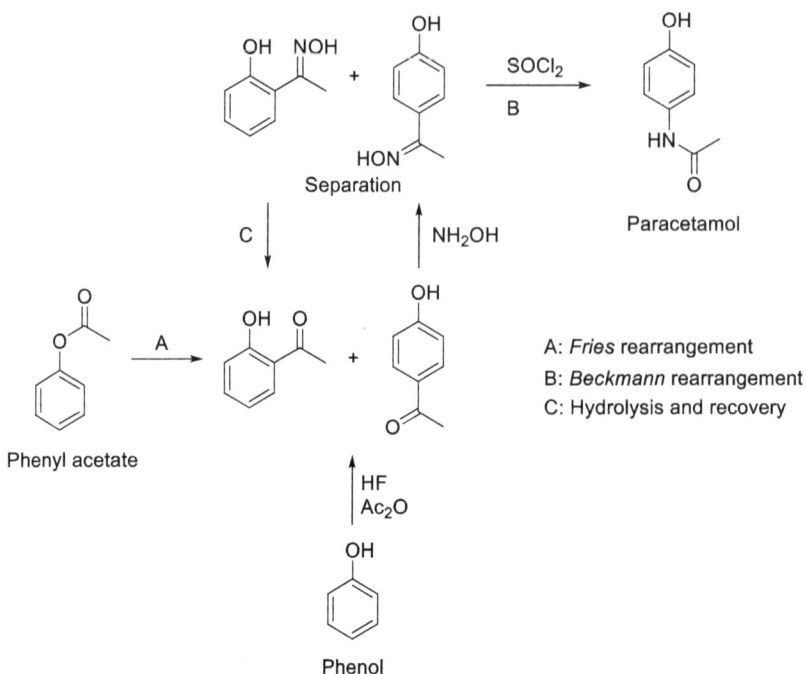

Scheme 8.4: Syntheses of paracetamol.

in an increased selectivity for *p*-hydroxy acetophenone, especially if phenylacetate is used neat [18]. An overview of routes to paracetamol is shown in Scheme 8.4.

Furthermore, strong acidic resins are industrially applied in etherification reactions of olefins and alcohols. An example is the production of methyl *t*-butyl ether starting from methanol and isobutene.

The industrial application of ion-exchange resins in esterification reactions is well known. Maleic anhydride or maleic acid is transformed into diethyl maleate in a solid acid-catalysed reaction with high conversion (Kvaerner process technology, DOW, 20,000 t/a, Rohm and Haas, BASF, Scheme 8.5). The hydrogenation of the diester results in unsaturated or saturated diols, which are ingredients in the manufacture of

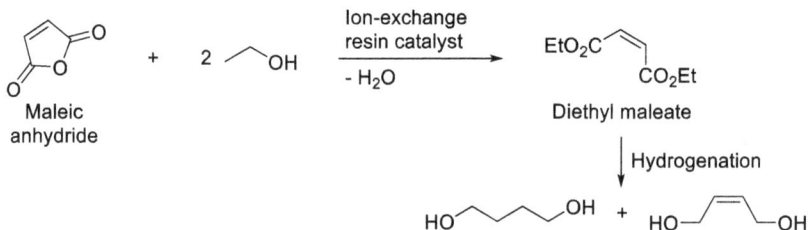

Maleic
anhydride

+ 2 OH

Ion-exchange
resin catalyst

- H$_2$O

EtO$_2$C

CO$_2$Et

Diethyl maleate

Hydrogenation

HO OH + HO OH

Scheme 8.5: Synthesis of 1,4-butanediol and 1,4-butendiol from maleic acid anhydride.

detergents [19]. This technology was also described for the production of fatty acid methyl esters. In general, the ester formation is carried out with >99% conversion and >99% selectivity.

This technology is cost efficient and environmentally friendly because the waste streams are reduced; it is a good example of the replacement of sulphuric acid or methyl sulphonic acid (homogeneous catalyst) by a heterogeneous acid catalyst.

Strong acidic ion-exchange resins of the sulphonic acid type are also applied in the hydration of olefins, for example the hydration of propene or the synthesis of *t*-butanol from isobutene (Scheme 8.6). The application of these solid acid catalysts in hydration reactions has been implemented at production volumes of several 100,000 t/a in several plants. Beneficial aspects of these catalysts (e.g. AMBERLYST®-type resins) are the long life-time of the catalyst, excellent selectivity and lower waste formation [20].

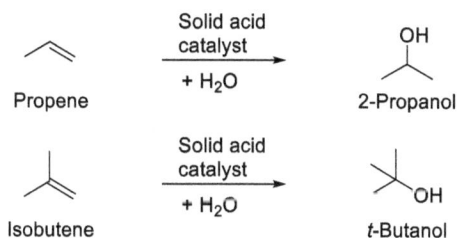

Typical catalysts: ion-exchange resins

Scheme 8.6: Solid acid-catalysed hydration reaction of olefins.

The dehydration of 1,4-butanediol into tetrahydrofuran is performed on an industrial scale of >20,000 t/a (Scheme 8.7). Another example of the use of macroporous ion-exchange resins in dehydration reactions and the replacement of mineral acid catalysts is the formation of bisphenol A (Scheme 8.1) and the synthesis of 2-vinyl-4-hydroxy-1,3-dioxolanes [21, 22].

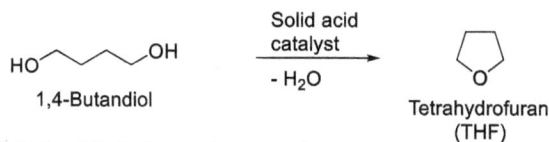

Typical catalysts: ion-exchange resins

Scheme 8.7: Application of acid ion-exchange resins in dehydration reactions.

The gasoline octane boosters such as the ethers methyl *tert*-butyl ether (MTBE), ethyl *tert*-butyl ether (ETBE) and *tert*-amyl methyl ether (TAME) are produced industrially on a large scale by addition of methanol to isobutene, ethanol to isobutene and methanol to 2-methyl-1(or 2)-butene, respectively, over an acid catalyst [23, 24]. These reactions take place in the liquid phase over microporous sulphonic acid resin catalysts, for example AMBERLYST®-15 and DOWEX® M32 [25]. The acid capacity is an important parameter for the activity of a resin catalyst, similar to other solid acid catalysts; a higher acid capacity results in an increased catalyst activity [26]. However, the surface area and pore diameter are less important. Microporous resins are preferred due to the better accessibility of the active sites [27]. The ethers, diisopropyl ether [28] and t-butyl propyl ether [29], have been considered as fuel additives and are also produced with cation-exchange resin-catalysed etherification reactions. The production volume of these ethers is up to several 10^3 to several 10^6 t/a.

Another example of the replacement of a *Brønsted* acid by a heterogeneous acid (here an ion-exchange resin) is the industrial manufacture of an intermediate of vitamin B$_6$ (Scheme 8.8). The water-soluble vitamin B$_6$ has several functions in organisms, for example amino-group transfer, and is produced in <10,000 t/a. In all industrial processes, a *Diels–Alder* reaction of a diene compound (here an oxazole derivative) and a dienophile (a dioxepine) are used [30, 31]. The dienophile is produced by a solid acid-catalysed reaction applying an ion-exchange resin; preferred are strong acidic ion-exchange resins containing sulphonic acid groups. In these processes, which can be carried out batch-wise or continuously, sulphuric acid was replaced by an ion-exchange resin. Advantages are the reduced waste formation and high selectivity.

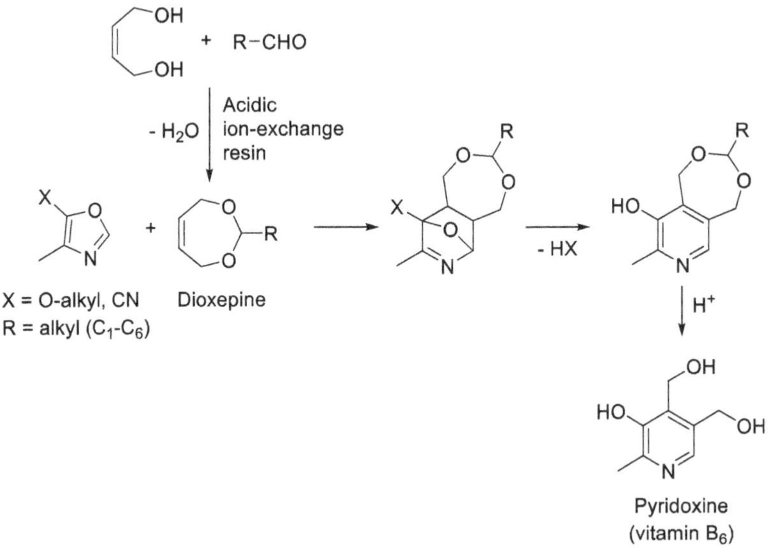

Scheme 8.8: Synthesis of vitamin B$_6$ key intermediate dioxepine.

Another fragrance compound based on pinene is α-terpineol (1-*p*-menthen-8-ol), which has a lilac odour and is applied as a mixture of isomers in soaps and cosmetics. The production volume is estimated to be several t/a. In the industrial process, an acid-catalysed reaction of water and α-pinene results in the formation of terpin hydrate, and the following partial dehydration gives rise to the final product (Scheme 8.9) [32].

Scheme 8.9: Synthesis of α-terpineol from α-pinene.

In the field of carotenoid synthesis and manufacture several concepts to form C_{40}-skeletons have been reported. One of the most important and widely used approaches is the combination of suitable $C_{15} + C_{10} + C_{15}$ building blocks (Scheme 8.10) (see Chapter 8.4: *Lewis* acids) [33, 34].

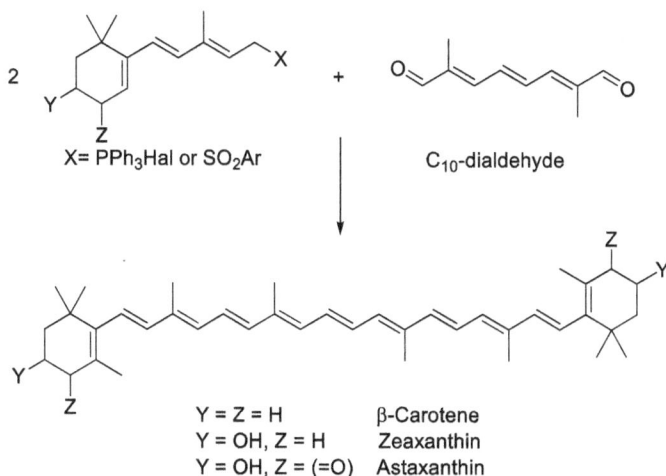

Scheme 8.10: Simplified $C_{15} + C_{10} + C_{15}$ concept applied in carotenoid synthesis.

The central key building block, C_{10}-dialdehyde can be prepared from furan. Treatment of furan with bromine in methanol results in the formation of 2,5-dimethoxydihydrofuran (DMDF), which is ring-opened under acid conditions to give the diacetal tetramethoxybutene (TMB); this undergoes condensation with two C_3-building blocks to form the C_{10}-dialdehyde (Scheme 8.11 and Scheme 8.27 in chapter 8.4).

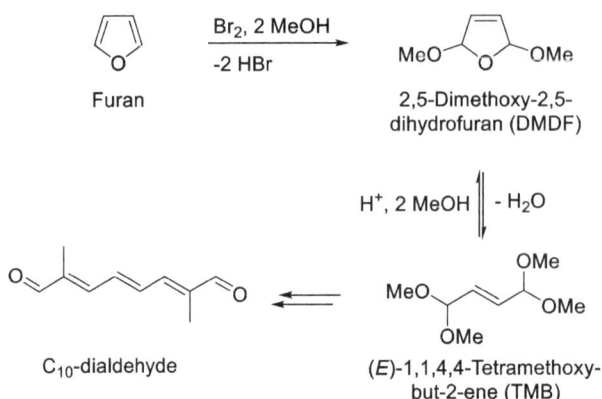

Scheme 8.11: Synthesis of C_{10}-dialdehyde via tetramethoxy butene (TMB).

During the bromination reaction, HBr is liberated which catalyses the ring opening reaction of DMDF into TMB. The main problems of this process are the formation of salt-waste by neutralisation of the mineral acid with base, the downstream processing of the product, the formation of side-products and the ecological footprint.

The application of a solid acid catalyst, especially an acidic ion-exchange resin (IER), allows improved control of the equilibrium of DMDF/TMB and thus results in an efficient process and a reduction of the amount of waste formed [35]. The reaction can be carried out in a continuous mode applying mild reaction temperature (around 20 °C) and the acidic ion-exchange resin has a long lifetime. The MeOH/DMDF ratio is in the range of 45–100, and a selectivity of >97% can be achieved. A simplified reaction set-up is depicted below (Figure 8.1).

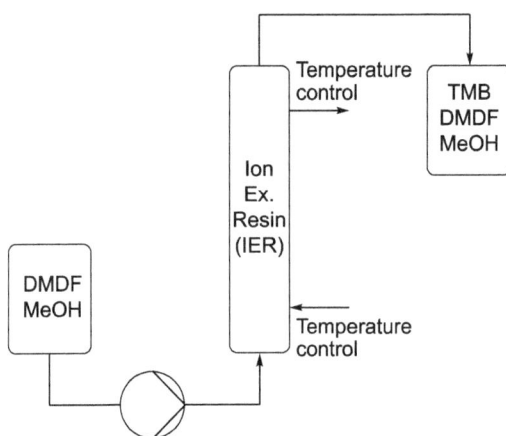

Figure 8.1: Simplified reaction set-up of an ion-exchange resin (IER) application in the synthesis of TMB.

The optical isomers of carvone (2-methyl-5-(prop-1-en-2-yl)cyclohex-2-en-1-one) differ in their organoleptic properties. The (S)-(+)-enantiomer is found in carawa seeds (*Carum carvi*) and the (R)-(-)-isomer is found in spearmint (*Mentha spicata*) and dill. The latter has a sweetish, minty smell and is manufactured in around 10 t/a [36, 37]. It can be separated from natural sources by fractionated distillation.

(-)-Carvone applied in the flavour industry is manufactured from (R)-(+)-limonene (Scheme 8.12) [38–40]. (+)-Limonene is reacted with gaseous nitrosyl chloride to produce (-)-carvone oxime. This is then converted into carvone in the presence of oxalic acid in acetone (s/c = 20).

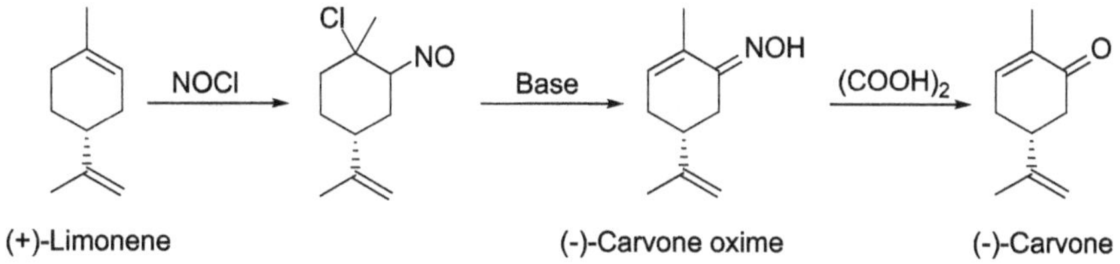

(+)-Limonene   NOCl   Base   (-)-Carvone oxime   (COOH)$_2$ (-)-Carvone

Scheme 8.12: Manufacture of (-)-carvone.

8.2 Zeolite-based catalysis

Zeolites are microporous crystalline materials with the ability to perform cation exchanges. They are insoluble in water and common organic solvents and have a sufficient thermal stability that allows the removal of all pore filling agents during their synthesis. Mesoporous solids, metal-organic frameworks (MOFs) and cationic and anionic classes are included in this group (International Zeolite Association). Zeolites are often aluminosilicates consisting of Si, Al and O, and metals such as Ti, Sn and Zn. They have hierarchical structures containing small (8-ring), medium (10-ring) or large (12-ring) entrances to the pores (Figure 8.2). Zeolites with a pore size < 2 nm are microporous materials, whereas mesoporous materials have a pore size of 2–50 nm and macroporous systems a pore size of >50 nm. Furthermore, zeolites are form-selective materials; the pore size and the pore architecture can influence the selectivity of heterogeneous reactions based on the specific substrate shape [41, 42]. Zeolite-based catalysis is well established in the petrochemical industry, especially for cracking and hydrocracking, but will not be discussed further here.

The application of H-ZSM-5 as a solid acid alkylation catalyst in the manufacture of ethylbenzene from ethene and benzene in a fixed-bed reactor system was industrialised 1973 and is known as the Mobil Oil and Bagger process (Scheme 8.13) [43]. The process is carried out in the gas phase at 420–430 °C, 15–20 bar and a yield of 99% is achieved. Ethylbenzene is produced in several million t/a by this process and is subsequently used for the production of styrene [44].

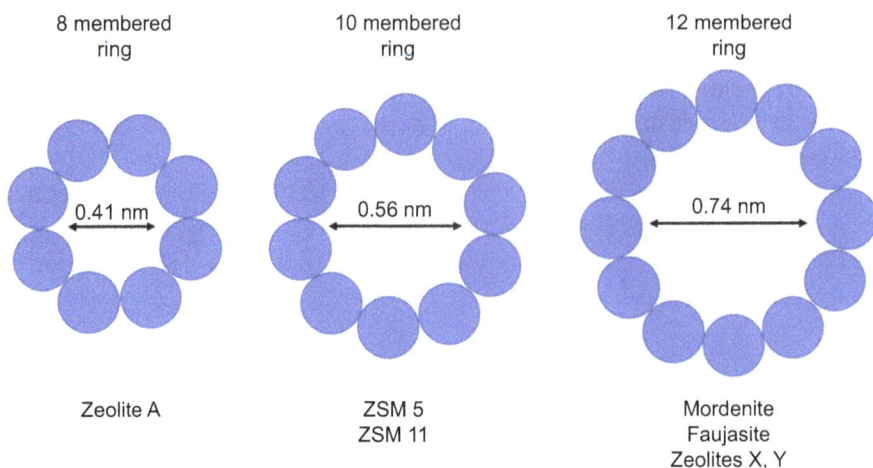

8 membered ring	10 membered ring	12 membered ring
0.41 nm	0.56 nm	0.74 nm
Zeolite A	ZSM 5 ZSM 11	Mordenite Faujasite Zeolites X, Y

Figure 8.2: Schematic ring size of selected zeolites.

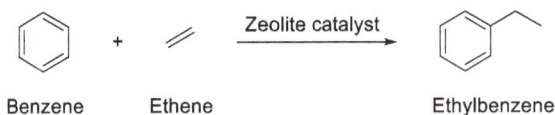

Benzene   Ethene   Ethylbenzene

Scheme 8.13: Synthesis of ethylbenzene from benzene and ethene.

The process can also be carried out in liquid phase in the presence of an MCM-22-based catalyst, with a benzene/ethene ratio of 3–5:1.

Similar alkylation processes are processes for the production of cumene, which is important for the manufacture of phenol, and processes for the production of alkyl benzene sulphonic acids (linear alkyl benzene sulphonates). These sulphonic acids are obtained by alkylation of benzene and C_{10}–C_{16} olefins, which are used as surfactants [44, 45].

Vanillin is an important compound in the flavour industry [46]. In nature, vanillin is found in its glycoside form and needs to be further treated before it can be used as a flavour material [47]. The natural product is manufactured by extraction of vanilla beans and further treatment with water, sunlight and enzymes. Applications of vanillin are sweet food, baked goods, medical applications (odours, taste), cleaning products, and as an intermediate for the synthesis of fine chemicals and pharmaceuticals [48, 49]. The demand for vanillin is so large that it cannot solely be met with material from natural sources, and approximately 80% of vanillin production is based on chemical routes. Therefore, a cost-efficient manufacturing procedure is needed.

Today, the chemical processes of vanillin production start from guaiacol (2-methoxyphenol) obtained from catechol (see Chapter 4: Oxidations). In the past, eugenol (also applied in the flavour industry) was the starting material of vanillin production. Later, lignin from waste streams in paper production was the main starting material for the synthesis of vanillin. Lignin treated under basic conditions followed by oxidation of the intermediate compound results in vanillin [50]. An overview of vanillin routes starting from guaiacol is found in [51].

The industrial production of vanillin (Rhône–Poulenc Process) starts from catechol (produced by phenol hydroxylation with H_2O_2, or glucose). The EniChem process performs the phenol hydroxylation on an industrial scale, using a titanium silicalite-1 catalyst (TS-1) at around 360 K (Scheme 8.14). The phenol conversion is approximately 20–30% and a selectivity of >90% is achieved [52]. The gas phase methylation with methanol in the presence of a lanthanide-phosphate at >500 K results in 2-methoxyphenol. The following C–C bond formation, hydroxy-methylation with formaldehyde, is catalysed by a zeolite and results in p-hydroxymethyl guaiacol. The alkylation reaction is performed at around 310 K in the presence of an H-mordenite with a Si/Al ratio of 18, the reaction is run at 30% conversion resulting in approximately 98% selectivity [53, 54]. Oxidation in the presence of a supported Pt or Pd catalyst results in the final product [55–57].

In general, it must be pointed out that in the reaction sequence, four different heterogeneous catalysed reactions are performed in excellent yield and selectivity, and that only water is produced as a by-product. From a point of sustainability and green chemistry, this process fulfils all criteria of modern chemistry and its industrial application.

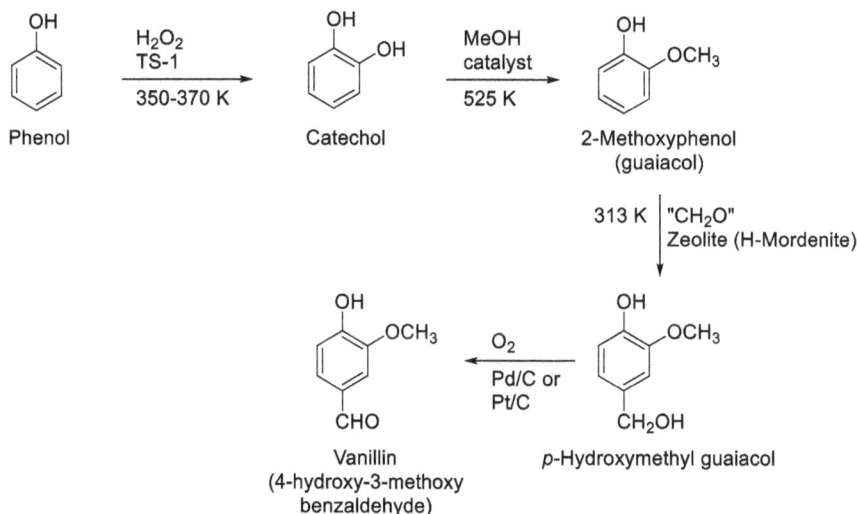

Scheme 8.14: Scheme Rhône–Poulenc process to vanillin.

p-Acetyl-anisole (4-methoxy-acetophenone, acetanisole) is used in the flavour and fragrance industry, as a flavouring in the food industry and in the pharmaceutical industry (a starting material for the synthesis of sulfarlem (treatment of *Parkinson's* disease) and methyl-synephrine). In current processes, anisole is treated in a *Friedel–Crafts* acylation reaction with acetyl chloride or acetic anhydride in the presence of aluminium chloride or zeolite. The application of zeolites for the acylation of anisole has been studied in detail, and the main focuses are the catalyst activity, selectivity and recovery. Starting from acetic anhydride and anisole, 4-methoxyacetophenone is produced using a zeolite catalyst, such as a zeolite Y or Faujasite, in a fixed-bed reactor system (Scheme 8.15). This approach is simple, cost efficient and environmentally friendly [58].

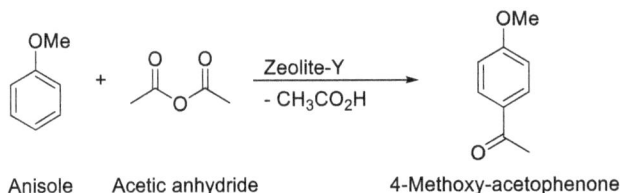

| Anisole | Acetic anhydride | | 4-Methoxy-acetophenone |

Scheme 8.15: Zeolite-catalysed synthesis of 4-methoxy-acetophenone synthesis.

The *Lewis* acid-catalysed diastereoselective cyclisation reaction of citronellal to isopulegol followed by a hydrogenation reaction is an important industrial process in the manufacture of *L*-menthol (Scheme 8.16) (see Chapter 2: Heterogeneous hydrogenations). The citronellal ring closure reaction is usually performed in the presence of $ZnBr_2$ giving 92% yield. The problems of high $ZnBr_2$ loading (nearly stoichiometric) and decreased selectivity in the presence of water (due to the hygroscopic nature of $ZnBr_2$) can be avoided by using Sn-β-zeolite. In acetonitrile, at 25 °C a conversion of >95% and selectivity of 98% was achieved [59]. The application of a tris(2,6-diarylphenoxy) aluminium catalysts can give a ratio of 99.7:0.3 of (-)-isopulegol to other isomers [60].

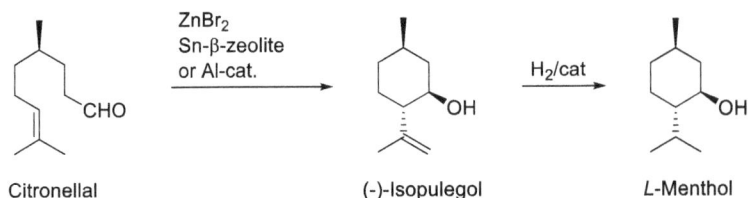

| Citronellal | (-)-Isopulegol | L-Menthol |

Scheme 8.16: Cyclisation reaction of citronellal.

An alternative variation of the citronellal ring closure to isopulegol is the Ru-salen-complex catalysed reaction performed in nitromethane at room temperature in 80% yield [61].

Sandalwood oil, *Oleum santali*, is one of the most expensive aroma chemicals, and is isolated *via* steam distillation from wood chips; its production volume is approximately 100 t/a [38]. Depending on the region, various trees can be used as the source of the sandalwood oil, such as *Santalum album*, in India, *Santalum austrocale-*

donicum in Vanatu, *Osyris lanceolata* and *Baphia nitida* in Africa [62, 63]. The main class of compounds found in sandalwood oil are sesquiterpenic alcohols such as α-santalol and β-santalol (Figure 8.3) [64, 65].

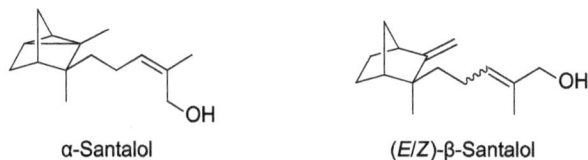

α-Santalol   (E/Z)-β-Santalol

Figure 8.3: α- and β-santalol.

Due to the limited availability of natural sandalwood oil and its high price, there have been a number of investigations into the production of synthetic alternatives. The key intermediate in most syntheses is campholenic aldehyde. (R)-Campholenic aldehyde is it-self found in fruits [66] and is a constituent of many essential oils of the *Juniper*, *Eucalyptus* and *Thyme* species. The industrial manufacture of synthetic campholenic aldehyde is by utilising a Lewis acid-catalysed rearrangement of α-pinene oxide (Scheme 8.17) [67, 68]. α-Pinene is converted into pinene oxide and then undergoes rearrangement reaction to campholenic aldehyde. The combined catalyst system zeolite/Brønsted acid results in a selectivity of 90% [69]. Formic acid and Na-type mordenite in dichloroethane at 35 °C were an alternative catalyst system. The application of heteropoly acids (HPAs, see below), such as $H_4SiW_{12}O_{40}$, for this reaction has a lower selectivity, based on the strong acid behaviour of the catalyst. In a comparative study, it was shown that Lewis acids such as $ZnCl_2$ are the preferred catalyst for the transformation of α-pinene oxide to campholenic aldehyde [70]. Other systems that have been reported are zeolites such as β-zeolite, or Mo-based catalysts [71, 72].

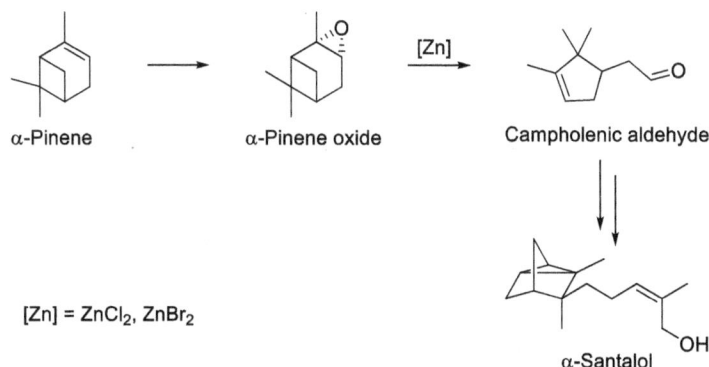

α-Pinene   α-Pinene oxide   Campholenic aldehyde

[Zn] = ZnCl₂, ZnBr₂

α-Santalol

Scheme 8.17: From pinene oxide to α-santalol.

8.3 Heteropoly acids (HPAs)

Polyoxo metal compounds such as polyoxy acids and heteropoly acids (HPAs) are interesting catalysts in the field of acid-catalysed reactions and oxidation reactions. The physical-chemical properties of HPAs result in the design and application of catalysts with outstanding performance. In particular, heteropoly acids used in solution or as solid acids are often supported on a carrier. The redox properties and acidic behaviour allow the creation of controllable multifunctional catalysts for applications in the bulk and fine chemical industry. Fundamental aspects of HPAs have been previously reviewed [73, 74].

The structure of HPAs is complex, dominated by the polyoxometalate anion having a metal-oxygen octahedron as its basic structural unit. Typically, the *Keggin* structure, $XM_{12}O_{40}^{n-}$, is representative of this type (Figure 8.4). The $XM_{12}O_{40}^{n-}$ anion contains $X = Si^{4+}$, P^{5+} and $M = Mo^{6+}$, W^{6+}; M can be substituted by V^{5+}, Co^{2+} or Zn^{2+}, and n is the overall charge of the anion. The *Keggin* anion can be described by a XO_4 tetrahedron as the central unit, surrounded by 12 edge and corner-sharing MO_6 units. The MO_6 units form four M_3O_{13} groups; each group is formed by three-octahedral sharing edges and have a shared O-atom with the XO_4 tetrahedron [75].

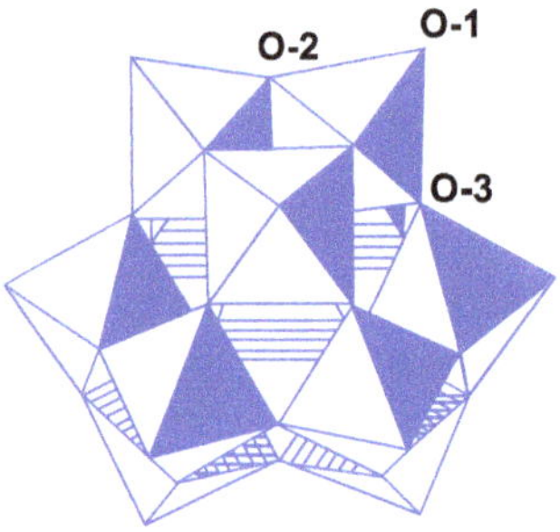

Figure 8.4: *Keggin* anion $XM_{12}O_{40}^{n-}$, O-1 = terminal, O-2 = edge, O-3 = corner oxygen atoms.

HPAs, especially of the *Keggin* type, are applied as catalysts in various reactions. Applications as strong *Brønsted* acids are well known; the protons are part of the formation of the crystal structure and packing. X-ray and neutron diffraction data show that the $PW_{12}O_{40}$-anion is a body-centred cubic structure, and the bulk protic sites are $H_5O_2^+$ ions, which are linked to four heteropolyanions (Figure 8.5).

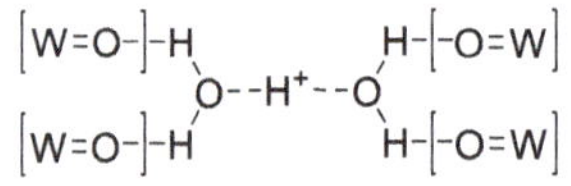

Figure 8.5: Bulk proton sides in *Keggin* anions.

HPAs are highly soluble in polar solvents such as water or alcohols but are insoluble in non-polar solvents. The dissociation constant in water and *Hammett* acidity function shows that HPAs such as $H_3PW_{12}O_{40}$ or $H_4SiW_{12}O_{40}$ or the equivalent Mo types are fully dissociated. The stability of HPAs in water is HSiW > HPW > SiMo > PMo. Usually HPAs are stronger acids than sulphuric acid or nitric acid. Their *Hammett* acidity depends on their composition [76]. HPAs in the *Keggin* form are among the most thermally stable forms of HPAs. For example, the stability of $H_3PW_{12}O_{40}$ (*Keggin* form) is higher than that of $H_6P_2W_{18}O_{62}$ [77, 78].

A carrier can influence the acidity and catalytic performance of HPAs. Industrialised catalytic applications of HPAs are found in petrochemical industry: hydration of olefins such as propene, isobutenes and *n*-butene, as oxidation catalysts in the production of methacrylic acid, and the oxidation of ethene to acetic acid. The application of HPA-based catalysts has the advantages of strong acids combined with a high oxidation potential in one material. A characteristic behaviour of HPA is the proton mobility and the stabilisation of (carbo)-cation intermediates [79]. The application of HPA on a carrier such as silica also has the advantage that the catalyst deactivation by coke formation on the surface can be avoided, and the catalyst can be reactivated by heat treatment at T > 500 °C.

An excellent example of a green process is the manufacture of *t*-butanol (Schemes 8.18 and 8.19). In the Asahi process, a mixture of butene isomers from petrochemical-crackers reacts in a 50% aqueous solution of HPAs. The butene mixture contains around 50% isobutene, and different reaction rates are observed compared to 1- and 2-butene. *t*-Butanol is obtained in >90% yield. The applied catalyst is a hetero-molybdic acid-based catalyst. The product occurs in the water phase and is separated by a decanter, followed by degassing of non-reacted butenes (small amounts) and distillation. The water phase containing the catalyst is directly reused, whereas the product is further used for industrial applications [80, 81].

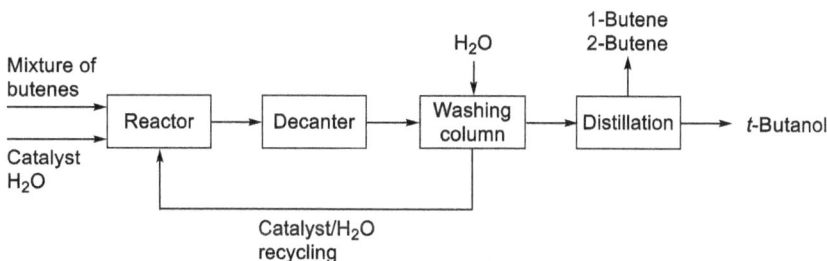

Scheme 8.18: Block-flow diagram of Asahi process for *t*-butanol synthesis (simplified).

The performance of the isobutene hydration is connected to the strong acidity of the catalyst and the good solubility of isobutene. The process is favourable with competitive sulphuric acid or ion-exchange resin-based processes because the HPA process has a high selectivity and less than 0.1% of isobutene is converted to polymers (an equivalent process using sulphuric acid in excess results in approximately 4–8% decrease of selectivity by polymer formation). Also, no waste treatment is required in the HPA process.

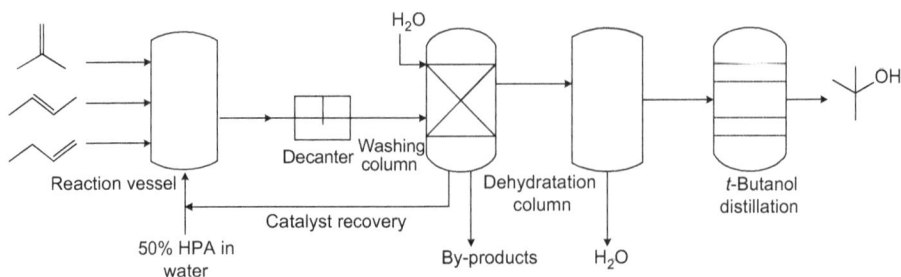

Scheme 8.19: Simplified industrial set up of the Asahi process for *t*-butanol.

The industrial application of HPA in the manufacture of ethyl acetate from ethene and acetic acid is performed on a several hundred thousand t/a scale (AVADA process from BP, AVADA = Advanced Acetates by Direct Addition). The ethyl acetate world market is around 1.5 million t/a, and ethyl acetate is used in the cosmetic and pharmaceutical industry and as a solvent. In the AVADA process, a silica supported HPA is applied; preferred catalysts are HSiW or HGeW which achieve a selectivity of 97% and an ethyl acetate purity of >99.5%. The process is carried out at approximately 180 °C at 10 bar, and a space time yield of 380 g $L^{-1} h^{-1}$ can be achieved with an ethene to acetic acid ratio of 12:1 (Scheme 8.20) [82].

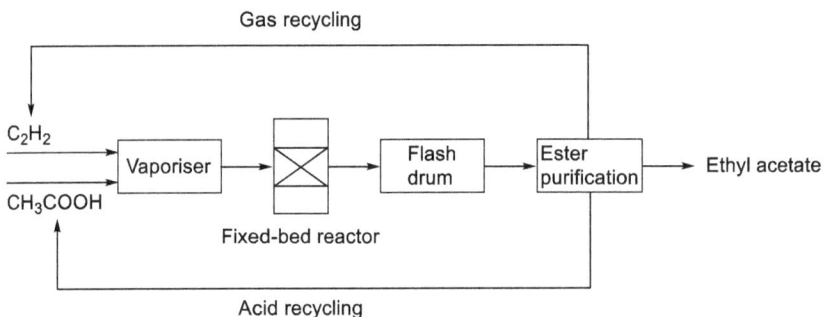

Scheme 8.20: Simplified flow chart of AVADA ethyl acetate process.

The AVADA process is superior based on low waste production, lower feedstock consumption, avoiding the need to go through the intermediate production of ethanol and a reduction in energy costs. In this process, a high atom economy, a very low e-factor and the application of a catalyst in the presence of non-toxic starting materials are applied.

The ethylene oxidation to acetic acid, commercialised in Japan, is catalysed by a palladium metal catalyst supported on a heteropoly acid such as silicotungstic acid [83]. Similar processes use the same metal catalyst on silicotungstic acid and silica [84].

Another HPA catalyst reaction of olefins is the *Prins* reaction. This is the electrophilic addition of an aldehyde or ketone to an alkene or alkyne which results in the formation of allyl alcohols or dioxanes, depending on reaction conditions. In the presence of water and a *Brønsted* acid, protic 1,3-diols are formed, whereas in absence of water, allylic alcohols are the main product. At low reaction temperatures and in the presence of an excess of formaldehyde, 1,3-dioxanes are formed [85–87]. The *Prins*-type reaction is used on an industrial scale in the synthesis of prenol. The synthesis of 1,3-dioxanes from alkenes, for example styrene, with formaldehyde catalysed by HPA shows superior results compared to H_2SO_4 or *p*-TsOH. The activity differed only slightly in a series of different HPAs [88].

The fibre and elastomer compound polytetramethylene ether glycol (PTMG) (poly-tetrahydrofuran or poly(tetramethylene oxide)), a polymer of tetrahydrofuran, is manufactured (60 kt/a) by the ring-opening polymerisation reaction of THF in a liquid–liquid two-phase reaction system at around 60 °C (Scheme 8.21). The water-catalyst phase and the THF-product phase are separated, and the catalyst-phase is reused. HPAs ($H_3PW_{12}O_{40}$ or $H_4SiW_{12}O_{40}$) are applied as catalyst, and they perform well in the reaction (Scheme 8.22). The polymer (molecular weight 250–300 Daltons) is separated from THF and is obtained in narrow molecular weight distribution [89, 90]. This process is applied in Japan by Asahi Chemical Industry since 1987.

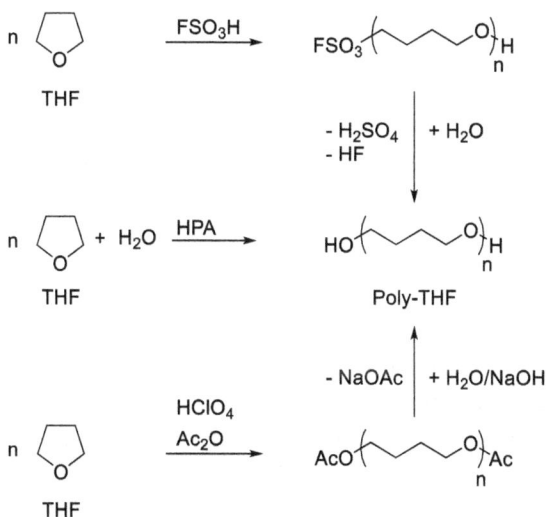

Scheme 8.21: Syntheses of poly-THF by polymerisation.

The application of HPAs in this process allows the synthesis of the polymer in a one-step process, with a simple workup and downstream processing. Furthermore, this green process results in very low levels of waste and avoids the use of strong acids like perchloric acid or fluorosulphuric acid in combination with acetic acid, which

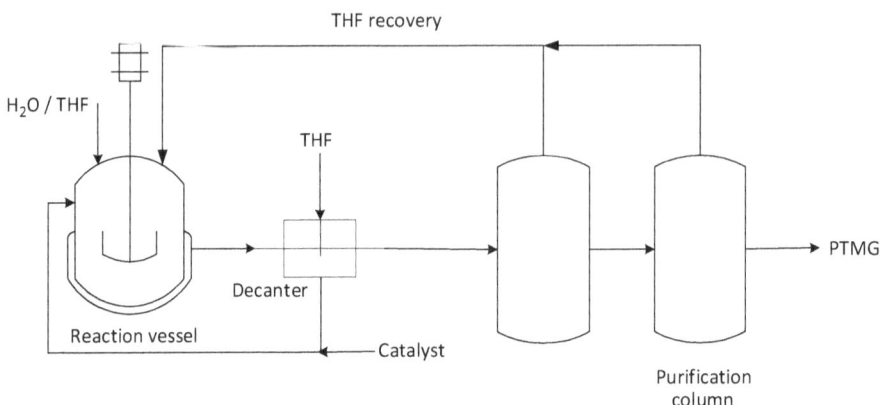

Scheme 8.22: HPA-catalysed poly-THF manufacture.

were used on the old two-step processes. An additional economic advantage is the lower investment required because corrosion problems are reduced.

The application of HPA such as $H_3PW_{12}O_{40}$ or $H_4SiW_{12}O_{40}$ as catalysts for *Friedel–Crafts*-type reactions in the homogeneous phase has been demonstrated by I. Kozhevnikov [91]. The reaction of trimethyl-1,4-benzoquinone and isophytol under homogeneous conditions, in solvents such as butyl acetate, results in α-tocopherol. If the reaction is performed in liquid–liquid two-phase solvent systems, an excellent yield of >95% of α-tocopherol can be achieved [92]. An advantage of the application of heterogeneous system is the mild reaction conditions, the easy recovery and reuse of the catalyst, and the fact that the reaction can be carried out in a 1:1 ratio of trimethylhydroquinone-isophytol (TMHQ-IP) or in an excess of TMHQ, whereas under homogeneous conditions an excess of IP is necessary to obtain high enough yields and selectivity.

HPAs have been applied in the alkylation of *p*-cresol with isobutene and was commercialised in Russia (Scheme 8.23) [93]. The product 2,6-di-tert-butyl-4-methylphenol (BHT) is applied as an antioxidant (see Chapter 5: Gas-phase reactions). The other commonly applied catalyst for these alkylation reactions is concentrated sulphuric acid. The applied *p*-cresol should contain only minor amounts of water (<0.1 wt%). The reaction is usually carried out at ca. 70 °C and 1 bar partial pressure of isobutene. The main by-products are monobutylated phenols [94]. The use of HPA provides an increase in selectivity of 7–10% compared to the corresponding reaction with sulphuric acid. Furthermore, toxic water pollution is almost eliminated in the process.

m-/*p*-Cresol mixtures which cannot be separated by distillation are also used as starting materials in the synthesis of BHT. The mixture is dibutylated to yield 2,6-di-*tert*-butyl-4-methylphenol (boiling point of 147 °C at 27 mbar) and 4,6-di-*tert*-butyl-3-methylphenol (boiling point of 167 °C at 27 mbar). The latter is subsequently converted back to *m*-cresol with catalytic amounts of oleum at 160–200 °C. The procedure allows

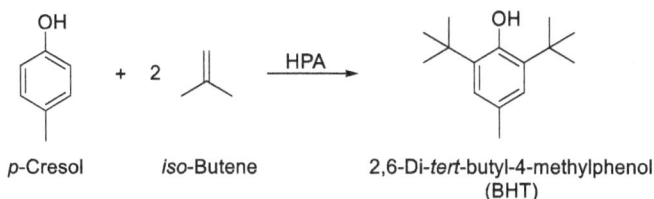

Scheme 8.23: Alkylation of p-cresol in the presence of HPA.

the separation of *m*-cresol and *p*-cresol and the synthesis of BHT. The annual production of BHT is around 30,000–45,000 t/a [95].

2-Methyl-1-naphthol can be oxidised to menadione at 50 °C in the presence of oxygen, resulting in around 90% yield at full naphthol conversion (Scheme 8.24) [96]. The
process is carried out in a two-phase solvent system (water-hydrocarbon) and is a promising alternative to the conventional oxidation process using CrO_3. The preferred catalyst
is $H_5PMo_{10}V_2O_{40}$ or its salts. The produced menadione is a key intermediate in the synthesis of vitamins K_1 and K_2. 2-Methyl-1-naphthol can be produced in a high yield by a
heterogeneously catalysed gas phase alkylation of naphthalene with methanol.

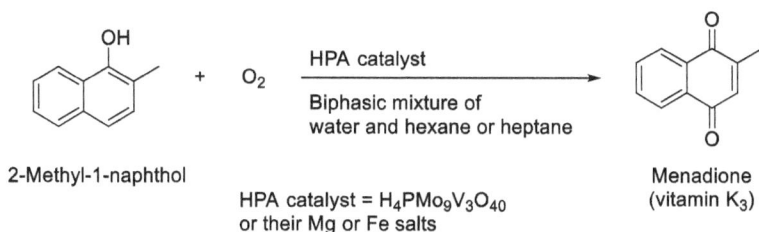

Scheme 8.24: Vitamin K_3 by P-Mo-V-catalysed oxidation of 2-methyl-1-naphtol.

8.4 *Lewis* acids

Lewis acid-catalysed reactions are well described in the literature. Here, we concen-
trate on their applications in the synthesis of selected fine chemicals. In the area of
O–C rearrangements, aryl-prenyl ether undergoes rearrangement catalysed by BF_3,
forming a prenyl substituted hydroquinone. The two-step procedure, O-alkylation and
rearrangement reaction, is higher yielding then the direct *Friedel–Crafts* alkylation.
These types of compounds are useful in the synthesis of quinones, especially in the
manufacture of Coenzyme Q_{10} (Scheme 8.25) [97].

Another BF_3-catalysed rearrangement reaction in the field of isoprenoid chemistry
is the 1,2-methyl shift obtained in the nootkatone skeleton synthesis (Scheme 8.26) [98].
The reaction is carried out in halocarbon solvents at room temperature and yields of

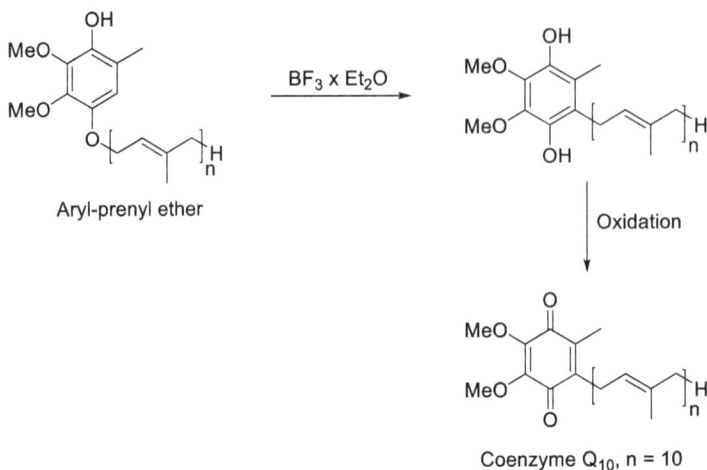

Scheme 8.25: BF$_3$ x Et$_2$O-catalysed O–C rearrangement.

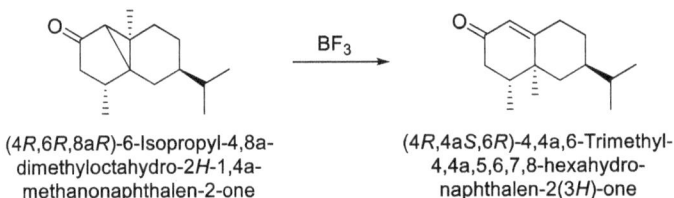

(4R,6R,8aR)-6-Isopropyl-4,8a-dimethyloctahydro-2H-1,4a-methanonaphthalen-2-one

(4R,4aS,6R)-4,4a,6-Trimethyl-4,4a,5,6,7,8-hexahydro-naphthalen-2(3H)-one

Scheme 8.26: BF$_3$-catalysed approach to the nootkatone skeleton.

around 45% can be achieved. The nootkatone class of compounds, such as nootkatone itself, are applied in the flavour and fragrance industry (see Chapter 4: Oxidations).

The *Friedel–Crafts* reaction of primary allyl alcohols and aromatic compounds is of great importance in the manufacture of fine chemicals, for example vitamins E, K$_1$ and K$_2$. The BF$_3$-catalysed reaction of 2-methyl-1,4-naphthohydroquinone-1-monoacetate (the monoester derivative of menadiol) is of industrial relevance for the manufacture of vitamin K$_1$ (Scheme 8.27) [99]. The alkylation is performed at elevated temperatures (>70 °C)

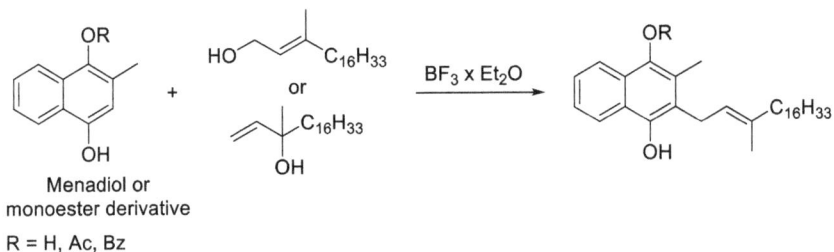

Menadiol or monoester derivative

R = H, Ac, Bz

Scheme 8.27: BF$_3$-catalysed *Friedel–Crafts* reaction.

in dioxane for >1 h reaction time. A yield of up to 65% of vitamin K$_1$ was achieved; an improved version applies isophytol and the monoacetate-protected aromatic precursor resulting in high yields [100].

A key intermediate for thiamine (vitamin B$_1$) production on an industrial scale is *Grewe* diamine, a 4-aminopyrimidine derivative [101]. Thiamine is produced in several thousand t/a. A new catalytic approach to *Grewe* diamine avoids the drawbacks of handling carcinogenic reagents (dimethylsulphate, *o*-chloroaniline) and the related formation of sodium methylsulphate waste and loss of *o*-chloroaniline due to incomplete recycling, which are issues in the traditional synthesis. This first catalysed, and scalable access to 4-aminopyrimidines via direct condensation of protected enolates with amidinium chlorides is applicable for the large-scale industrial production of 4-aminopyrimidines (Scheme 8.28) [102]. (See Chapter 2: Heterogeneous hydrogenation for an alternative synthesis of *Grewe* diamine).

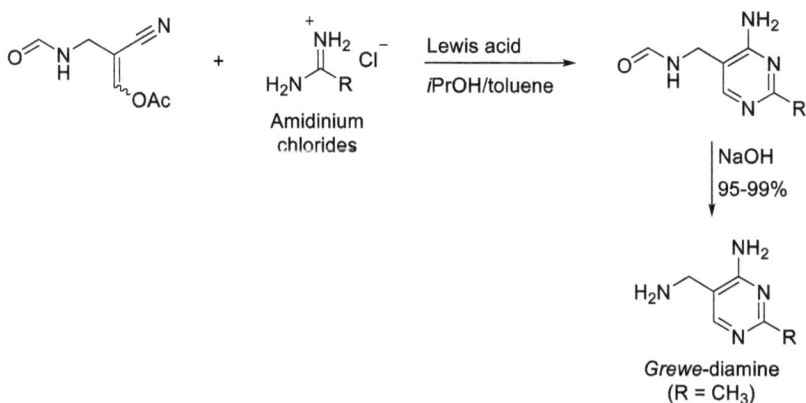

Scheme 8.28: Synthesis of *Grewe* diamine.

Reaction of the β-cyano enolate derivative with a small excess of amidine hydrochlorides at around 80 °C in the presence of *Lewis* acids such as CuCl, FeCl$_2$ or ZnCl$_2$ (0.2 – 0.3 equiv.) in isopropanol – toluene furnished the corresponding aminopyrimidines in around 85% yield. Commercially available formamidine hydrochloride, acetamidine hydrochloride, guanidine hydrochloride and *o*-methyl urea hydrochloride were successfully applied in the condensation reaction. Removal of the *N*-formyl protecting group of the corresponding pyrimidines was easily accomplished in high yields following standard basic conditions in a two-phase solvent system [103].

This procedure also works especially well for the synthesis of *N*-((2,4-diaminopyrimidin-5-yl)methyl)-formamide, a key building block in the synthesis of the antibiotic trimethoprim (Figure 8.6).

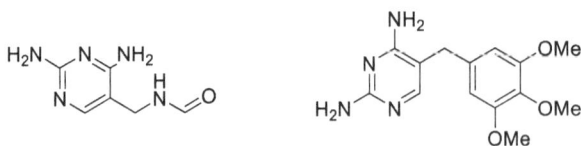

Figure 8.6: *N*-((2,4-diaminopyrimidin-5-yl)methyl)-formamide (left) and trimethoprim (right).

Lewis acids can also be used in the condensation reaction of dienes and enol ethers (Scheme 8.29). Suitable catalysts for this reaction type are BF_3, $ZnCl_2$ or Fe(III)-compounds. The synthesis of intermediates of carotenoids such as astaxanthin, canthaxanthin or apoester follows this approach [104]. The carotenoids are natural compounds that are produced synthetically and used in colourisation of food. For instance, astaxanthin is used in fish farming to ensure the flesh of salmon and lobster maintains its pink colour. It is produced in several hundred t/a. Canthaxanthin and apoester are produced on a smaller scale and used in chicken feed for egg yolk and broiler pigmentation.

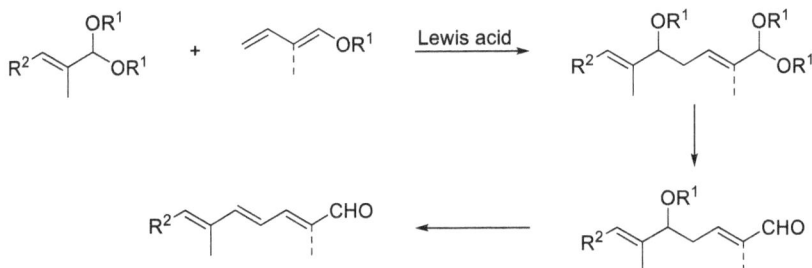

Scheme 8.29: *Lewis* acid-catalysed enol ether condensation.

In general, the enol ether reaction of an alkyl enol ether and an acetal is a suitable alternative for the aldol reaction. An advantage is that the reaction is not limited to C_2- or C_3-enol ether building blocks. The s/c is usually around 20–100, and yields of around 95% at full conversion can be achieved in inert non-polar solvents.

The catalytic enol ether condensation can be applied in the synthesis of polyenes, such as astaxanthin (Scheme 8.30) as a suitable alternative for the *Wittig*, or *Wittig*-type reactions [104]. This avoids the problem of the generation of large quantities of phosphorus waste (phosphine oxides and phosphates).

The enol ether condensation in the presence of *Lewis* acids, such as $FeCl_3$, BF_3 or $ZnCl_2$, is also applied in the manufacture of 2,7-dimethyl-octa-2,4,6-trienedial (C_{10}-dialdehyde), a building block in carotenoid chemistry [105–108]. Starting from furan, the dialdehyde is produced in approximately 1,000 t/a. The enol ether condensation is performed in s/c of 500 to 1,000 at room temperature (Scheme 8.31 and Scheme 8.11 in chapter 8.1) [108].

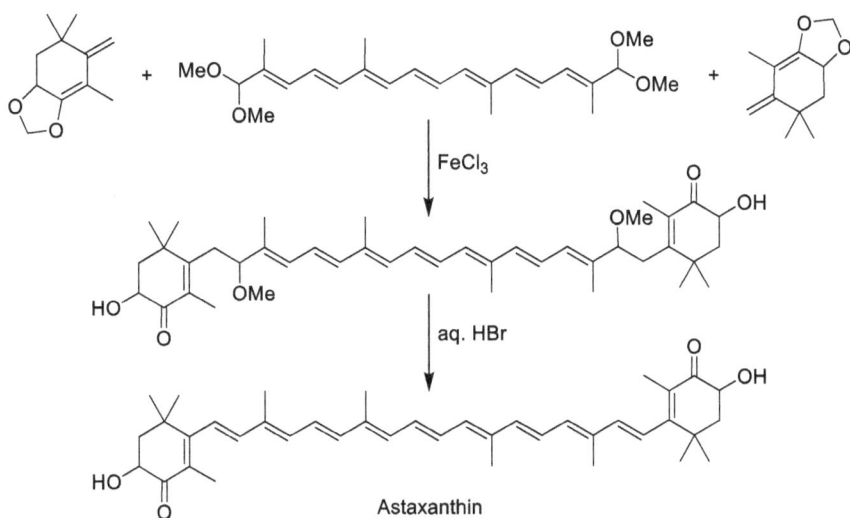

Scheme 8.30: Astaxanthin synthesis by *Lewis* acid catalysis of enol ether condensation.

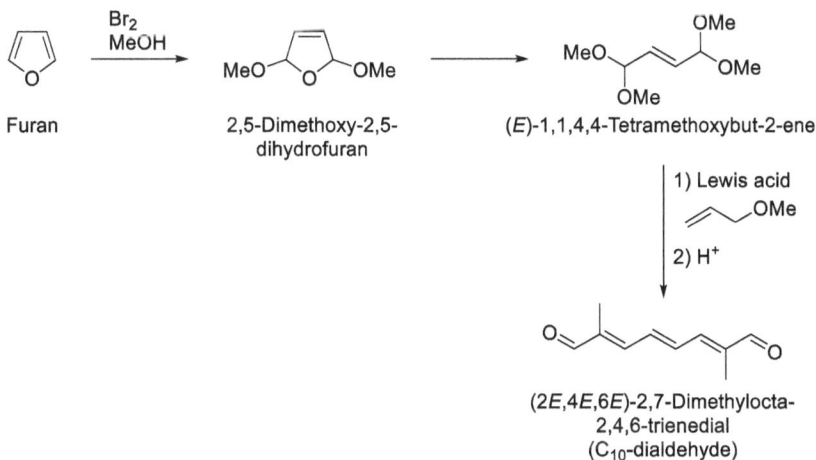

Scheme 8.31: *Lewis* acid-catalysed double enol ether condensation.

In the field of isoprenoid chemistry, vitamin E (α-tocopherol, commercial form: α-tocopheryl acetate) is one of the most important substances. α-Tocopheryl acetate is produced in around 70,000 t/a and applied in animal- and human-nutrition, food industry, and cosmetics. The main product for the animal nutrition market (around 75%) is the tocopheryl acetate adsorbate on silica. There are two different types of vitamin E on the market: (all-R)-α-tocopherol acetate from natural sources, approximately 3,000 t/a, and (all-rac)-α-tocopherol acetate from chemical synthesis. Natural vitamin E is extracted from palm, soy, sunflower and pine nut oil as mixtures of chromanols, a mixture of tocols such as α-, β-, γ- and δ-tocopherol, all compounds with an (R)-configuration at the stereo centres, and tocotrienols, which have C = C bonds in position 3′, 7′ and 11′ of the isoprenoid side chain (Figure 8.7). Tocopherol from natural sources has a limited and variable availability, depending on harvest, and the different isomers have different biological activities [109, 110]. The methylation of non-α-tocopherol to natural identical α-tocopherol is well described and performed on an industrial scale [111].

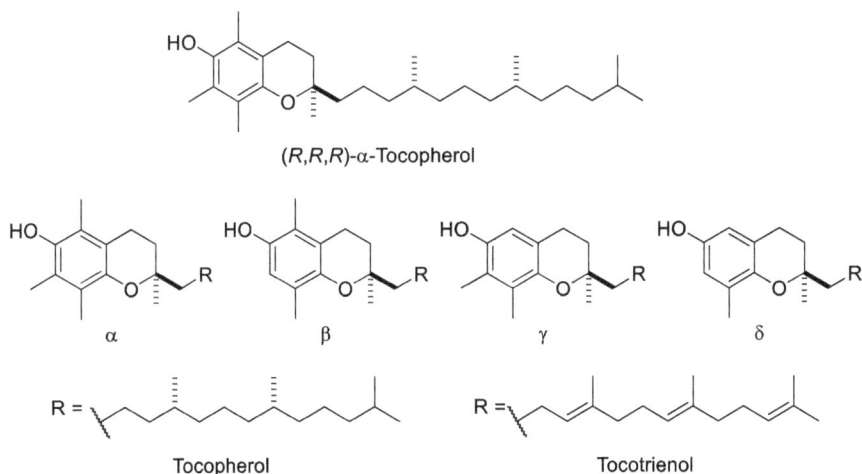

(R,R,R)-α-Tocopherol

α   β   γ   δ

Tocopherol   Tocotrienol

Figure 8.7: Tocopherol and tocotrienol compounds.

The *Friedel–Crafts* condensation reaction of trimethylhydroquinone (TMHQ) and isophytol is an important step in the manufacture of *(all-rac)*-α-tocopherol (Scheme 8.32). For many decades, a catalyst combination based on a *Lewis* acid and a *Brønsted* acid, such as ZnCl$_2$/HCl, was applied. The disadvantages of all these procedures are corrosion problems, the formation of by-products, the lack of selectivity and often a high catalyst loading was required. Modern approaches which avoid these problems are to perform the *Friedel–Crafts* reaction of isophytol and trimethylhydroquinone, or derivatives thereof, in a two-phase liquid–liquid solvent system, that means an aliphatic solvent (e.g. hexane, heptane or octane) and a polar solvent such as a carbonate (e.g. ethylene or propylene carbonate or an ester). The two-phase solvent approach produced tocopherol in high yield (>95%) and selectivity, applying only a *Lewis* or *Brønsted* acid alone as catalyst with a s/c of 1,000 to 10,000 depending on reaction conditions [13]. An advantage of the two-phase solvent approach is the easy recovery and reuse of the catalyst in the polar solvent phase. Following this concept, *Lewis* acids such as metal-triflates, metal-methides or metal-triflimides have been used as efficient catalysts [112–114].

RO.

+

R^{1}O

Ethylene carbonate/heptane | Catalyst -HOR^1 | Conventional catalyst: AlCl$_3$, ZnCl$_2$, HCl, BF$_3$, etc. Modern catalyst: M(OTf)$_3$, R$_f$SO$_3$H, M(NTf$_2$)$_3$

RO.

O

R = H
R = Ac

RO.

O

(all-*rac*)-α-Tocopheryl acetate

R = H, Ac, Bz

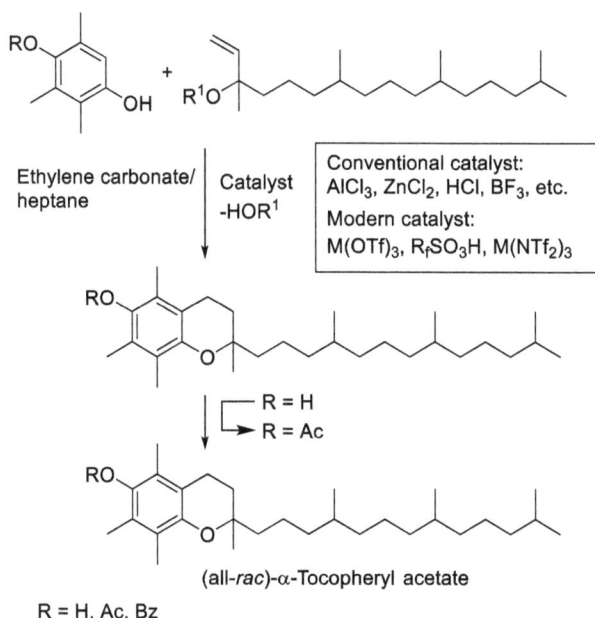

Scheme 8.32: Liquid–liquid two-phase approach in the synthesis of α-tocopherol.

The discovery of alternative and efficient *Brønsted* acids for the synthesis of (all-*rac*)-α-tocopherol not only improved the condensation reaction considerably, but also had a strong impact on other key steps in vitamin E production, such as the synthesis of tri-methylhydroquinone (Scheme 8.33) [115–117]. In particular, methane trisulphonic acid or sulphur(VI) containing *Brønsted* acids are efficient catalysts which allow the synthe-

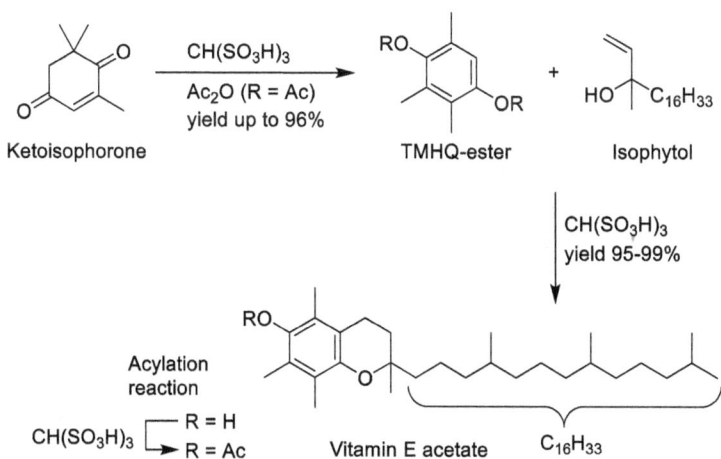

Ketoisophorone

CH(SO$_3$H)$_3$
Ac$_2$O (R = Ac)
yield up to 96%

RO.

OR

TMHQ-ester

+

HO C$_{16}$H$_{33}$

Isophytol

CH(SO$_3$H)$_3$
yield 95-99%

Acylation reaction

CH(SO$_3$H)$_3$

R = H
R = Ac

RO.

O

Vitamin E acetate

C$_{16}$H$_{33}$

Scheme 8.33: Synthesis of tocopherol from ketoisophorone and isophytol.

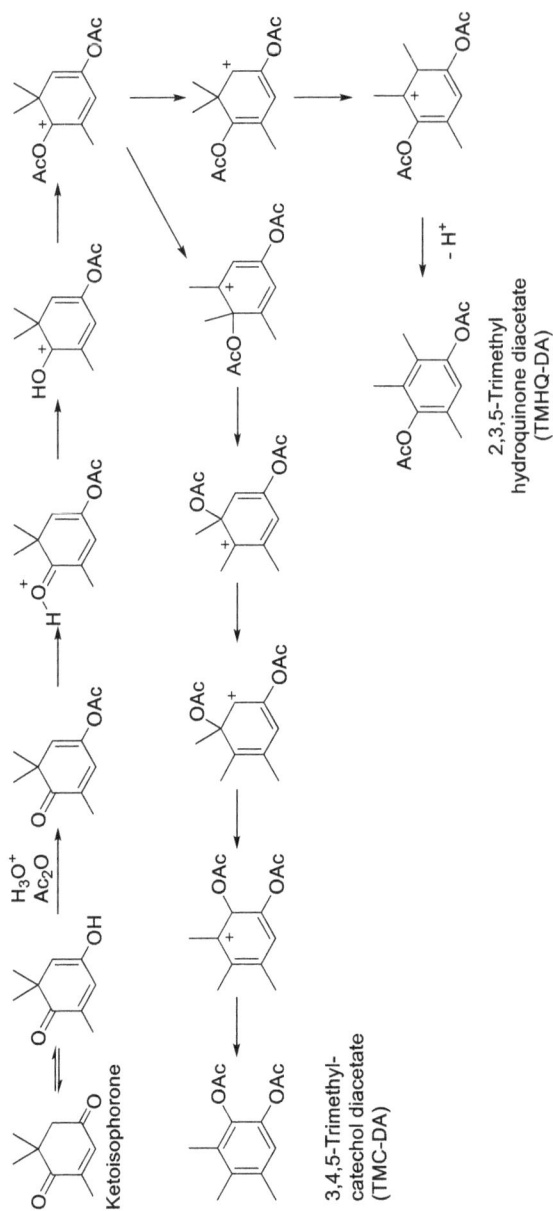

Scheme 8.34: Mechanism scheme for *Wagner–Meerwein* rearrangement of ketoisophorone.

sis of tocopherol with high selectivity and yield (>95%) [118]. Methane trisulphonic acid also performs very well in the synthesis of TMHQ-esters by the *Wagner–Meerwein* rearrangement (see Chapter 7: Rearrangement reactions) from ketoisophorone (KIP). The benefit of this approach is that the catalyst can be used in the three different reaction types: the *Friedel–Crafts* alkylation, the *Wagner–Meerwein* rearrangement and the *O*-acylation reaction. The catalyst loadings for all three reactions are from 100 to 5,000. Furthermore, the catalyst is produced from acetone, which is also a starting material for ketoisophorone and isophytol.

The proposed mechanism for the *Wagner–Meerwein* rearrangement of KIP (ketoisophorone) into TMHQ-DA and TMC-DA is shown in Scheme 8.34. The main by-product formed is 3,4,5-trimethylcatechol diacetate.

In the field of synthetic aroma chemicals, compounds with a musk-like odour are of great commercial interest. The compound HHCB (4,6,6,7,8,8-hexamethyl-1,3,4,6,7,8-hexahydrocyclopenta[y]benzopyran) (Scheme 8.35), a viscous oil, is widely applied as a mixture of stereoisomers in cosmetics, cleaning industry and soaps; the production volume is >1,000 t/a. The starting materials for manufacture of HHCB are *t*-amylene and α-methylstyrene. A cycloaddition reaction of these starting materials results in the formation of 1,1,2,3,3-pentamethylindane, which is converted in a *Friedel–Crafts* reaction with propylene oxide catalysed by aluminium trichloride into 1,1,2,3,3-pentamethyl-5-(β-hydroxyisopropyl)indane. The ring closure and treatment with formaldehyde or its corresponding acetal results in the formation of HHCB, which is also known by its trade-name of Galaxolide[TM] (Scheme 8.35) [119, 120].

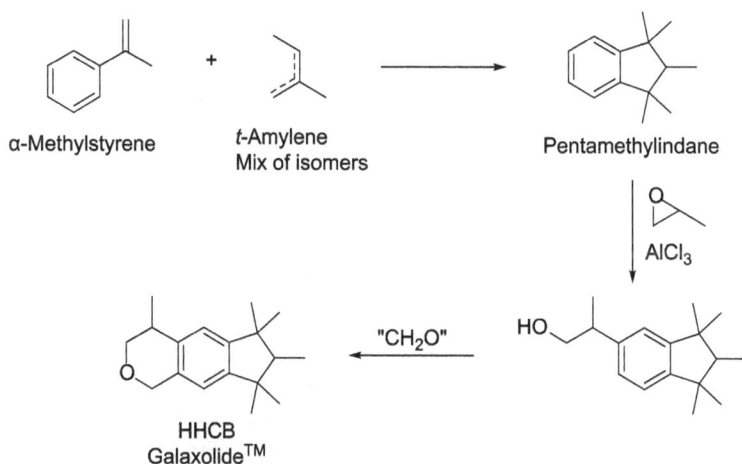

Scheme 8.35: Synthesis of HHCB (Galaxolide[TM]).

Myrcenol (2-methyl-6-methylene-7-octene-2-ol) is an isomer of geraniol that is used in the F + F industry because of its top note in citrus and lavender compositions. Myrcenol is found in nature, e.g. in Chinese lavender oil, and is produced from myrcene [121, 122]. Myrcene is treated with diethylamine and the resulting mixture of geranyl-, and neryldiethylamine is hydrated in the presence of dilute acid to form the corresponding hydroxydiethylamines. The deamination is performed in the presence of a Pd-phosphine complex (Scheme 8.36).

For the application of Lewis acids in oxidation (dehydrogenation) reactions of alkanes, see Chapter 4: Oxidations.

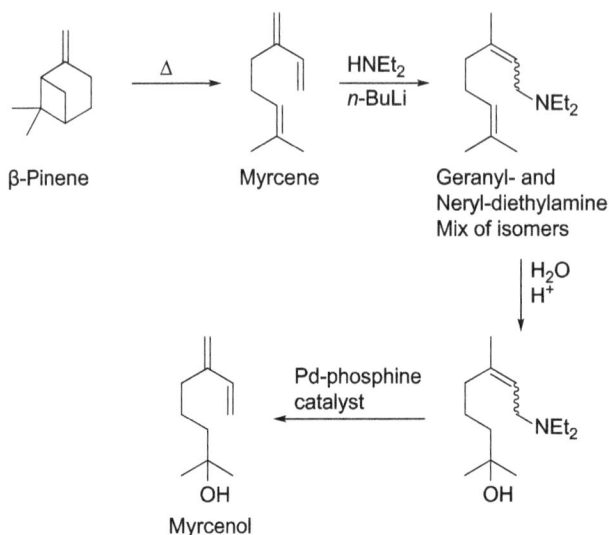

Scheme 8.36: Manufacture of myrcenol.

8.5 Base catalysis

In the introduction (Chapter 1), we pointed out that industrial base catalysed reactions, especially in the fine chemical industry, are limited. Aldol-type reactions are an important exception; several processes are known where this reaction type is used on an industrial scale.

The self-aldol condensation of acetaldehyde to give 3-hydroxybutanal ($\Delta H = -113$ kJ/mol) is an important process because the reaction product is used for the production of crotonaldehyde. This compound in turn is used for the manufacture of 2-ethylhexanol (aldol reaction and hydrogenation) and crotonic acid (by liquid-phase oxidation) (Scheme 8.37). The aldol reaction of acetaldehyde is performed in a tubular reactor catalysed by sodium hydroxide at around 20 °C with partial conversion of 50–60% (to avoid by-product formation) in 85% selectivity. The main by-product is crotonaldehyde. Crotonaldehyde itself is produced from 3-hydroxybutanal by acid

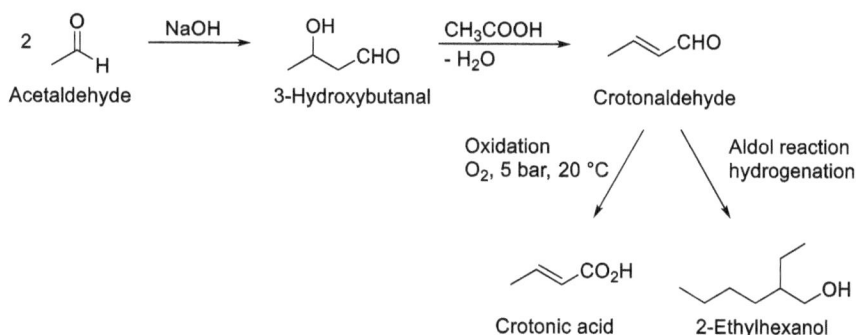

Scheme 8.37: Synthesis of crotonaldehyde by aldol reaction.

treatment in 95% selectivity. An alternative procedure for the production of crotonal-dehyde is the hydroformylation of propene.

2-Ethylhexanal from butanal by aldol condensation and hydrogenation is produced in several thousand t/a. The main applications of 2-ethylhexanal are as an intermediate in the production of 2-ethylhexanol, 2-ethylhexanoic acid and 2-ethylhexylamine, as a solvent and as an intermediate in pharmaceutical and flavour industries. 2-Ethylhexanol is applied as plasticiser and in the manufacture of surfactants (reaction with ethylene oxide) [123].

The chemical industry is interested in the application of acetone from renewable resources. An interesting product from acetone is its aldol addition product diacetone alcohol, 4-hydroxy-4-methyl-2-pentanone, which is important for the manufacture of mesityl oxide (Scheme 8.38). Further applications of diacetone alcohol are in the fra-grance industry [124, 125]. The aldol reaction of acetone to diacetone alcohol is base-catalysed; it is performed under homogenous or heterogeneous conditions, and the major technical challenges are the separation of the reaction product and overheating effects [126–128].

Modern procedures for the synthesis of diacetone alcohol are based on solid base catalysts, mainly alkali or earth alkali metals based in the presence of a metal binder, or solid catalysts based on alumina silicates [129, 130]. Furthermore, basic ion-exchange resins have been studied in the aldol reaction, and in particular, kinetic studies have been performed for the acetone condensation [131, 132]. Basic macroreticular ion-exchange resins, containing alkylamine functionalities were studied batch-wise and in continuous reaction mode in a temperature range of 10–50 °C [133–135]. Acid-catalysed aldol reactions carried out in the presence of solid acid catalysts are described in [136, 137]; however, usually higher by-product formation is observed compared to the base-catalysed versions [132].

The aldol reaction of acetone results in the formation of diacetone alcohol and mesityl oxide as main products. Overall, the di- and trimerisation of acetone results in a very complex reaction network. Depending on the reaction conditions, other prod-

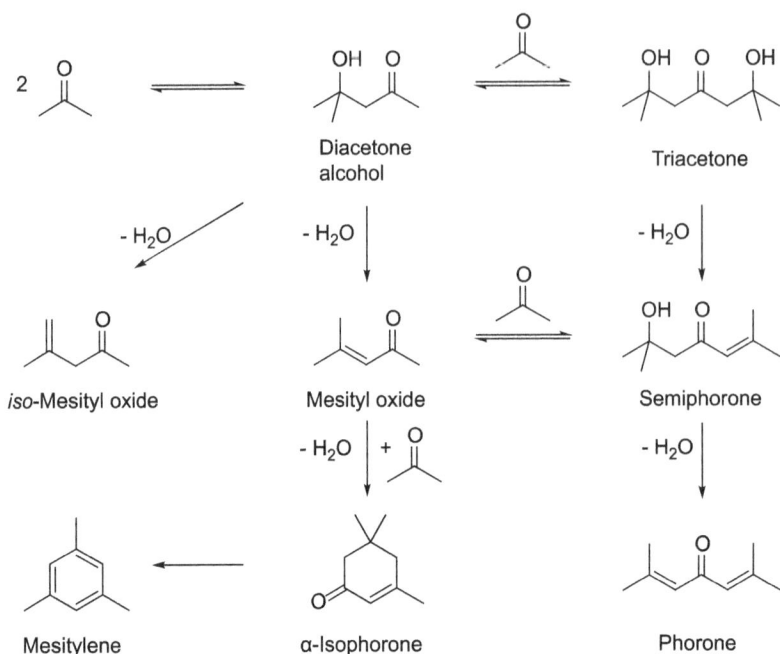

Scheme 8.38: Reaction network of aldol reaction of acetone.

ucts such as *iso*-mesityl oxide, phorone, semiphorone, triacetone, mesitylene and α-isophorone can be formed [132] (Scheme 8.38).

The main reaction products mesityl oxide and isophorone are applied as large volume products, e.g. as solvents. The main application of mesityl oxide is found in the manufacture of isobutyl methyl ketone (produced on a 100,000 t/a scale) by hydrogenation and is commonly used as a solvent for extractions, in resins, printing inks, nitrocellulose, paints, and gums.

Isobutyl methyl ketone is produced in a one-pot procedure where acetone is transformed to diacetone alcohol which is followed by water elimination and hydrogenation (Scheme 8.39). Suitable catalyst precursors are based on Rh, Ni- or Co- salts. Usually, the hydrogenation reactions are performed under pressure, up to 200 bar, in the presence of iodine and/or carbon monoxide. In the temperature range 140–200 °C

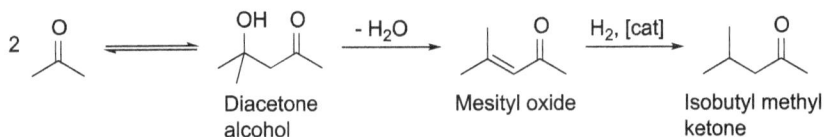

Scheme 8.39: Manufacture of isobutyl methyl ketone.

a selectivity of 98% at 50% conversion can be achieved. An alternative stepwise procedure is also known [138].

High-purity isobutyl methyl ketone can also be prepared using the *Uhde-Technology*, applied by *Sasol*. A bi-functional catalyst based on a Pd-doped ion-exchange resin is used in a tubular reactor. The mesityl oxide hydrogenation is performed at around 30 bar and 120 °C. Selectivity of 90% and a purity of >99% can be achieved; the conversion is approximately 30% per pass, resulting in >80% yield [139]. The unconverted acetone is recycled and re-used.

As mentioned earlier, another main product of the acetone condensation reaction is α-isophorone. The production volumes of this compound doubled between 2000 (50,000 t/a) and 2005 (100,000 t/a) [140]. As Evonik started the operation of a new isophorone plant with a 50,000 t/a capacity in 2012 [141], the yearly production volumes are assumed to be significantly higher than in 2005. Applications of isophorone are in the polymer industry for dissolving polymers, oils, in the coating industry or in printer inks, paints, adhesives and pesticides [142, 143]. An important application is the synthesis of isophorone diisocyanate (IPDI), which is used to produce special polyurea foams resistant to light and abrasion [142, 143]. The isophorone is additionally transformed into 3,3,5-trimethylcyclohexanone, 3,3,5-trimethyl-1-cyclohexanol and 3,5 dimethylphenol, important intermediates for the manufacture of 2,4,4- and 2,2,4-trimethyladipic acid, and 2,2,4-trimethylhexane-1,6-diamine. Furthermore, from a viewpoint of fine chemical industry, the manufacture of 2,3,6-trimethylhydroquinone (a key intermediate of vitamin E, α-tocopherol) from α-isophorone is of great importance (Scheme 8.40) [144–148] (see Chapter 7.2: *Wagner-Meerwein* rearrangements).

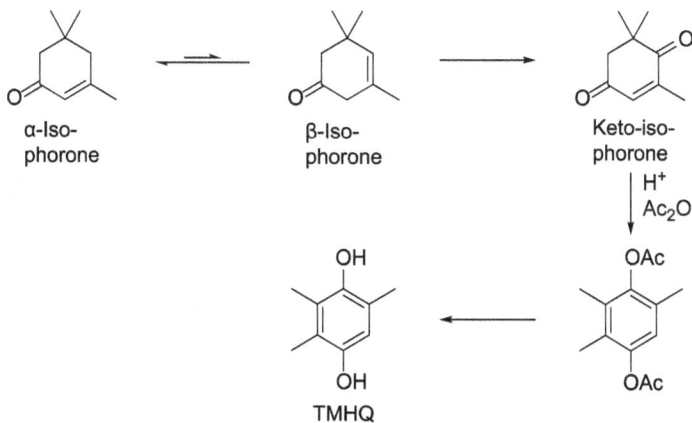

Scheme 8.40: Simplified reaction sequence from isophorone to TMHQ.

The manufacture of isophorone can be carried out in the liquid- or gas-phase; this trimerisation of acetone is usually a base-catalysed condensation. The liquid phase process has been operated on industrial scale since 1941 utilising KOH or NaOH as catalysts in aqueous phase, and is known as the *Scholven-process*, resulting in a selectivity of around 70% [142, 143, 149–152]. This process usually operates in a temperature range of 200–250 °C, at around 35 bar.

The gas-phase condensation is performed in the presence of a heterogeneous solid base catalyst. Advantages of the gas-phase process are the higher selectivity and avoidance of basic wastewater, but disadvantages are the higher energy consumption [142, 143, 149–151]. The reaction mechanism of isophorone synthesis has been studied for several decades and it can be described as a three-step process. Acetone undergoes a self-condensation to form mesityl oxide, which is followed by a Michael addition resulting in 4,4-dimethylheptane-2,6-dione. This then undergoes an intramolecular aldol reaction/cyclisation/elimination to yield α-isophorone (Scheme 8.41) [149, 153–157].

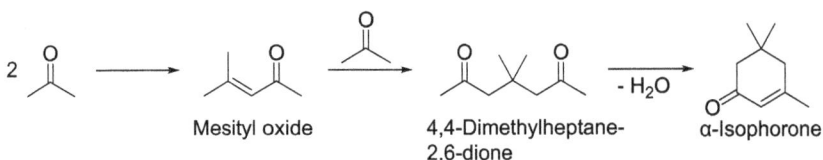

Scheme 8.41: From acetone to α-isophorone.

In recent studies the effect of the catalyst (alkali) concentration on the yield and selectivity has been studied [155, 158, 159]. Using a KOH concentration of 0.75% in water and a temperature of 200 °C, a selectivity of >85% can be achieved [158, 159]. In the presence of higher NaOH concentrations (e.g. 4% in water) a yield of 52% and an 80% selectivity can be obtained after a 4 h reaction time at around 220 °C [155]. If the reaction is performed under continuous conditions applying KOH or NaOH (10% in water) under supercritical conditions at 320 °C, a yield of around 90% can be achieved [160]. For a recent review on the topic see [140].

ψ-Ionone (pseudoionone), a key intermediate in the manufacture of carotenoids, flavour compounds, vitamins A and E, is produced on around 30,000 t/a. For the production on an industrial scale, two main approaches are followed. The acid-catalysed *Saucy–Marbet* reaction of dehydrolinalool, followed by a base-catalysed rearrangement of the intermediate allene-ketone, results in an excellent yield of ψ-ionone (see Chapter 7: Rearrangement reactions). Alternatively, the base-catalysed cross-aldol condensation of citral and acetone is industrially carried out with an excess of acetone (Scheme 8.42). Yields of 85–90% based on citral can be achieved [30]. The hydrogenation of ψ-ionone results in the formation of hexahydro-ψ-ionone, an intermediate in the vitamin E manufacture (see Chapter 2: Heterogeneous hydrogenations).

The synthesis of ψ-ionone from citral and acetone is usually performed in the presence of strong bases, such as sodium ethoxide. Quaternary ammonium salt bases are also efficient catalysts for this reaction [161, 162]. On an industrial scale, citral is

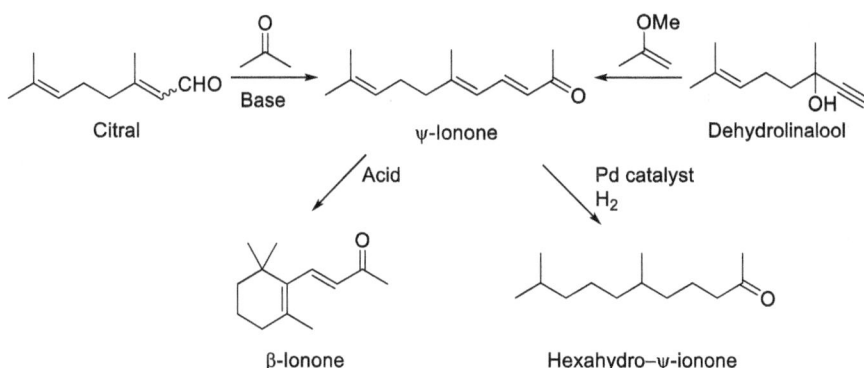

Scheme 8.42: ψ-Ionone synthesis and main products thereof.

reacted with acetone to form ψ-ionone through an aldol condensation in a liquid bi-phasic process by using a homogeneous base in the aqueous phase. This process has several disadvantages related to corrosion, disposal of the spent basic materials, and the relatively low yield and selectivity (approximately 85%) [163]. A modern approach for the cross-aldol condensation of various aldehydes and ketones, for example the condensation of acetone and citral to ψ-ionone, is the use of anionic (basic) ion-exchange resins resulting in good yield and selectivity. The reactions were run in batch and continuous modes with no significant loss in activity and selectivity; in par-ticular, in continuous mode a stable performance was observed. A variety of alde-hydes and ketones were reacted successfully to the corresponding condensed aldol products [164]. Solid bases, such as lanthanum salts, have also been applied in the liq-uid phase for the aldol reaction of citral and acetone. The catalyst was prepared from lanthanum oxide by heat treatment at around 900 °C. The catalysts have a surface area of 5–10 m^2/g and were applied at s/c of 100 to 1,000 resulting in conversions of 95% and a selectivity of 85% [165]. The spent catalyst can be recycled and reused with-out a loss of activity or selectivity.

The application of aldol condensation reactions in flavour and fragrance chem-istry is a useful tool in the manufacture of 1-(2,2,6-trimethylcyclohexyl) hexan-3-ol (Timberol® by *Symrise*, Norlimbanol® by *Firmenich*), a compound with an amber-woody aroma. Applications are in the cosmetic and aroma industry. For the synthe-sis of 1-(2,2,6-trimethylcyclohexyl) hexan-3-ol or its precursor timberone, various synthetic protocols have been described. The olfactory active *trans*-forms were syn-thesised and further applied in the fragrance industry. The industrially applied pro-cess starts from cyclocitral obtained from β-ionone by ozonolysis, followed by aldol condensation with 2-pentanone and heterogeneously catalysed hydrogenation (Pd/C) [166, 167]. A main disadvantage of this route is that the direct route from citral to cyclocitral on an industrial scale is not known.

An alternative route has been described in the literature applying heterogeneous catalysts in each of the three steps starting from citral to timberone, avoiding the formation of large amounts of waste [168]. Solid basic catalysts were applied in the aldol reaction between 2-pentanone and citral, since separation of the product from the catalyst is simple (Scheme 8.43). Basic ion-exchange resins such as Ambersep® 900 OH were the preferred catalysts, achieving 98% selectivity at 97% conversion. The acid-catalysed cyclisation of the aldol condensation product can be performed using a homogeneous or heterogeneous catalyst since the position of the formed C=C double bond (α- or β-form) is less important, as the resulting cyclic compound is hydrogenated to timberone in the presence of a supported Pd-catalyst. Efficient systems for such type of hydrogenation reactions are Pd/alumina, which allows full conversion at 10 bar in less than 2 h and yields a selectivity of >92%.

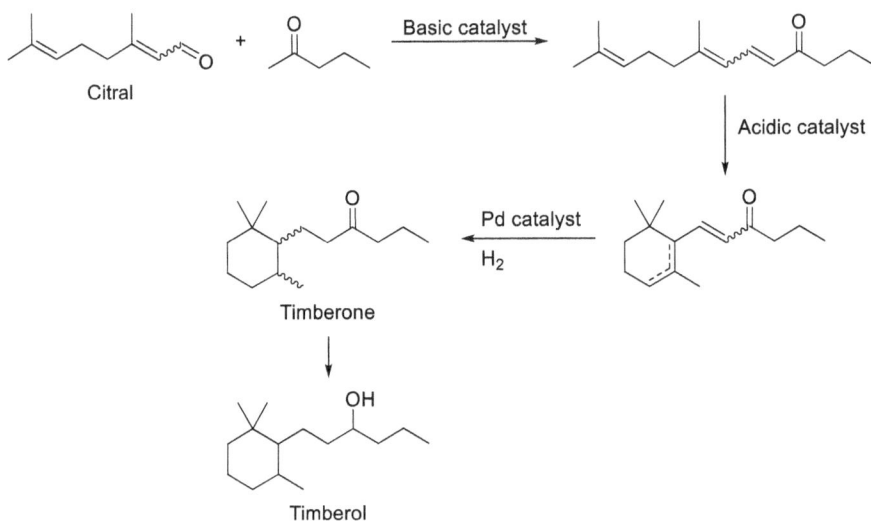

Scheme 8.43: Synthesis of timberol by aldol condensation.

In addition to known compounds from sandalwood oil (see Section 8.2 Acid-Base-Catalysed Reactions), a number of other derivatives have been developed with interesting aroma profiles. For example, 5-(2,2,3-trimethyl-3-cyclopenten-1-yl)-3-methylpentan-2-ol, known as Sandalore™, which does not occur in nature. It is manufactured by an aldol condensation of methyl ethyl ketone with campholenic aldehyde in the presence of potassium *tert*-butoxide, followed by hydrogenation. Sandalore™ can be used in pure state or as mixture with its regio-isomer (Scheme 8.44) [169]. Similar concepts of aldol type condensation starting with campholenic aldehyde are also applied in the manufacture of 2-ethyl-4-(2,2,3-trimethyl-3-cyclopenten-1-yl)-2-but-(*E*)-en-1-ol, which is commercialised under several trade names (e.g. Bacdanol®, Sandranol®) [38].

α-Pinene

Cat.

KO*t*Bu

OHC

Campholenic-
aldehyde

Cat. = Lewis acid,
e.g. ZnCl₂, or zeolite or BenzCp₂Mo(CNMe)₂

+

H₂
Catalyst

O=

OH

Bacdanol®
Sandranol®

OH

Sandalore™

+

HO

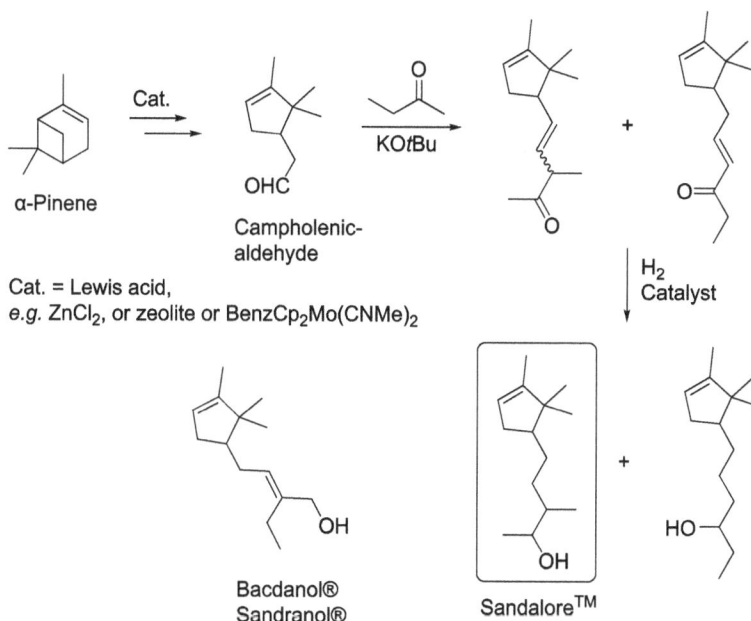

Scheme 8.44: Synthesis of Sandalore™.

In the application of jasmine aroma compounds, aldol reactions of benzaldehyde and aliphatic aldehydes such as heptanal or octanal are applied on industrial scale. The products α-amyl cinnamaldehyde and hexyl cinnamaldehyde are applied in soaps because of their stability under basic conditions; the compounds are produced in several t/a. The aldol condensation reaction is catalysed by a base, preferentially NaOH, resulting in excellent yields and high selectivity (Scheme 8.45) [170, 171]. Yields up to 85% can be achieved at s/c of 100.

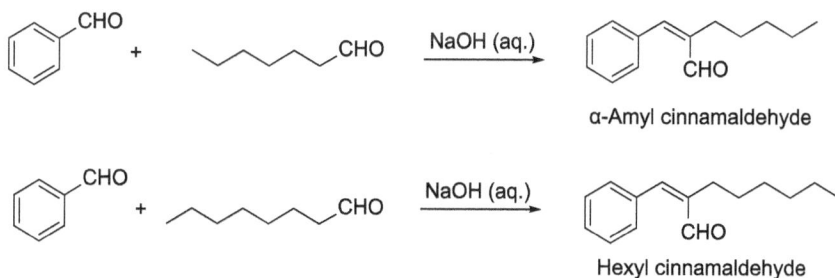

CHO

+

CHO

NaOH (aq.)

CHO

α-Amyl cinnamaldehyde

CHO

+

CHO

NaOH (aq.)

CHO

Hexyl cinnamaldehyde

Scheme 8.45: Manufacture of α-amyl cinnamaldehyde and hexyl cinnamaldehyde.

The base-catalysed condensation of a C–H acidic compound with a carbonyl compound followed by dehydration results in the formation of a new C=C double bond.

This is a type of aldol reaction called the *Knoevenagel* condensation. Suitable C–H acidic compounds are diethyl malonate and ethyl acetoacetate. The *Knoevenagel* reaction is applied in the pharmaceutical industry, for example, in the manufacture of lumefantrine, an anti-malarial drug (Scheme 8.46) [172]. It is synthesised in a base-catalysed (NaOH/NaOMe) one-pot *Knoevenagel* reaction, starting from 2-dibutylamino-(2,7-dichloro-9-H-fluoren-4-yl)ethanol.

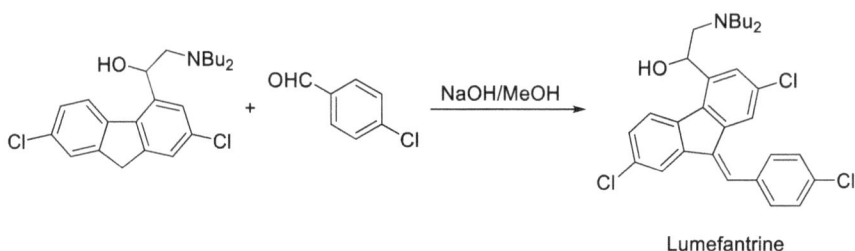

Scheme 8.46: *Knoevenagel* condensation for the anti-malarial drug, lumefantrine.

The base-catalysed reaction of an α-haloester and a carbonyl compound (aldehyde or ketone) to form an α,β-epoxy ester (glycidic ester) is called the *Darzens* reaction (Scheme 8.47). The *Darzens* glycidic ester product reacts further by hydrolysis and decarboxylation, followed by a rearrangement reaction to a carbonyl compound, a C_1-elongated compound from view of the starting material. The reaction is initiated by strong bases. When the starting material is an ester, the alkoxide corresponding to the ester side-chain is commonly chosen in order to prevent complications due to potential acyl exchange side reactions [173].

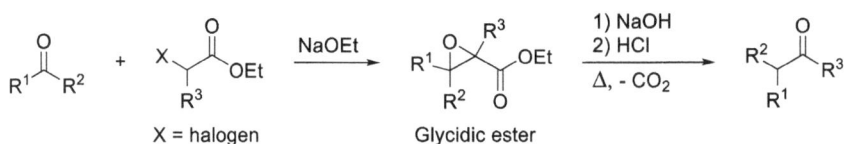

Scheme 8.47: General concept of *Darzens* reactions.

An application of the *Darzens* reaction is the C_1-elongation of isoprenoids in the synthesis of a C_{14}-aldehyde from β-ionone (Scheme 8.48). This reaction is important in the manufacture of vitamin A by the *Isler* route and is performed in a several thousand t/a scale [30].

In the field of flavour and fragrance chemistry, the *Darzens* reaction is applied in the synthesis of various aldehydes such as 6-methoxy-2,6-dimethylheptanal, 2,6-dimethylheptenal, 2,6,10-trimethylundeca-5,9-dienal, 2,6,10-trimethylundeca-9-enal, 2-methyl-4-(2,6,6-trimethyl-cyclohexene-1-yl)-2-butanal and 2-methyl-4-(2,6,6-trimethyl-2-cyclohexen-1-yl)-2-butenal (Scheme 8.49) [174–178].

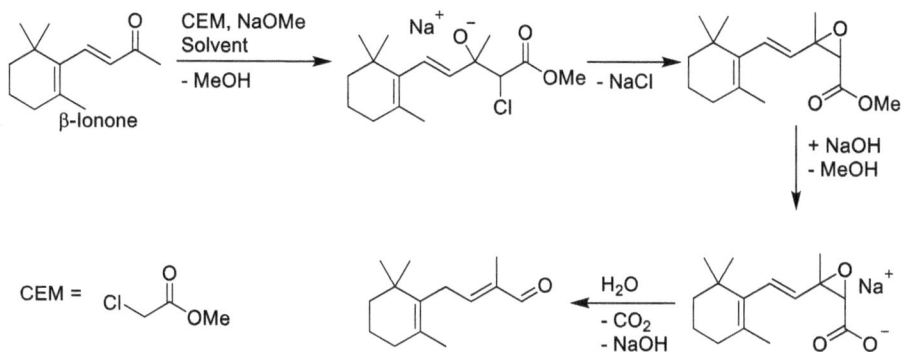

Scheme 8.48: *Darzens* reaction for the manufacture of a vitamin A intermediate.

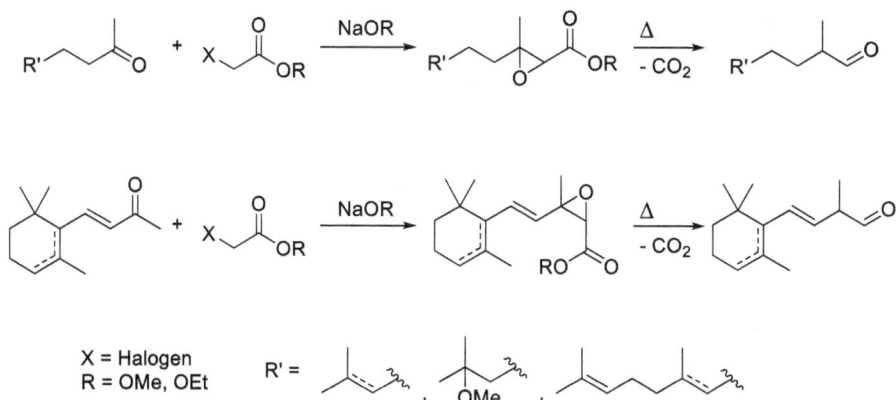

X = Halogen
R = OMe, OEt

R' =

Scheme 8.49: *Darzens* reaction applied in the synthesis of aliphatic aldehydes.

8.6 Ethynylation

The synthesis of α-alkynols from carbonyl compounds and 1-alkynes is of special interest in industry, especially the C_2-elongation with ethyne; the reaction of carbonyl compounds by ethynylation reactions is catalysed by a base. For simple ketones such as acetone, the catalytic ethynylation has been known since the 1960s [179, 180]. However, C_2-elongation reactions of α,β-unsaturated carbonyl compounds were more difficult, resulting in lower selectivities and yields; this was partly solved by using LiC_2H or special reaction conditions [181, 182].

The application of catalytic methods in the manufacture of acetylenic compounds such as α-alkynols by ethynylation of carbonyl compounds is performed at s/c >200 (Scheme 8.50). The reaction products are key intermediates in the production of carotenoids, isoprenoids and fat-soluble vitamins [183]. The main advantage of this approach is the high selectivity and low by-product formation, for example aldol condensation products or the so-called dimer formation by addition of the alkynol to a second equivalent of the carbonyl compound. Usually the reactions are performed at 0–50 °C and pressures of >10 bar. Acetylene undergoes deflagration at 1.6 bar; therefore, the presence of ammonia allows the reaction to be performed safely.

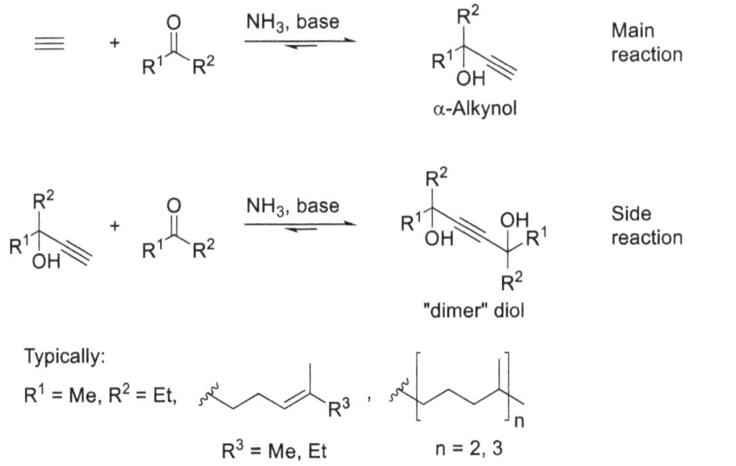

Scheme 8.50: Catalytic ethynylation reaction of different ketones involved in the isoprenoid chemistry.

The application of solid bases in a fixed-bed reactor system for the manufacture of α-alkynols from carbonyl compounds is a further example of modern reaction technology. Macroporous anion-exchange resins, preferably featuring a polystyrene matrix, can be used in the presence of ammonia to perform ethynylation reactions in high selectivity [184]. A typical reactor set-up is shown in Scheme 8.51. After deactivation, the catalyst can easily be recovered and reused. The feed stream of ethyne and carbonyl compound is transferred to a second fixed-bed reactor, and the first is washed with a base/water mixture.

8.7 *Mannich* reactions

The synthesis of β-amino-carbonyl compounds by an amino alkylation reaction of non-enolisable aldehydes or ketones in the presence of a primary or secondary amine is known as the *Mannich* reaction; it can also be used in asymmetric synthesis (see Chapter 6: C–C-bond and C–N-bond forming reactions). In the field of alkylation reactions, the *Mannich* reaction is applied in the synthesis of alkylated aromatic compounds, for example in the synthesis of trimethylhydroquinone (see Chapter 4: Oxidations and Chapter 2: Heterogeneous hydrogenations) [185].

Crude
Product

Waste
(regeneration and
wash solvent)

Reactors in
up-flow mode

Regeneration
Solvent

Wash
Solvent

Ammonia +
Acetylene

Substrate

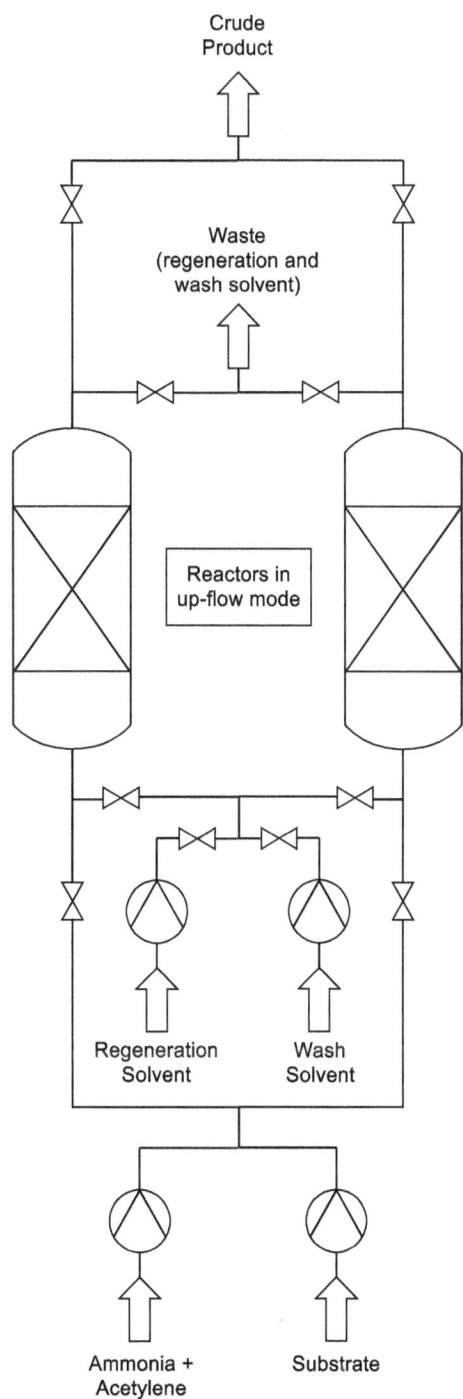

Scheme 8.51: Application of solid ion-exchange resins in ethynylation reactions.

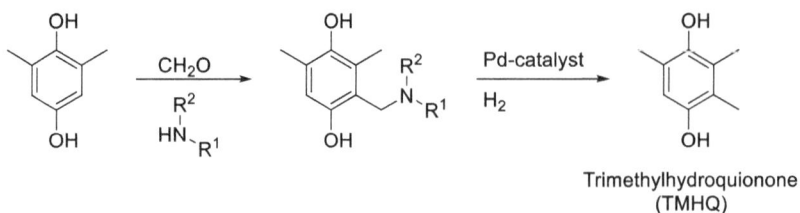

Scheme 8.52: *Mannich* reaction in the manufacture of a trimethylhydroquinone derivative.

The *Mannich* reaction is also applied on industrial scale in the methylation of mixed tocopherols to α-tocopherol starting from soy bean oil (Scheme 8.52) [111, 186]. The process is carried out on several thousand t/a. Alternative approaches including hydroxymethylation, or the application of methanol or formaldehyde in the presence of a mixed metal oxide or hydrotalcite under supercritical or near supercritical conditions, or a boric acid-catalysed methylation have also been described (Scheme 8.53). However, they are not carried out on an industrial scale [187–189].

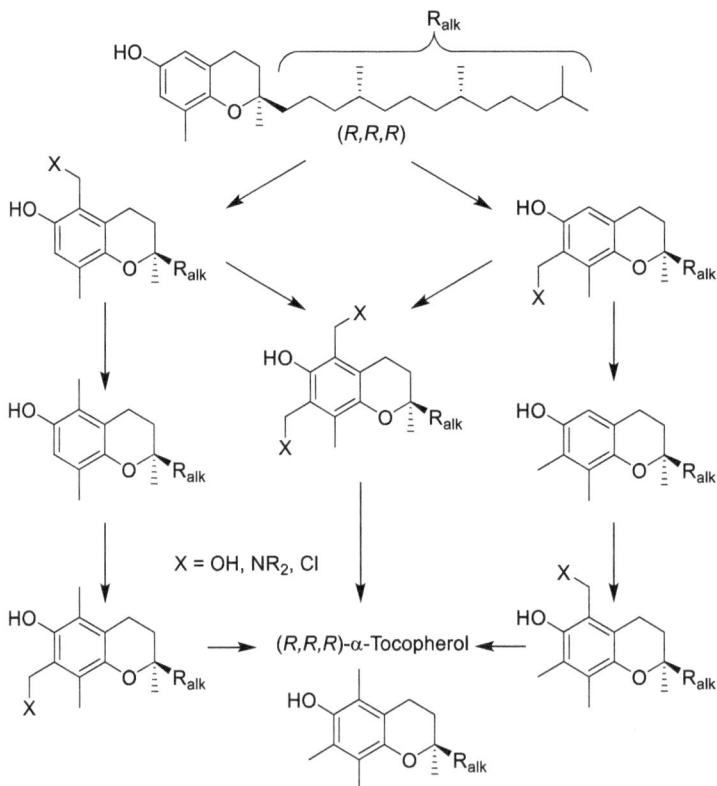

Scheme 8.53: Methylation of mixed tocopherols to α-tocopherol.

References

[1] Bonrath W, Netscher T. Catalytic processes in vitamins synthesis and production. Appl Catal A Gen 2005, 280, 55–73.

[2] Bonrath W, Eggersdorfer M, Netscher T. Catalysis in the industrial preparation of vitamins and nutraceuticals. Catal Today 2007, 121, 45–57.

[3] Hara M, Yoshida T, Takagaki A, Takata T, Kondo J N, Hayashi S, Domen K. A carbon material as a strong protonic acid. Angew Chem 2004, 116, 3015–3018.

[4] Gillespie R J, Peel T E, Robinson E A. Hammett acidity function for some super acid systems. I. Systems H^2SO_4-SO_3, H_2SO_4-HSO_3F, H_2SO_4-HSO_3Cl, and H_2SO_4-$HB(HSO_4)_4$. J Am Chem Soc 1971, 93, 5083–5087.

[5] Olah G A, Prakash G K S, Sommer J. Superacids. New York, USA, John Wiley & Sons, 1985.

[6] Samms S R, Wasmus S, Savinell R F. Thermal stability of nafion in simulated fuel cell environments. J Electrochem Soc 1996, 143, 1498–1504.

[7] Harmer M A, Sun Q, Farneth W E. High surface area nafion resin/silica nanocomposites: A new class of solid acid catalyst. J Am Chem Soc 1996, 118, 7708–7715.

[8] Harmer M A, Sun Q. Solid acid catalysis using ion-exchange resins. Appl Catal A General 2001, 221, 45–62.

[9] Wüstenberg B, Müller M-A, Schütz J, Wyss A, Schiefer G, Litta G, John M, Hähnlein W. Vitamins, 2. Vitamin A (Retinoids). In Elvers, ed. Ullmann's Encyclopedia of Industrial Chemistry, Online ed, Weinheim, Germany, Wiley-VCH Verlag GmbH & Co, 2020.

[10] Isler O, Ruegg R, Schudel P. Synthetic carotenoids for food coloring. Chimia 1961, 15, 208–226.

[11] Bonrath W, Wehrmüller J. Process for Isomerizing a pent-1-en-3-ol derivatives in a multiphase system. WO 2008095724, DSM IP Assets B.V., June 2, 2009.

[12] Aquino F, Bonrath W. Process for the rearrangement of allyl alcohols. WO 2009144328, DSM IP Assets B.V., February 8, 2008.

[13] Bonrath W. New trends in (heterogeneous) catalysis for fine chemical industry. Chimia 2014, 68, 485–491.

[14] Clavey T, Herguijuela J R, Stubbe A. Process intensification with pervaporation of acetals – Membrane screening. Chem Ing Tech 2009, 81, 1471–1478.

[15] Clavey T. Process intensification with pervaporation of acetals-pilot plant study. Chem Ing Tech 2009, 81, 1473 and 1583–1590.

[16] Gelbard G. Organic synthesis by catalysis with ion-exchange resins. Ind & Eng Chim Res 2005, 44, 8468–8498.

[17] Gierke T D, Munn G E, Wilson F C. The morphology in nafion perfluorinated membrane products, as determined by wide- and small-angle x-ray studies. J Polm Sci Polymer Phys Ed 1981, 19, 1687–1704.

[18] Heidekum A, Harmar M A, Hoelderich W F. Highly selective fries rearrangement over zeolites and nafion in silica composite catalysis: A comparison. J Catal 1998, 176, 260–263.

[19] Lundquist E. Esterification process using acidic vinylaromatic polymer beads as catalysts. US 5426199, Rohm and Haas, December 14, 1992.

[20] Armor I N New catalytic technology in the USA during the 1980s. Appl Catal 1991, 78, 141–173.

[21] Chauvel A, Delmon B, Hölderich W F. New catalytic process developed in Europe during the 1980's. Appl Catal A General 1994, 115, 173–217.

[22] Wieland S, Panster P. Replacing Liquid Acids in Fine Chemical Synthesis by Sulfonated Polysiloxanes as Solid Acids and as Supports for Precious Metal Catalysts. In: Blaser H U, Baiker A, Prins R, eds. Heterogeneous Catalysis and Fine Chemicals IV, Amsterdam, The Netherlands, Elsevier, 1997, 67–74.

[23] Brockwell H L, Sarathy P R, Trotta R. Synthesize ethers. Hydrocarbon Process Int Ed 1991, 133–141.

[24] Parra D, Izquierdo J F, Cunill F, Tejero J, Fite C, Iborra M, Vila M. Catalytic activity and deactivation of acidic ion-exchange resins in methyl tert-butyl ether liquid-phase synthesis. Ind Eng Chem Res 1998, 37, 3575–3581.

[25] Panneman H J, Beenackers A A C M. Synthesis of methyl tert-butyl ether catalyzed by acidic ion-exchange resins. Influence of the proton activity. Ind Eng Chem Res 1995, 34, 4318–4825.

[26] Chu W, Yang X, Shan Y, Ye X, Wu Y. Immobilization of the heteropoly acid (HPA) $H_4SiW_{12}O_{40}$ (SiW_{12}) on mesoporous molecular sieves (HMS and MCM-41) and their catalytic behaviour. Catal Lett 1996, 42, 201–208.

[27] Buttersack C. Accessibility and catalytic activity of sulfonic acid ion-exchange resins in different solvents. React Poly 1989, 10, 143–164.

[28] Heese F P, Dry M E, Moller K P. Single stage synthesis of diisopropyl ether – An alternative octane enhancer for lead-free petrol. Catal Today 1999, 49, 327–335.

[29] Calderon A, Tejero J, Izquierdo J F, Iborra M, Cunill F. Equilibrium constants for the liquid-phase synthesis of isopropyl tert-butyl ether from 2-propanol and isobutene. Ind Eng Chem Res 1997, 36, 896–902.

[30] Eggersdorfer M, Laudert D, Letinois U, McClymont T, Medlock J, Netscher T, Bonrath B. One hundred years of vitamins – A success story of natural sciences. Angew Chem Int Ed 2012, 51, 12960–12990.

[31] Bonrath W, Peng K, Zhang Q-M, Pauling H, Weimann B-J. Vitamins 10. Vitamin B_6. In Elvers, ed. Ullmann's Encyclopedia of Industrial Chemistry, Online ed, Weinheim, Germany, Wiley-VCH Verlag GmbH & Co, 2020.

[32] Hoyer H, Keicher G, Schubert H. Verfahren zur Herstellung von Terpineol durch Wasserabspaltung aus Terpin oder Terpinhydrate. DE 955499, Hoechst, January 3, 1957.

[33] Soukup M, Spurr P, Widmer E. General Strategies. In: Britton G, Liaaen-Jensen S, eds. Carotenoids Volume 2: Synthesis, Basel, Boston, Berlin, Birkhäuser Verlag, 1996, 10.

[34] Wüstenberg B, Stemmler R, Letinois U, Bonrath W, Hugentobler M, Netscher T. Large-scale production of bioactive ingredients as supplements for healthy human and animal nutrition. Chimia, 2011, 65, 420–428.

[35] Bonrath W, Goy R, Joray M. Sustainable process for the manufacture of 1,1,4,4-tetramethoxy-2-butene. WO 2021170864, DSM, September 02, 2021.

[36] Hornok L. Cultivation and Processing of Medical Plants. Chichester, UK, John Wiley & Sons, 1992.

[37] De Carvalho C C C R, Da Fonseca M M R. Carvone: Why and how should one bother to produce this terpene. Food Chem, 2006, 95, 413–422.

[38] Fahlbusch K-G, Hammerschmidt F-J, Paten J, Pickenhagen W, Schatkowski D, Bauer K, Garbe D, Suburg H. Flavors and Fragrances. In: Elvers, ed. Ullmann's Encyclopedia of Industrial Chemistry, Online ed., Weinheim, Germany, Wiley-VCH Verlag GmbH & Co, 2012.

[39] Rothenberger S O, Krasnoff S B, Rollins R B. Conversion of (+)-Limonene to (-)-Carvone: An organic laboratory sequence of local interest. J Chem Edu, 1980, 57, 741–742.

[40] Derfer J M, Kane B J, Young D G. Preparation of carvone. US 3293301, Glidden, December 20, 1966.

[41] Everett D H. Manual of Symbols and Terminology for Physicochemical Quantities and Units, Appendix II: Definitions, Terminology and Symbols in Colloid and Surface Chemistry. Pure Appl Chem 1972, 31, 579–638.

[42] Weitkamp J, Ernst S, Puppe L. In: Weitkamp J, Puppe L, eds. Catalysis and Zeolites, Berlin, Germany, Springer, 1999, 327–376.

[43] Welch V A, Fallon K J, Gelbke H-P. Ethylbenzene. In: Elvers B, ed. Ullmann's Encyclopedia of Industrial Chemistry, Online ed., Weinheim, Germany, Wiley-VCH Verlag GmbH &; Co, 2005.

[44] Degnan T F, Smith C M, Venkat C R. Alkylation of aromatics with ethylene and propylene: Recent developments in commercial processes. Appl Catal A Gen 2001, 211, 283–294.

[45] Bipin V V, Kocal J A, Barger P T, Schmidt R J, Johnson J A. Alkylation. In: Kirk-Othmer, ed. Kirk Othmer Encyclopedia of Chemical Technology, Online ed., Weinheim, Germany, Wiley-VCH Verlag GmbH; Co, 2003.

[46] Matheis G. Flavour Modifiers. In: Ziegler E, Ziegler H, eds. Flavourings, Weinheim, Germany, Wiley-VCH Verlag GmbH & Co, 1989, 324.

[47] Walton N J, Mayer J M, Narbad A. Vanillin. Phytochemistry 2003, 63, 505–515.

[48] Esposito L J, Formanek K, Kientz F, Mauger V, Maureaux G, Robert G, Truchet F. Vanillin. In: Kirk-Othmer Encyclopedia of Chemical Technology, 3rd edition, New York, USA, John Wiley & Sons, 1997, 812–825.

[49] Sinha A K, Sharma U K, Sharma N. A comprehensive review on vanilla flavor: Extraction, isolation and quantification of vanillin and other constituents. Int J Food Sci Nutr 2008, 59, 299–326.

[50] Hocking M B. Vanillin: Synthetic flavoring from spent sulfite liquor. J Chem Educ 1997, 74, 1055–1059.

[51] Van Ness J H. Vanillin. In: Kirk-Othmer Encyclopedia of Chemical Technology, 3rd ed, New York, USA, John Wiley & Sons, 1997, 704–717.

[52] Romano U, Esposito A, Maspero F, Neri C, Clerici M G. Selective oxidation with titanium silicalite. Chim Ind (Milan) 1990, 72, 610–616.

[53] Moreau C, Razigade-Trousselier S, Finiels A, Fazula F, Gilbert L. Method for hydroxyalkylating carbocyclic aromatic ethers, WO 9637452, Rhone-Poulenc Chimie, November 28, 1996.

[54] Moreau C, Fajula F, Finiels A, Razigade S, Gilbert L, Jacquot R, Spagnol M. Selective Hydroxymethylation of Guaiacol to Vanillic Alcohols in the Presence of H-Form Mordenites. In: Sowa J R, ed. Catalysis of Organic Reactions, New York, USA, Marcel Dekker Inc., 1998, 51–62.

[55] Maliverney C, Jouve I, Metivier P. Preparation of 3-alkoxy-4-hydroxybenzaldehyde. WO 9626175, Rhone-Poulenc Chimie, February 14, 1996.

[56] Metivier P. Preparation of 3-carboxy-4-hydroxybenzaldehydes. WO 9637454, Rhone-Poulenc Chimie, May 24, 1996.

[57] Metivier P, Malivernay C, Denis P. Method for selective preparation of a 2-hydroxybenzoic acid and a 4-hydroxybenzaldehyde and derivatives. WO 9816493, Rhodia Chimie, October 13, 1997.

[58] Spagnol M, Gilbert L, Benazzi E, Marcilly C. Aromatic ether acylation process using zeolite catalysts, and application to the acetylation of anisole. WO 9635656, Rhone-Poulenc Chimie, May 10, 1996.

[59] Corma A, Renz M Sn-Beta zeolite as diastereoselective water-resistant heterogeneous lewis-acid catalyst for carbon-carbon bond formation in the intramolecular carbonyl-ene reaction. Chem Commun 2004, 550–551.

[60] Hori Y, Iwata T, Okeda Y. Process for producing isopulegol. EP 1225163, Takasago International Corporation, July 24, 2002.

[61] Ellis W W, Odenkirk W. Homogeneous catalysis. Use of a ruthenium(II) complex for catalyzing the ene reaction. Chem Commun 1998, 1311–1312.

[62] Buettner A. Springer Handbook of Odor. Cham, Switzerland, Springer International Publishing, 2017, 67.

[63] Kapoor L D Handbook of Ayurvedic Medicinal Plants. Herbal Reference Library Series, Vol. 2, Boca Raton, USA, CRC Press, 2001.

[64] Eintrag zu Sandelholzöl. (Accessed December 12, 2023, at https://roempp.thieme.de/lexicon/RD-19-00318?context=lexiconOverview.)

[65] Krotz A, Helmchen G Total synthesis, optical rotations and fragrance properties of sandalwood constituents (-)-(Z)- and (-)-E-β-santalol and their enantiomers, ent-β-santalene. Liebigs Ann Chem, 1994, 601–609.

[66] Fahlbusch K G, Hammerschmidt F-J, Panten J, Pickenhagen W, Schatkowski D, Bauer K, Garbe D, Surburg H. Flavors and Fragrances. In: Elvers, ed. Ullmann's Encyclopedia of Industrial Chemistry, Online ed. Weinheim, Germany, Wiley-VCH Verlag GmbH & Co, 2003.

[67] Arbusov B. Isomerization of terpene oxides. I. Isomerization of α-pinene oxide in the reformatsky reaction. Chem Ber 1935, 1430.

[68] Lewis J B, Hedrick G W. Reaction of α-pinene oxide with zinc bromide and rearrangement of 2,2,3-trimethyl-3-cyclopentene products derived there from. J Org Chem 1965, 30, 4271–4275.

[69] Fujiwara Y, Nomura M, Igawa K. Preparation of camphorene aldehyde as a perfume intermediate. JP 62019549, Toyo Soda Mfg. Co., July 19, 1985.

[70] Kaninska J, Schwegler M A, Hoefnagel A J, Bekkum H V. The isomerization of α-pinene oxide with brønsted and lewis acids. Recl Trav Pays-Bas 1992, 111, 432–437.

[71] Bruno A M, Valente A A, Pillinger M, Amelse J, Romao C C, Goncalves I S. Efficient isomerization of α-pinene oxide to campholenic aldehyde promoted by a mixed-ring analogue of molybdenocene. ACS Sustainable Chem Eng, 2019, 7, 13639–13645.

[72] Kunkeler P J, van der Waal J C, Bremmer J, Zuurdeeg B J, Dowing R S, van Bekkum H. Applications of zeolite titanium Beta in the rearrangement of α-pinene oxide to campholenic aldehyde. Catal Lett, 1998, 53, 135–138.

[73] Kozhevnikov I V. Catalysis by heteropoly acids and multicomponent polyoxometalates in liquid-phase reactions. Chem Rev 1989, 98, 171–198.

[74] Mizuno N, Misono M. Heterogeneous catalysis. Chem Rev 1989, 98, 199–217.

[75] Keggin J F. The structure and formula of 12-phosphotungstic acid. Proc R Soc London 1934, A144, 75.

[76] Kozhevnikov I V. Advances in catalysis by heteropolyacids. Russ Chem Rev 1987, 56, 1417–1443.

[77] Tsigdinos G A. Heteropoly compounds of molybdenum and tungsten. Top Curr Chem 1978, 76, 1–64.

[78] Fournier M, Jantou C-F, Rabia C, Herve G, Launay S. Polyoxometalates catalyst materials: X-ray thermal stability study of phosphorus-containing heteropolyacids $H_{3+x}PM_{12-x}V_xO_{40}$ x 13–14 H_2O (M = molybdenum, tungsten; X = 0–1). J Mater Chem 1992, 2, 971–978.

[79] Okuhara T, Mizuno N, Misaona M. Catalytic chemistry of heteropoly compounds. Adv Catal 1996, 41, 113–252.

[80] Misono M, Nojiri N. Recent progress in catalytic technology in Japan. Appl Catal 1990, 64, 1–30.

[81] Misono M, Ono I, Koyano G, Aoshima A. Heteropolyacids Versatile green catalysts usable in a variety of reaction media. Pure Appl Chem 2000, 72, 1305–1311.

[82] Ng F T T, Mure T, Jiang M, Sultan M, Xie J-H, Gayraud P, Smith W J, Hague M, Watt R, Hodge S. Avada – A New Green Process for the Production of Ethyl Acetate. Chemical Industries, Boca Raton. In: Sowa J R, ed. Catalysis of Organic Reactions, New York, USA, Marcel Dekker Inc, 2005, 104, 251–260.

[83] Ken-ichi S, Uchida H, Wakabayashi S. A new process for acetic acid production by direct oxidation of ethylene. Catal Surv Japan 1999, 3, 66–60.

[84] Makoto M. Recent progress in the practical applications of heteropolyacid and perovskite catalysts: Catalytic technology for the sustainable society. Catal Today 2009, 144(3–4), 285–291.

[85] Arundale E, Mikeska L A. The olefin-aldehyde condensation. The prins reaction. Chem Rev 1952, 51, 505–555.

[86] Miles R B, Davis C E, Coates R M. Syn- and anti-selective prins cyclizations of unsaturated ketones to 1,3-Halohydrins with Lewis Acids. J Org Chem 2006, 71, 1493–1501.

[87] Overman L E, Velthuisen E J. Scope and facial selectivity of the prins-pinacol synthesis of attached rings. J Org Chem 2006, 71, 1581–1587.

[88] Izumi Y, Urabe K, Onaka M. Zeolite. In: Clay and Heteropoly Acid in Organic Reactions, Tokyo, VCH Publishing, 1993, 99.

[89] Aoshima A, Tonomura S, Mitsui R. Polyether glycol. EP 126471, Asahi Chemical Industry, May 18, 1984.

[90] Aoshima A, Tonomura S, Yamamatsu S. New synthetic route of polyoxytetramethyleneglycol by use of heteropolyacids as catalyst. Polym Adv Technol 1990, 2, 127–132.

[91] Kozhevnikov I V, Kulikov S M, Chukaeva N G, Krisanov A T, Letunova A B, Blinova V I. Syntheses of vitamins E and K1 catalyzed by heteropoly acids. React Kinet Catal Lett 1992, 47, 59–64.

[92] Aquino F, Bonrath W. Manufacture of d,l-α-tocopherol. EP 970953, F. Hoffmann-La Roche, June 26, 1999.

[93] Kozhevnikov I V. Heteropoly acids as catalysts for organic reactions. Stud Surf Sci Catal 1994, 90, 21–34.

[94] Zundel C L, Chron L. Alkylation of phenols or thiophenols with olefins. DE 1518460, October 22, 1963.

[95] Fiege H, Voges H-W, Hamamoto T, Umemura S, Iwata T, Miki H, Fujita Y, Buysch H-J, Garbe D, Paulus W. Phenol Derivatives. In: Elvers, ed. Ullmann's Encyclopedia of Industrial Chemistry, Online ed. Weinheim, Germany, Wiley-VCH Verlag GmbH & Co, 2012.

[96] Matveev K I, Zhizhina G E, Odyakov V F. New catalytic for the synthesis of vitamin K. React Kinet Catal Lett 1995, 55, 47–50.

[97] Yoshizawa T, Toyofuku H, Tachibana K, Kuroda T. Regioselective polyprenyl rearrangement of polyprenyl 2,3,4,5-tetrasubstituted phenyl ethers promoted by boron trifluoride. Chem Lett 1982, 1131–1134.

[98] Caine D, Graham S L. The acid-catalyzed rearrangement of a cyclopropyl ketone related to 10-epieudemane. Tetrahedron Lett 1976, 2521–2524.

[99] Hirschmann R, Miller R, Wendler N L. The synthesis of vitamin K_1. J Am Chem Soc 1954, 76, 4592–4594.

[100] Isler O, Doebel K. Syntheses in the vitamin K series. I. The total synthesis of vitamin K_1. Helv Chim Acta 1954, 37, 225–233.

[101] Letinois U, Moine G, Hohmann H-P. Vitamins, 6. Vitamin B_1 (Thiamin). In: Elvers, ed. Ullmann's Encyclopedia of Industrial Chemistry, Online ed. Weinheim, Germany, Wiley-VCH Verlag GmbH & Co, 2020.

[102] Létinois U, Schütz J, Härter R, Stoll R, Huffschmidt F, Bonrath W, Karge R. Lewis acid-catalyzed synthesis of 4-aminopyrimidines: A scalable industrial process. Org Process Res Dev 2013, 17, 427–431.

[103] Bonrath W, Fischesser J, Giraudi L, Karge R. Process for the manufacture of a precursor of vitamin B1. WO 2006079504, DSM IP Assests, January 24, 2006.

[104] Rüttimann A. Dienolether condensations – A powerful tool in carotenoid synthesis. Pure Appl Chem 1999, 71, 2285–2293.

[105] Makin S M, Gabrielyan S M, Chebotarev A S, Vladimirskaya E K, Morlyan N M. Chemistry of unsaturated ethers. XXXVI. Synthesis and study of higher α,β-unsaturated ene and diene aldehydes. J Org Chem USSR 1974, 10, 2061–2065.

[106] Labler L, Rüttimann A, Giger A. Enol Ether and Aldol Condensation. In: Carotenoids Volume 2, Chapter 2, Part I, Switzerland, Birkhäuser Verlag Basel, 1996, 27–54.

[107] Bernhard K, Jäggli S. Enol Ether Condensation. In: Carotenoids Volume 2, Birkhäuser Verlag Basel, Switzerland, 1996, 301–302.

[108] Ernst H, Klaus H, Keller A. Verfahren zur Herstellung von 2,7-dimethyl-octa-2,4,6-trien-dial. DE 102004006579, BASF, February 10, 2004.

[109] Weiser H, Vecchi M. Stercoisomers of α-tocopheryl acetate II. Biopotencies of all eight stereoisomers, individually or in mixtures, as determined by rat resorption-gestation tests. J Int Nutr Res 1982, 52, 351–370.

[110] Weimann B J, Weiser H. Functions of vitamin E in reproduction and in prostacyclin and immunoglobulin synthesis in rats. Am J Clin Nutr 1991, 53, 1056S–1060S.

[111] Netscher T. Synthesis and Production of Vitamin E. In: Gustone F D, ed. Lipids: Synthsis and Manufacture, UK, Sheffield Academic Press ltd, 1999, 250–267.

[112] Bonrath W, Haas A, Hoppmann E, Netscher T, Pauling H, Schager F, Wildermann A. Synthesis of (all-rac)-α-tocopherol using fluorinated NH-acidic catalysts. Adv Synth Catal 2002, 344, 37–39.

[113] Netscher T, Bonrath W, Haas A, Hoppmann E, Pauling H. Perfluorinated bronsted "superacids": Powerful catalysts for the preparation of vitamin E. Chimia 2004, 58, 153–155.

[114] Bonrath W, Haas A, Hoppmann E, Netscher T, Pauling H. Process for manufacturing all-rac-α-tocopherol in the presence of a tris(perfluoroalkanesulfonyl or pentafluorobenzenesulfonyl) methane catalyst. EP 1134218, F. Hoffmann-La Roche, March 10, 2001.

[115] Aquino F, Bonrath W, Pace F. Acylation process for the preparation of 2,3,5-trimethylhydroquinone diacylates in the presence of methanetrisulfonic acid catalyst. WO 2005044775, DSM IP Assets, October 26, 2004.

[116] Bonrath W, Hoppmann S, Haas A, Netscher T, Pauling H. Manufacture of α-tocopherol from the reaction of trimethylhydroquinone with isophytol or phytol in the presence of methane trisulfonate. WO 2004046127, DSM IP Assets, September 30, 2003.

[117] Bonrath W, Haas A, Hoppmann S, Netscher T, Pauling H. Process for the manufacture of tocyl and tocopheryl acylates. WO 2004096790, DSM IP Assets, April 19, 2004.

[118] Baak M, Bonrath W, Kreienbühl P. Process for manufacturing d,l-α-tocopherol in a carbonate solvent and in the presence of a sulfur-containing acid catalyst. EP 949255, F. Hoffmann-La Roche, March 30, 1999.

[119] Sanders J M, Michael L H. Process for production of isochromanes. US 3910964, Int Flavors & Fragrances Inc, October 7, 1975.

[120] Finkelmeier H, Hopp R. Verfahren zur Herstellung von Isochroman-Derivaten. EP 120257, Haarmann & Reimer GmbH, October 3, 1984.

[121] Dorn H, Tian N, Qiao H, Lu X. Chemical components of lavender oil. Chin J Chromatogr, 1985, 2, 75–79.

[122] Mitsuhashi S, Kumobayashi H, Akutagawa S. Process for producing myrcenol. EP 112727, Takasago Perfumery Co Ltd, July 4, 1984.

[123] Schulz R P, Blumenstein J, Kohlpaintner C Crotonaldehyde and Crotonic Acid. In: Elvers, ed. Ullmann's Encyclopedia of Industrial Chemistry, Online ed. Weinheim, Germany, Wiley-VCH Verlag GmbH & Co, 2015.

[124] Tietze L F, Eicher T. Reaktionen und Synthesen im organisch-chemischen Praktikum und Forschungslaboratorium, 2nd ed. New York, USA, Georg Thieme Verlag, 1991.

[125] Sell C. The Chemistry of Fragrances: From Perfumer to Consumer, 2nd ed. Cambridge, UK, RCS popular science, RCS Publisher, 2006.

[126] Hawkins C, Yeomans B Process for the production of diacetone alcohol. GB 1527033, BP Chem, October 04, 1976.

[127] Guthrie J P. Equilibrium constants for a series of simple aldol condensations, and linear free energy relations with other carbonyl addition reactions. Can J Chem, 1978, 56, 962–973.

[128] Craven E C. Alkaline condensation of acetone. J Appl Chem, 1963, 13, 71–77.

[129] Yokoyama K, Iwade S, Isogai S. Process for producing diacetone alcohol. WO 2004101585, Mitsubishi, November 25, 2004.

[130] Osei-Twum E Y, Hassan N, Elwaer N, Bhat G S, Seoane C G, Al-Assaf K H. Synthesis of diacetone alcohol and mesityl oxide. WO 2016012974, SABIC, January 28, 2016.

[131] Bonrath W, Fleischhauer H, Hoelderich W F, Schütz J. Aldol condensation reaction and catalysts therefore. WO 2008145350, DSM, December 04, 2008.

[132] Eisenacher M, Venschott M, Dylong D, Hölderich W F, Schütz J, Bonrath W. Upgrading bio-based acetone to diacetone alcohol by aldol reaction using Amberlyst A26-OH as catalyst. React Kinet Mech Catal, 2022, 135, 971–986.

[133] Podrebarac G G, Ng F T T, Rempel G L. The production of diacetone alcohol with catalytic distillation: Part I: Catalytic distillation experiments. Chem Eng Sci, 1998, 53, 1067–1075.

[134] Podrebarac G G, Ng F T T, Rempel G L. The production of diacetone alcohol with catalytic distillation: Part II: A rate-based catalytic distillation model for the reaction zone. Chem Eng Sci, 1998, 53, 1077–1088.

[135] Kim Y K, Hatfield J D. Kinetics and equilibrium data of the dehydration-hydration reaction between diacetone alcohol and mesityl oxide in phosphoric acid. J Chem Eng Data, 1985, 30, 149–153.

[136] Tanabe K, Misono M, Ono Y, Hattori H New Solid Acid and Bases: Their Catalytic Properties. Studies in Surface Science and Catalysis, Vol. 51. Amsterdam, Oxford, New York, Tokyo, Elsevier, 1989.

[137] Tanabe K, Hoelderich W F. Industrial application of solid acid-base catalysts. Appl Catal A, 1999, 181, 399–434.

[138] Isogai N, Okawa T, Wakui N. Verfahren zur Herstellung von Methylisobutylketon. DE 3021764, Mitsubishi Gas Chemical Corp., December 18, 1980.

[139] Technology Profil, Methyl isobutyl ketone (MIBK). (Accessed October 18, 2021, at https://web.ar chive.org/web/20131203010302/http:/www.uhde-ftp.de/cgi-bin/byteserver.pl/pdf/technologies/TP_ MIBK_2005.pdf.)

[140] Ruther T, Müller M-A, Bonrath W, Eisenacher M. The production of isophorone. Encyclopedia, 2023, 3, 224–244.

[141] Luckenbach O. Groundbreaking ceremony for Evonik's new isophorone plants in Shanghai – Evonik Industries (Accessed September 28 2023, at https://corporate.evonik.com/en/investor-relations /groundbreaking-ceremony-for-evoniks-new-isophorone-plants-in-shanghai-107041.html).

[142] Braithwaite J. Ketones. In: Kirk-Othmer, ed. Kirk-Othmer Encyclopedia of Chemical Technology, Online ed., Weinheim, Germany, Wiley-VCH Verlag GmbH & Co, 2000.

[143] Siegel H, Eggersdorfer M Ketones. In: Elvers, ed. Ullmann's Encyclopedia of Industrial Chemistry, Online ed. Weinheim, Germany, Wiley-VCH Verlag GmbH & Co, 2000, 187–205.

[144] David Raju B, Rama Rao K S, Salvapathi G S, Sai Prasad P S, Kanta Rao P. Aromatization of isophorone to 3,5-xylenol over Cr_2O_3/SiO_2 catalysts. Appl Catal A Gen, 2000, 193, 123–128.

[145] Kirichenko G N, Glazunova V I, Kirichenko V Y, Dzhemilev U M. Promising process for synthesis of 3,5-xylenol from isophorone. Pet Chem, 2006, 46, 434–438.

[146] Tungler A, Máthé T, Petró J, Tarnai T. Enantioselective hydrogenation of isophorone. J Mol Catal, 1990, 61, 259–267.

[147] Xu L, Sun S, Zhang X, Gao H, Wang W. Study on the selective hydrogenation of isophorone. RSC Adv, 2021, 11, 4465–4471.

[148] Zhang X, Wang R, Yang Y, Yu J. Central composite experimental design applied to the catalytic aromatization of isophorone to 3,5-xylenol. Chemom Intell Lab Syst, 2007, 89, 45–50.

[149] Mei J, Chen Z, Yuan S, Mao J, Li H, Yin H. Kinetics of isophorone synthesis via self-condensation of supercritical acetone. Chem Eng Technol, 2016, 39, 1867–1874.

[150] Vaughn T H, Jackson D R. Process for preparing isophorone. US2183127, Union Carbide, December 12, 1939.

[151] Li Y, Lü J, Jin Z. Research progress in the synthesis of isophorone by acetone self-condensation. Chem Ind Eng Prog, 2016, 35, 1190–1196.

[152] Franck H G, Turowski J, Eurnlu K, Storch G, Zander M, Lemke R. 3,3,6,8-Tetramethyl-tetralon-(1), ein Kondensationsprodukt des Acetons. Justus Liebigs Ann Chem, 1969, 724, 94–101.

[153] Salvapati G S, Ramamamurty K V, Janardanarao M. Selective catalytic self-condensation of acetone. J Mol Catal, 1989, 54, 9–30.

[154] Canning A S, Jackson S D, McLeod E, Vass E M. Aldol condensation of acetone over $CsOH/SiO_2$: A mechanistic insight using isotopic labelling. Appl Catal A: Gen, 2005, 289, 59–65.

[155] Li G, Dong Y, Fan Z, Ma J, Yang L, Zhang X, Teng Z. Kinetics model of synthesis of isophorone. Lanzhou Ligong Daxue Xuebao, 2009, 35, 61–63.

[156] Podrebarac G G, Ng F T T, Rempel G L. A kinetic study of the aldol condensation of acetone using an anion exchange resin catalyst. Chem Eng Sci, 1997, 52, 2991–3002.

[157] Sliepcevich A, Moscatelli D, Gelosa S. Kinetic study of the polycondensation of acetone to produce isophorone adopting alumina and magnesia as catalyst. Chem Eng Trans, 2007, 11, 605–610.

[158] Li L, Li Q, Yin D. The liquid phase condensation of acetone to isophorone by one step. Acta Sci Nat Univ Norm Hunanensis, 2003, 23, 67–71.

[159] Walton J R, Yeomans B. Isophorone production using a potassium hydroxide. US 3981918, BP Chemicals International, September 21, 1976.

[160] Chen Z, Li H, Yin H, Xu Y, Wang C. Method for preparing α-isophorone from acetone. CN 101633610, Zhejiang NHU, January 27, 2010.

[161] Galera E, Zabza A. Insect growth regulators. II. C-15 derivatives of ethyl 6,7-dihydrofarnesoate. Bull Acad Pol Sci Ser Sci Chim 1977, 25, 615–625.

[162] Huang S, Wen H, Liu C, Wu Z. Tianran Chanwu Yanjiu Yu Kaifa 1993, 5, 21–25.

[163] Dobler W, Bahr N, Breuer K, Kindler A. Continuous process for producing pseudoionones and ionones. WO 2004041764, BASF, October 28, 2003.

[164] Bonrath W, Pressel Y, Schütz J, Ferfecki E, Topp K-D. Aldol condensations catalyzed by basic ion-exchange resins. Chem Cat Chem 2016, 8, 3584–3591.

[165] Hölderich W F, Ritzerfeld V, Russbüldt B M E, Fleischhauer E H, Bonrath W, Karge R, Schütz J. Process for preparing pseudoionone. WO 2012022562, DSM IP Assets, July 29, 2011.

[166] Schulte-Elte K-H. An isomeric composition of 1-(2,6,6-trimethylcyclohexyl)hexan-2-ol, use of this composition, and intermediate carbinols. EP 0118817, Firmenich, February 24, 1984.

[167] Ohmoto T, Shimada A, Yamamoto T. Preparation of cyclohexanepropanol derivatives as perfume components. EP 456932, Takasago International, October 5, 1990.

[168] Schütz J, Bonrath W. A novel approach to timberwood fragrances. Catal Sci Technol 2012, 2, 2037–2038.

[169] Bajgrowicz J A, Frater G. Preparation of optically pure isomers of campholenic aldehyde derivatives for use as detergent fragrances. EP 841318, Givaudan-Roure, May 13, 1998.

[170] Richmond H R. Preparation of cinnemal aldehyde. US2529186, United States Rubber Co, November 07, 1950.

[171] Heynderickx P M. Activity coefficients for liquid organic reactions: Towards a better understanding of true kinetics with the synthesis of jasmin aldehyde as showcase. Int J Mol Sci, 2019, 20, 3819.

[172] Beutler U, Fuenfschilling P C, Steinkemper A. An improved manufacturing process for the antimalaria drug coartem. Part II. Org Proc Res Dev 2007, 11, 341–345.

[173] Ballester M. Mechanisms of the darzens and related condensation. Chem Rev 1955, 55, 283–300.

[174] Beumer R, Bonrath W, Fischesser J. Process for the manufacture of a mixture of 2-methyl-4-(2,6,6-trimethyl-1-cyclohexen-1-yl)-2-butenal and 2-methyl-4-(2,6,6-trimethyl-2-cyclohexen-1-yl)-2-butenal. WO 2018069452, DSM IP Assets, October 12, 2017.

[175] Beumer R, Bonrath W, Eggertswyler C, Fischesser F. Process for the manufacture of 2,6,10-trimethylundeca-5,9-dienal. WO 2018069457, DSM IP Assets, October 12, 2017.

[176] Beumer R, Bonrath W, Eisele F, Fischesser J, Wehrli C. Process for the manufacture of 2,6-dimethyl-5-hepten-1-al. WO 2018069456, DSM IP Assets, October 12, 2017.

[177] Beumer R, Bonrath W, Fischesser F. Process for the manufacture of 6-methoxy-2,6-dimethylheptanal. WO 2018069458, DSM IP Assets, October 12, 2017.

[178] Beumer R, Bonrath W, Fischesser J. Process for the manufacture of 2,6,10-trimethylundec-9-enal. WO 2018069454, DSM IP Assets, October 12, 2017.

[179] Tedeschi R J, Casey A W, Clark J G S, Huckel R W, Kindley L M, Russell J P. Base-catalyzed reaction of acetylene and vinylacetylenes with carbonyl compounds in liquid ammonia under pressure. J Org Chem 1963, 28, 1740–1743.

[180] Tedeschi R J. The mechanism of base-catalyzed ethynylation in donor solvents. J Org Chem 1965, 30, 3045–3049.

[181] Wiederkehr H. Examples of process improvements in the fine chemicals industry. Chem Eng Sci 1988, 43, 1783–1791.

[182] Bonrath W, Schütz J, Netscher T, Wüstenberg B. Ethynylation of polyunsaturated aldehydes and ketones. WO 2016059151, DSM IP Assets, October 15, 2015.

[183] Bonrath W, Scheer P L, Tschumi J, Zenhaeusern R. Ethynylation process. WO 2004018400, DSM IP Assets, August 9, 2003.

[184] Bonrath W, Englert B, Karge R, Schneider M. Ethynylation process. WO 2003029175, Roche Vitamins, September 29, 2002.

[185] Bonrath W, Schütz J, Netscher T, Wüstenberg B. Process for manufacturing TMHQ. WO 2012/025587, DSM IP Assets, April 29, 2013.

[186] Müller R K, Schneider H. Aminomethylierung von Tocopherolen. EP 0735033, F. Hoffmann-La Roche, March 21, 1996.

[187] Swanson R R. Vapour/liquid phase methylation of non-alpha-tocopherols. EP 176690, Henkel, July 29, 1985.

[188] Breuninger M. Preparation of α-tocopherol by catalytic methylation of other tocopherols. EP 882722, F. Hoffmann-La Roche, May 28, 1998.

[189] Brüggemann K, Herguijuela J R, Netscher T, Riegel J. Hydroxymethylierung von Tocopherolen. EP 0769497, F. Hoffmann-La Roche, October 9, 1996.

9 Phase transfer catalysis (PTC)

Phase transfer catalysis (PTC) is a catalytic process in which the reactants are in non-miscible phases (a minimum of two phases) and the addition of a further compound, the phase transfer catalyst (PT-cat), facilitates the transfer of compounds from one phase to another, allowing/accelerating the chemical reaction. The term *phase transfer catalysis* was introduced by Stark in 1971 [1]. PT-cats are often (but not exclusively) ionic compounds that have aliphatic groups attached and so can be partially solvated in multiple different phases, including those of different polarity. This nature is generally the origin of the accelerating effect that they produce. Herein we describe selected examples of PTC to demonstrate the fundamental and broad applicability of this methodology. Further examples and applications are discussed in other chapters.

The power of this concept was demonstrated in the nucleophilic substitution reaction of alkyl halides. 1-Bromooctane and sodium cyanide were dissolved in ether and water, respectively. Without a PT-cat, no conversion to the desired product was observed, even after 2 weeks, and the only reaction that occurred was the slow hydrolysis of NaCN to sodium formate. Whereas, with the addition of 1.5 mol% of a phosphonium salt >95% conversion was obtained after only 90 min (Scheme 9.1) [1].

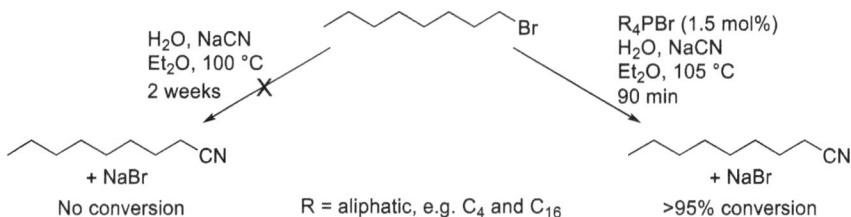

Scheme 9.1: Synthesis of octyl nitrile by PTC.

PTC often occurs in biphasic systems containing water and an organic solvent which is immiscible with water. The nature of the catalyst and that of the solvent are important factors in PTC. Usually, one of the reactants is insoluble in the organic phase, typically an anion such as OH^- or CN^- or an ionic oxidant such as permanganate. This reactant is transferred from the water into the organic phase by the PT-cat. PT-cats for anions are often quaternary ammonium or phosphonium salts, whereas for cations generally crown ethers are used (Scheme 9.2) [2].

In inversed PTC, the reactant, for example a carboxylic acid halide, is transferred into the aqueous phase with the aid of a PT-cat, for example a pyridine derivative (Scheme 9.3) [3]. Although this concept is most often applied for liquid–liquid reactions, liquid–solid or gas–liquid reactions are also known [4, 5]. The advantages of using PTC are often higher reaction rates and increased reaction yields. Avoiding or using reduced

https://doi.org/10.1515/9783111102672-009

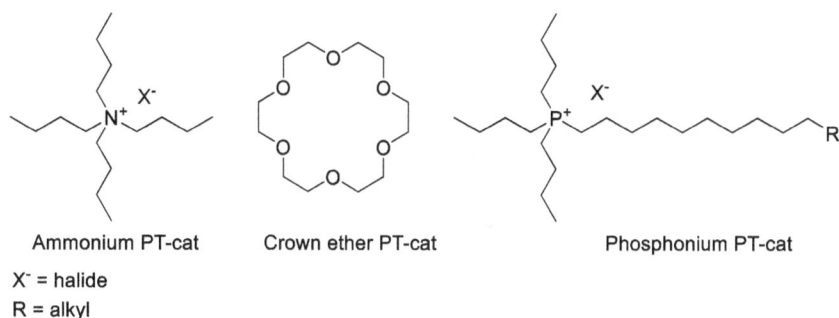

Ammonium PT-cat   Crown ether PT-cat   Phosphonium PT-cat

X⁻ = halide
R = alkyl

Scheme 9.2: Examples of PT-catalysts.

amounts of organic solvents can be an additional benefit [2, 6]. Typically, PTC is carried out with a catalyst loading of around 1%.

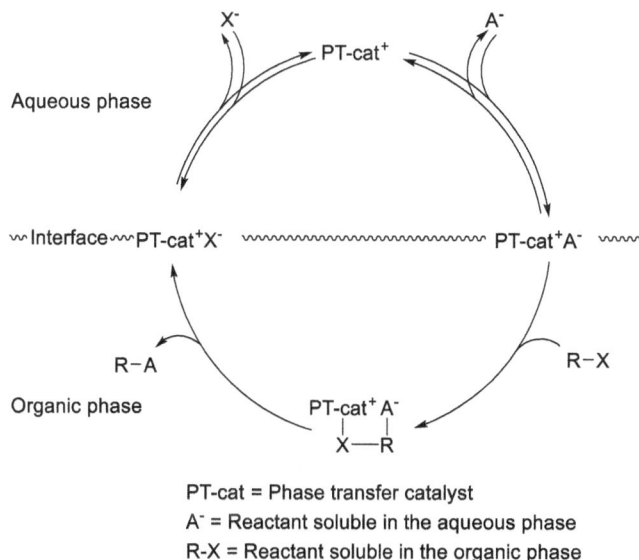

PT-cat = Phase transfer catalyst
A⁻ = Reactant soluble in the aqueous phase
R-X = Reactant soluble in the organic phase

Scheme 9.3: General principle of phase transfer catalysis.

In industrial applications, PTC is exploited in polymer chemistry (e.g. the manufacture of bisphenol A) and in the synthesis of pesticides. Quaternary ammonium salts derived from the chiral cinchona alkaloids are used in asymmetric PTC [7, 8]. PTC can also be applied for various base-induced reactions, for example carbanion chemistry like *Darzens, Wittig, Julia* or *Michael* reactions [9].

Bisphenol A is an important building block for various polymeric materials and is among the most manufactured chemicals worldwide (estimated $>2.2 \times 10^6$ t/a year in 2009) [10]. It is synthesised via the condensation of two equivalents of phenol with acetone, and an excess of phenol is used to ensure full conversion. The following transformations to achieve the synthesis of polycarbonates or epoxy resins are usually conducted using phase transfer conditions (Scheme 9.4). In the production of polycarbonates, phosgene is usually used to combine the bisphenol A units. Phosgene itself is highly toxic and large amounts of HCl are generated during the course of the reaction. The biphasic mixture consists of an aqueous phase containing the phenol. A base (e.g. NaOH) is then added, which deprotonates the phenol, this results in increased nucleophilicity but also increased solubility in the aqueous phase. The additional role of the base is to quench the HCl generated in the condensation reaction. Phosgene is added to the organic phase (e.g. dichloromethane), and the PT-cat aids the transfer of the phenolate to the organic phase and reaction with phosgene to produce the polymer. A side reaction is the hydrolysis of phosgene in water to carbonic acid, and it is accepted due to the high advantages in terms of process safety. The production of epoxy resins can be conducted in a similar fashion. The phenolate of bisphenol A is present in the basic aqueous phase and an epoxide, such as epichlorohydrin, is added to the organic layer.

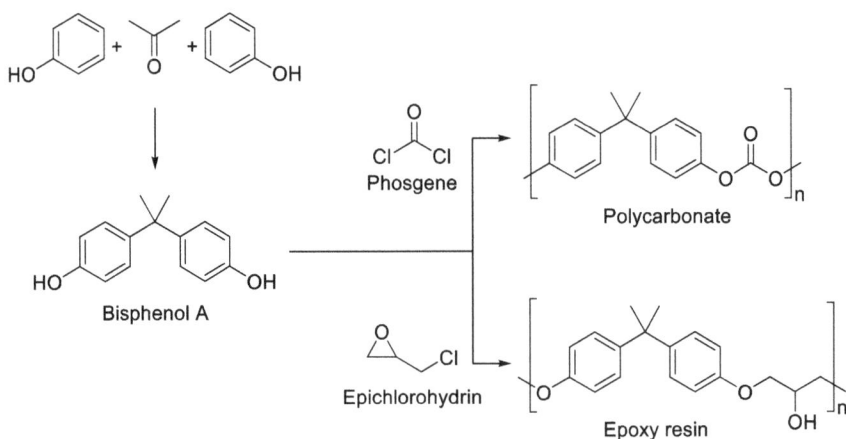

Scheme 9.4: Synthesis of bisphenol A and its conversion to polymeric materials.

The reaction mechanism for the formation of epoxy resin with epichlorohydrin is base-catalysed (Scheme 9.5). After the first deprotonation step, the epoxide is opened via nucleophilic attack of the phenolate. The generated species undergoes a subsequent intramolecular substitution to generate a second epoxide with a chloride as a leaving group. On first glance, an intermolecular substitution could also be envisioned; however, this is significantly slower. The newly formed epoxide is attacked by a second deprotonated bisphenol A molecule. In the next step, a proton is transferred from a phenol and the chain reaction can proceed.

In the production of fluoroaromatics, PTC is applied for the halogen exchange of Br or Cl, the so-called *Halex* reaction [11]. These reactions are performed in polar

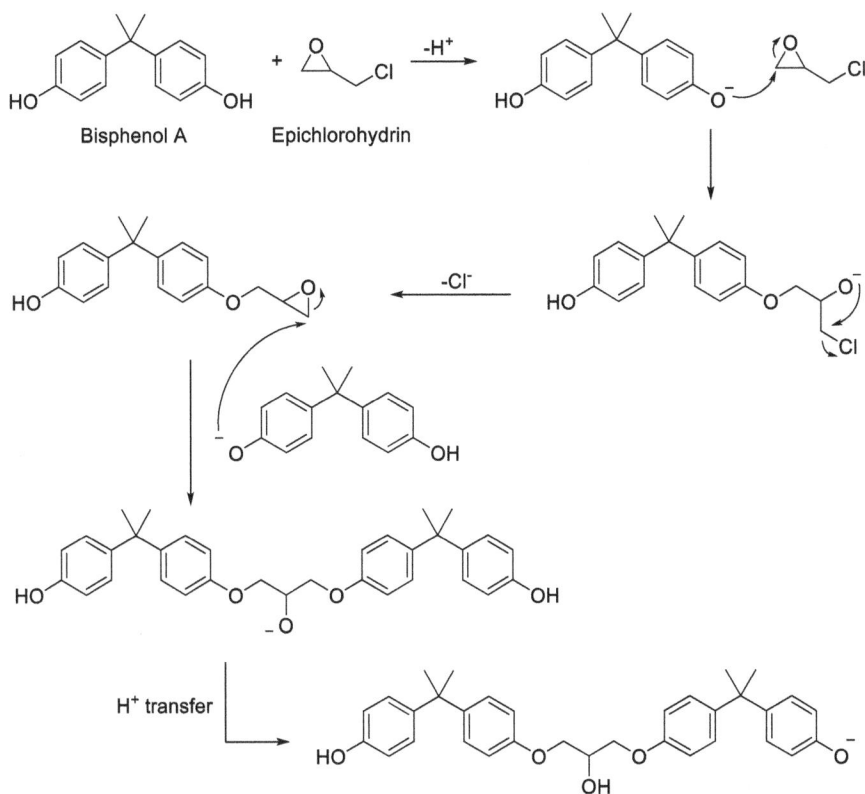

Scheme 9.5: Mechanism of epoxy resin formation of bisphenol A and epichlorohydrin.

aprotic solvents, for example sulfolane with KF as the fluorine source. Yields in the range of 80–90% can be achieved under reflux conditions in a reaction time of approximately 2 h. The use of potassium fluoride, especially in its spray-dried form, in combination with PTC enabled the main breakthrough in the *Halex* chemistry [12, 13].

The intermediate 2,6-difluorobenzonitrile, precursor of the insecticide difluben-zuron (which inhibits the production of chitin) (Scheme 9.6), and the herbicide precursor 2,4-difluoroaniline are synthesised by the *Halex* reaction on an industrial scale [11].

Alkylation reactions can also be accelerated by PTC. For example, the reaction of phenyl acetonitrile with alkyl halides and sodium hydroxide under phase transfer conditions using quaternary ammonium salts has been described in the 1960s (Scheme 9.7) [14].

Further examples applying this concept at room temperature have been demonstrated using tetrabutylammonium bromide (NBu$_4$Br) or methyltrioctylammonium chloride (MeOct$_3$NCl) as catalysts [14].

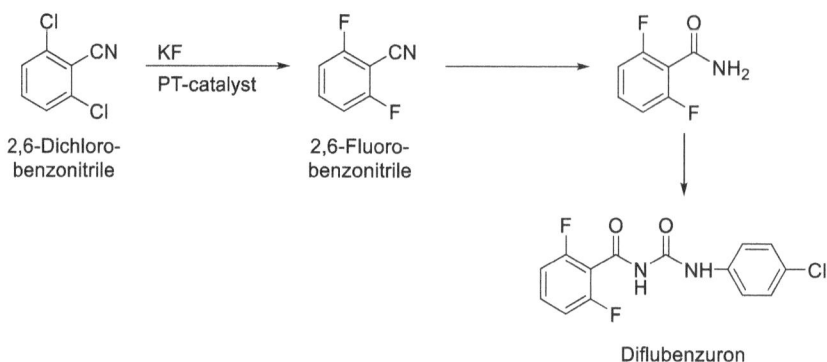

Scheme 9.6: Synthesis of diflubenzuron.

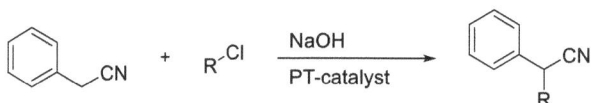

Scheme 9.7: PTC in alkylation reactions of nitriles.

The application of asymmetric PTC is well known for the synthesis of drug intermediates [8]. In particular, ammonium salts are applied in this variation. It is also known that some of the phase transfer limitations of multiphase catalysis can be reduced by the use of sonication. For example, the yield in the alkylation of imine ester derivatives in toluene with KOH (50% in water) at 0 °C (Scheme 9.8) was increased when ultrasound was applied [15].

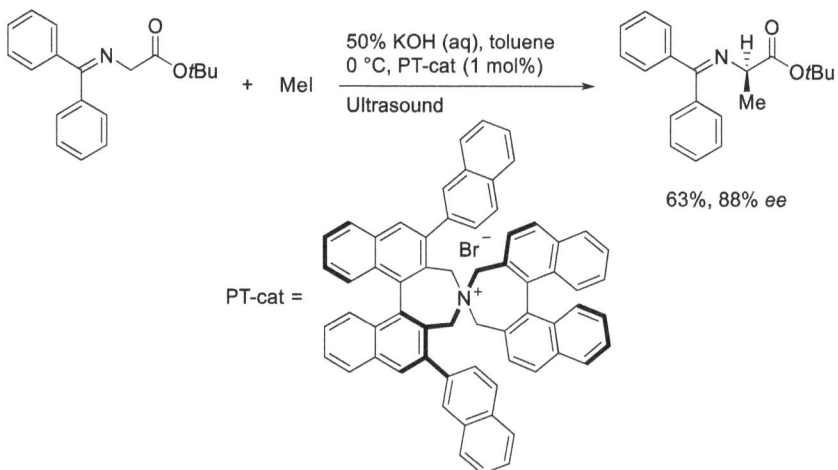

Scheme 9.8: Chiral PTC alkylation reaction of an imine ester derivative under ultrasound irradiation.

Glyoxylic imides were also alkylated using a chiral PT-cat [16]. The method allows the synthesis of α-H-α-amino acids based on the alkylation of imino acid esters or their corresponding amides. The alkylation of imines with various electrophiles results in >95% yield and >95% *ee* under chiral PTC conditions using a cinchona alkaloid deriva- tive Scheme 9.9).

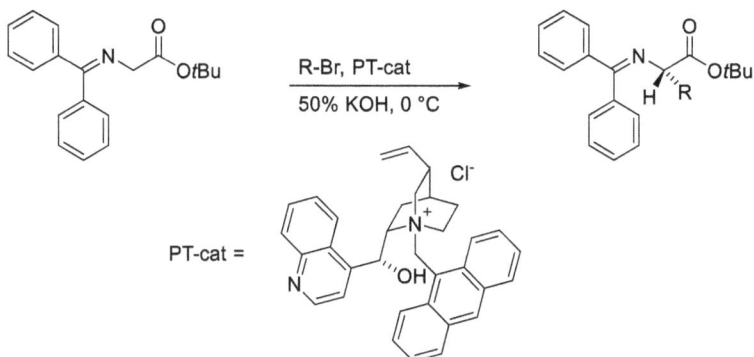

Scheme 9.9: Glyoxylic imines as glycine *Schiff*-base equivalents in PTC.

In the field of radiopharmaceuticals, [^{18}F]-fluoro-L-Dopa is applied in positron emission tomography (PET). This compound can be synthesised in an asymmetric PTC alkylation reaction from 2-[^{18}F]-fluoro-4,5-dimethoxybenzyl bromide. Caesium hydroxide is used as the base, and the stereochemistry is controlled by the chiral PT-cat (an excess is re- quired) (Scheme 9.10) [17]. The reaction is performed in toluene at 0 °C and *ee* values of >95% and a yield of around 90% can be achieved.

Alkylation reactions under PTC conditions are used for the asymmetric synthesis of intermediates of the anti-diabetes drug (−)-ragaglitazar. The reaction also uses CsOH as a base in a mixture of CH$_2$Cl$_2$-hexane (1:1) (Scheme 9.11) [18, 19]. A related cinchona alkaloid PT-cat is used, but it contains a different nitrogen substituent. The application of a solvent mixture CH$_2$Cl$_2$-hexane increases the *er* to 93:7, whereas in toluene and KOH (aqueous) an *er* of 90:10 was achieved.

In the synthesis of naproxen (see Chapter 8: Acid–base-catalysed reactions), PTC is applied in the asymmetric alkylation of an aliphatic or arylacid ester under basic conditions (KO^{t}Bu). A solvent mixture of toluene-CH$_2$Cl$_2$ at −50 °C is used and a yield of around 80% and 60% *ee* of the desired product is obtained (Scheme 9.12) [20].

The *Strecker* reaction applying acetone cyanohydrin under asymmetric PTC con- ditions is an efficient method to replace KCN as the cyanide source, via an imine inter- mediate (Scheme 9.13) [21]. The reaction can be carried out using a catalyst loading of 10 mol% in toluene at −20 °C resulting in 85–95% yield and 68–88% *ee* depending on the substrate structure.

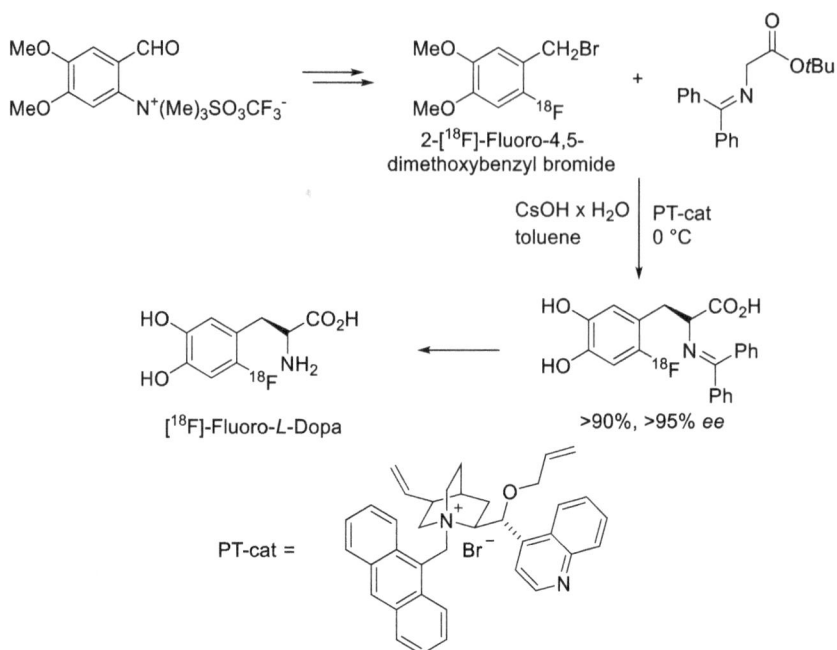

Scheme 9.10: PTC in the synthesis for radiopharmaceutical compound [^{18}F]-fluoro-L-Dopa.

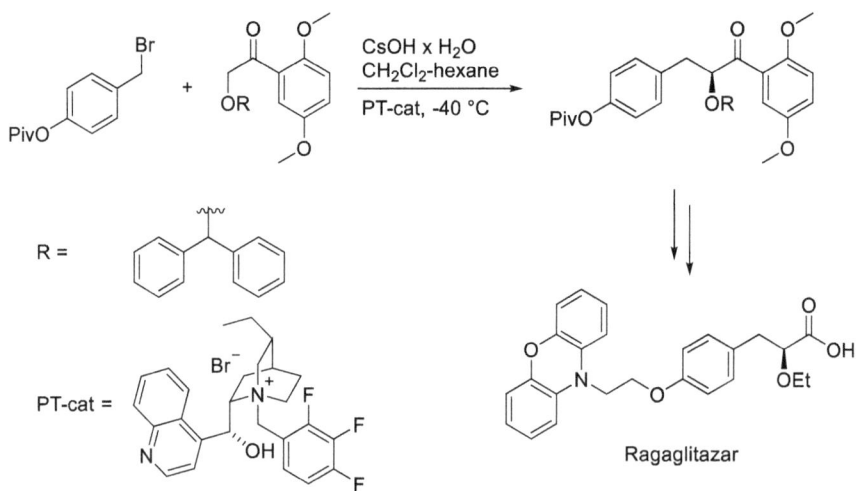

Scheme 9.11: Synthesis of an anti-diabetes intermediate.

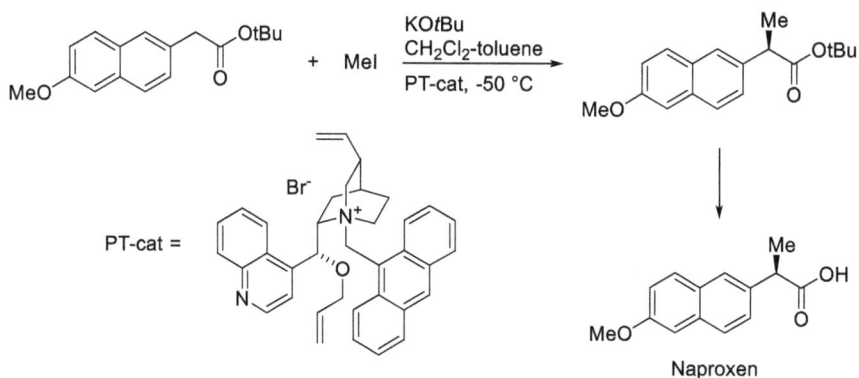

Scheme 9.12: Synthesis of naproxen under PTC conditions.

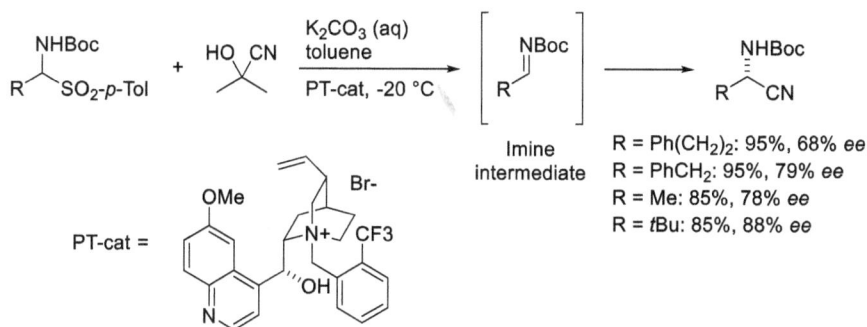

R = Ph(CH₂)₂: 95%, 68% ee
R = PhCH₂: 95%, 79% ee
R = Me: 85%, 78% ee
R = tBu: 85%, 88% ee

Scheme 9.13: Asymmetric *Strecker* reaction under PTC conditions.

In the pharmaceutical industry, methods for *O*- and *N*-alkylation are well established in the preparation of heterocycles such as azipine or 2 H-chrome synthesis [22, 23]. The application of PTC in large-scale alkylation of the carboxylic acid of benzylpenicillin was the first application in this field (Scheme 9.14) [24, 25].

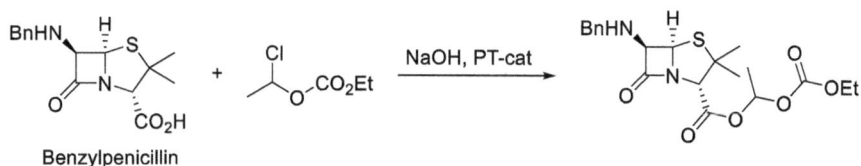

Scheme 9.14: Esterification of benzylpenicillin.

The use of *para*-substituted *N*-benzylcinchonium bromides as PT-cat allows the synthesis of chiral products, for example indacrinone with an *ee* value of 97% and a yield of 95%. The reaction is carried out in toluene with a 50% solution of NaOH in water at

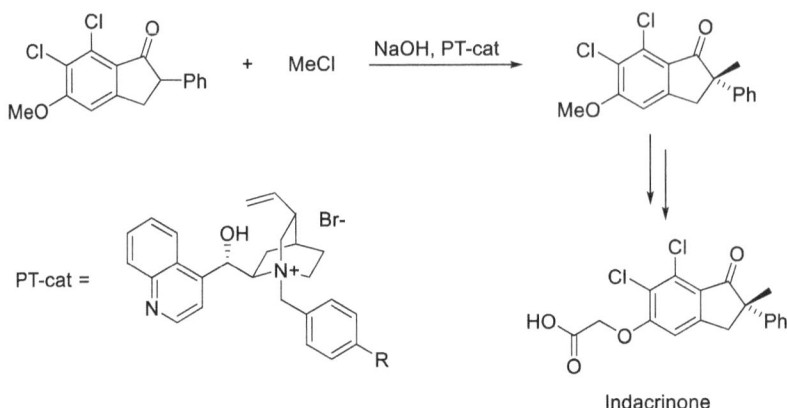

Scheme 9.15: (+)-Indacrinone synthesis applying PTC.

room temperature in a reaction time of 18 h. These compounds are applied as anti-hypertensive drugs (Scheme 9.15) [26–28].

For further examples of the application of PTC in pharmaceutical industry, see [29].

The production of pesticides, for example pyrethroids such as fenvalerate and cy-permethrin (Scheme 9.16), is carried out in the presence of PTC for C-alkylation and esterification. Several patent applications by Sumitomo, Shell and ICI show the importance of PTC [30, 31]. Cypermethrin is applied as insecticide against *Haematobia* spp. and *Musca autumnalis*; Fenvalerate is used as an insecticide. Both compounds are sold as a mixture of stereoisomers. The production of fenvalerate starts from *p*-chlorotoluene and *m*-phenxytoluene, which is oxidised to the corresponding aldehyde and treated with 2-isopropyl-(4-chlorophenyl)-acetic acid [32].

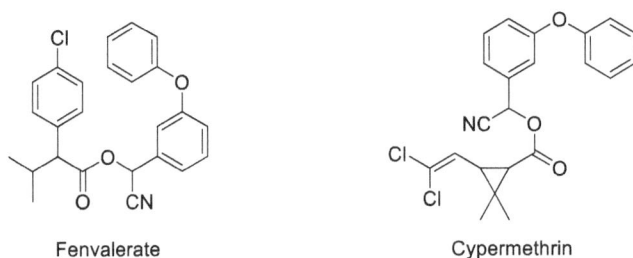

Fenvalerate

Cypermethrin

Scheme 9.16: Fenvalerate and cypermethrin.

The concept of two-phase catalysis in water-organic solvent mixtures is applied in reactions such as reductive dehalogenation of allyl and benzyl halides. Here, a PdL_2Cl_2 complex (L = o-$(KO_3SC_6H_4)PPh_2$), synthesised from $PdCl_2$ and diphenyl(2-potassium sulfonatobenzyl)phosphine in CH_3CN is used (Scheme 9.17). This complex is a PT-cat applied for the production of polyethers [33].

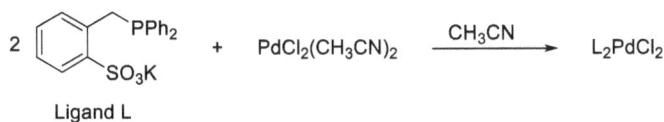

Scheme 9.17: Pd-complex for PTC.

The concept of PTC for allylic nucleophilic substitution was demonstrated by the use of a Pd(0)-catalyst, generated from Pd(dibenzylidene acetone)$_2$ or Pd(OAc)$_2$ and TPPTS (trisodium 3,3′,3″-phosphinetriyltribenzenesulfonate), in a water–nitrile mixture. These reactions result in >93% yield (Scheme 9.18) [34].

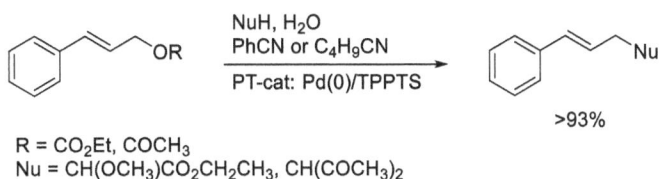

R = CO$_2$Et, COCH$_3$
Nu = CH(OCH$_3$)CO$_2$CH$_2$CH$_3$, CH(COCH$_3$)$_2$

Scheme 9.18: Allylic substation under PTC conditions.

The manufacture of chloroprene by dehydrohalogenation of 3,4-dichlorobut-1-ene is carried out under phase transfer conditions with NaOH (aqueous, 20%) and a quaternary ammonium salt (Scheme 9.19). The mixture of 1,4-dichlorobut-2-ene and 3,4-dichlorobut-1-ene obtained by butadiene chlorination is isomerised to the 3,4-dichlorobut-1-ene derivative and the dehydrohalogenation is performed. This process carried out in a cascade-type reactor set-up is the leading process for chloroprene production and it replaces older processes based on ethyne. The chlorination process is carried out in up to 25% conversion with a selectivity of 95%. The preferred catalyst is a quaternary ammonium chloride, for example bis(hydroxypropyl)benzylhexyl ammonium chloride. Chloroprene is produced in several 10,000 t/a and is the monomer for polychloroprene (neoprene). The by-product, 1-chlorobutadiene, is separated by distillation. The yield based on chloroprene is >99% [35, 36].

A similar approach is applied for the mono-dehydrohalogenation of 1,2-dihalo-1-phenylethane. Catalytic amounts of C$_{10}$–C$_{18}$ alkylbenzyldimethyl ammonium chloride (*Katamin AB*) are used. The corresponding olefin is obtained in 85% yield and the mono-halogenated by-product is observed in 5% yield (Scheme 9.20) [37].

Dehydrohalogenation is also used in the industrial production of 3,5-dichloro-α-methylstyrene, an intermediate for agricultural products. The dehydrobromination of 1-(2-bromopropan-2-yl)-3,5-dichlorobenzene (Scheme 9.21) is carried out at 110 °C in the presence of tetrabutylammonium bromide as PT-cat (s/c around 100). Using an excess of NaOH (45% aq.), a 98% yield can be obtained after a reaction time of 8 h [38].

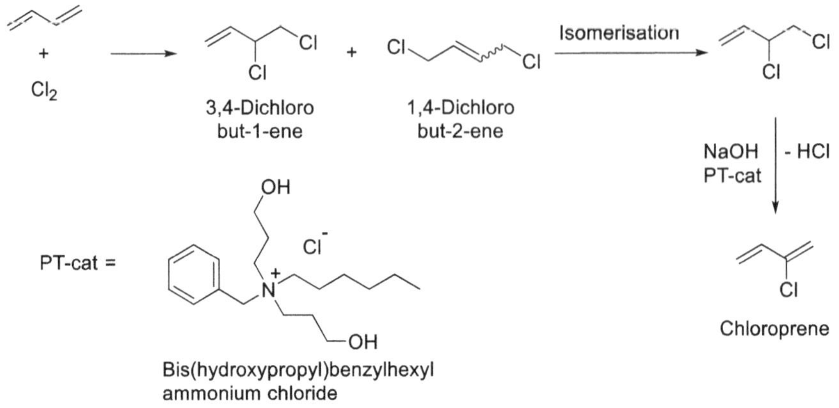

Scheme 9.19: PTC in the production of chloroprene.

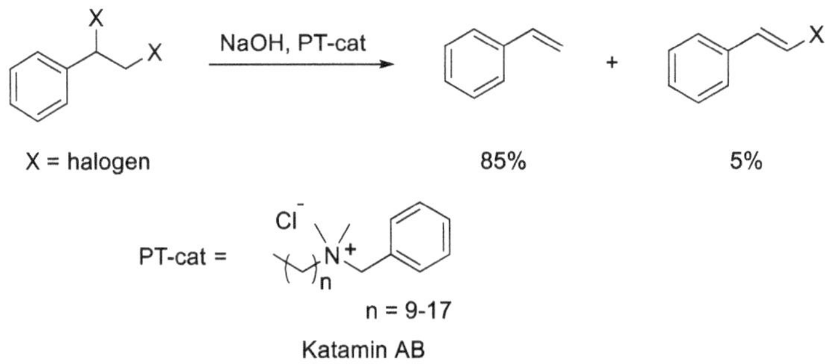

Scheme 9.20: Selective dehydrohalogenation of 1,2-dihalo-1-phenylethane.

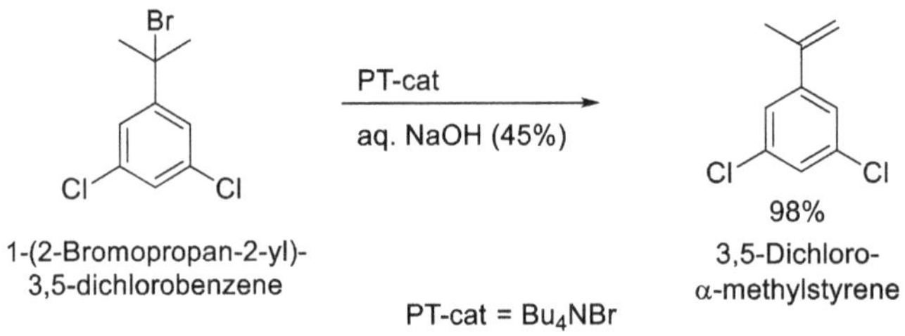

Scheme 9.21: Dehydrobromination.

The advantages of two-phase solvent systems in hydroformylation reactions using Rh-complexes are described in more detail in Chapter 6 of C–C-bond formation. The reaction of C_6–C_{20} olefins with CO/H_2 in the presence of Rh-TPPTS at a temperature of around 125 °C and 25 bar in the presence of a PT-cat, such as $Me(CH_2)_{13}NMe_3SO_4Me$, results in a linear to branched ratio (n to i) of 98:2 of the corresponding aldehyde, and it is conducted at 41% conversion (Scheme 9.22) [39].

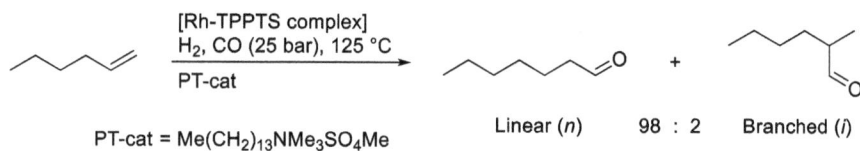

PT-cat = Me(CH$_2$)$_{13}$NMe$_3$SO$_4$Me

Linear (*n*)   98 : 2   Branched (*i*)

Scheme 9.22: Hydroformylation under phase transfer conditions.

For some large-scale applications, oxidising agents such as hypochlorite (10% in water) are the oxidising reagents of choice. The oxidation of hydroquinone and catechols are carried out with hypochlorite under PTC conditions. Yields >90% are obtained for the hydroquinone and catechol oxidation in the presence of NBu$_4$HSO$_4$ as PT-cat (Scheme 9.23) [40].

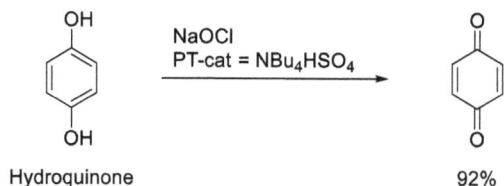

Hydroquinone   92%

Scheme 9.23: Hydroquinone oxidation under PTC.

The oxidation of alcohols with NaOCl in the presence of NBu$_4$HSO$_4$ as PT-cat results in the formation of aldehydes and ketones, whereas amines form the corresponding *N*-chloroimines, which can be hydrolysed later to form ketones. Yields up to 76% for aldehydes and 98% for ketones can be achieved (Scheme 9.24) [41].

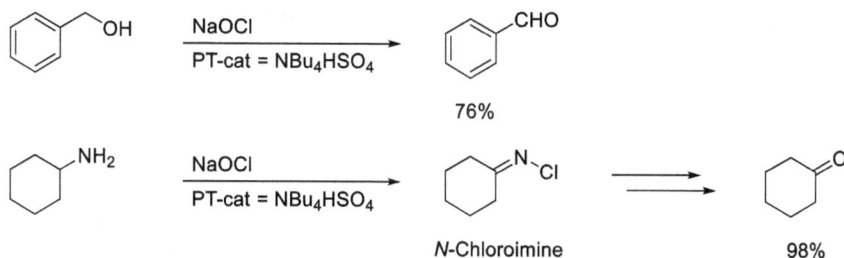

76%

N-Chloroimine   98%

Scheme 9.24: Hypochlorite (bleach) oxidation under PTC condition.

In a new approach to (+)-biotin (see Chapters 2 and 3: Heterogeneous and Homogeneous hydrogenations), the synthesis of ethyl 5-iodopentanoate was of interest [42, 43]. Ethyl 5-iodopentanoate was required to introduce the side chain of (+)-biotin by a Finkelstein reaction in the form of a zinc reagent. The iodo-compound is sensitive to handle and traces of impurities, such as solvents, result in a difficult scale-up of the reaction. It was found that in inert solvents in the presence of a PT-catalyst, for example tetra-*n*-butylammonium bromide in water, ethyl 5-bromopentanoate reacts with sodium iodide to give ethyl 5-iodopentanoate in excellent yield. The reaction was performed in toluene at room temperature with >99% conversion and 88% yield (Scheme 9.25). The reaction of ethyl 5-iodopentanoate with thiolactone, resulted in a 90% yield of the (+)-biotin intermediate. The reaction was carried out in THF/DMF at 40 °C and has the potential for further optimisation.

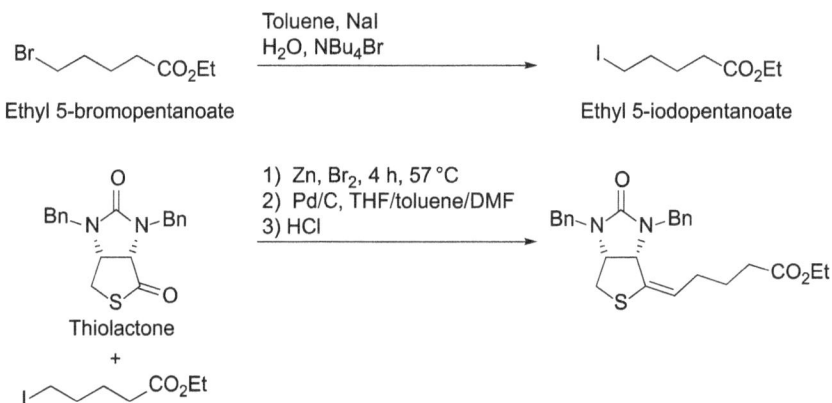

Scheme 9.25: Finkelstein reaction to ethyl 5-iodopentanoate.

In conclusion, PTC can be successfully applied in a broad range of applications, especially in asymmetric synthesis. Mass-transport limitations can be overcome by the use of technologies such as ultrasound. During the past decades, the application of PTC has been established as a core technology in catalysis and application in the production of organic compounds. The main advantages are the use of mild reaction conditions, and cheap and environmentally friendly reagents and solvents.

References

[1]   Starks C M. Phase-transfer catalysis. I. Heterogeneous reactions involving anion transfer by quaternary ammonium and phosphonium salts. J Am Chem Soc 1971, 93, 195–199.
[2]   Metzger J O. Phase-transfer catalysis. A general green methodology in organic synthesis. Pure Appl Chem 2000, 72, 1399–1403.
[3]   Dehmlow E V, Dehmlow S S. Phase Transfer Catalysis, 3rd ed. Weinheim/ VCH Publishers, New York, VCH Verlagsgesellschaft, 1993, 499.
[4]   Mąkosza M, Fedoryński M. Phase transfer catalysis. Catal Rev – Sci Eng 2003, 45, 321–367.
[5]   Tundo P, Venturello P. Synthetic resections by gas-liquid phase-transfer catalysis. Polym Sci Technol 1984, 24, 275–290.
[6]   Metzger J O. Free organic syntheses. Angew Chem Int Ed 1998, 21, 2975–2978.

[7]   Halpern M. Phase-Transfer Catalysis. In: Elvers, ed. Ullmann's Encyclopedia of Industrial Chemistry, Online ed. Weinheim, Germany, Wiley-VCH Verlag GmbH & Co, 2002.

[8]   Hashimoto T, Maruoka K. Recent development and application of chiral phase-transfer catalysts. Chem Rev 2007, 107, 5656–5682.

[9]   Makosza M, Fedorynski M. Catalysis in two-phase systems: Phase transfer and related phenomena. Adv Catal 1987, 35, 375–422.

[10] Kelland K. Experts demand European action on plastics chemical. Reuters. 22 June 2010. https://www.reuters.com/article/us-chemical-bpa-health/experts-demand-european-action-on-plastics-chemical-idUSTRE65L6JN20100622?loomia_ow=t0:s0:a49:g43:r3:c0.084942:b35124310:z0

[11] Langlois B, Gilbert L, Forat G. Industrial chemistry library, fluorination of aromatic compounds by halogen exchange with fluoride anions ("halex" reaction). Ind Chem Lib 1996, 8, 244–292.

[12] Pleschke A, Marhold A, Schneider M, Kolomeitsev A, Röschenthaler G-V. Halex reactions of aromatic compounds catalysed by 2-azaallenium, carbophosphazenium, aminophosphonium, and diphosphazenium salts: A comparative study. J Fluorine Chem 2004, 125, 1031–1038.

[13] Adams D J, Clark J H. Nucleophilic routes to selectively fluorinated aromatic. Chem Soc Rev 1999, 28, 225–231.

[14] Herriott A W, Picker D. Phase-transfer catalysis. Evaluation of catalysis. J Am Chem Soc 1975, 97, 2345–2349.

[15] Ooi T, Tayama E, Doda K, Takeuchi M, Maruoka K. Synthesis of Kurasoin B using phase-transfer-catalyzed acylimidazole alkylation. Synlett 2000.

[16] Hyett D J, Didone M, Milcent T J A, Broxtermann Q B, Kaptein B. A new method for the preparation of functionalized unnatural α-H-α-amino acid derivatives. Tetrahedron Lett 2006, 47, 7771–7774.

[17] Lamaire C, Gillet S, Guillouet S, Plenevaux A, Aerts J, Luxen A. Highly Enantioselective synthesis of no-carrier-added 6-[^18F]Fluoro-L-dopa by chiral phase-transfer alkylation. Eur J Org Chem 2004, 2899–2904.

[18] Andrus M B, Hicken E J, Stephens J C. Phase-transfer-catalyzed asymmetric glycolate alkylation. Org Lett 2004, 6, 2289–2292.

[19] Ooi T, Miki T, Maruoka K. Asymmetric synthesis of funktionalized aza-cyclic amino acids with quaternary stereocenters by a phase-transfer-catalyzed alkylation strategy. Org Lett 2005, 7, 191–193.

[20] Kumar S, Ramachandran U. A simple catalytic route to naproxen. Tetrahedron Asymmetry 2005, 16, 647–649.

[21] Herrera R P, Sgarzani V, Bernardi L, Fini F, Petersen D, Ricci A. Phase transfer catalyzed enantioselective strecker reactions of α-amido sulfones with cyanohydrins. J Org Chem 2006, 71, 9869–9872.

[22] Gozlan I, Halpern M, Rabinovitz M, Avnir D, Ladkani D. Phase transfer catalysis in N-alkylations of pharmaceutical intermediates 5-H-dibenz[b,f]azepinw and 5H-10,11-dihydrobenz[b,f]azepine. J Heterocycl Chem 1982, 19, 1569–1571.

[23] Reisch J, Dharmartne H, Ranjith W. Natural product chemistry. 97. A convenient synthesis of the 2-dimethyl-2H-chromen system. Z Naturforsch B 1985, 40B, 636–638.

[24] Lindblom L, Elander M. Phase-transfer catalysis in the production of pharmaceuticals. Pharm Tech 1980, 4, 59–69.

[25] D'Incan E, Viout P, Gallo R. Phase transfer catalysis and selectivity in organic synthesis. Israel J Chem 1985, 26, 277–281.

[26] Dolling U-H, Davis P, Grabowski E J. Efficient catalytic asymmetric alkylations. 1. Enantioselective synthesis of (+)-indacrinone via chiral phase-transfer catalysis. J Am Chem Soc 1984, 106, 446–447.

[27] Dolling U H. Use of an Achiral Co-catalyst Promoter in a Chiral Phase Transfer Alkylation Process for Preparation of a Substituted Fluorenyloxyacetic Acid. US 4605761, Merck, 1985.

[28]   Vlasses P H, Rotmensch H, Swanson P N, Irvin J D, Johnsen C, Ferguson R K. Indacrinone: Natriuretic and uricosuric effects of various ratios of its enantiomers in healthy men. Pharmacotherapy 1984, 4, 272–277.

[29]   Freedman H H. Industrial applications of phase transfer catalysis (PTC): Past present, future. Pure Appl Chem 1986, 58, 857–868.

[30]   Reinink A, Sheldon R A. Preparation of Esters. US 4175094, Shell Oil Cooperation, 1979.

[31]   Wood D A. Preparation of Cyanobenzyl Cyclopropane Carboxylates. US 4409150, Shell International Research, 1983.

[32]   Franck H-G Herstellung und Verwendung von Benzol-Derivaten and Herstellung und Verwendung von Toluol-Derivaten. In: Franck H-G, Stadelhofer J W, ed. Industrielle Aromatenchemie, Berlin, Heidelberg, Germany, Springer-Verlag, 1987, 175–176, and 269–270.

[33]   Paetzold E, Oehme G, Costisella B. Synthesis of diphenyl(2-potassium sulfonatobenzyl) phosphine: A new water soluble ligand for catalytic active transition metal complexes. Z Chem 1989, 29, 447–448.

[34]   Safi M, Sinou D. Palladium(0)-catalyzed substitution of allylic substrates in a two-phase aqueous-organic medium. Tetrahedron Lett 1991, 32, 2025–2028.

[35]   Maurin L J. Dehydrohalogenation Process. US 4418232, E.I. Du Pont de Nemours, 1983.

[36]   Rossberg M, Lendle W, Pfleiderer G, Tögel A, Dreher E-L, Langer E, Rassaerts H, Kleinschmidt P, Strack H, Cook R, Beck U, Lipper K-A, Torkelson T R, Löser E, Beutel K, Mann T. Chlorinated Hydrocarbons. In: Elvers, ed. Ullmann's Encyclopedia of Industrial Chemistry, Online ed. Weinheim, Germany, Wiley-VCH Verlag GmbH & Co, 2006, 233.

[37]   Kurginyan K A. Elimination under phase-transfer catalysis conditions and industrial synthesis of unsaturated compounds. Zhurnal Vsesoyuznogo Khimicheskogo Obshchestva Imeni D I Mendeleeva 1986, 31, 164–169.

[38]   Henneke K-W, Diehl H, Wedemeyer K. Preparation of 3,5-dichloro-α-methylstyrene. US 4594467, Bayer AG, 1986.

[39]   Bahrmann H, Cornils B, Konkol W, Lipps W. Verfahren Zur Herstellung von Aldehyden. EP 0157316, Ruhrchemie AG, 1985.

[40]   Ishii F, Kishi K. Oxidation of hydroquinone and catechols with aqueous sodium hypochlorite under phase-transfer catalysis. Synth 1980, 706–708.

[41]   Lee G E, Freedman H H. Phase transfer catalyzed oxidations of alcohols and amines by aqueous hypochlorite. Tetrahedron Lett 1976, 20, 1641–1644.

[42]   Seki M, Takahashi Y. Practical synthesis of (+)-biotin key intermediate by calcium borohydride reduction and temperature-dependent purity upgrade during crystallization. Org Process Res Dev 2021, 25, 1950–1955.

[43]   Takahashi Y, Seki M. Finkelstein reaction in -non-polar organic solvents: A streamlined synthesis of organic iodides. Org Process Res Dev 2021, 25, 1974–1978.

10 Biocatalysis

10.1 Introduction to biocatalysis

Biocatalytic processes and procedures have gone beyond being an object of mere curiosity and are nowadays applied in many industrial processes. Enzymes are the catalysts of nature that have evolved to catalyse reactions in living organisms. Like all catalysts (e.g. metal catalysts), they lower the activation energy of the overall reaction, alter the reaction rate and make feasible transformations that do not normally proceed. However, like all catalysts, they never change the equilibrium of a reaction, that is, enzymes work reversibly. Enzymes are usually very selective due to the specific shape of the active side where the catalysis takes place and exhibit additional positive features desired in industrial production (Table 10.1). In addition to the catalysis of chemical reactions, enzymes also play a crucial role in cell movement, signalling, transportation of metabolites or in ion pumps.

Table 10.1: Some advantages and disadvantages of biocatalysis [1].

Advantages	Disadvantages
High substrate-, regio- and chemo-selectivity.	Prone to substrate or product inhibition.
Usually mild reaction conditions	
(pH 5–8 and temperature 20–40 °C).	Harsher reaction conditions such as higher temperature, extreme pH values or organic solvents can lead to deactivation.
Can be completely degraded in the environment.	Often require additional co-factors for correct functioning.
Reuse is possible.	Enzymes in nature exist only in one enantiomeric form (and are often substrate specific).
	Usually water is used as solvent or traces of water are required.

Most enzymes are proteins which are built up from amino acids. Exceptions are, for example catalytically active DNA or ribosomes. Proteins are polar macromolecules in which the different amino acids are connected via amide bonds (Figure 10.1).

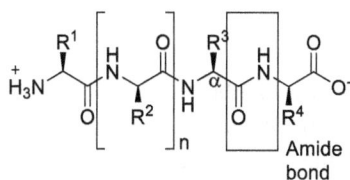

Figure 10.1: Primary protein structure built up from different amino acids.

https://doi.org/10.1515/9783111102672-010

The individual properties of an enzyme are closely connected to its amino acid sequence. In total, twenty different amino acids can be used in protein biosynthesis. The number of amino acids in typical proteins increases going from Archaea to Bacteria to Eukaryotes due to the higher number of protein domains [2]. The median protein length of yeast is 379 amino acids and of *Homo sapiens*, 375 [3]. A protein consisting of 200 amino acids therefore has 20^{200} possible combinations, which is significantly higher than the number of protons in the observable universe [4]! An important factor in the functionality of a protein is how these macromolecules orient themselves in space. The amide bond has a partial double-bond character with restricted rotation, but free rotation occurs around the bonds connected to the α-carbon atom. The folded structure of an enzyme is stabilised predominately by numerous non-covalent interactions such as hydrogen bonds, ionic bonds and *Van der Waals* interactions. In the folded structure, the hydrophobic parts of the enzymes are preferably oriented to the inner side of the protein, minimizing the interaction with the aqueous media. Therefore, water molecules are frequently located at defined positions, if they are present in the hydrophobic pockets where chemical reactions take place.

The specific interaction of an enzyme and a substrate can be explained by the *Fischer* lock-and-key model postulated first, in 1894, by Emil Fischer (Scheme 10.1) [5, 6]. This is a simplified model that assumes that the substrate simply fits into the active side of the enzyme, and an enzyme-substrate complex is formed. Only the right size and shape of a substrate (key) can fit into the enzyme (lock).

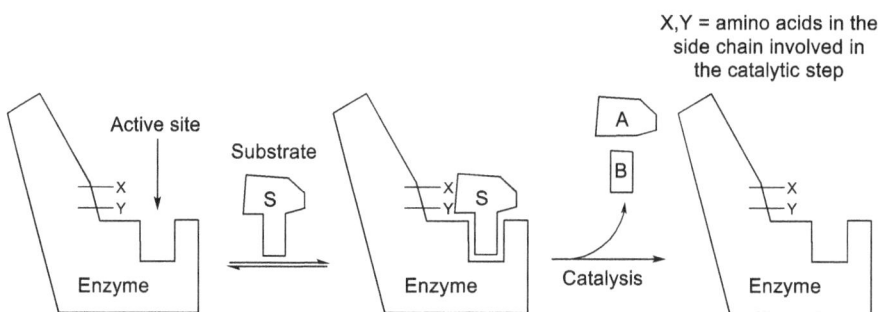

Scheme 10.1: Lock-and-key model postulated by Emil Fischer in 1894.

However, considering the enzyme only as a rigid assembly of amino acids is not correct. Enzymes have a certain degree of flexibility in their structure, which is adopted by the induced-fit model. This postulated model is an extension of the lock-and-key model and was reported by Koshland in 1958 (Scheme 10.2) [7]. This extension explains why compounds can bind to the enzyme but do not react, for example because they are too small to induce a conformational change in the enzyme or because the enzyme is too distorted and is not capable of introducing the right alignment of the reacting groups.

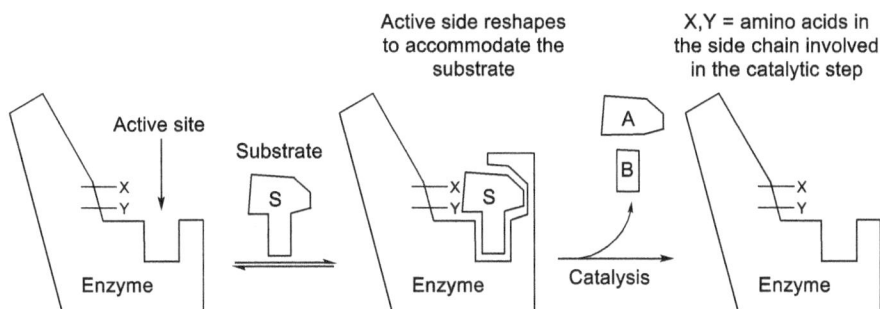

Scheme 10.2: Induced-fit model postulated by Koshland in 1958 to extend the lock-and-key model.

An example of how enzymes can catalyse reactions is illustrated by the serine protease catalytic hydrolytic cleavage of a peptide bond (Scheme 10.3). In the first step, a proton is transferred to histidine from the serine residue. The activated serine residue can undergo a nucleophilic attack on the carbonyl-carbon of a peptide bond. The negative charge in the transition state is stabilised by the oxyanion-hole, a pocket in the active site that can stabilize a negative charge. The hydrogen atom which was previously transferred protonates the nitrogen of the peptide bond, which triggers the cleavage of the C–N bond forming a primary amine group. This amine can diffuse out of the protein and is replaced by a water molecule. The water molecule is activated by interaction with the histidine residue and can attack the carbonyl-carbon attached to the serine residue. This results in the generation of the carboxylic acid, and the catalytic cycle is closed [8].

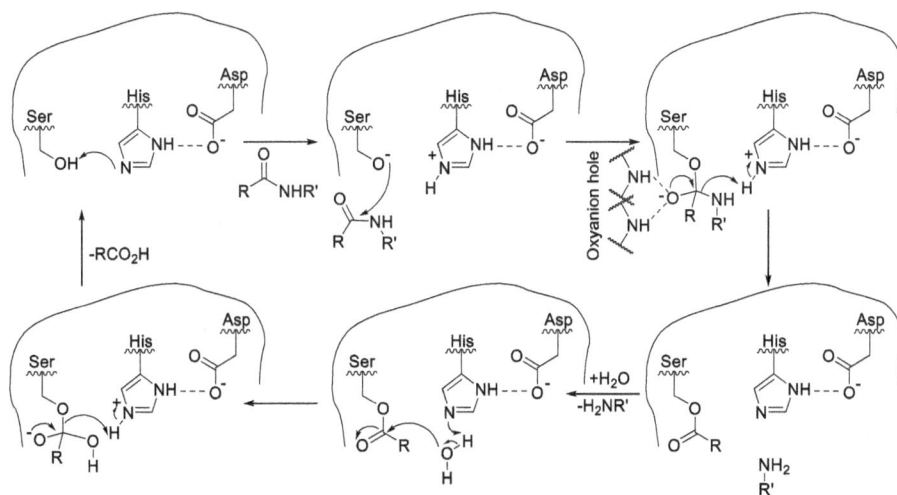

Scheme 10.3: Mechanism for the hydrolytic cleavage of a peptide bond by a serine protease [8].

The activity of such enzymatic reactions is usually dependent on the reaction temperature and the pH value of the reaction media (Figure 10.2). In the depicted graph, almost no reaction takes place below 20 °C, whereas at ~48 °C, denaturation of the enzyme is observed. The activity of an enzyme is maximum at a specific temperature that is unique to the enzyme. In this case, the optimal temperature and pH values are around 36 °C and at a pH value of ~8.4. At pH values <6 or >10, often, denaturation of the enzyme occurs. Denaturation is when enzymes lose their structure/orientation present in their native state while they are in space. This can be caused by heat, inorganic salts, radiation, solvent, or when a residue in the active side obtains a different charge due to acids or bases; in all cases, this results in the deactivation of the enzyme. Therefore, in most enzymatic reactions, control of the reaction temperature and the pH value is essential for high-yielding reactions.

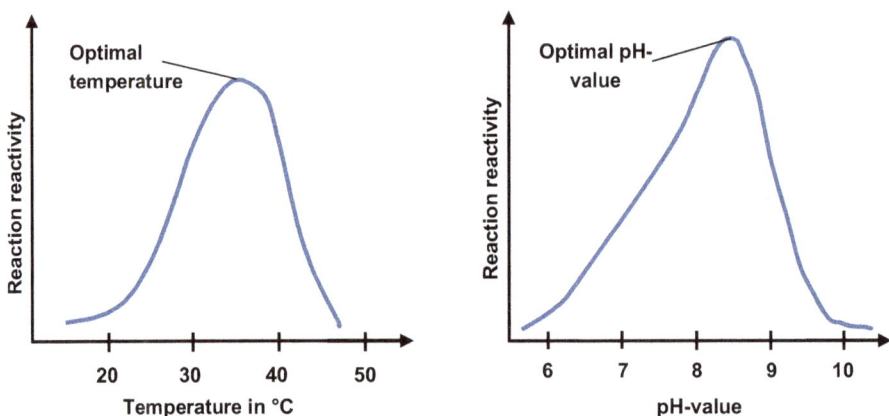

Figure 10.2: Dependency of enzyme activity on temperature and pH value.

Biotransformations or biocatalysis can be performed as a whole-cell process or using isolated enzymes. Depending on the specific application, the use of isolated enzymes or a whole-cell fermentation might be preferred (Table 10.2). Transformations with whole cells are usually advantageous if the employed enzyme is not stable enough on its own, or a change in the oxidation state of the substrate is involved. In the latter case, co-factor recycling is necessary to achieve a cost-competitive process required by the fine chemical industry.

As previously mentioned, the stoichiometric use of expensive co-factors such as $NAD^+/NADH$ and $NADP^+/NADPH$ (Figure 10.3), are not possible on industrial scale. Thus, methods for their regeneration are essential [10]; various strategies have been developed (Figure 10.4).

In the past, electrochemical methods were applied in industrial processes but were rather limited in terms of inadequate mediator and electrode combinations. The main problems of direct electrochemical methods are that they are limited to sub-

Table 10.2: Comparison of transformations using isolated enzymes and whole cells [9].

	Advantages	Disadvantages
Isolated enzymes	Variable catalyst concentration	Limited stability
	Usually clean reactions	Co-factor regeneration required
	Simple product recovery	
	No transport limitations	
Whole cells	Multi-step conversion possible	Side reactions frequently occur (due to other enzymes present)
	Co-factor recycling by cellular machines	Transport limitations into/out of the cell

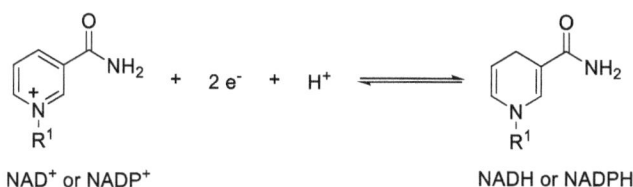

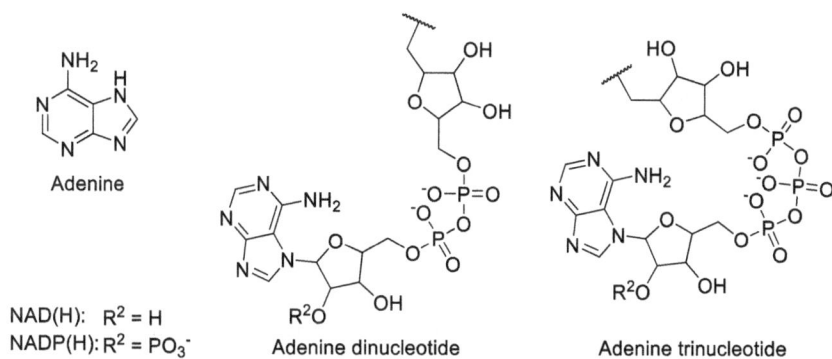

Figure 10.3: Co-factors NAD(H) and NADP(H).

strates or products that are stable under the applied conditions, especially the high oxidation potential. With indirect electrochemical reduction methods, an organic or metal-organic molecule acts as a mediator, thus allowing milder conditions. The chemical or photochemical co-factor regeneration is also performed in the presence of mediator molecules. Photochemical methods have been developed but have not been applied on industrial scale. For enzymatic co-factor regeneration methods, a second enzyme is applied, combined with a cheap organic compound such as isopropanol, ethanol or glucose.

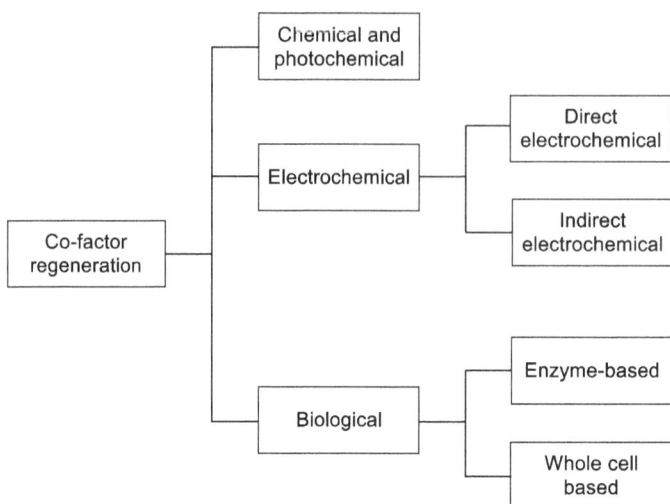

Figure 10.4: Various strategies to achieve co-factor regeneration.

From an industrial point of view, the recovery of the catalyst and its reuse is a major topic. Therefore, if an isolated enzyme is used, immobilisation of the enzyme is a preferred method. Additionally, the advantages of immobilisation are that it can be used in fixed-bed or fluidised-bed reactor systems that permit continuous processing, are often associated with an increased enzyme stability and allow higher temperatures to be used. The disadvantages of immobilised systems are not only mass transport limitations, but also the loss of activity that cannot be neglected. Various immobilisation techniques such as cross-linking, covalent adsorption on a carrier or forming of alginate beads are described in the literature. An example of lipase immobilisation with an outstandingly high activity, especially for *Candida antartica lipase B* and *Rhizomucor miehei*, is the use of functionalised methacrylic resins [11]. If a whole-cell system is applied, the cells can be separated by centrifugation or ultrafiltration [12, 13].

With the constantly increasing number of known enzymes, it is sometimes hard to keep the whole overview. The NC-IUBMB (Nomenclature Committee of the International Union of Biochemistry and Molecular Biology) classifies enzyme-catalysed reactions (and therefore enzymes) into seven main classes sorted by the EC number [14], six of which are relevant for industrial processes. This number contains 4 digits (EC A.B.C.D) which are separated by a dot and progressively refine the classification further. Therefore, non-identical enzymes could receive the same EC number, and this number identifies a group of enzymes that catalyse the same reaction. Preliminary EC numbers contain an "n" as part of the last digit.

A: stands for the main type of reaction of an enzyme
B: indicates the substrate type or the type of transformed molecule

C: gives the nature of the substrate or co-substrate specificity

D: represents the individual enzyme number

For example, an enzyme with the number EC 1.1.1.1 is a dehydrogenase, that catalyses the oxidation of an alcohol to an aldehyde.

The main classes are oxidoreductases, transferases, hydrolases, lyases, isomerases and ligases. There is a seventh class, translocases, which catalyse the movement of ions and molecules across or within membranes, but these are not relevant for the synthesis of chemicals (Table 10.3):

Table 10.3: Overview on the enzyme classes.

Class number	Name	Function	Reaction type (example)
EC 1	Oxidoreductase	Redox reactions	$A + O \rightarrow AO$
EC 2	Transferase	Intermolecular transfer of functional groups	$AB + C \rightarrow AC + B$
EC 3	Hydrolase	Hydrolysis	$AB + H_2O \rightarrow AOH + BH$
EC 4	Lyase	Double-bond formation or adding to a double bond	$RCOCO_2H \rightarrow RCOH + CO_2$
EC 5	Isomerase	Isomerisation	$ABC \rightarrow ACB$
EC 6	Ligase/synthase	Connection of substrates	$A + B \rightarrow AB$
EC 7	Translocases	Movement of ions/molecules	–

– Oxidoreductases catalyse oxidation/reduction reactions and act in hydrogen transfer reactions or shuffling of electrons between molecules. Dehydrogenases, oxidases, oxygenases and peroxidases are examples of this class (Scheme 10.4).

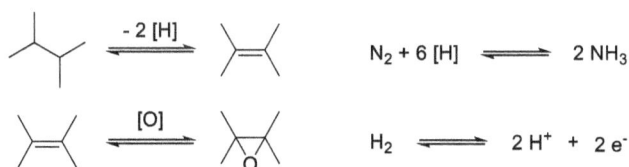

Scheme 10.4: Oxidoreductase-type reactions.

– The transfer of amino-, acetyl-, phosphoryl-, glycosyl-groups and other similar groups is catalysed by transferases (Scheme 10.5).

R–X + Y ⟶ R–Y + X

Scheme 10.5: Simplified scheme of a transferase-catalysed reaction.

– Hydrolases catalyse hydrolytic bond cleavage. From the technical application view, the most important are acylases, amylases, esterases, lipases and proteases (Scheme 10.6).

Scheme 10.6: Hydrolase-catalysed reaction types.

– Non-hydrolytic bond cleavage by elimination reactions or adding groups to double bonds are catalysed by lyases (Scheme 10.7). Examples are aldolases, aspartases, decarboxylases and dehydratases.

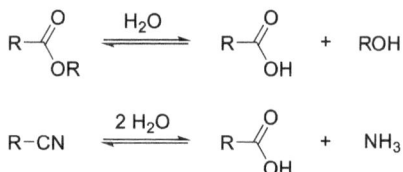

Scheme 10.7: Examples of lyase-performed reactions.

– In a catalytic isomerisation or functional group transfer reactions performed within one molecule, isomerases are the responsible enzymes (Scheme 10.8).

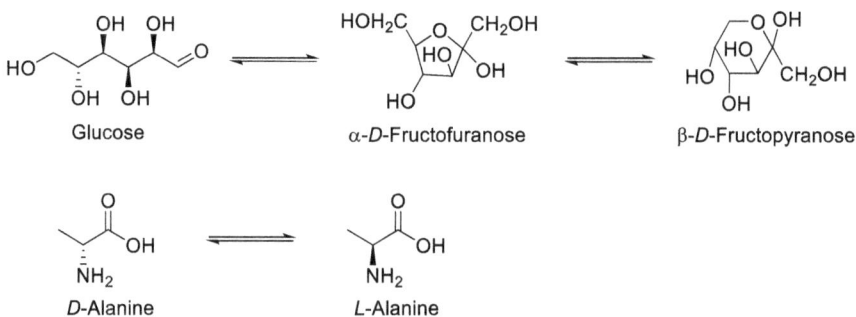

Glucose   α-D-Fructofuranose   β-D-Fructopyranose

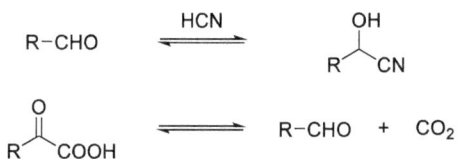

D-Alanine   L-Alanine

Scheme 10.8: Examples of isomerisation reactions.

– Ligases catalyse the condensation of two molecules via a covalent bond (Scheme 10.9). The required energy is obtained via cleavage of one or two of phosphate residues for example adenosine-5'triphosphate (ATP).

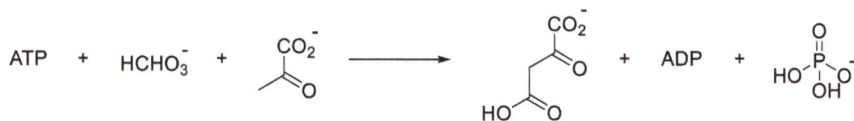

$$ATP \quad + \quad HCHO_3^- \quad + \quad \text{[structure]} \quad \longrightarrow \quad \text{[structure]} \quad + \quad ADP \quad + \quad \text{[phosphate structure]}$$

Scheme 10.9: Example of a ligase-catalysed reaction.

The main applications of biocatalytical processes are the production of chiral molecules, the application of hydrolases and redox biocatalysis (Figure 10.5). Other enzyme classes such as isomerases or ligases are used only rarely.

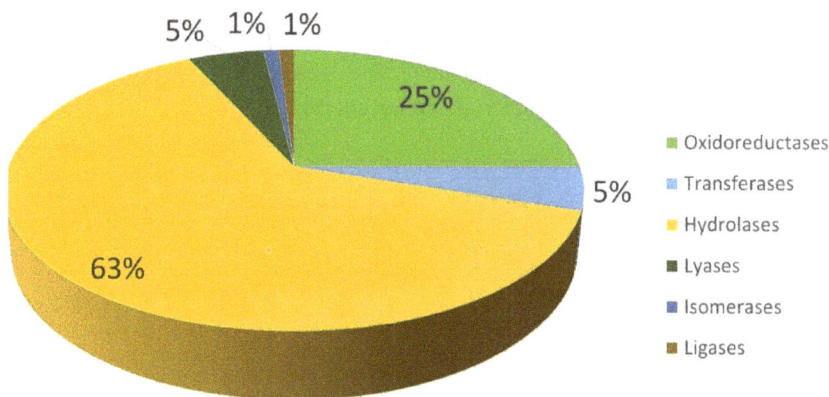

Figure 10.5: Relative use of the different enzyme classes in industry [15].

Biotechnological processes can be classified into different areas, each represented by a colour code [16]:
– Red biotechnology: medicine and human health, for example vaccines and antibiotics.
– Green biotechnology: agricultural, environmental biotechnology, for example biofuels, plant cell cultures or transgene plants.
– Yellow biotechnology: nutrition science and food technology, for example wine or beer making.
– Blue biotechnology: marine organisms and for microbiological processes.
– White biotechnology: application of biocatalysts in industrial processes, for example biodegradable polymers. This field is considered to be the biggest branch of biotechnology.
– Dark biotechnology: bioterrorism, biowarfare and so on.
– Brown biotechnology: desert zones and dry regions.

- Grey biotechnology: environmental applications.
- Gold biotechnology: bioinformatics, nanobiotechnology.
- Violet biotechnology: deals with publications, patents, law, ethical and philosophic issues.

From an industrial point of view, the development of a biocatalytic process must fulfil various criteria and must progress through several stages to be an economically attractive process (Figure 10.6). Compared to the pharmaceutical industry, where the time-to-market criteria are dominant, in the fine chemical industry, productivity is more important, requiring a productivity of 15–20 gl/h and a product concentration of around 120 g/l to be economically attractive [17]. To establish a biocatalytic process on an industrial scale, several development tools and cycles of development are necessary. First, a suitable biocatalyst needs to be selected (screening phase, cell/enzyme), followed by the determination of the enzyme kinetics and optimisation of the reaction conditions. The next step of development is engineering of the enzyme – or cell engineering (if needed), followed by process engineering. In the application phase, immobilisation, enzyme stability and co-factor regeneration are investigated. The development of the downstream processing is the final step before industrial application [18].

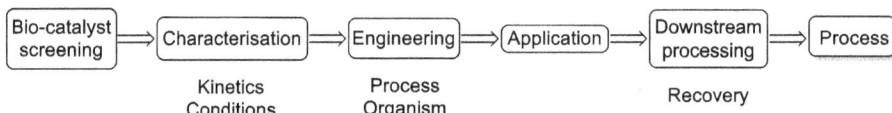

Figure 10.6: Development of a (bio)catalytic process.

In the development of biocatalytic processes, a breakthrough was achieved by implementing liquid–liquid two-phase catalysis. In the water-phase, the biocatalyst such as growing cells is present, and the non-polar phase contains the substrate and the product. This approach circumvents the problem of toxicity of solvent or substrate/product for the biocatalyst [19]. Furthermore, bioprocesses are often performed batch-wise and under aqueous conditions, which result in low product concentrations. Recent developments in the engineering of biocatalysts and extraction methods and more efficient downstream processing or working under solvent-free conditions are a way forward to minimise or avoid this disadvantage. In general, biocatalytic processes generate wastewater, salts and biomass. If organic solvents are used, that is, during downstream processing, they need to be recovered and reused. Processes where recombinant organisms are used must include a final biomass treatment to deactivate the engineered cells.

If whole-cell systems are used, the cells may need to be disrupted for product separation so that the product can be isolated by solvent extraction. The cells can be disrupted by physical, chemical or enzymatic methods. Several mechanical and non-mechanical methods for cell disruption are known and applied (Figure 10.7). During mechanical cell disruption, control of the temperature is necessary to avoid degradation of the desired product. Applying ball mills, the cells are levigated (ground to a powder or paste), whereas ultrasound treatment results in cell disruption, based on cavity effects. The cavitation is a mechanical effect, also obtained by hydrodynamic cavitation, which can be produced by high-speed stirring. Pressure-based systems use the effect of nitrogen accumulation in cells under high pressure, and the cell disruption occurs spontaneously under decompression. Shear force-based methods, often in the presence of hypotonic buffers, are used under freeze–thaw conditions. In general, non-mechanical methods are beneficial based on their mild conditions. These methods are applied to cells with very stable cell walls such as yeast cells. Extraction methods are applied for non-plant cells, whereas yeast cells are treated with toluene and lyases to destroy the cell membrane. The denaturing of bacteria cells depends on the type of bacteria. Bacteria are usually treated with ethylenediaminetetraacetate (EDTA), lysozyme and triton-X; pH adjustment can also be used for denaturation [20–22].

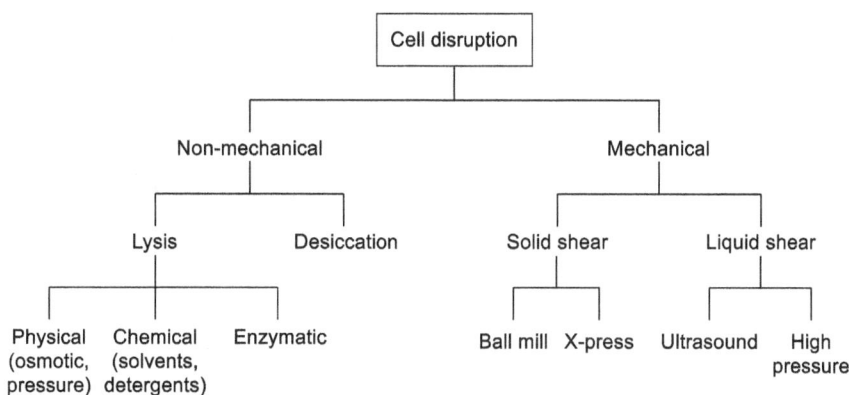

Figure 10.7: Methods of cell disruption.

Industrial biocatalytical processes comprise the stages of raw material preparation (see, e.g. upstream processes), fermentation medium, fermentation (product formation) and purification (downstream processing). The source of carbohydrates, nitrogen and growth factors often limit the economy of a process (Table 10.4) [23].

The utilisation of genetic resources is highly regulated by the *Nagoya protocol* that was signed in October 2010 and came into force on 12.10.2014. It deals with access to genetic resources and the fair and equitable sharing of benefits arising from their utilisation. Consequently, the country where the genetic resource originated from has to be compensated by the countries in which a benefit arises from the utilisation of genetic resources as well as subsequent applications and commercialisation [24]. So

Table 10.4: Source of starting materials.

Type	Source
Glucose	Hydrolysed starch
Lactose	Pure lactose
Starch	Barley, soybean meal, rye flour
Sucrose	Beet molasses, cane molasses, brown sugar, white sugar
Several vitamins (examples)	
Biotin	Corn meal
Pantothenic acid	Beet molasses, corn steep
Pyridoxal	Corn steep liquor, wheat grain, yeast

far, the protocol has been ratified by 127 parties, including China and the European Union. The United States of America have not signed the protocol [25].

As discussed previously in this chapter, the folded structure of proteins is a crucial parameter to understand their activity and identify sites/areas for directed evolution and protein engineering. Recently, computer-based programs that allow the prediction of 3D-structures have significantly increased their accuracy and influence. The most well-known of these is AlphaFold which uses artificial intelligence (AI) for the prediction of the crystal structures of proteins based on the corresponding amino acid sequence. The open access software is available in the second edition and is seen as a game changer in the world of protein structure determination, especially in the field of drug discovery [26–28].

10.2 Industrial applications

Fermentative and enzyme-catalysed processes have been established in various industrial areas. The use of biocatalysis was first applied in the preparation of food and drinks, although it was never explicitly called that. Traditional methods, known for generations, were then modified and applied on an industrial scale. Based on a deeper understanding of biochemical reactions, processes using enzymes, including genetically modified enzymes, are now key technologies used in the pharmaceutical industry and food production. Here, we discuss several historical applications (including the use in food production) and then modern applications for the synthesis of individual molecules.

10.2.1 Sauerkraut and ethanol: Applications with a long history

The manufacture of food by fermentation has been known since the times of the Greeks and the Romans. From Roman times, the preservation of cabbages and turnips using salt is known. The Tartars brought the recipe of the fermentation of cabbages from

China to Europe, and it became very popular in central and eastern European countries. Also, in Korea, a version of fermented cabbages is known (Togil Kimchi). James Cook (eighteenth-century British explorer) always had a stock of Sauerkraut on his sea voyages, since experience had taught him it prevented scurvy [29].

Sauerkraut has been industrially produced since around 1860. The process is divided into three major phases. In phase one, acetic acid-producing bacteria consume oxygen-producing ethanol, acids and esters. In phase two, *Lacotobacillus* species, for example *Leuconostoc mesenteroides* produce lactic acid, carbon dioxide and mannitol. The ethanol is consumed, producing esters (which form part of the taste component), and the lactic acid content is 1–2%. Each phase one and two needs around three days for completion. In the three- to six-week period of phase three, *Lactobacillus brevis* finalises the fermentation process (Scheme 10.10) [29]. Based on the applied process, different types of Sauerkraut are commercially produced.

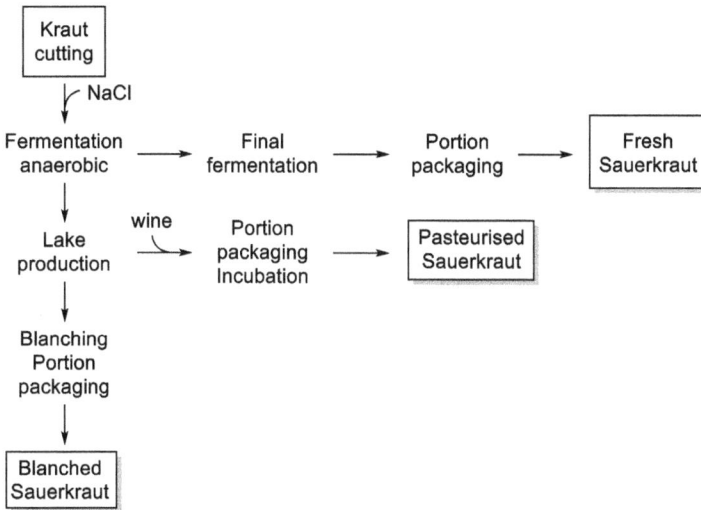

Scheme 10.10: Methods of Sauerkraut manufacture, simplified flow sheet.

The rationale behind the application of microorganisms such as bacteria or yeast is their high productivity, that is, the protein production of microorganisms is several times greater when compared to higher plants (e.g. soya beans). The energy source for microorganisms are, like for other organisms, oligosaccharides such as starch, which are degraded into smaller units (monosaccharides such as glucose) by enzymes (Scheme 10.11). For biotechnological processes, yeasts, fungi, bacteria and actinomycetales (generally gram-positive bacteria, which are often rod-shaped) are applied. Biotechnology research and development deals with several aspects of microorganisms, the organism itself (e.g. yeast), the macromolecules produced from microorganisms (such as enzymes), the primary metabolites (e.g. organic acids or amino acids) and inhibitors (such as antibiotics or alkaloids). Usually, during metabolism, the energy source (e.g. glucose) is transferred into smaller molecules by oxidation reactions. The resulting energy is used as heat energy inside the organism or for further cell processes such as the building of cell material or cell growth. The oxidative metabolism of glucose into water, carbon dioxide and adenosine-5'triphosphate (ATP) is a unique characteristic of cells. Organisms that can work with a limited amount of oxygen produce ATP, carbon dioxide and ethanol; these processes are called fermentation. Such processes are known to produce ethanol, lactic acid, acetone, butanol and methane. Organisms such as *Saccharomyces* can switch between various fermentation processes. Furthermore, not all microbiological products are typically fermentation products, for example acetic acid is produced from ethanol in the presence of excess oxygen.

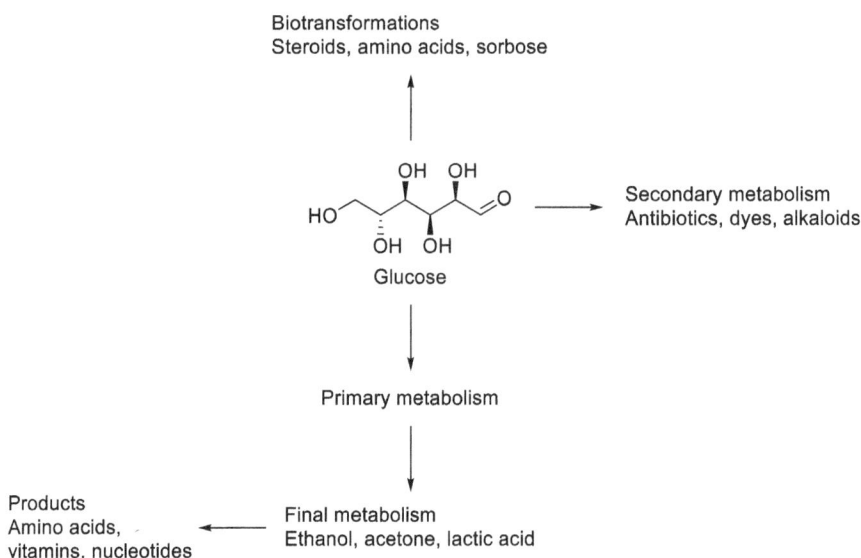

Biotransformations
Steroids, amino acids, sorbose

Secondary metabolism
Antibiotics, dyes, alkaloids

Glucose

Primary metabolism

Products
Amino acids,
vitamins, nucleotides

Final metabolism
Ethanol, acetone, lactic acid

Scheme 10.11: A general scheme on microbiological products.

The oxidative fermentation of glucose to ethanol is amongst the oldest and most important fermentation processes. It can be simplified into a few important steps; however, the detailed process is based on twelve enzyme catalysed steps (*Embden–Meyerhof* pathway). In the simplified process (Scheme 10.12), glucose is split and converted to pyruvate via hydrogen transfer, and after decarboxylation acetaldehyde is formed, which is reduced to ethanol.

Based on the sugar source and the flavour/fragrance molecules present in them, various alcoholic drinks that exhibit different flavours can be produced (Table 10.5).

Glycolysis is the central process for the degradation of carbohydrates in eukaryotic cells and is also common for bacteria and archaea. In glycolysis, the energy is stored as ATP. The pathway starts with the formation of fructose-1,6-diphosphate with the consumption of two ATP units, which are later reformed in the process from the dehydrogenation of glyceraldehyde-3-phosphate. Using NAD^+ as a co-factor (acting as a hydrogen acceptor), pyruvate is formed (*Embden–Meyerhof* pathway). The enzyme catalysed decarboxylation that ensues results in carbon dioxide and acetaldehyde. Acetaldehyde is transformed into ethanol with the consumption of NADH and reformation of NAD^+ (Scheme 10.13) [30].

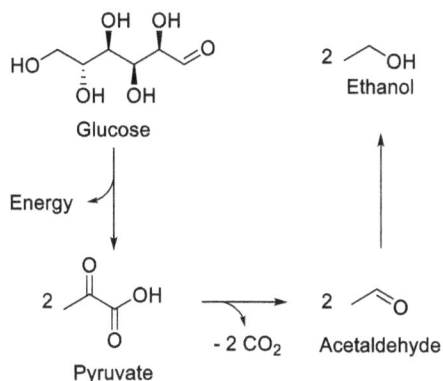

$$C_6H_{12}O_6 + 2\ NAD^+ \longrightarrow 2\ NADH + 2\ H^+ + 2\ CH_3COCOOH$$

Scheme 10.12: Simplified ethanol process.

Table 10.5: Manufacture of various alcoholic drinks.

Alcoholic drink	Source	Distilled
Beer	Barley	No
Cider	Apple juice	No
Sake	Rice	No
Wine	Grape	No
Gin	Rye, corn, juniper berry	Yes
Tequila	*Agave tequilana*	Yes
Vodka	Potatoes, grain	Yes
Whiskey	Barley, wheat, corn	Yes

10.2.2 Sugars and oligosaccharides

10.2.2.1 Glucose/fructose isomerisation

Glucose/fructose isomerisation is a major industrial application of isomerase enzymes (Scheme 10.14). This process has been applied for many decades, especially in the USA, for the industrial production of sweeteners, because fructose has approximately 30% increased sweetness compared to glucose. At room temperature, the equilibrium between β-pyranose, α-furanose and β-furanose is 76:4:20. The main starting material for fructose and the high fructose corn syrup (HFCS) is maize. In the first step, starch is broken down

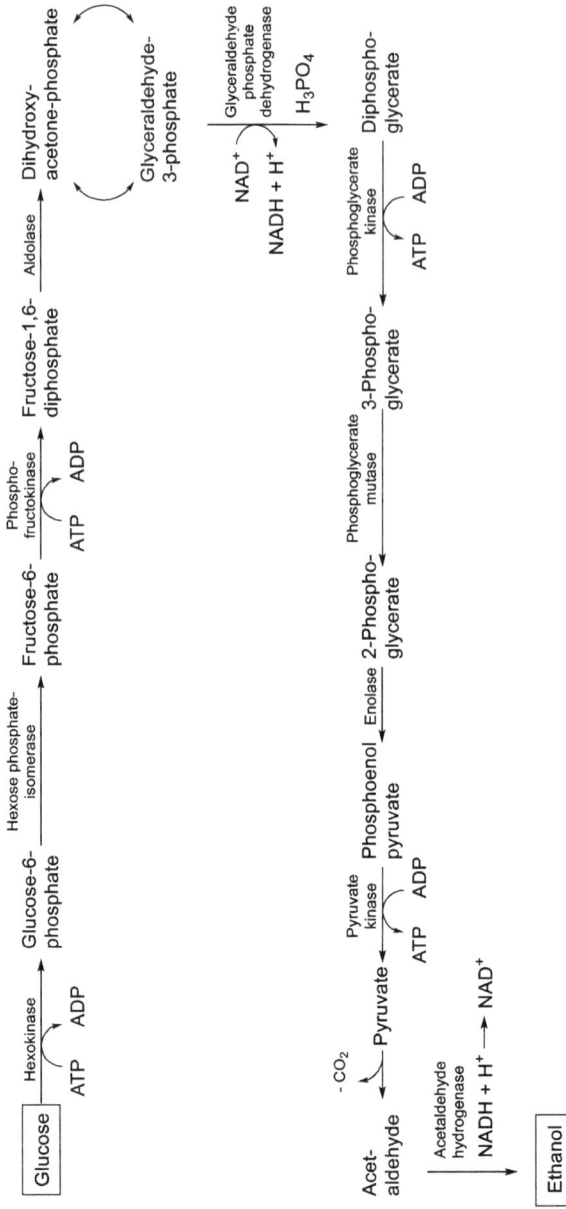

Scheme 10.13: *Embden–Meyerhof* pathway for the synthesis of ethanol from glucose.

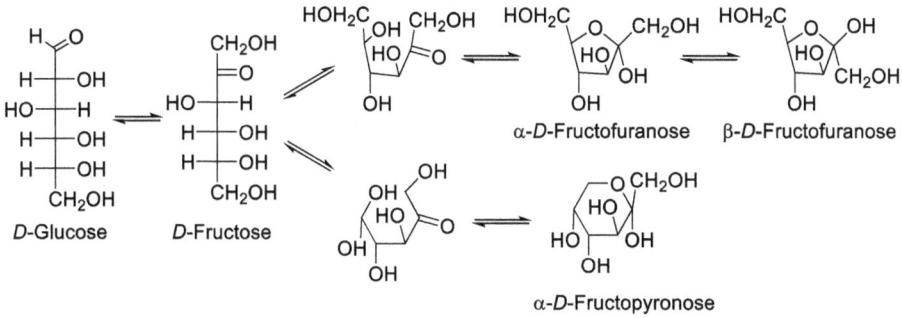

Scheme 10.14: Glucose isomerisation.

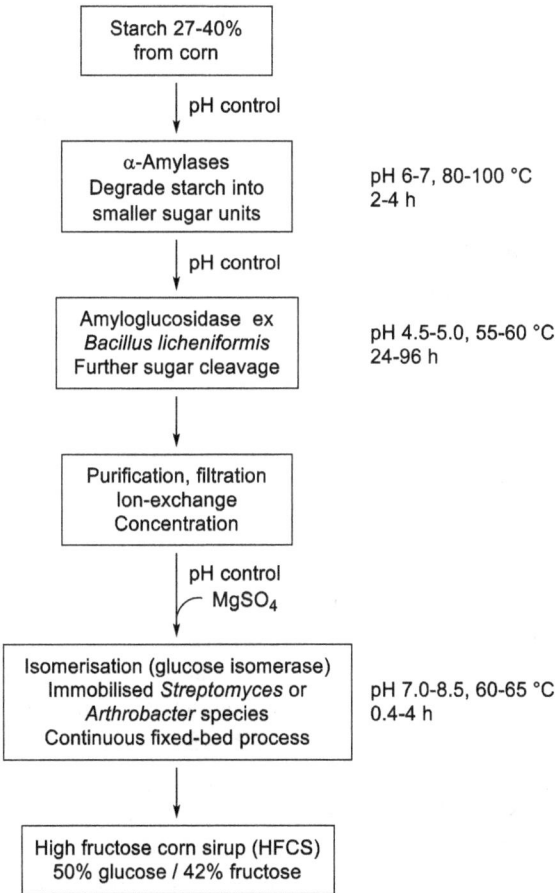

Scheme 10.15: Glucose-fructose isomerase flow sheet [33].

into shorter chains. Then, it is converted by various amylases into glucose, which is purified and isomerised into a mixture of glucose and fructose [31]. HFCS is used in processed foods, breakfast cereals and soft drinks. The process is carried out on a scale of $>10^7$ t/a.

Fructose production from corn starch is industrially carried out with amylases and glucose isomerases (Scheme 10.15). First, glucose is produced (see below), which is then treated with the immobilised enzyme xylose isomerase to convert the sugars into a mixture of glucose and fructose. A fructose content of around 42–55% can be achieved. HFCS with 90% content of fructose needs further treatment by chromatography. For application in the food industry, HFCS with 42% fructose is used, whereas HFCS with 55% fructose is used in soft drinks [32].

The manufacture of glucose from starch is also an enzyme-catalysed reaction. This process is carried out in $>10^7$ t/a scale and starts from grain. In the process, two enzymes are used: α-*amylase* (E.C. 3.2.1.1) and *amyloglucosidase* (E.C. 3.2.1.3) [34].

Another enzyme-catalysed isomerisation/epimerisation reaction is the transformation of N-acetyl-D-glucosamine into N-acetyl-D-mannosamine performed in the presence of *N-acylglucosamine 2-epimerase* (EC 5.1.3.8) (Scheme 10.16). The reaction is carried out in such a manner that the N-acetyl-D-mannosamine formed is further transformed in the presence of pyruvate, because the equilibrium of the epimerisation reaction is on the side of the starting material. N-Acylglucosamine 2-epimerase is cloned from the porcine kidney and overexpressed in *Escherichia coli*. N-Acetyl-D-mannosamine is a building block of neurotransmitters and is also applied as the starting material in the synthesis of sialic acid derivatives [35].

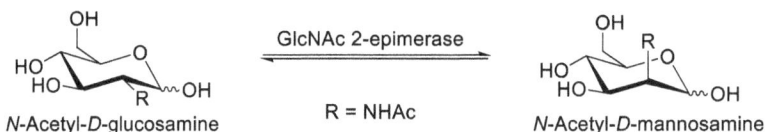

Scheme 10.16: Epimerisation of N-acetyl-D-glucosamine to N-acetyl-D-mannosamine.

10.2.2.2 Maltose and beer brewing

Starch degradation results in maltose ($C_{12}H_{22}O_{11}$), a crystalline disaccharide, which is found in barley, beer and potatoes and is used in the production of whiskey and beer. Maltose (Figure 10.8) is formed from starch and glycogens with α/β-amylase, and derives its name from malt since this process is found in germinating seeds. During beer brewing, around 80% of the starch is transformed into maltose. The maltose fermentation during beer brewing is carried out in the temperature range of 40–70 °C for 2–4 h. The further processing of beer production includes the following steps: addition of hops or hop-extract followed by heating at 80 °C for denaturisation of α-amylase, concentration and filtration to remove proteins and insoluble hop particles. After cooling to around 5 °C in a counter-current cooler, the addition of yeast and fermentation follows. For 5–8 days, the maltose is transformed into ethanol and carbon dioxide

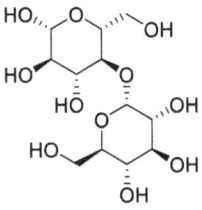

Figure 10.8: Maltose.

at a yield of 60–70%, which is separated and reused during the bottling of the beer. During storage of between two weeks and a month, the remaining maltose is transformed into ethanol and carbon dioxide, which is not separated. All insoluble materials settle out during this time. After filtration, the beer is bottled or stored in barrels under oxygen-free conditions [36]. Maltose is produced in around 1.5×10^7 t/a [37].

10.2.2.3 Fucosyllactose

Human milk oligosaccharides (HMOs) play an important role in the field of infant nutrition, with a constantly increasing market size expected to reach $100 million in 2027 [38]. Approximately 200 different HMOs are present in human mother's milk – the most prevalent is 2′-fucosyllactose (2′FL) making up about 30% of all HMOs present. 2′-Fucosyllactose (2′FL) protects against infectious diseases, such as by preventing epithelial-level adhesions of toxins and pathogens [39]. 2′FL also offers protection against pathogens such as *Campylobacter jejuni, Salmonella enterica serovar Typhimurium* and *Helicobacter pylori. Helicobacter pylori* harms the stomach and is responsible for gastritis, and cancer in the stomach and pancreas. *Campylobacter jejuni* is responsible for food poisoning, diarrhoea and fever. *Salmonella enterica* is mainly responsible for gastroenteritis [40]. 2′FL is manufactured using *Escherichia coli* [41, 42].

Fucose is elongated by the sequential addition of N-acetylglucosamine (GlcNAc), galactose and sialic acid; this is catalysed by fucosyltransferases which are located in the endoplasmic reticulum and glycosyltransferases which are active in the Golgi apparatus (Scheme 10.17).

10.2.3 Organic acids

10.2.3.1 Acetic acid for vinegar

The world market for acetic acid is around 13 million tons at a price of 1200 to 1600 US $/t in 2015 [43]. The main production process is the carbonylation of methanol [44]. The biological route reflects only around 10% of the worldwide production [45] and is especially important for the production of vinegar. It is produced by *Acetobacter, Lactobacillus* and *Polyporus* species. The main substrate used to produce acetic acid by *Acetobacter* species is ethanol. Beer, wine, cider and other alcoholic beverages or distilled ethanol can be used as starting material. The fermentation to vinegar is one of

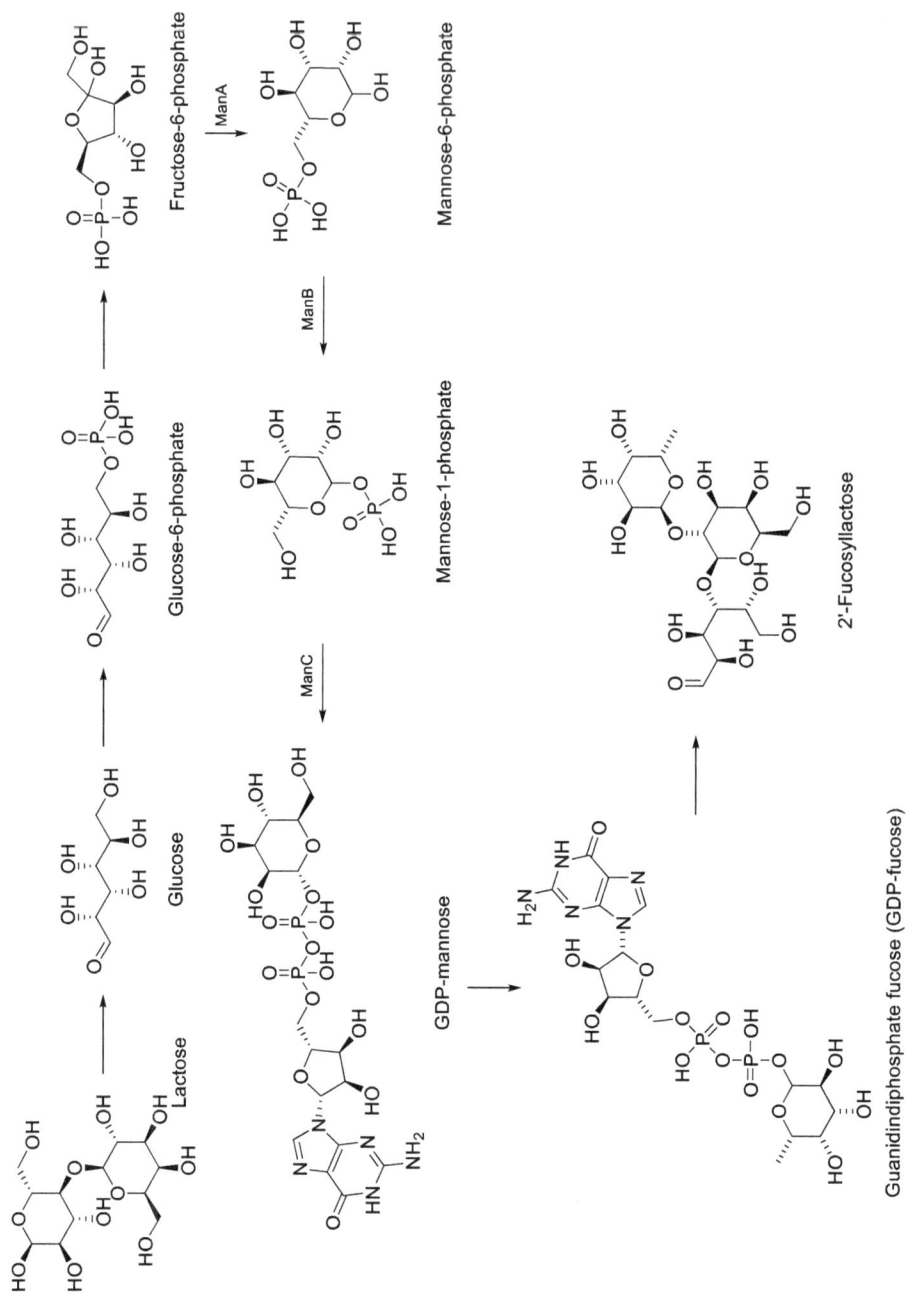

Scheme 10.17: Synthesis of 2′-fucosyllactose.

the oldest fermentation processes. When a pure ethanol/water/acetic acid mixture is used, additions of nutrients such as vitamins are necessary. Depending on the source, several types of vinegar are known. The acetic acid concentration in vinegar is 5–6%, and only traces of residual ethanol can be detected. For example, the popular vinegar, *Aceto balsamico di Modena*, is manufactured from grape juice. In the production process, amylose is transformed into maltose and glucose by enzymes. The sugars are transformed into carbon dioxide and ethanol by *Saccaromyces cerevisae*. The quality of the vinegar depends on the ratio of fermented to non-fermented sugars (aroma compounds). Vinegar fermentation must be carried out in the presence of oxygen. Applications of vinegar are in the food industry (conservation), and as cosmetics (e.g. *vinaigre de toilette*) and cleaning agents.

Acetic acid production by fermentation is performed with *Acetobacter* species or *Gluconobacter* species. The oxidation of ethanol is carried out with a membrane-associated alcohol dehydrognase (ADH) and an aldehyde dehydrogenase (ALDH). The electrons are transferred to an oxidase by ubiquinone. The oxidation performs well if there is an efficient oxygen supply to the fermentation broth. *Clostridium* bacteria can transform sugars directly into acetic acid [46].

10.2.3.2 Citric acid

The worldwide production of citric acid in 2007 was 1.6 million tonnes [47], and it is used in many different fields such as the food and beverage, pharmaceutical and cosmetics industries [48], and in detergents. The manufacture of citric acid starts from corn or molasses by fermentation in the presence of *Aspergillus niger*. High titres and yields can be achieved with a high content of oxygen and glucose in the fermentation broth. The fermentation must be performed under pH control at pH <3, with the aim of inhibiting the further metabolism of citric acid. Efficient production of citric acid also requires a very low content of Fe^{2+} ions (<5 mg/l) and ions such as Zn^{2+}, and Cu^{2+}; this is because Fe^{2+} is necessary as co-factor for the enzyme aconitase, which is responsible for the reversible conversion of citrate or isocitrate to aconitate and should be minimised. The fermentation has to be carried out under air for several days at a sugar concentration of around 25% (Scheme 10.18). Yields of up to 85% are achieved, based on the carbohydrate supplied. Usually, *Aspergillus niger* is grown on a solid nutrient media. The crude product is purified by treatment with an ion exchange resin, followed by concentration, crystallisation and centrifugation. After drying and packing, the final product is obtained [49].

Applications of citric acid are in cleaning products (water softeners), in the cosmetic industry (cremes and lotions) and in the food industry (preserving agents). Citric acid is marketed as a crystalline anhydrous material, as a monohydrate and as sodium salt.

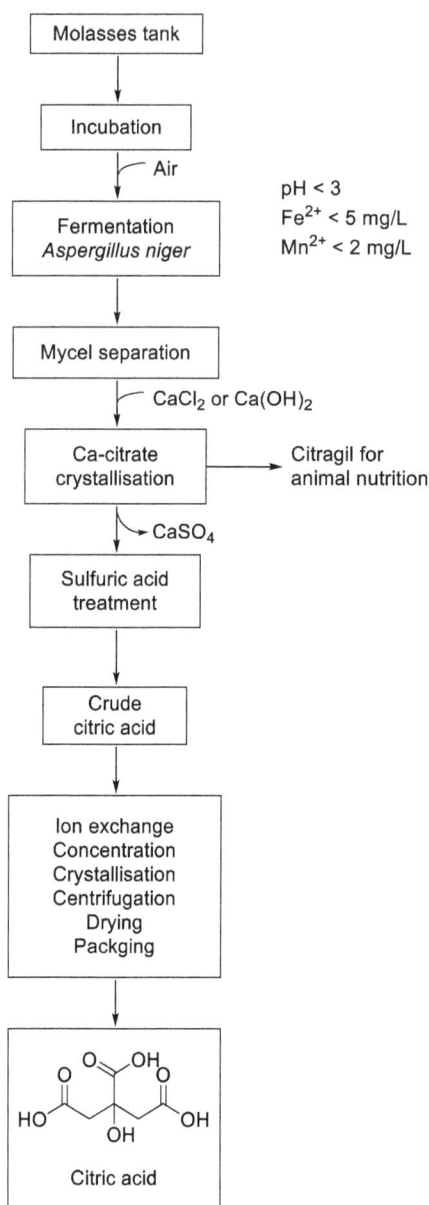

Scheme 10.18: Block-flow diagram for the manufacture of citric acid.

10.2.3.3 Lactic acid

L-(+)-Lactic acid is found in several biological sources such as kidney serum and bile; the racemate of D-, and L-lactic acid is found in the juice of tomatoes, beer and dairy products. In general, it has been found that the D–L-ratio of lactic acid products depends on the purification step and the bacteria used during the fermentation. Fungi such as *Rhizopodus, Al-*

lomyces and *Blastocladiella* are also able to produce lactic acid [50]. The lactic acid world market is >1 Mio t/a, which means that lactic acid is considered a bulk chemical. More than 90% of the world market of lactic acid is produced by fermentation of carbohydrates. The industrial processes are carried out using *Lactobacillus acidophilus, Lactobacillus casei, Lactobacillus delbrueckii subsp. bulgaricus* (*Lactobacillus bulgaricus*) and *Lactobacillus helveticus*. Furthermore, *Streptococcus salivarius* subsp. *thermophilus* (*Streptococcus thermophilus*) and *Lactococcus lactis* are also used [51, 52].

In general, lactic acid is produced from glucose, maltose, sucrose or lactose. Starches from corn or potatoes are hydrolysed, preferably by enzymes from *Lactobacillus delbrueckii* or *Lactobacillus bulgaricus,* or by using acids such as sulphuric acid to form maltose and glucose, followed by lactic acid fermentation. The sugar concentration during fermentation is around 12%, the fermentation is carried out under pH control (pH 5.5–6.5), and the produced acid is neutralised ($Ca(OH)_2$) at 45–50 °C. Fermentation times of approximately 72 h are required for complete sugar consumption, and yields of 95% are obtained.

Lactic acid by chemical synthesis is performed by the addition of hydrocyanic acid to acetaldehyde. The lactonitrile that is formed is hydrolysed with HCl, producing lactic acid (Scheme 10.19) [53].

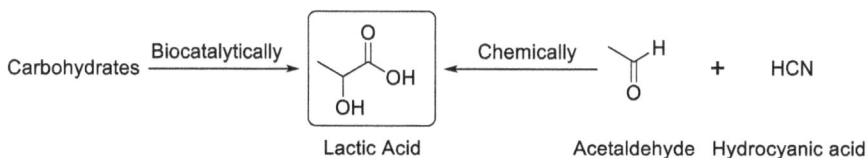

Scheme 10.19: Lactic acid production.

10.2.3.4 Itaconic acid

The production volume of itaconic acid (Figure 10.9) is >80,000 t/a, and the fermentation is performed at around 40 °C with *Aspergillus terreus* and *Aspergillus itaconicus* starting from molasses [54]. A starting concentration of sugar of 150 g/l is used, and the final content of itaconic acid in the fermentation broth is 85 g/l. During the fermentation, the produced acid is neutralised by the addition of ammonia or sodium hydroxide, which allows a constant fermentation rate and a final acid concentration of 18%. Itaconic acid fermentations resemble those of citric acid, and Fe^{2+} ions reduce the accumulation of itaconic acid.

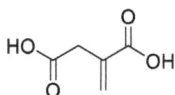

Figure 10.9: Itaconic acid.

Applications of itaconic acid are in the polymer industry as a co-monomer for the synthesis of polyacrylates and rubber, in food industry and as a thickening agent.

10.2.3.5 Gluconic acid

Gluconic acid and its derivatives, such as salts and esters, occur widely in nature, for example in fruits, honey and wine. The acid itself is the free form of D-glucono-δ-lactone, produced from D-glucose by dehydrogenation. The reaction is performed with *Aspergillus niger* at glucose concentrations of around 200–350 g/l (Scheme 10.20). The lactone-acid equilibrium is controlled by pH (pH 6.5) and temperature (around 30 °C). Alternatively, sodium gluconate can be produced since the solubility of the sodium salt is much higher.

A modified process is the *Glucose oxidase*-catalysed transformation of glucose to the corresponding lactone, followed by product formation under *Catalase*-catalysed hydrogen peroxide transformation to D-gluconic acid. The process is characterised by high yield and selectivity (Scheme 10.20) [55].

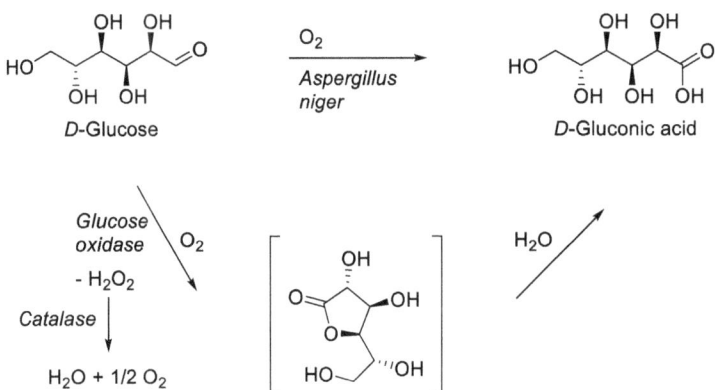

Scheme 10.20: Enzymatic oxidation of D-glucose to D-gluconic acid.

Applications of gluconic acid are in the fields of food additives (E574), in medical applications such as calcium gluconate injections and in the treatment of hypocalcaemia. Quinine gluconate, a salt between gluconic acid and quinine, is used for intramuscular injection in the treatment of malaria. Zinc gluconate injections are used to neuter male dogs. Ferrous gluconate injections have been proposed in the past to treat anaemia and in cleaning products [56, 57].

10.2.3.6 Tartaric acid

Tartaric acid is commonly used as an additive in the food industry (E334), but also finds technical applications as an anti-microbial agent or in the degumming of silk fabrics

[58]. It contains two stereogenic centres and has three stereoisomers: a pair of enantiomers (the *D*- and *L*-forms) and the *meso*-form having a mirror plane (Figure 10.10). *L*-(+)-Tartaric acid is isolated from natural sources by treatment of potassium hydrogen tartrate (tartar) with calcium hydroxide and sulphuric acid. The by-products of this process are calcium sulphate and yeast waste. In an alternative process, potassium hydrogen tartrate is treated with potassium hydroxide, and after ion-exchange resin treatment, tartaric acid is isolated. *D*-(−)-tartaric acid is produced from the racemic tartaric acid with *Penicilliun glaucum* via fermentation. In this process, the *L*-(+)-tartaric acid is degraded and *D*-(−)-tartaric remains untouched [59].

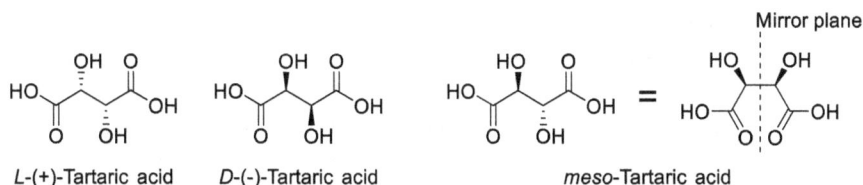

Figure 10.10: The different stereoisomers of tartaric acid.

10.2.3.7 Mandelic acid

The manufacture of mandelic acid by the kinetic resolution of racemic mandelonitrile is carried out on a scale of several t/a. The process is carried out by applying a nitrilase of type E.C. 4.2.1.84, which selectively hydrolyses the (*R*)-nitrile to yield (*R*)-mandelic acid. The remaining (*S*)-configurated enantiomer is racemised and reused (Scheme 10.21) [60]. Mandelic acid is used in the treatment of urinary tract infections [61]. Furthermore, mandelic acid can be used as a resolving agent for other chiral molecules such as amines [62].

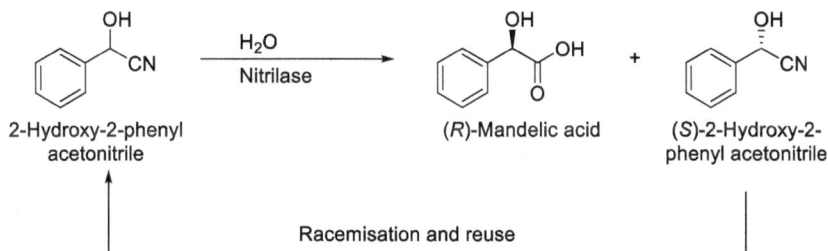

Scheme 10.21: Mandelic acid synthesis by enzymatic kinetic resolution.

10.2.4 Amino acids

Amino acids, especially in the *L*-configuration, are the building blocks for proteins and are essential for both humans and animals. They are usually obtained from the diet but can also be obtained from nutritional supplements. Amino acids are applied in the food industry (Glu, Gly, Cys), animal nutrition (Trp, Thr, Met, Lys), medical applications (acetyl-cys, *L*-DOPA) and in cosmetics. In the last few decades, the manufacture of amino acids by biotechnological processes has increased, in the number of amino acids that are produced, the production volume and the number of processes that are applied (Table 10.6). For the manufacture of amino acids, several different industrial methods are applied. One method is extraction, which can be used if a protein hydrolysate has different physical–chemical behaviour, for example separation by crystallisation. Cystine and tyrosine can be produced based on their limited solubility in water from a protein hydrolysate. The manufacture of amino acids by chemical synthesis is applied for large volume and low-price products which require very cheap starting materials such as acrolein. Furthermore, chemical production is favoured in the manufacture of glycine (no stereogenic centre) and methionine (where both *D*- or *L*-forms have similar nutritional values). An advantage of the chemical production of amino acids is the possibility of continuous processing; however, a disadvantage is the need to resolve the mixture of enantiomers, which is an additional production step. *D,L*-Valine, *D,L*-methionine and *D,L*-cysteine are produced by *Strecker*-synthesis. The separation of the enantiomers can be performed by crystallisation of diastereomeric salts, crystallisation with seed crystals or enzymatic methods. Amino acids such as the production of *L*-aspartic acid from fumaric acid are carried out with an aspartase enzyme from *Escherichia coli* or *Brevibacterium flavum*. *tert-L*-leucine and *L*-DOPA are produced by enzymatic methods. Resolution of *D,L*-amino acids in the presence of enzymes is used in the production of *L*-methionine (acylase, see later), valine, alanine, phenylalanine and tryptophan. Glucose is fermented in the manufacture of glutamic acid. Furthermore, the *L*-amino acids leucin, *iso*-leucine, lysine tyrosine, alanine, valine, phenylalanine, tryptophan and threonine are manufactured by fermentation [63, 64].

Table 10.6: Amino acids produced by biotechnology.

Amino acid	Abbreviation*	Production volume t/a	Organism
L-Glutamic acid	Glu or Q	2.0×10^6	*Corynebacterium glutaminicum*
L-Lysine	Lys or K	1.5×10^6	*Brevibacterium flavum*
L-Threonine	Thr or T	3.1×10^5	Fermentation
L-Phenylalanine	Phe or F	1.2×10^5	Chemistry
L-Aspartic acid	Asp or D	1.5×10^5	*Escherichia coli* or *Brevibacterium flavum*
L-Cysteine	Cys or C	3.0×10^3	*Escherichia coli*
L-Tryptophan	Trp or W	1.2×10^3	Tryptophan synthase (*Escherichia coli*)
L-Arginine	Arg or R	1.2×10^3	*Brevibacterium flavum*
L-Alanine	Ala or A	10^3	Aspartate-β-decarboxylase
L-Methionine	Met or M	10^3	Acylase
L-DOPA		10^3	*Erwinia herbicola*
tert-L-Leucine		5×10^3	*Streptomyces bottropensis*
D- and *L*-Valine		10^2	*Bacillus species, Corynebacterium acetoacephilium*

*Only for proteinogenic amino acids given.

10.2.4.1 L-DOPA

The amino acid L-3,4-dihydroxyphenylanaline (L-DOPA) plays an important role in the biology of humans, animals and plants. Blue mussels (*mytilus edulis*) use L-DOPA as an intercalation compound in the protein chain of its adhesive for fixing. The role of L-DOPA is to increase the hydrophilicity of the adhesive [65]. Furthermore, in humans, L-DOPA is a precursor of neurotransmitters such as adrenaline, noradrenaline (catechol amines), melamine and L-dopamine. L-DOPA itself mediates neurotrophic factors (molecules that support the growth and survival of neurons) [66]. The treatment of *Parkinson's* disease is the main application of L-DOPA which has been produced and manufactured (Monsanto) using asymmetric hydrogenation as a key technology. This transformation is discussed in more detail in Chapter 3: Homogeneous hydrogenations. Biotechnological approaches by microbial fermentation using lyases with suspended whole cells were industrially applied by Ajinomoto (Japan). Starting from catechol, L-DOPA was produced in a fed-batch process (Scheme 10.22) [67–69]. More than 250 t/a are produced by the Ajinomoto process.

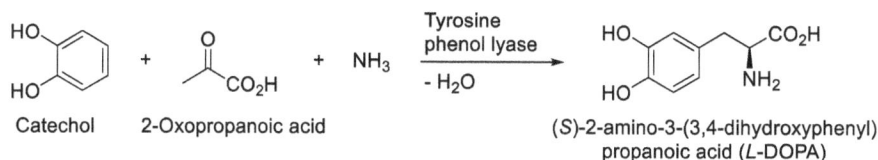

Scheme 10.22: Ajinomoto process for the fermentative L-DOPA production.

10.2.4.2 L-Methionine

L-Methionine is produced in several hundred t/a by kinetic resolution with an amino acylase from *Aspergillus oryzae*. The reaction performs well in the presence of Co^{2+} and, therefore, cobalt salts are added during the reaction. The non-converted acetyl-D-methionine is racemised and reused. The racemisation is performed chemically by acetic anhydride treatment under basic conditions. The L-methionine is separated by crystallisation. The undesired enantiomer is recovered and racemised. This allows a yield > 80% and an *ee* value of >99.5% for the continuous process (Scheme 10.23) [70]. The enzyme can be separated using a membrane process [71]. The amino acids L-alanine, L-phenyl alanine, L-valine, L-norvaline and L-homophenylalanine are produced at Evonik in a similar production set up.

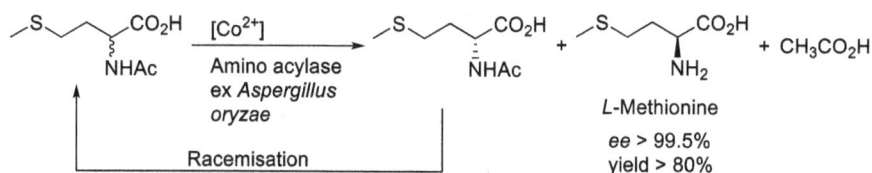

Scheme 10.23: L-Methionine by resolution with *Aspergillus oryzae*.

α-H-amino acids are industrially relevant compounds (used as food and feed additives and in the pharmaceutical industry) and can be manufactured in the presence of *Pseudomonas putida* (E.C. 3.4.1.11) in a whole-cell fermentation process (Scheme 10.24) of various α-amino amides. *D*-phenyl-glycine, 4-hydroxyphenylalanine and *L*-homophenylalanine are produced by applying the procedure, and the recovery and racemisation of the unwanted enantiomer allows a nearly 100% approach. An additional benefit is that in vivo protein engineering has enlarged the number and type of substrates that can be used [72, 73]. The starting material for the whole-cell fermentation is obtained via a *Strecker* reaction (addition of cyanide to an aldehyde) followed by saponification. The resulting α-amino amides can be further chemically modified.

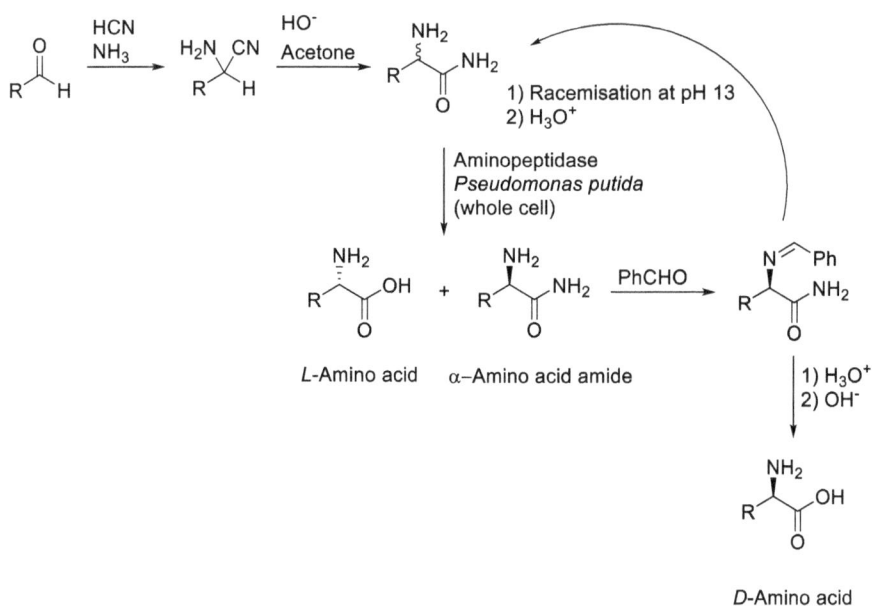

Scheme 10.24: *L*- and *D*-α-amino acids by kinetic resolution of α-amino acid amides.

10.2.4.3 *tert*-Leucine

tert-Leucine (terleucine), a chiral non-proteogenic α-amino acid, occurs in *L*- and *D*-forms. It occurs in nature as the building block of the antibiotic Bottromycin, formed by the bacterium *Streptomyces bottropensis*. *tert*-*L*-leucine is produced in several thousand t/a by fermentation and used in pharmaceutical applications and drug synthesis. *tert*-*L*-Leucine is produced from trimethyl pyruvic acid by reductive amination in the presence of leucine dehydrogenase (LDH) and formate dehydrogenase, with continuous co-factor regeneration (Scheme 10.25) [74–76].

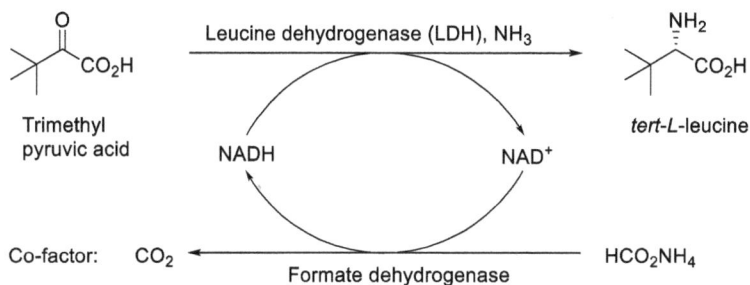

Scheme 10.25: Synthesis of *tert-L*-leucine [74].

10.2.4.4 Cbz-*D*-proline

An interesting combination of chemical and biotechnological methods is used in the process for Cbz-*D*-proline (Cbz stands for carboxybenzyl), a drug intermediate for migraine treatment. The proline racemisation is performed chemically in water, followed by treatment with benzoyl chloroformate (Scheme 10.26). The resulting racemic *N*-Cbz-proline is hydrolysed with the proline acylase *Arthrobacter* sp. in an efficient manner, with *ee* value of 99.5%. Titres of 70 g/l can be achieved. The final product is separated by extraction, whereas (*S*)-proline is recovered and reused [77–79].

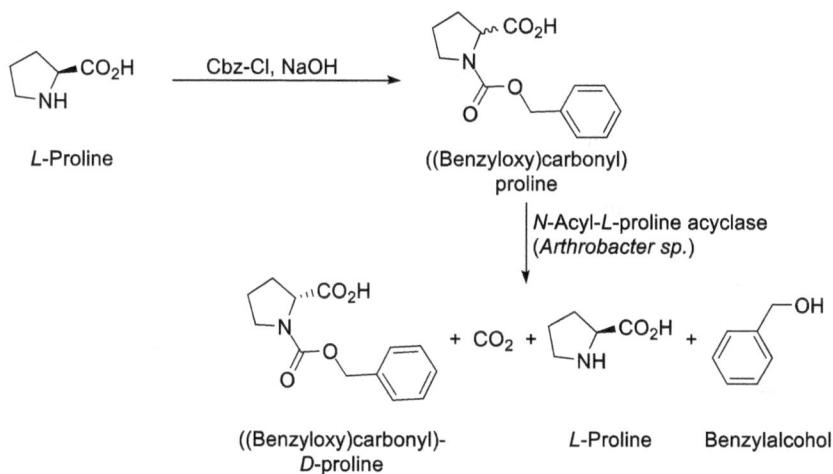

Scheme 10.26: Synthesis of Cbz-*D*-proline.

10.2.4.5 *L*-Carnitine

The growth factor *L*-carnitine can be manufactured by chemical or biotechnological processes. The chemical route starts from epichlorohydrin or diketene [80–82]. However, the disadvantages of the chemical routes are the low yield and the need to recover and reuse the resolving agent. Comparison of the two processes favours the biotechnical approach because of the significantly lower amounts of all waste streams (salt waste, total organic carbon waste, waste for incineration and wastewater) [83].

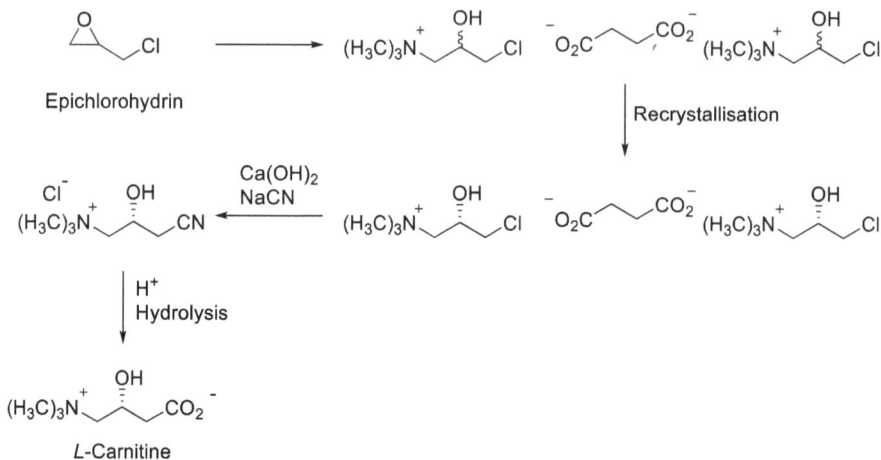

Scheme 10.27: Chemical route to *L*-carnitine from epichlorohydrin [84].

For the biotechnological synthesis, two processes have been developed. A more efficient process of *L*-carnitine manufacture starts from achiral precursors and involves the resolution of racemic carnitine (or its derivatives) [85–87] (Scheme 10.27). However, they still suffer from some disadvantages, for example less stable starting materials, low productivity and limited yield (a maximum yield of 50%, *ee* value usually <100%).

The most efficient methodology of *L*-carnitine manufacture is the whole-cell fermentation starting from γ-butyrobetaine (Scheme 10.28). The process is carried out with growing cells because the betaine and *L*-carnitine must be transported through the membrane, which is energy-intensive. Biocatalysts from *Agrobacterim* and *Rhizobium* (E.C. 4.2.1.89) allow the selective introduction of the stereogenic centre by crotonobetainyl-CoA-hydrolase. The production of *L*-carnitine is achieved in >99% conversion, an *ee*-value of >99% and a titre of 80 g/l [88].

10.2.4.6 *L*-Cysteine

L-Cysteine is an important amino acid with multiple uses in the food and pharma industry, and it is produced on several kt/a. In the past, *L*-cysteine was prepared by HCl treatment of keratin sources from animal materials such as feathers or dog hair [89]; recently, processes based on fermentation have been established. *L*-Cysteine and its analogues are synthesised using the gene-modified bacteria *Escherichia coli* [90–92]. The hydrolysis of *DL*-2-amino-Δ^2-thiazoline-4-carboxilic acid (ATC, Figure 10.11) available from thiourea and α-chloroacrylate in the presence of *Pseudomonas thiazoliophi-*

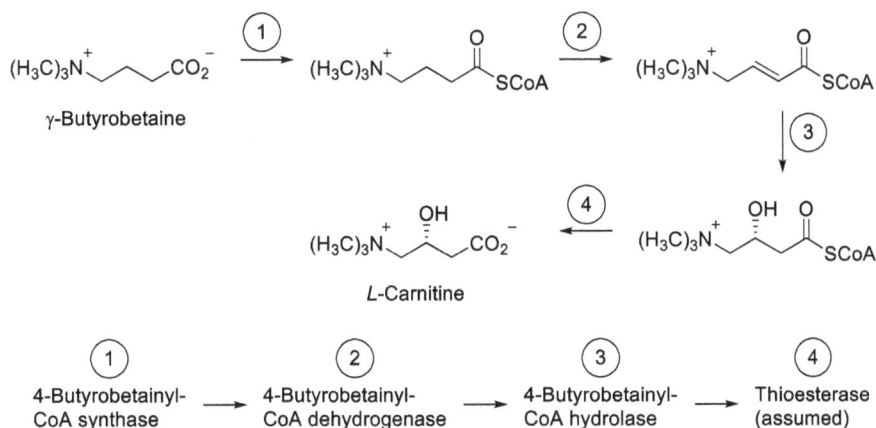

Scheme 10.28: Biotransformation process to *L*-carnitine.

lum was developed and a yield 30 g/l of cysteine can be achieved [93]. *L*-Cysteine and *L*-cystine are manufactured from plant materials and inorganic materials (vegan production). The fermentation process was introduced by Wacker and is carried out in Leon (Spain) using a modified *E. coli* [94, 95]. The class of E.C. 2.I.9.2 shows an increased amount of *L*-cysteine compared to the wild-type enzyme. Furthermore, the enzyme activity of E.C. 2.7.9.2 (KEGG data base) of the wild type is reduced (especially preferred) by 70%. Furthermore, the synthesis of *L*-cysteine from glucose is known. Starting from 20 mM glucose, 10 mM cysteine can be synthesised during a fermentation time of 12 h [96].

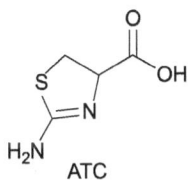

Figure 10.11: *DL*-2-Amino-Δ^2-thiazoline-4-carboxilic acid (ATC).

It should be mentioned that cysteine can also be synthesised from cystine by electrochemical reduction (Scheme 10.29) [97]. In a batch-type set-up using Pb-electrodes with a surface area of 10–100 cm^2, a product yield of around 95% can be achieved (Scheme 10.29). The electric energy consumption is 0.6 kWh/kg (*L*-cysteine hydrochloride).

$$\text{L-Cystine} \xrightarrow[\substack{\text{MMO anode, H}_2\text{SO}_4 \\ \text{metal cathode, HCl}}]{\substack{\text{2-3 F} \\ \text{20-400 mA/cm}^2}} 2 \ \text{L-Cysteine}$$

L-Cystine

L-Cysteine

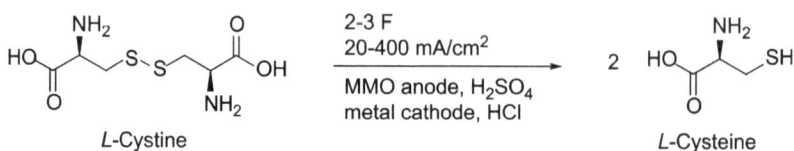

MMO = Mixed metal oxide anode (IrO$_2$-RuO$_2$/Ti)

Scheme 10.29: Electrochemical reduction of cystine to cysteine.

10.2.5 Vitamins and carotenoids

10.2.5.1 Vitamin A acetate

Vitamin A refers to a group of diterpenes found in animal tissue, such as vitamin A alcohol (retinol) and its derivatives such as the corresponding acetate, propionate or palmitate. There are also provitamin A compounds known, for example β-carotene, α-carotene or cryptoxanthin, which are oxidatively degraded into vitamin A. Vitamin A is important for cell differentiation and growth and its involvement in the vision process [98]. The production volume of vitamin A is more than 5,000 t/a [99].

In the industrial vitamin A process of DSM, the dihydroxy-intermediate is partially acetylated and transformed to vitamin A acetate [(all-*E*)-retinyl acetate] (Scheme 10.30). However, the chemical route gives a mixture of mono- and bis-acylated productions. An alternative method for the selective synthesis of the mono-acetylated intermediate is a lipase-catalysed reaction of the primary allylic alcohol [100]. In a screening of various lipases, it was found that enzymes of the group of E.C. 3.1.1.3 show good selectivity with high activity. In particular, immobilised lipases such as *Candida antarctica* and *Chirazyme L-2* (on C-2 carrier) proved to be suitable for large-scale application in view of the conversion rate and the high selectivity. Moreover, an advantage of a heterogeneous (bio)catalyst is that it allows the transformation to be carried out in a continuous manner. Vinyl acetate displayed the highest activity as an acyl donor, compared to other acetates [101]. This can be explained by the formation of acetaldehyde in the transesterification step that shifts the equilibrium of the reaction to the product side.

10.2.5.2 β-Carotene

β-Carotene is a C$_{40}$-carotenoid with an orange-red colour and is applied in food and drink colouration, in cosmetics and in the animal feed industry, such as in egg and broiler pigmentation. Industrially, β-carotene is produced by chemical synthesis in several hundred t/a [102]. The main synthetic routes to β-carotene are using a *Wittig* reaction, a C$_{15}$ + C$_{10}$ + C$_{15}$ approach or a C$_{20}$ + C$_{20}$ approach using a vitamin A waste stream (Scheme 10.31). In addition to the chemical routes for β-carotene, microorganisms such as *Blakeslea trispora* [103], *Rhodotorula glutinis* [104], *Sphingomonas* sp. [105] and *Phyco-*

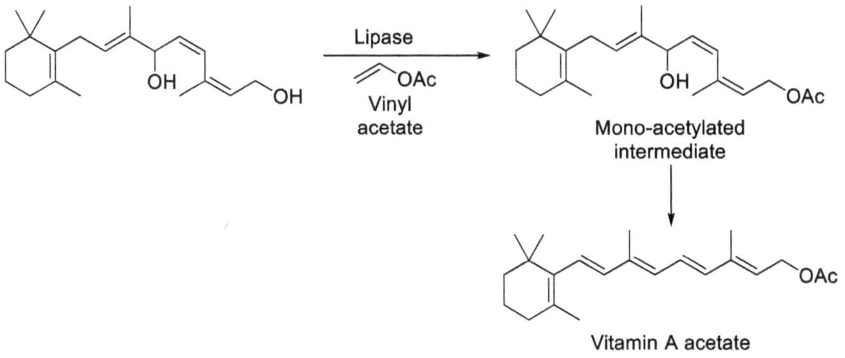

Scheme 10.30: Lipase-catalysed synthesis of a vitamin A precursor.

myces blakesleeanus [106] are also known to synthesise this carotenoid; the most promising commercially sources of the microbiological production are *Blakeslea trispora*.

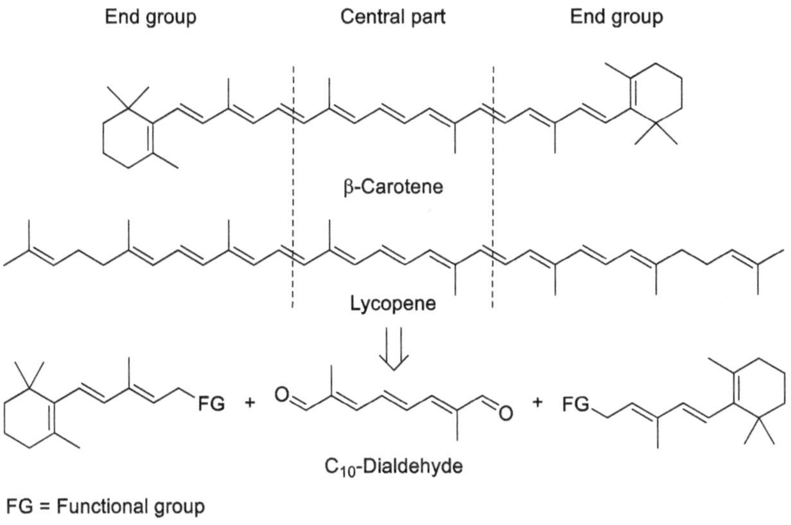

FG = Functional group

Scheme 10.31: β-Carotene chemical routes, with the preferred $C_{15} + C_{10} + C_{15}$ approach.

In *Blakeslea trispora*, β-carotene biosynthesis follows the mevalonate pathway [107] (Scheme 10.32) involving at least five crucial enzymatic steps, which are catalysed by enzymes encoded by the genes hmgR, ipi, carG, carRA and carB [108–112].

The production process starts with seed production, where glucose is consumed. Male and female strains are cultured and after cell growth, the fermentation process of β-carotene starts with the addition of oils such as soy oil. In the industrial process, a content of around 4% β-carotene can be achieved. The downstream processing involves the steps of biomass separation and drying, natural β-carotene (nBC) isolation

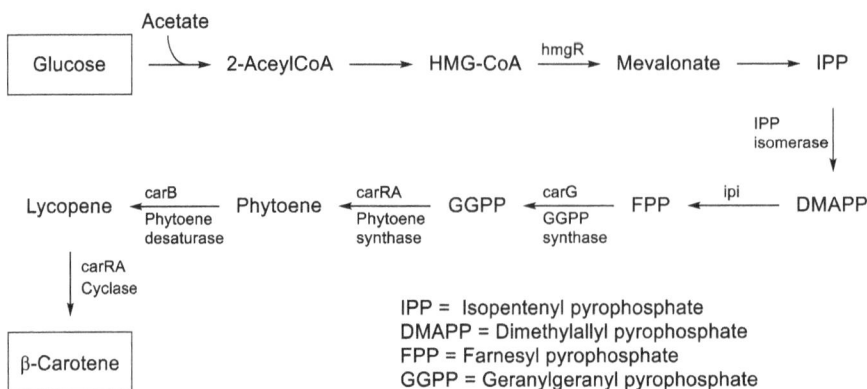

Scheme 10.32: Biosynthetic pathway to β-carotene in *Blakeslea trispora*.

by extraction, crystallisation, filtration and drying of the final product (Scheme 10.33 and Figure 10.12), which is packaged and ready for sale [107].

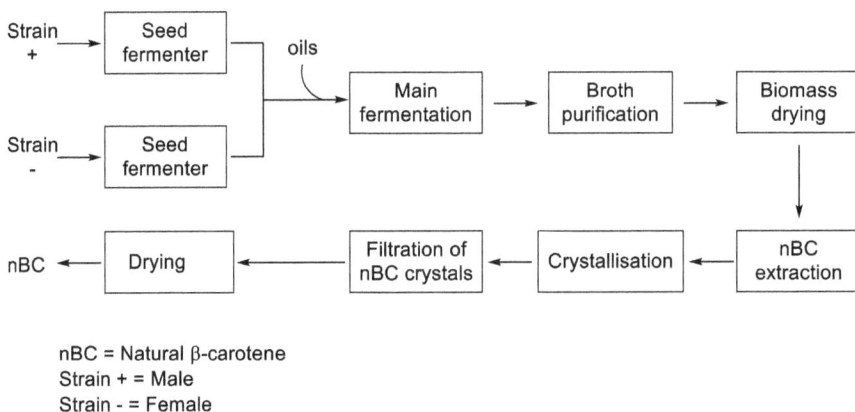

nBC = Natural β-carotene
Strain + = Male
Strain - = Female

Scheme 10.33: Natural β-carotene process flow sheet.

10.2.5.3 Riboflavin (Vitamin B$_2$)

The replacement of a chemical process by a biochemical one can be demonstrated with several examples, one of which is the manufacture of riboflavin (vitamin B$_2$). For decades, the manufacture of riboflavin was carried out by a chemical process (Scheme 10.34) and later by a combination of chemical and biochemical processes. In the original process, 3,4-xylidine was treated in a reductive coupling with ribose and further reacted with phenyldiazonium chloride and barbituric acid to form the final product. Ribose was originally synthesised from glucose by oxidation and later produced by fermenta-

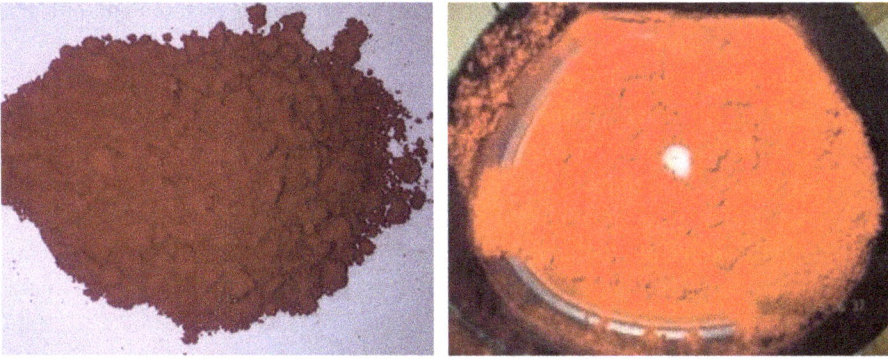

Figure 10.12: nBC from DSM's process. Left: after biomass drying, approx. 5% nBC. Right: final crystalline product (copyright photos: Antonio Estrella, DSM).

tion. Nowadays, only biochemical procedures are applied for the industrial-scale production of riboflavin (Scheme 10.35) [113].

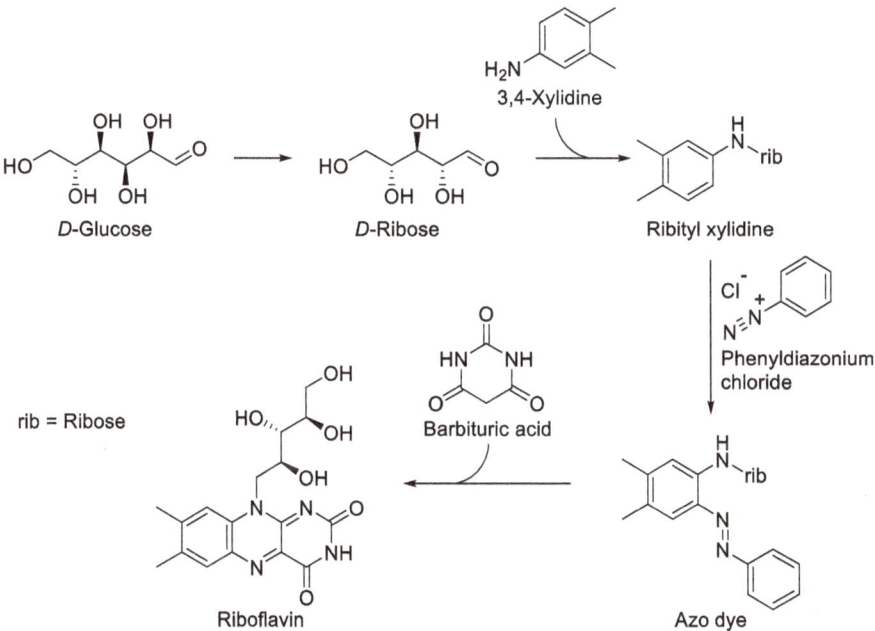

Scheme 10.34: Chemical process of riboflavin production.

The current microbiological processes are a result of systematic strain optimisation, for example the optimisation of the genetically modified fungus *Ashbya gossypii* [114] or *Bacillus subtilis* [115]. The titres of riboflavin could be increased from 10 g/l to >20 g/l, in water. The main advantages of these processes are the improved environmental and economic benefits, due to lower waste formation, cheap starting material and efficient downstream processing.

The riboflavin process based on *Ashbya gossypii* uses vegetable oils as the fermentation feedstock. An active transport process in *Ashbya gossypii* deposits the produced riboflavin in its vacuoles. After completion of the fermentation, the downstream processing starts with a heat-induced lysis of the biomass, followed by crystallisation of the product, separation by decantation and final purification by crystallisation.

The manufacture of riboflavin using a high performing *Bacillus subtilis* strain (RB50 mutant) was designed, isolated, optimised for industrial application and implemented on an industrial scale by DSM Nutritional Products (since 2000). The strain over-produces and secretes riboflavin. Genetic engineering was applied to further enhance the expression of the rib gene by making use of strong, constitutive promoters and by increasing the dosage of the rib gene [116]. Riboflavin accumulates during the fermentation and crystallises in the fermentation broth, and is separated from the biomass by decanting, washing and packaging. The material obtained is 96% pure and can be recrystallised for food and pharma applications (to >99% purity). The riboflavin world market is estimated to be approximately 100,000 t/a.

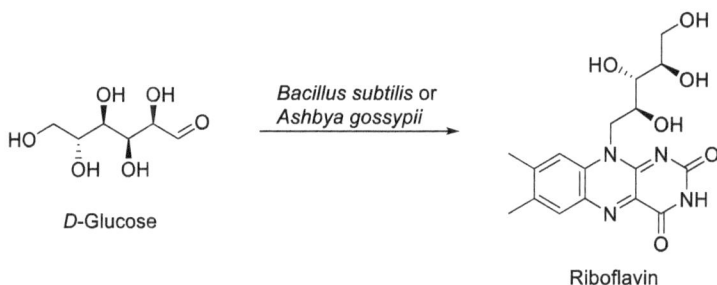

Scheme 10.35: Synthesis of riboflavin by fermentation processes.

10.2.5.4 Cobalamin (Vitamin B$_{12}$)

The group of cobalamins, compounds with corrinoid structure, includes adenosyl cobalamin, methyl cobalamin and hydroxy cobalamin, and is known as vitamin B$_{12}$ (Figure 10.13). These compounds have a tetrapyrrole ring, like porphyrins, which is coordinated to a cobalt ion. Nowadays, it is produced by fermentation using *Pseudomonas denitrificans*. This transforms glucose into vitamin B$_{12}$ at pH 7 and around 30 °C, and has been optimised to achieve >150 mg/l with a two-day fermentation time [117, 118]. Cobalamin plays a crucial role in the metabolism of every cell of the human body, is a co-factor in DNA synthesis and is important for the function of the nervous system.

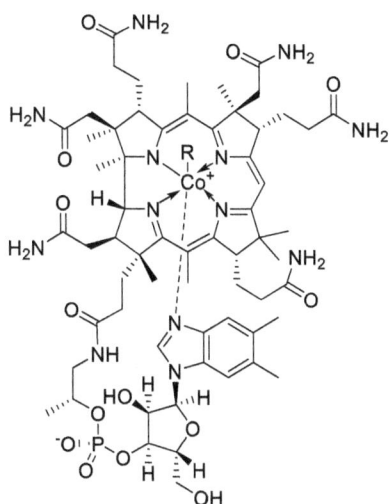

Figure 10.13: Cobalamin with R = –C≡N.

10.2.5.5 Vitamin C (ascorbic acid)

L-Ascorbic acid or vitamin C, a water-soluble carbohydrate acid, has several functions such as antioxidant properties, and it is an essential nutrient for humans, non-human primates and a few other mammals. The discovery of ascorbic acid is related to scurvy, a disease that was prevalent until the twentieth century, especially among sailors, with the symptoms of tiredness, shortness of breath, bleeding of mucous membranes, anaemia and eventually death. *L*-Ascorbic acid has a production volume of approximately 150,000 t/a at prices of 1.5–8 €/kg depending on the application and grade [119].

The manufacturing processes for ascorbic acid are based on a mixed chemical/fermentation procedure, which was initially developed by *Reichstein* and *Grüssner* in 1933. Most commercially produced *L*-ascorbic acids are produced by a variety of processes, which are generally variations of the Reichstein process. In the classical *Reichstein–Grüssner* process, *D*-glucose is converted in five steps into *L*-ascorbic acid [113].

In the *Reichstein–Grüssner* process, *D*-glucose is hydrogenated over a nickel-alloy catalyst to afford sorbitol (see Chapter 2: Heterogeneous hydrogenations), followed by the microbiological oxidation to *L*-sorbose; this was originally carried out with *Acetobacter xylinum* and later with *Gluconobacter oxydans* (Scheme 10.36). The acid-catalysed reaction with acetone results in the formation of 2,3,4,6-di-*O*-isopropylidene-α-*L*-sorbofuranose (also known as diacetone-*L*-sorbofuranose), which is oxidised in high yield and selectivity to di-*O*-isopropylidene-2-ketogulonic acid. Removing the ketal protecting groups by acid treatment and rearrangement (acid or base catalysed) results in ascorbic acid. The oxidation of 2,3,4,6-di-*O*-isopropylidene-α-*L*-sorbofuranose can be affected by treatment with hypochlorite in the presence of a nickel salt, by electrochemistry, or by air-oxidation catalysed by palladium or platinum on a carrier [113].

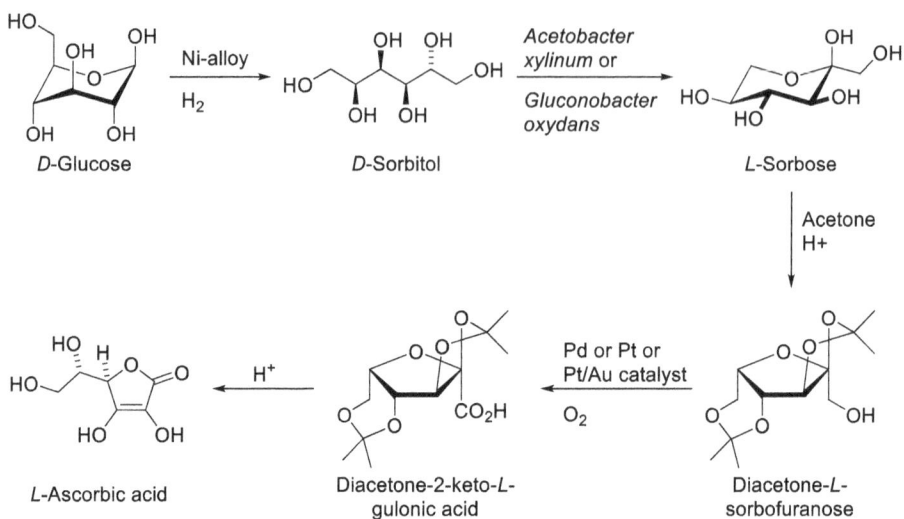

Scheme 10.36: Classical *Reichstein–Grüssner* process of ascorbic acid synthesis.

In general, several options for industrial production of ascorbic acid using fermentation processes have been described. The direct fermentation of sorbitol to ascorbic acid from glucose, or the one-step or two-step 2-KGA (2-ketogulonic acid) process from sorbitol and further treatment to ascorbic acid are known (Scheme 10.37) [120].

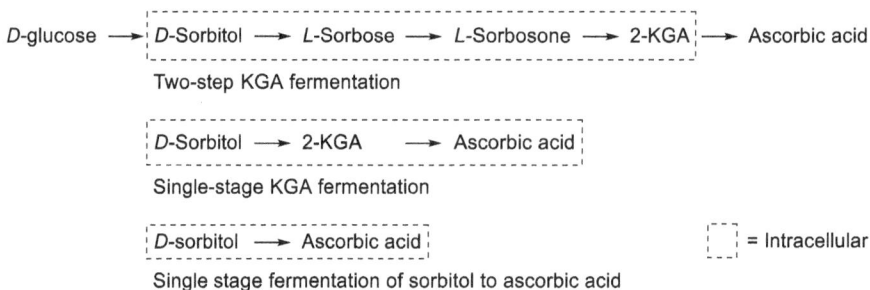

Scheme 10.37: Options for industrial production of vitamin C by fermentation processes.

Today, most industrial processes used for vitamin C production are based on 2-KGA (Scheme 10.38) – manufactured from *D*-sorbitol (from glucose). The chemical oxidation to 2-KGA was replaced by a fermentation step using a dehydrogenase from *Ketogulonicigenium vulgare* and a "helper strain" of *Bacillus megaterium*. The intermediate *L*-sorbosone is obtained by sorbose dehydrogenase oxidation (SDH) and further oxidised to 2-KGA by sorbosone dehydrogenase (SNDH) [120]. The concomitant bacterium is required for the growth of *Ketogulonicigenium*. The selective oxidation of sugar alcohols in the presence of acetic acid bacteria is well known [121]. *Gluconobacter*, formerly named as *Acetobacter suboxidants*, performs best in such oxidations. The D-sorbitol oxidation to L-sorbosone is carried out by this organism in nearly full conversion and high selectivity on an industrial scale [122].

The fermentation processes for 2-KGA production are mainly batch processes at pH 7 and at temperatures <30 °C. The fermentation time is between 40 and 65 h and 90–130 g/l 2-KGA can be achieved. Developments on continuous fermentation processes for both fermentation steps are described in the literature, especially the application of *Ketogulonicigenium vulgare* DSM 4025 at a steady-state concentration of 112 g/l 2-KGA, operating at >90% conversion. The space-time yield of 2.15 g/l/h demonstrated that the process is suitable for vitamin C production [123].

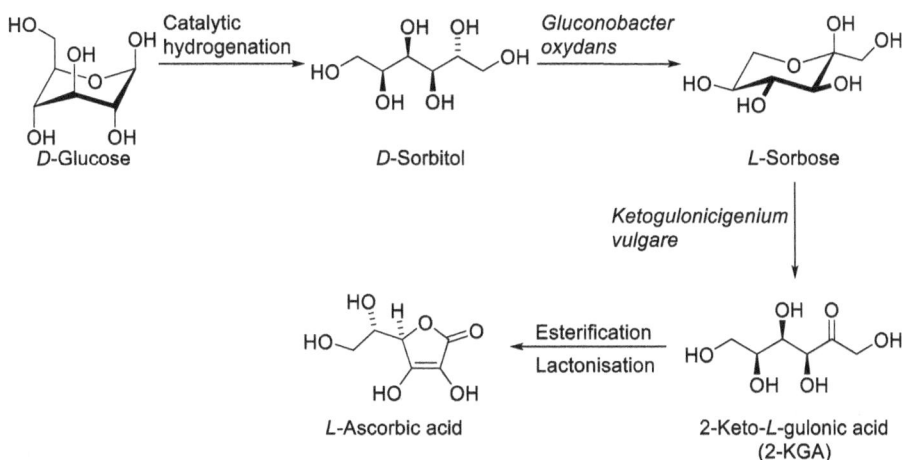

Scheme 10.38: Ascorbic acid process by two-step fermentation.

10.2.5.6 *(R)*-Pantolactone

(R)-Pantolactone is the key intermediate in the manufacture of pantothenic acid (vitamin B_5) and its derivatives. This vitamin is essential for the synthesis of co-enzyme A, acyl group transfer, and the metabolism of carbohydrates and fatty acids. Pantothenic acid and its derivatives are produced in capacities of around 10,000 t/a. The industrial manufacture is mainly carried out by aldol condensation of isobutyraldehyde and formaldehyde, followed by the addition of HCN under acidic conditions and subsequently by hydrolysis to form *(R/S)*-pantolactone (Scheme 10.41). The racemate of *(R/S)*-pantolactone is separated by resolution with chiral amines, the *(S)*-enantiomer is isomerised into the *(R/S)*-mixture and reused. The *(R)*-enantiomer is further converted into the *(R)*-pantothenic acid and its derivatives [124]. For the synthesis of *(R)*-pantolactone by enantioselective hydrogenation, see Chapter 3: Homogeneous hydrogenations.

A possible alternative method for the resolution of pantolactone is the lipase-catalysed kinetic resolution in the presence of vinyl acetate. In this process, the (S)-alcohol is selectively acetylated and separated from the (R)-alcohol. The recycling of the unwanted (S)-acetate proceeds via saponification and racemisation, but the requirement of additional steps mean that it is not economically beneficial compared to the classical resolution (Scheme 10.39) [125].

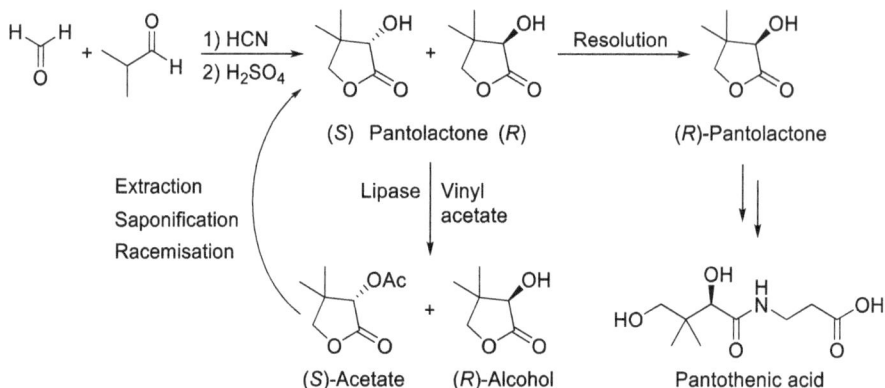

Scheme 10.39: Lipase-catalysed resolution of (R/S)-pantolactone.

An alternative enzymatic resolution approach is the enzyme-catalysed lactone opening. This approach is used by a number of manufacturers in Asia for the synthesis of (R)-pantolactone. It was first developed by the Daiichi Fine Chemical Company [126, 127]. Racemic pantolactone is treated with a lactonase from *Fusarium oxysporum*, which selectively hydrolyses the (R)-pantolactone to the ring-opened (R)-hydroxy acid. The (R)-acid can be separated from the unreacted (S)-pantolactone and ring-closed to produce (R)-pantolactone. This is further reacted to produce pantothenic acid derivatives (Scheme 10.40). The (S)-isomer is racemised and resubmitted to the process. There is a small background rate of hydrolysis of the (S)-isomer, so slightly lower enantioselectivities are obtained compared to the chemical resolution process. There have also been reports that an (S)-selective enzyme has been developed for this hydrolysis approach [128, 129].

All these approaches are based on resolution and so require the separation and racemisation of the unwanted enantiomer, else a maximum yield of 50% is possible. A more elegant enzymatic approach is the direct synthesis of the (R) enantiomer from achiral starting materials. This is achieved by the addition of hydrogen cyanide (HCN) to an aldehyde in an oxynitrilase catalysed reaction [130]. For the synthesis of (R)-pantolactone, the procedure is optimised using a modified oxynitrilase from *Prunus amygdalus* or *Pichia pastoris* [131]. Under buffer conditions at pH value <3, *ee*-values of >96% and a yield of around 90% of the (R)-dihydroxynitrile intermediate can be

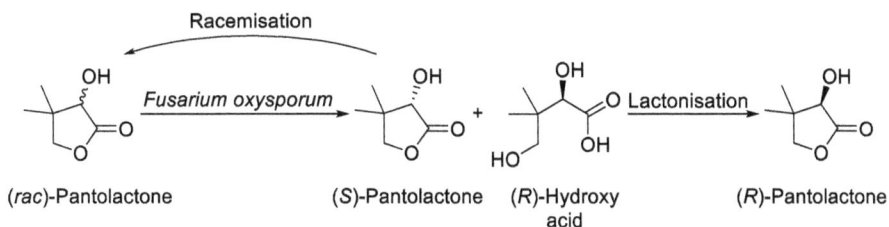

Scheme 10.40: Lactonase-catalysed resolution of (R/S)-pantolactone.

achieved in a 20 h reaction time. The unselective background reaction caused by the addition of HCN, resulting in a mixture of (R)- and (S)-isomers can be suppressed. The oxynitrilase catalysed reaction needs the hydroxy pivalaldehyde monomer, so the available dimer is cleaved at elevated temperatures or in strongly acidic conditions (Scheme 10.41).

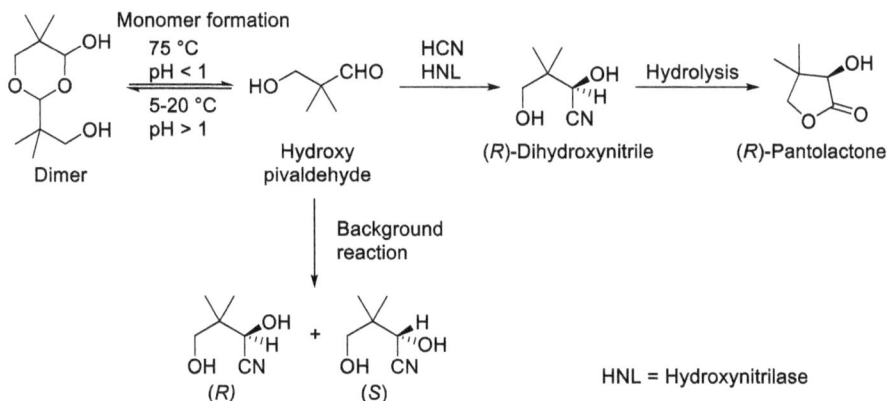

Scheme 10.41: Oxynitrilase application in the synthesis of (R)-pantolactone.

10.2.5.7 Nicotinamide (vitamin B₃)

The essential nutrient, nicotinamide (known as vitamin PP (pellagra preventing) or vitamin B$_3$), cures the skin disease pellagra and promotes animal growth. It is naturally available in a range of foods such as vegetables and plants and is generally consumed as part of normal diets and from supplements [132]. Nicotinamide is manufactured by a chemical process. The key intermediate is 3-picoline (3-methylpyridine), which is synthesised from ammonia, acetaldehyde and formaldehyde. The ammoxidation results in 3-cyanopyridine, which is converted into nicotinamide by treatment with an alkali. The process has the disadvantage of low selectivity, not only in the first step in which pyridine is also formed, but nicotinic acid formation also cannot be avoided [133]. A modified manufacturing process for nicotinamide starts from 2-methyl-5-ethypyridine, obtained from acetaldehyde and ammonia, oxidised with nitric acid and purified by crystallisation (Chapter 5: Gas-phase reactions).

In a new continuous process (>3,000 t/a), chemical and biocatalytic methods are combined (Scheme 10.42). The route starts from 2-methyl-1,5-diaminopentane (which is available from 5-methylglutaronitrile, a by-product of adiponitrile production). This diamine is cyclised and dehydrogenated to form 3-methylpyridine and converted into 3-cyanopyridine. The final hydrolysis step is carried out biocatalytically with polyamide immobilised *Rhodococcus rhodochrous* cells and is highly selective [134]. This procedure is environmentally friendly and safe, takes place in water and the hydrogen and ammonia are recycled [88].

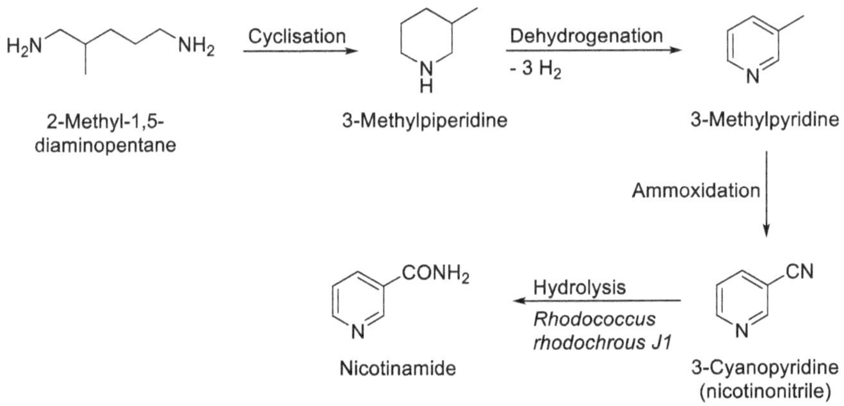

Scheme 10.42: Nicotinamide manufacture by a combined chemical-biotransformation approach (Lonza).

10.2.5.8 Calcifediol
7-Dehydrodesmosterol (or cholesta-5,7,24-trien-3-ol) is a cholesterol intermediate and is a key intermediate in the manufacture of 25-hydroxy-vitamin D_3 (calcifediol, Scheme 10.43) [135]. The fermentation process, developed at AMOCO, using a yeast organism from *Saccharomyces cerevisiae ATC 1562* produces cholesta-5,7,24-trien-3-ol in an efficient manner [136]. Further transformation of the trienol derivate into 25-hydroxy-vitamin D_3 is carried out by chemical transformations [137]. The alternative chemical route to 25-hydroxy-vitamin D_3 starts from 25-hydroxycholesterol, which is obtained from desmosterol by chemical steps.

The biotechnological approach to cholesta-5,7,24-trien-3-ol started with a strain construction by the mating of an erg5 strain and an erg6 strain M610-12B. A generated diploid was sporulated and a germinated spore called ATC05403mu was selected. The strain was mated with the lab strain *S. cerevisiae* DBY745 to give improvements in growth properties, sterol content and/or transformation efficiency. In analogy, researchers at AMOCO have also constructed a truncated version of the yeast HMG1 [138].

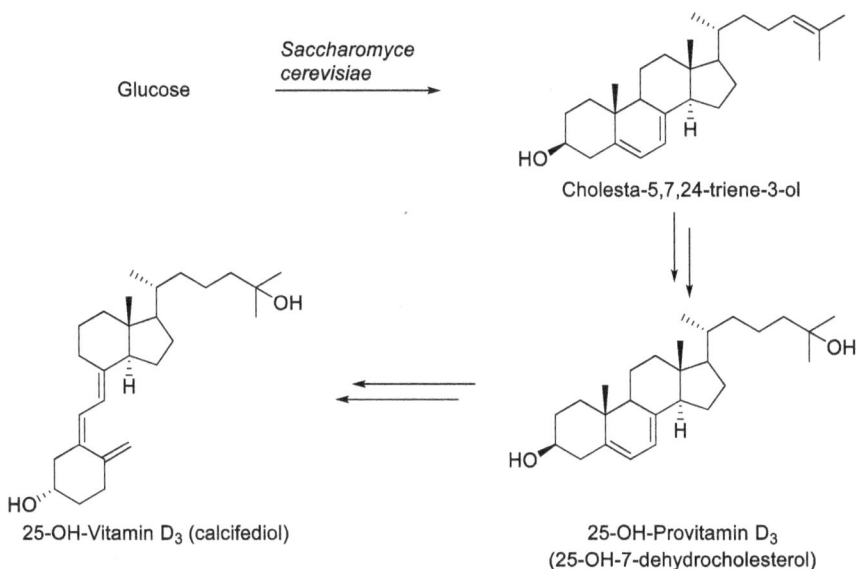

Scheme 10.43: 25-Hydroxy-vitamin D₃ from 25-hydroxycholesterol.

10.2.6 Active pharmaceutical ingredients

10.2.6.1 *(R)*-Phenylethylmethoxyamide
Enantiomerically pure compounds by lipase-catalysed reactions are well described in the literature, for example *Pseudomonas* species lipases are used in the manufacture of chiral amines or alcohols [139, 140]. The kinetic resolution of 1-phenyl-ethylamine and derivatives with an immobilised enzyme (polyacrylate carrier) allows the synthesis of *(R)*-phenylethylmethoxyamide, which is hydrolysed to *(R)*-phenylethylamine, an impor-tant building block for the pharma industry, in high *ee*-values (Scheme 10.44). The *(S)*-enantiomer can be isomerised with a Pd-catalyst and reused. The process is performed in MTBE and the enzyme activity can be increased by freeze-drying of the lipase in the presence of a fatty acid. The process is used in capacities of several 100 t/a, using a li-pase from *Burkholderia plantarii* (E.C. 3.1.1.3).

10.2.6.2 Diltiazem intermediate
For the manufacture of an intermediate of the calcium channel blocker diltiazem, a lipase-catalysed resolution is used in industrial settings. Various processes have been developed using either a *Rhizomucor miehei* lipase (RML) or a lipase from *Serratia mar-cescens* (Scheme 10.45). The latter is inhibited by an aldehyde, which can be formed via decarboxylation. The inhibition can be avoided if the transformation is carried out in

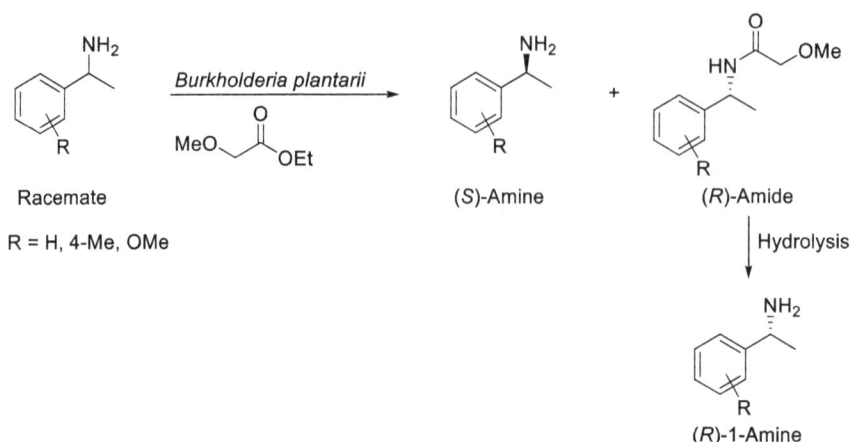

Scheme 10.44: Enantiomerically pure amines by lipase-catalysed resolution.

the presence of bisulfite. This process is carried out in a membrane reactor, resulting in a product with high purity and around 45% yield [141].

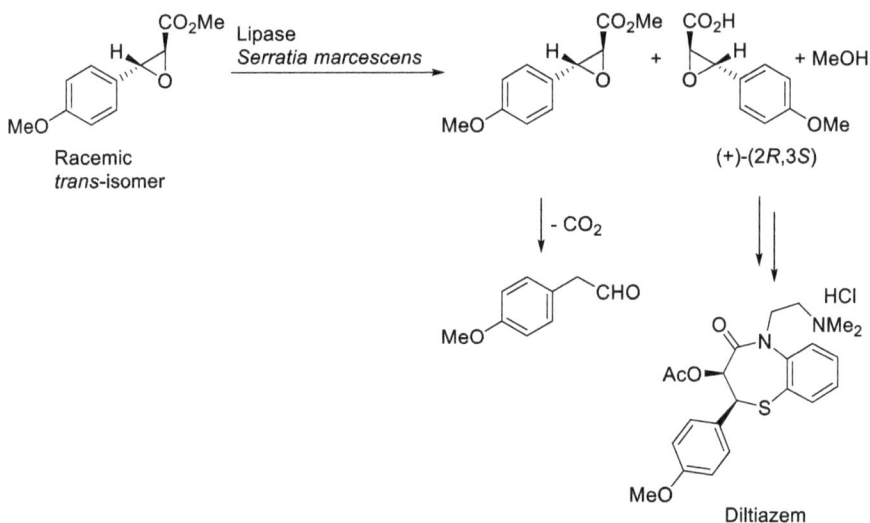

Scheme 10.45: *Serratia marcescens* catalysed kinetic resolution for a diltiazem intermediate (Tanabe Pharmaceutical).

10.2.6.3 (1S,2S)-*trans*-2-Methoxycyclohexanol

The manufacture of the secondary alcohol (1S,2S)-2-methoxycyclohexanol by resolution of the racemic mixture of the *trans*-alcohol is carried out by *Candida antarctica lipase B* (CAL-B) in scales of several t/a (Scheme 10.46). The alcohol is used for the manufacture of tricyclic β-lactam antibiotics such as Sanfetrinem. The acetate source is vinyl acetate and the reaction is carried out in cyclohexane/triethylamine with a reaction time of 6–8 h [142, 143].

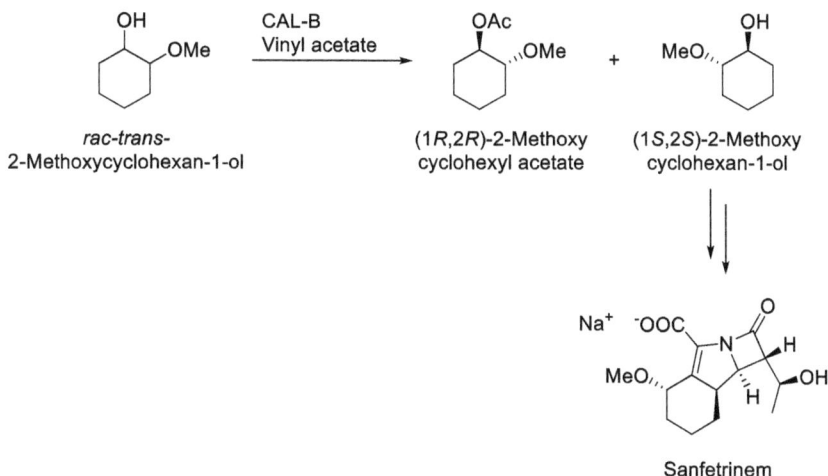

Scheme 10.46: Lipase CAL-B in the synthesis of an antibiotic intermediate.

10.2.6.4 Abacavir®
The enantioselective synthesis of a carbocyclic nucleoside precursor is carried out in capacities of several t/a using a γ-lactamase from *Bradyrhizobium japonicum*, expressed in *E. coli* (Scheme 10.47). The intermediate lactam, synthesised in >98% *ee*, is an intermediate of Abacavir®, a drug used in a combination therapy for the treatment of HIV-1 (reverse transcriptase inhibitor) [144].

10.2.6.5 Montelukast
Montelukast is a leukotriene receptor antagonist used in the treatment of asthma [145] and was developed by Merck in the early 1990s [146]. Since then, many efficient synthetic routes have been reported. Several different approaches to introduce the stereogenic centre have been developed and one of the most efficient approaches is via an engineered ketoreductase (Scheme 10.48). This approach was superior compared to the previously used stoichiometric reduction using the chiral boron reagent (−)-DIP-Cl. Enzyme activity was improved 3,000-fold using directed evolution technology. The described process is robust and gave the desired intermediate >99.9% *ee*, >95% yield, >98.5% purity at a >200 kg scale at concentrations of 100 g/l. IPA was used for cofactor (NADH or NADPH) regeneration [147].

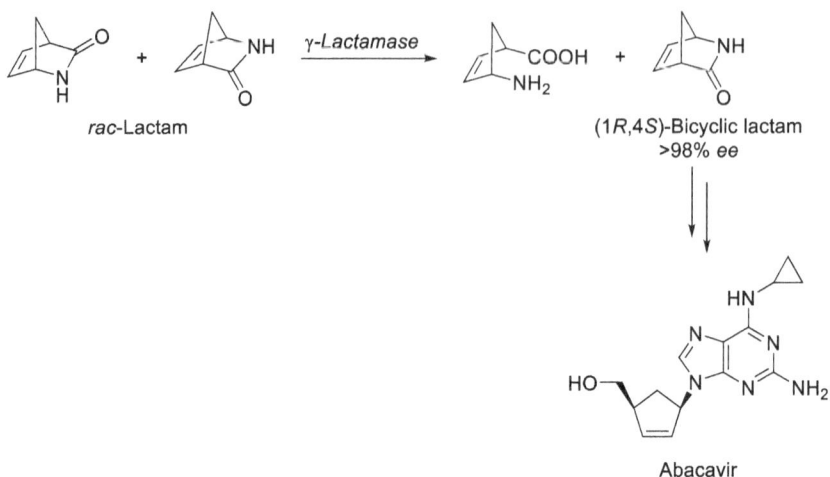

Scheme 10.47: Lactamase-catalysed synthesis of a HIV-1 inhibitor intermediate.

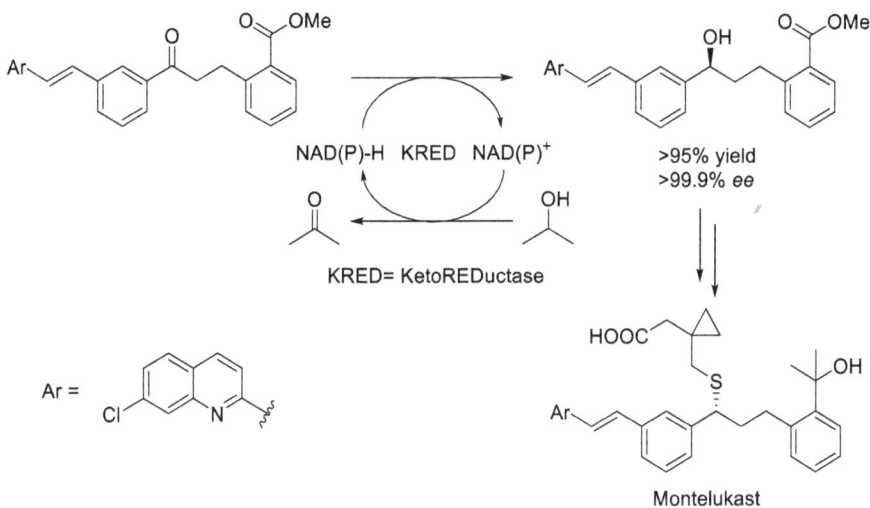

Scheme 10.48: The efficient application of a ketoreductases in the synthesis of montelukast.

10.2.6.6 *(R)*-Ethyl-4,4,4-trifluoro-3-hydroxybutanoate

(R)-Ethyl-4,4,4-trifluoro-3-hydroxybutanoate, an intermediate in the synthesis of the an-tidepressant monoamine oxidase-A inhibitor Befloxatone, is manufactured in a two-phase solvent system butyl acetate/water using a whole-cell fermentation (*Escherichia coli*) (Scheme 10.49). The organism contains two plasmids for the aldehyde reductase to reduce ethyl-4,4,4-trifluoroacetate and a glucose dehydrogenase to generate NADPH from NADP⁺. Titres of >300 g/l and an *ee* value of 99% can be achieved [148].

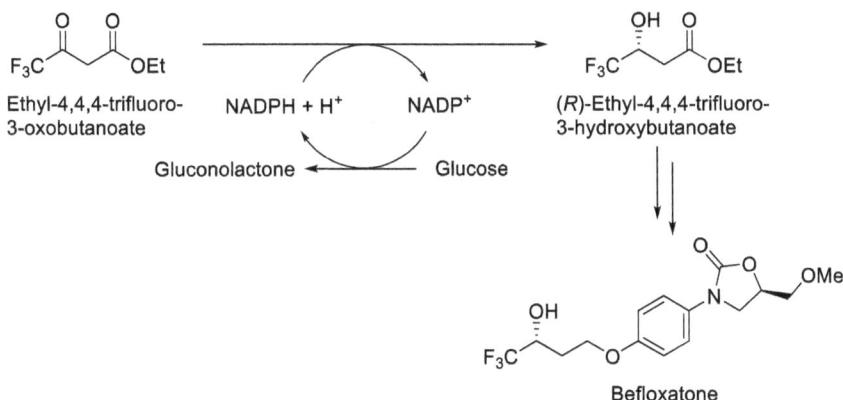

Scheme 10.49: Fermentative synthesis of (R)-ethyl-4,4,4-trifluoro-3-hydroxybutanoate.

10.2.6.7 Sitagliptin

Sitagliptin is an orally active pharmaceutical ingredient that was discovered by Merck and is used for the treatment of type 2 diabetes mellitus in Januvia® [149]. This compound is an impressive showcase of how processes develop over time in the pharmaceutical industry, where time to market is one of the most critical aspects. In the first- and second-generation syntheses, the stereogenic centre was introduced via homogeneous asymmetric hydrogenation of a β-ketoester or an enamine (Chapter 3: Homogeneous hydrogenations) [150]. In the current synthesis, the stereogenic centre is introduced by a transaminase (Scheme 10.50).

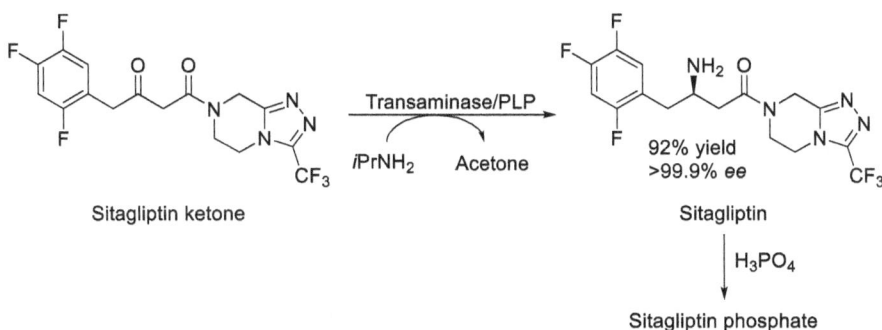

Scheme 10.50: Introduction of the stereogenic centre by an evolved transaminase.

The evolved enzyme contains 27 mutations, not only in the active side but also in the dimer interfacial region; it is assumed here that mutations enhance the enzyme stability. Under optimal conditions, 200 g/l sitagliptin ketone is converted to sitagliptin in >99.95% ee and 92% yield by using 6 g/l enzyme in 50% DMSO. The undesired enantiomer was never observed. The biocatalytic process provides sitagliptin with a 10–13%

increased overall yield, a 53% increase in productivity (kg/l per day) and a 19% reduction in total waste compared to the asymmetric enamine reduction. Additional benefits are the elimination of all heavy metals, a reduction in total manufacturing cost and the benefit that the enzymatic reaction can be performed in multipurpose vessels, avoiding specialised high-pressure hydrogenation equipment [151]. These improvements were awarded with the Presidential Green Chemistry Challenge Award in 2010 for Greener Reaction Conditions. This is the second award of this type for this molecule, after the recognition of the asymmetric enamine hydrogenation in the earlier synthesis.

10.2.6.8 6-Aminopenicillanic acid

The manufacture of 6-aminopenicillanic acid (6-APA), a backbone for a variety of semisynthetic antibiotics, starts from penicillin G, which is produced by fermentation using *Penicillium chrysogenum* strains. The penicillin G is converted to 6-APA at 37 °C in water, using *Penicillium acylase* (E.C.3.5.1.11, Scheme 10.51) [152]. Immobilised penicillin acylases can also be used in the forward/synthetic direction to transform 6-APA into ampicillin and amoxicillin, and also the related antibiotics cephalexin and cefadroxil, all in more than 1,000 t/a [153].

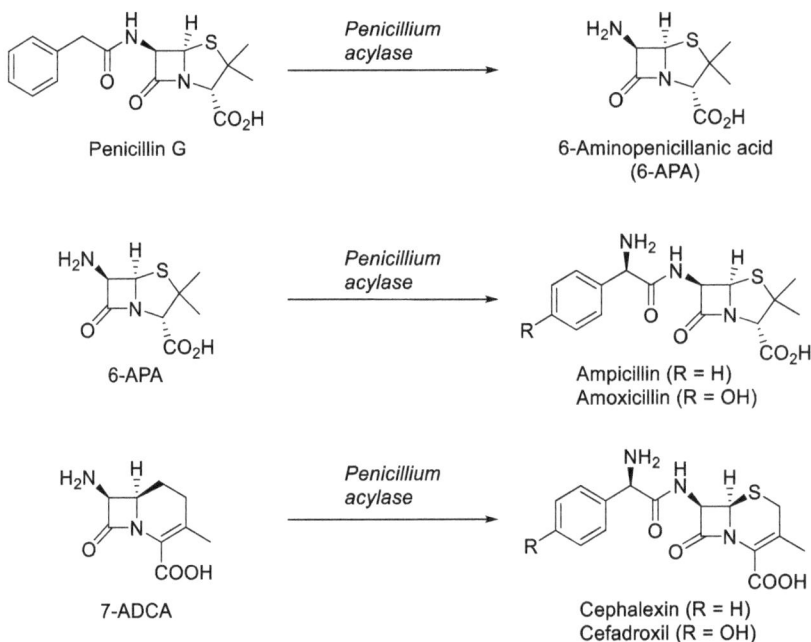

Scheme 10.51: Manufacture of 6-APA and other antibiotics using penicillin acylases.

Manufacture of Penicillin G, production volume >15 kt/a (Glaxo, former Gist Brocades) starts from phenyl acetic acid. The reaction temperature is around 25 °C for 200 h, followed by H_2SO_4 addition and adjustment to pH 2 (Scheme 10.52). Recovery yields for Penicillin G are >90%. The industrial production of Penicillin V uses phenoxyacetic acid as starting material [154].

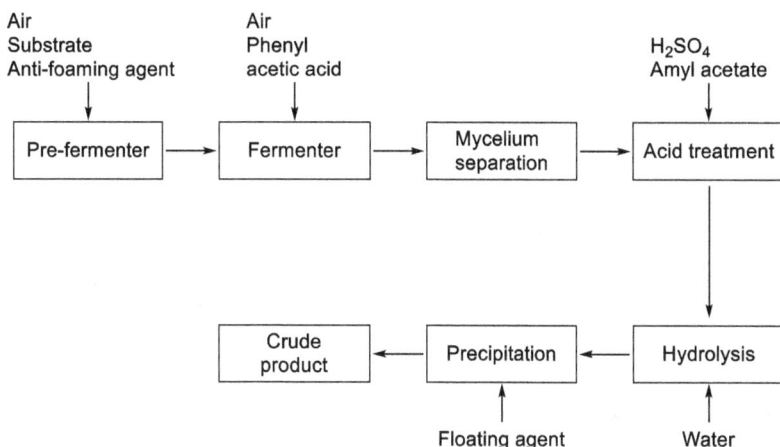

Scheme 10.52: Simplified flow scheme of penicillin G production.

10.2.6.9 5-Methylpyrazine-2-carboxylic acid, 2-(6-methylpyridyl-3)acetic acid and pyridyl-3-acetic acid

The manufacture of 5-methylpyrazine-2-carboxylic acid, an intermediate for the anti-diabetic drug Glipizide, is carried out by oxidation of 2,5-dimethylpyrazine with *Pseudomonas putida*, with living cells grown from xylene [155]. 2,5-Dimethylpyrazine is fed in amounts of 2 g/l into the reactor in the continuous process. After acidification, the product is separated by precipitation. The final product is obtained in >95% purity, 80% yield and a concentration of >20 g/l (Scheme 10.53). An advantage of the process is the high selectivity since oxidations using metal oxides are usually unselective [156].

A similar biocatalytic oxidation of alkyl groups is the use of *Pseudomonas oleovorans*, which is grown on *n*-octane. 5-Ethyl-2-methylpyridine can be oxidised to the corresponding acetic acid derivative 2-(6-methylpyridin-3-yl) acetic acid in high selectivity and the same transformation is used in the manufacture of 3-pyridylacetic acid, an intermediate in the synthesis of the drug Risedronate, which is used in the treatment of osteoporosis (Scheme 10.54) [155, 157].

10.2.6.10 Pipecolic acid and piperazine carboxylic acid

Hydrolysis of nitriles under mild reaction conditions can be carried out in the presence of nitrile hydrolysing enzymes. These enzymatic hydrolyses follow two different

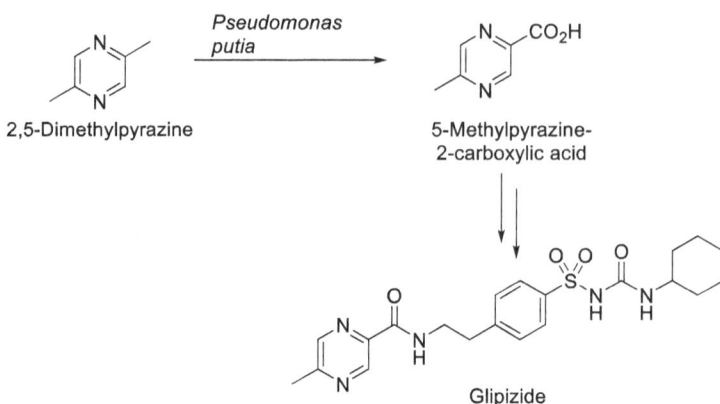

Scheme 10.53: Oxidation of 2,5-dimethylpyrazine.

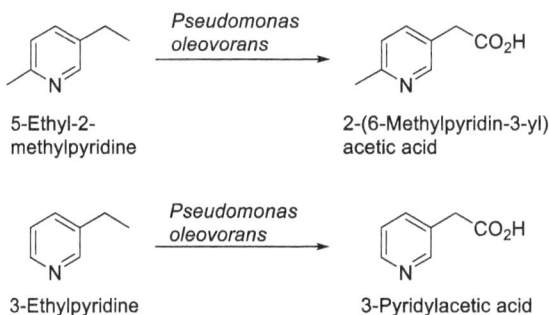

Scheme 10.54: Oxidation of alkyl groups in presence of *Pseudomonas oleovorans*.

pathways. In the direct conversion of nitriles, the corresponding acid and ammonia
are formed in nitrilase-catalysed reactions. Nitrile hydratase-catalysed reactions result in the formation of the amide, which can be transformed in the presence of an
amidase into acid and ammonia (Scheme 10.55).

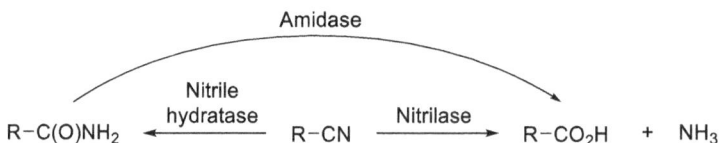

Scheme 10.55: Different pathways of nitrile hydrolysis.

The pharmaceutical drug building block (*S*)-pipecolic acid (piperidine 2-carboxylic acid)
is produced by a combined chemical and biotransformation approach in *ee*-value of
99%. Picolinonitrile is converted to the corresponding amide using a nitrile hydratase,

followed by hydrogenation to *rac*-piperidine-2-carboxamide. The whole-cell amidase-catalysed resolution with *Pseudomonas fluorescens* DSM9924 and precipitation under pH control (pH 7) results in the final product (Scheme 10.56). This technology also allows the production of (*R*)- and (*S*)-piperazine-2-carboxylic acid, where different enzymes hydrolyse different enantiomers of the starting amide [79, 158–160].

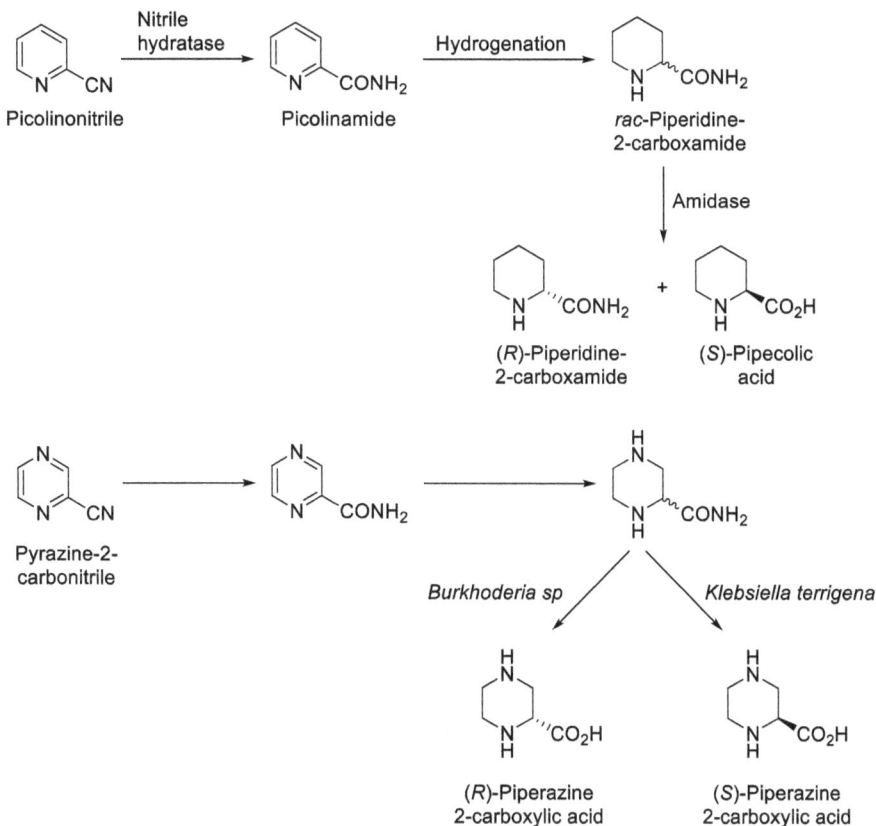

Scheme 10.56: Pipecolic acid and piperazine-2-carboxylic acid processes.

10.2.6.11 Pregabalin

Pregabalin is a lipophilic γ-aminobutyric acid (GABA) analogue that was developed by Pfizer for the treatment of several central nervous system disorders (e.g. anxiety and epilepsy). The first-generation manufacturing process was based on a racemic synthesis followed by a resolution by the formation of a salt with mandelic acid as the final step. Although this route was cost-effective, a more efficient route was desired that allowed a resolution earlier in the synthesis and recycling of the undesired enantiomer (Scheme 10.57).

First generation route:

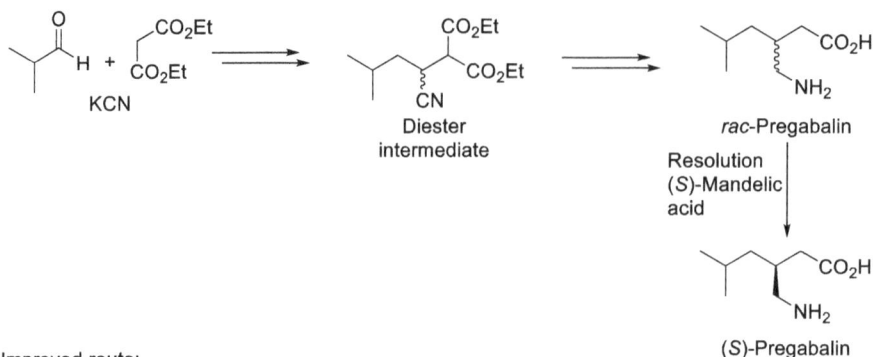

Improved route:

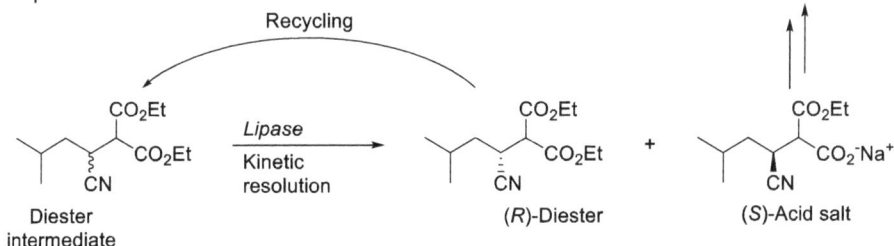

Scheme 10.57: Manufacturing processes for pregabalin.

The improved route used the same diester intermediate as the original route, but then used a kinetic resolution with a lipase at this stage. The resolution produced the salt of the (S) acid; the (R)-diester was left unreacted and could be easily recycled using a base [161]. A wide variety of lipases were screened and two were found to give the correct (S)-isomer with high selectivity, a lipase from *Rhizopus delemar* and another from *Thermomyces lanuginosus*. The latter was found to be more active, cheap and commercially available (as Lipolase from Novozymes), so was selected for further development. The enzyme was expressed in a genetically modified *Aspergillus oryzae* strain and used in the kinetic resolution in less than 24 h at high concentration (~3 molar), with a turnover of almost 100,000. Overall, this improved route resulted in the use of one-fifth of the amount of chemicals and one-eighth of the amount of solvent, compared to the first-generation process (Chapter 3: Homogeneous hydrogenations).

10.2.6.12 Artemisinin

Artemisinin is a sesquiterpene with high anti-malarial activity that was first isolated from the plant *Artemisia annua*. It is a key component of a combination therapy for the treatment of malaria recommended by the WHO (World Health Organization) in 2004. Originally, artemisinin was extracted from natural sources, but this resulted in very variable prices and availabilities, therefore an alternative source was required. Artemisinin has a complicated structure including an unstable peroxide bridge; therefore, a complete total synthesis of the drug substance is unlikely to be cost-effective. An ingenious solution was found by the combination of biotechnological and chemical steps, a so-called semi-synthetic route.

The synthesis of sesquiterpenes is known to proceed via the mevalonate pathway, producing FPP (farnesene pyrophosphate) (see Section 10.2.9.4 below on farnesene synthesis). Originally, the over-expression of the enzymes used to produce FPP and artemisinic acid was investigated in *E. coli*; titres of 25 g/l could be achieved for an artemisinic acid precursor, but only 1 g/l of artemisinic acid was achieved. The switch of host to baker's yeast (*Saccharomyces cerevisiae*) was more successful and allowed increased production of artemisinic acid (>25 g/l). A further advantage was that the produced artemisinic acid precipitated out of the growth medium and could be isolated by extraction (Scheme 10.58) [162–164].

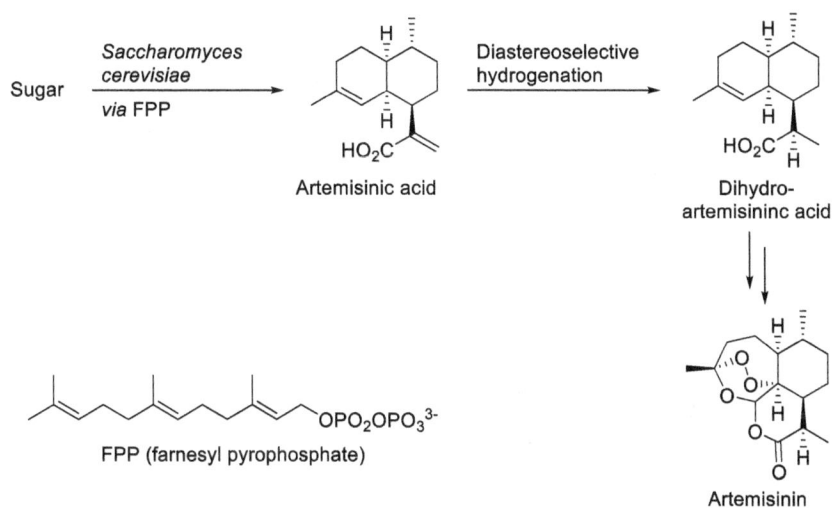

Scheme 10.58: Production of artemisinin by fermentation.

The semi-synthetic production of artemisinin was completed by a diastereoselective homogeneous hydrogenation (Chapter 3: Homogeneous hydrogenations), formation of a mixed-anhydride and a photo-oxidation (Chapter 4: Oxidations) [165]. The whole route (fermentation and further elaboration) was used to produce 60 tonnes of artemisinin in 2014.

10.2.6.13 LSD1-Inhibitor GSK2879552
In addition to alcohol and keto-reductases, more recently a related class of enzymes, imine reductases, have been shown to be efficient catalysts for the production of

pharmaceutical intermediates. These are different to the transaminases used, for example in the production of Sitagliptin (see Chapter 10.2.6.7: Sitagliptin).

GlaxoSmithKline developed the compound GSK2879552 as a lysine-specific demethylase-1 (LSD-1) inhibitor, which was applied in clinical trials for lung cancer and leukaemia. The original chemical route involved the resolution of a chiral cyclopropylamine followed by a reductive amination with an aldehyde intermediate. While successful, the resolution required the formation of a chiral salt and the reductive amination had to be performed at low temperature using sodium borohydride as reductant, generating stoichiometric amounts of boron waste [166]. A new process was developed using an imine reductase enzyme that had reduced cost, fewer chemical steps, lower solvent usage and a lower process mass intensity for a key intermediate of GSK2879552 (Scheme 10.59).

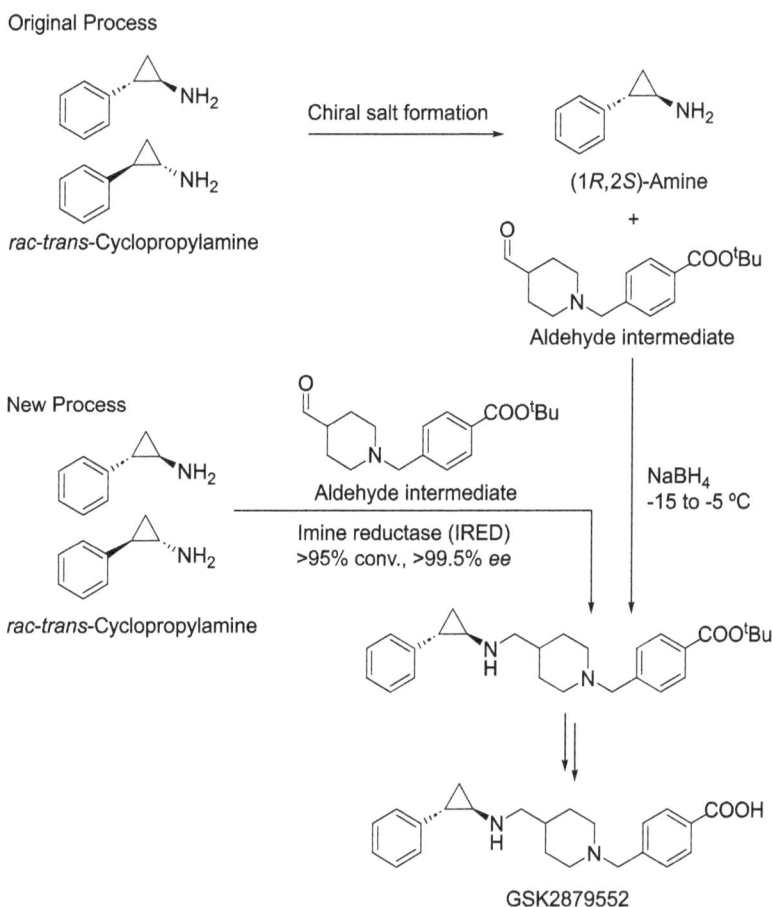

Scheme 10.59: Imine reductase approach to an intermediate of GSK2879552.

Initially, a panel of imine reductase enzymes was screened, and several hits were identified that would catalyse the reductive amination with the single enantiomer (1R,2S)-amine in up to 95% conversion. Three rounds of directed evolution were performed on the enzyme to obtain the final mutant that could perform a kinetic resolution using 2.2 equivalents of the *racemic-trans*-amine giving 84% isolated yield of the reductive amination product in >99.5% *ee*.

10.2.6.14 Abrocitinib

Another use of imine reductases is the production of Abrocitinib, a Janus kinase (JAK) inhibitor developed by Pfizer that is used in the treatment of atopic dermatitis. One of the key features of this compound is the *cis*-diaminocyclobutane unit in the centre of the molecule. Several different approaches were investigated to synthesise the key amino-cyclobutane intermediate. Imine formation with methylamine, followed by chemical reduction required low temperatures and resulted in an 80:20 mixture of *cis:trans* diastereoisomers. Although the desired (*cis*)-isomer could be obtained by multiple crystallisations, a more efficient route was required. Transamination of the starting ketone was successful in producing the primary amine; however, mono-alkylation was not straightforward (Scheme 10.60) [167].

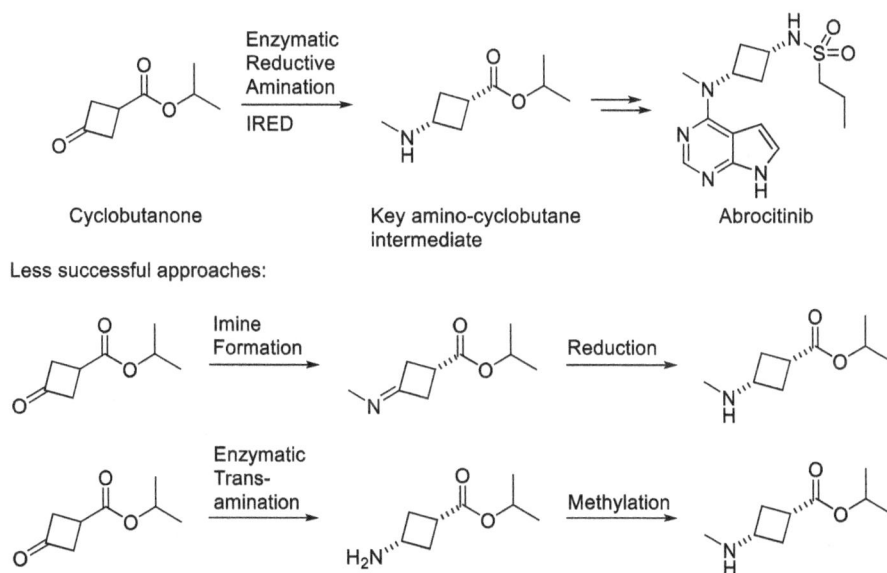

Scheme 10.60: Imine reductase approach to an intermediate of Abrocitinib.

An initial screen of a Pfizer library of 80 wild-type imine reductase (IRED) enzymes identified three hits and scaling up the reaction with the best enzyme from *Streptomyces purpureus* showed high selectivity for the *cis*-product (>99:1). However, repeating the reaction under conditions that would be required for a viable process showed very low conversion (<1%), but the high selectivity was maintained. In addition, the wild-type enzyme only operated in a narrow pH and temperature window. Three rounds of enzyme engineering were performed using a mixture of computational-based and bioinformatics-based approaches coupled with the identification of hot spots and synergistic recombination. The best mutant was then tested with multiple cofactor regeneration systems, with glucose dehydrogenase being found to be optimal. Under optimised conditions, batch sizes of >200 kg were run, with >91% conversion being obtained in 48 h. In total >3.5 t of material were produced in >99% purity and >99:1 Z:E selectivity.

10.2.7 Flavour compounds

10.2.7.1 Jasmonate
The enzyme-catalysed kinetic resolution of a jasmonate intermediate for the flavour in-dustry was developed by using an allylic alcohol and dimethyl malonate in the presence of an immobilised *Candida antarctica* species from Novozyme in the presence of $KHCO_3$. At conversions of 50%, an *ee* value of >97% was achieved. The unwanted isomer was recovered and racemised [168, 169] (Scheme 10.61). Silylation and *Claisen* rearrangement, followed by decarboxylation, epoxidation and epoxide opening results in the formation of the jasmonate derivative.

10.2.7.2 Aspartame ex Holland Sweetener Company
Aspartame is an artificial sweetener and is used in the food industry for soft drinks, bakery ingredients, corn flakes, bubble gums and puddings. It is 200 times sweeter than sucrose. The name aspartame is the combination of *L*-aspartic acid and *L*-(S)-phenylalanine, which is combined as a dipeptide. Aspartame is the methyl ester deriv-ative of this dipeptide and is produced by chemical or enzymatic methods.

In the chemical route, (S)-aspartic acid is treated with phosphorus oxychloride and (S)-phenylalanine methyl ester (Scheme 10.62). However, a by-product can be formed (the bitter-tasting β-aspartame) from the reaction of the phenylalanine methyl ester with the other carbonyl group of the anhydride and this must be removed from the final product.

The enzymatic route starts from racemic phenylalanine methyl ester (Scheme 10.63). This racemic mixture reacts with protected aspartic acid to form a mixture of protected aspartame and (R)-phenylalanine methyl ester. After filtration, acid treatment and re-moval of the protecting group by hydrogenation with Pd/C, it results in the final product (see Chapter 2: Heterogeneous hydrogenation). The (R)-amino acid is recovered, racem-ised and reused [170]. Around 25% of the world market of aspartame (around 15,000 t/a) is produced by this enzymatic process.

H₂SO₄

MeO OMe

Novozyme 435, KHCO₃
43%

HO

rac-2-Pentylcyclopent
-2-en-1-ol

(S)-Alcohol
OH

+

(R)-Malonate
OMe

NaH
TMSCl

CO₂Me

NMP
NaCl/H₂O

Decarboxylation

TMSO₂C
CO₂Me

THF
50-65 °C

Claisen
rearrangement

Me₃SiO OMe

CF₃CO₃H
Epoxidation

CO₂Me

AlCl₃

CO₂Me

(+)-Methyl
dihydroepijasmonate

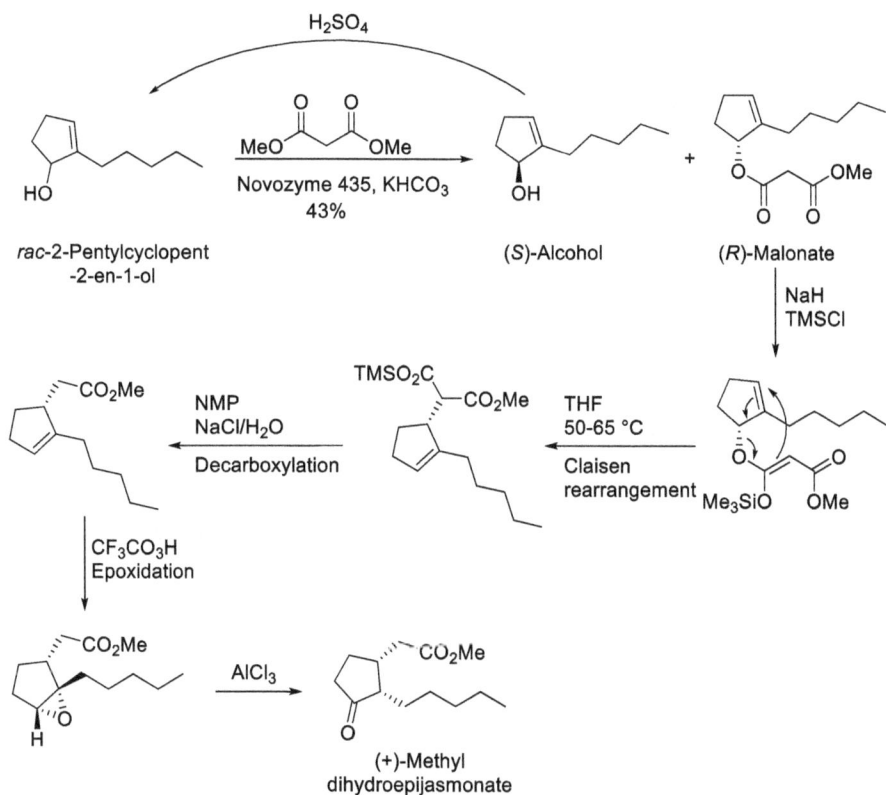

Scheme 10.61: Enzyme-catalysed resolution in the jasmonate synthesis.

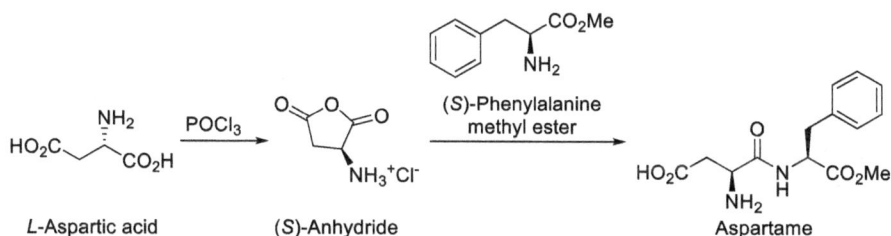

CO₂Me

NH₂

(S)-Phenylalanine
methyl ester

NH₂

HO₂C CO₂H

L-Aspartic acid

POCl₃

O O O

NH₃⁺Cl⁻

(S)-Anhydride

HO₂C N CO₂Me
H
NH₂

Aspartame

Scheme 10.62: Chemical route to aspartame.

10.2.8 Pesticides

10.2.8.1 *(S)*-2-Hydroxy-2-(3-phenoxyphenyl) acetonitrile

Another industrialised oxynitrilase-catalysed hydrocyanation reaction is the manufacture of (S)-2-hydroxy-2-(3-phenoxyphenyl) acetonitrile. The starting material is 3-phenoxybenzaldehyde, an intermediate in the synthesis of pyrethroids (a group of pes-

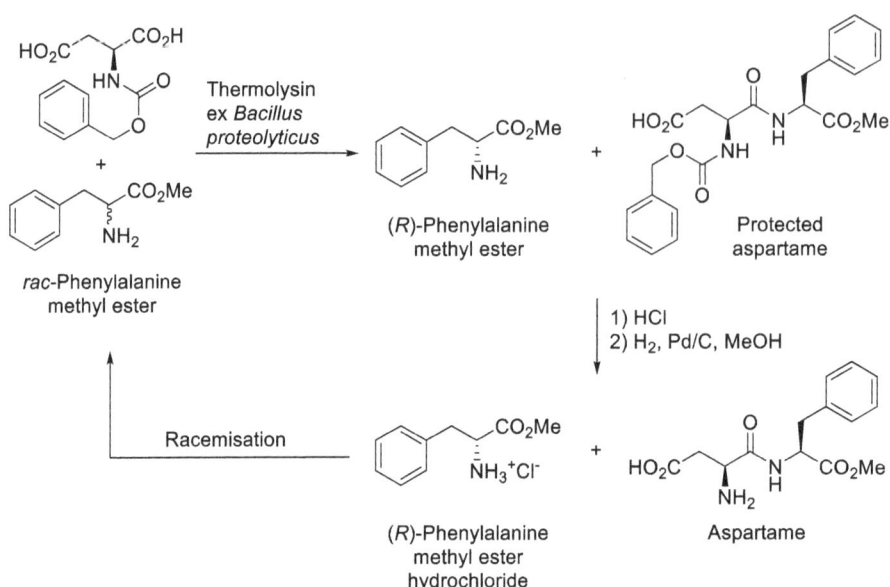

Scheme 10.63: Aspartame process of Holland Sweetener Company.

ticides). The reaction is carried out in a two-phase solvent system of water/*t*-butyl methyl ether. The oxynitrilase comes from *Hevea brasiliensis* (E.C. 4.1.2.39). The product is obtained with an *ee* value of >95% at a conversion of 46% (Scheme 10.64) [171].

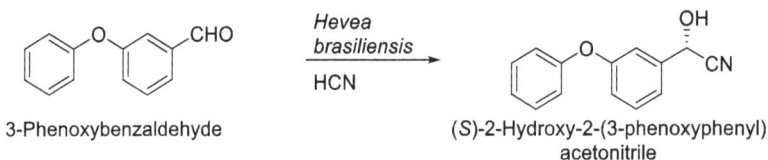

Scheme 10.64: *Hevea brasiliensis* catalysed synthesis of (S)-2-hydroxy-2-(3-phenoxyphenyl) acetonitrile.

10.2.9 Platform chemicals

10.2.9.1 Acrylamide

The transformation of acrylonitrile to acrylamide is carried out with *Rhodococcus* (nitrile hydratase, E.C. 4.2.1.84) immobilised on poly(acrylamide) gel at pH 8.0–8.5 in a semi-batch reaction mode. The process is carried out at Nitto Chemicals, Japan and was first industrialised in 1985. The substrate concentration is constant at 3% and the original production scale was around 10,000 t/a. In the meantime, this was increased to a worldwide production volume of approximately 50,000 t/a (Scheme 10.65) [172].

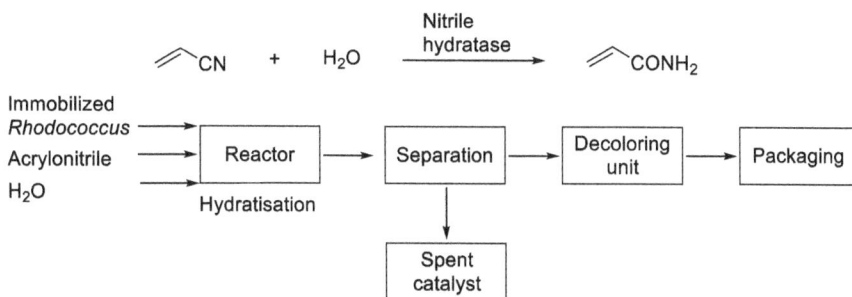

$$\begin{array}{ccccc} \nearrow CN & + & H_2O & \xrightarrow{\text{Nitrile hydratase}} & \nearrow CONH_2 \end{array}$$

Immobilized *Rhodococcus*

Acrylonitrile

H_2O

→ Reactor → Separation → Decoloring unit → Packaging

Hydratisation

Separation → Spent catalyst

Scheme 10.65: Block-flow diagram for *Rhodococcus*-based acrylamide process.

Concentrations of up to 600 g/l can be achieved. This process is the world's largest process of a commodity chemical (non-chiral) manufactured with an enzyme (non-fermentative). The main advantages of this process are the high selectivity at high conversion. The process runs at full conversion and as compared to the classical Cu-based chemical process, the recovery of non-reacted acrylonitrile and the copper catalyst is avoided [173].

10.2.9.2 Bioethanol

The transformation of biomass to ethanol, which is used as fuel, is called bioethanol. This ethanol is mixed with mineral oil and used in biofuel (sometimes, together with methanol). Natural materials such as wood, plants or straw are suitable starting materials; in general, materials with a high content of sugar or starch, such as corn (North America), wheat (EU), sugar beet route (EU) *triticale sorghum* or manioc (Asia) can be used. The pre-treatment that is needed depends on the starting material. If the material is first broken-up and then fermented or directly fermented (molasses) – materials like straw need an acidic and enzymatic treatment. The fermentation to bioethanol is carried out with a yeast of the *Saccharomyces cerevisiae* type. Ethanol concentrations of around 12% can be achieved. Further purification by azeotropic distillation results in a product of >99.5% content (Scheme 10.66). Also, membrane processes are used in the concentration/purification of bioethanol. Nowadays, several 10^6 t/a bioethanol are produced worldwide.

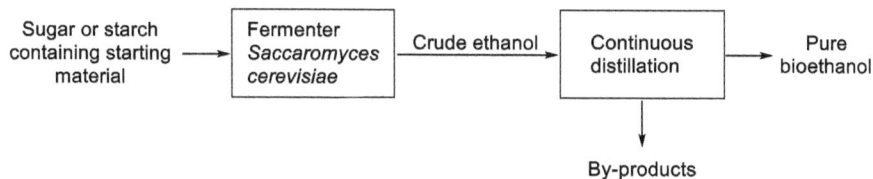

Sugar or starch containing starting material → Fermenter *Saccaromyces cerevisiae* —Crude ethanol→ Continuous distillation → Pure bioethanol

Continuous distillation → By-products

Scheme 10.66: Block flow diagram of bioethanol production.

10.2.9.3 Biogas

Microorganisms can be used in the processing of waste and wastewater, resulting in the production of biogas (Scheme 10.67). The fermentation of organic materials under anaerobic conditions results in the production of a mixture of carbon dioxide and methane. The organic materials are converted at yields >90%. The produced methane can be used for energy generation (co-generation of heat and power) or can be purified and sold as biomethane [174].

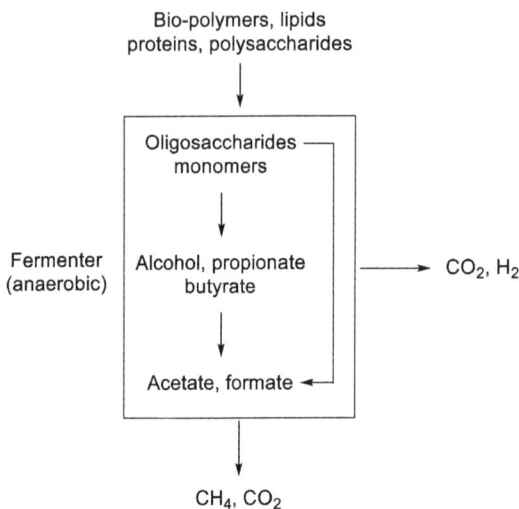

Bio-polymers, lipids
proteins, polysaccharides

↓

Fermenter (anaerobic)	Oligosaccharides — monomers
	↓
	Alcohol, propionate butyrate → CO_2, H_2
	↓
	Acetate, formate ←

↓

CH_4, CO_2

Scheme 10.67: Flow sheet of biogas formation.

The methanogenic bacteria (*archaebacteria*) convert formic acid, methanol and acetic acid into methane. These organic molecules are produced by the hydrolysis of amino acids or by acidogenic bacteria, converting fatty acids into butyric and/or propionic acids and alcohols.

10.2.9.4 Farnesene

Terpenoids are an important class of natural products and are constructed from multiple units of the isoprenoid (C_5) building block. Over the last decade, within the class of sesquiterpenes (C_{15}-isoprenoids), the family of farnesene ($C_{15}H_{24}$), six closely related chemical compounds have gained increasing industrial relevance. This was originally as a potential biofuel and more recently as a bio-based intermediate for the synthesis of more complicated molecules. The isomers, α-farnesene (3,7,11-trimethyl-1,3,6,10-dodecatetraene) and β-farnesene (7,11-dimethyl-3-methylene-1,6,10-dodecatriene), differ only in the location of one double bond (Scheme 10.68). The most common isomer (*E,E*)-α-farnesene is found in the coating of apples and other fruits. It is responsible for the characteristic green apple smell [175]. β-Farnesene occurs in nature in the (*E*)-

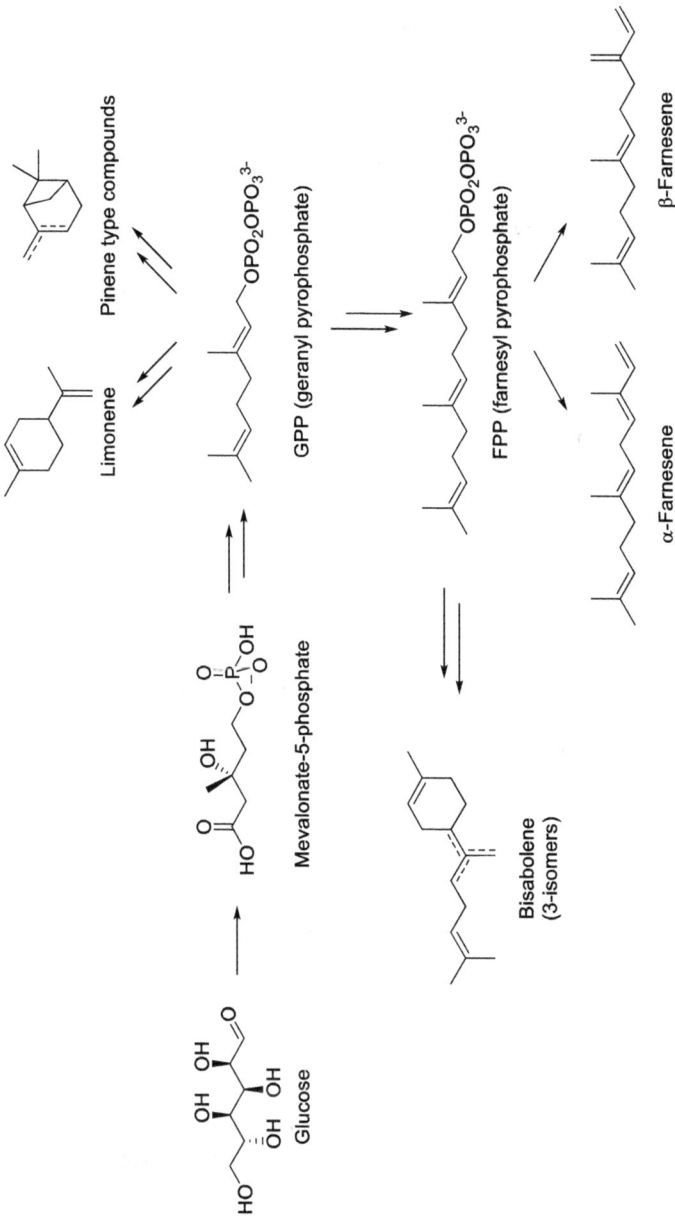

Scheme 10.68: Isoprenoids from sugar.

configuration in various essential oils and is also released by aphids as an alarm pheromone [176]. It is also known that potato species synthesize β-farnesene as a natural insect repellent [177].

There are two main biosynthetic pathways to produce isoprenoids, via mevalonate (MVA pathway) and via methylerythritol phosphate (MEP pathway). The MVA pathway has been successfully used for the production of isoprenoid compounds such as farnesene, and bisabolene as potential diesel and jet fuel, as it has produced higher titres than the MEP pathway [178–180]. The production of farnesene and bisabolene from FPP (farnesene pyrophosphate) by farnesene synthase has been reported. FPP overproduction in *Escherichia coli* and yeast strains led to a farnesene titres of >100 g/l in yeast fermentation [181]. The manufacture of β-farnesene is currently carried out in Brazil due to the low sugar price.

References

[1]   Milnera E, Maguire A R. Recent trends in whole cell and isolated enzymes in enantioselective synthesis. ARKIVOC 2012, i, 321–382.

[2]   Kozlowski L P. Proteome-pI: Proteome isoelectric point database. Nucleic Acids Res 2017, 45, D1112–D1116.

[3]   Brocchieri L, Karlin S. Protein length in eukaryotic and prokaryotic proteomes. Nucleic Acids Res 2005, 33, 3390–3400.

[4]   Koonin E V, Galperin M Y. Sequence – Evolution – Function: Computational Approaches in Comparative Genomics. Boston, Kluwer Academic, 2003, Chapter 8, Genomes and the Protein Universe. Available from: https://www.ncbi.nlm.nih.gov/books/NBK20267/.

[5]   Lemieux R U, Spohr U. How Emil Fischer was led to the lock and key concept for enzyme specificity. Adv Carbohydr Chem Biochem 1994, 50, 1–20.

[6]   Eschenmoser A. One hundred years lock-and-key principle. Angew Chem Int Ed 1995, 33, 2363.

[7]   Koshland D E Jr. The key–lock theory and the induced fit theory. Angew Chem Int Ed 1995, 33, 2375–2378.

[8]   Dodson G, Wlodawer A. Catalytic triads and their relatives. Trends Biochem Sci 1998, 23, 347–352.

[9]   Waldmann H, Drauz K. Enzyme Catalysis in Organic Synthesis, I. 2nd ed. Weinheim Germany, Wiley-VCH Verlag GmbH, 2002.

[10]   Uppada V, Bhaduri S, Noronha S B. Co-factor regeneration – An important aspect of biocatalysis. Curr Sci 2014, 106, 946–957.

[11]   Basso A, Hesseler M, Serban S. Hydrophobic microenvironment optimization for efficient immobilization of lipases on octadecyl functionalised resins. Tetrahedron 2016, 72, 7323–7328.

[12]   Keim W, Driessen-Hölscher B. Heterogenization of Complexes End Enzymes. In: Ertl G, Közinger H, Weitkamp J, eds. Handbook of Heterogeneous Catalysis, Wiley-VCH, Weinheim, 231–240.

[13]   Raser P. Chiral Catalyst Immobilization and Recycling, De Vos D E, Vankelecom I F J, Jacobs P A, eds. Wiley-VCH, 2000, 96–104.

[14]   http://www.chem.qmul.ac.uk/iubmb/enzyme/http://www.chem.qmul.ac.uk/iubmb/enzyme/. (21.06.2019).

[15]   Faber K. Biotransormations in Organic Chemistry: A Textbook. 3rd ed. Berlin, Germany, Springer-Verlag, 1997.

[16]   Kafarski P. Rainbow code of biotechnology. Chemik 2012, 66, 811–816.

[17] Straathof A J, Panke S, Schmid A. The production of fine chemicals by biotransformations. Curr Opin Biotechnol 2002, 13, 548–556.

[18] Schmid A, Dordick J S, Hauer B, Kiener A, Wubbolts M G, Witholt B. Industrial biocatalysis today and tomorrow. Nature 2001, 409, 256–268.

[19] Wubbolts M G, Favre-Bulle O, Witholt B. Biosynthesis of synthons in two-liquid phase media. Biotechnol Bioeng 1996, 52, 301–308.

[20] Freifelder D, Davison P F. Studies on the sonic degradation of deoxyribonucleic acid. Biophys J 1962, 2, 235–247.

[21] Chomczynski P, Sacchi N. Single-step method of RNA isolation by acid guanidinium thiocyanate-phenol-chloroform extraction. Anal Biochem 1987, 162, 156–159.

[22] Birnboim H C, Doly J. A rapid alkaline extraction procedure for screening recombinant plasmid DNA. Nucleic Acids Res 1979, 7, 1513–1523.

[23] Rhodes A, Fletcher D L. Principles of Industrial Microbiology. Oxford, Pergamon, 1966.

[24] https://www.cbd.int/abs/about/ (20.09.2020).

[25] https://www.cbd.int/abs/nagoya-protocol/signatories/ (20.09.2020).

[26] Jumper J, Evans R, Pritzel A, Green T, Figurnov M, Ronneberger O, Tunyasuvunakool K, Bates R, Žídek A, Potapenko A, Bridgland A, Meyer C, Kohl S A A, Ballard A J, Cowie A, Romera-Paredes B, Nikolov S, Jain R, Adler J, Back T, Petersen S, Reiman D, Clancy E, Zielinski M, Steinegger M, Pacholska M, Berghammer T, Bodenstein S, Silver D, Vinyals O, Senior A W, Kavukcuoglu K, Kohli P, Hassabis D. Highly accurate protein structure prediction with AlphaFold. Nature 2021, 596, 583–598.

[27] Ireland T. Will AlphaFold change bioscience research? The Biologist, 4 December 2020. (Accessed May 2, 2023, at https://thebiologist.rsb.org.uk/biologist-features/how-will-alphafold-change-bioscience-research.)

[28] Callawy E. The entire protein universe: AI predicts of nearly every known protein. Nature 2022, 608, 15–16.

[29] Dellweg H. Biotechnologie Verständlich. Springer, 1994, 32.

[30] Wang J, Yan Y. Chapter 1: Glycolysis and its metabolic engineering applications. Eng microb metabolism Chem Synth Rev Perspect 2018, 1–33.

[31] Hobbs L. Sweeteners from Starch: Production, Properties and Uses. Starch: Chemistry and Technology. 3rd ed. London, Academic Press/Elsevier, 2009, 797–832.

[32] Byong H L. Fundamentals of Food Biotechnology. John Wiley and Sons, 332, 2014.

[33] Labout J J M. Conversion of liquefied starch into glucose using a novel glucoamylase system. Starch 1985, 37, 157–161.

[34] Holm J, Bjoerck I, Ostrawska S, Eliasson A C, Asp N G, Larsson K, Lundquist I. Digestibility of amylose-liquid complexes in vitro and in-vivo. Starch 1983, 35, 294–297.

[35] Kragl U, Gygax D, Ghisalba O, Wandrey C. Enzymatische zweistufige Synthese von N-Acetylneuraminsäure im Enzym-Membranreaktor. Angew Chem 1991, 103, 854–855.

[36] Narziß L, Back W, Gastl M, Zarnkow M. Abriss der Bierbrauerei, 8, Wiley-VCH, 2017.

[37] Wagemann K, Tippkötter N. Biorefineries. 1st ed., Springer, 2019.

[38] Newswire P R (USA). Human Milk Oligosaccharides Market Size Worth $101.5 Million By 2027: Grand View Research, Inc. February 11, 2020, 4.

[39] Taylor D E, Rasko D A, Sherburne R, Ho C, Jewel L D, Laurence D. Lack of correlation between Lewis antigen expression by Helicobacter pylori and gastric epithelial cells in infected patients. Gastroenterology 1998, 115, 1113–1122.

[40] Lee W-H, Pathanibul P, Quarterman P J, J-h J, Han N-S, Miller M J, Jin Y-S, Seo J-H. Whole cell biosynthesis of a functional oligosaccharide, 2'-fucosyllactose, using engineered Escherichia coli. Microb Cell Fact 2012, 11, 48.

[41] Chin Y-W, Kim J-Y, Lee W-H, Seo J-H. Enhanced production of 2′-fucosyllactose in engineered Escherichia coli BL21star (DE3) by modulation of lactose metabolism and fucosyltransferase. J Biotechnol 2015, 210, 107–115.

[42] Chin Y-W, Seo N, Kim J-H, Seo J-H. Metabolic engineering of Escherichia coli to produce 2′-fucosyllactose via Salvage pathway of guanosine 5′-diphosphate (GDP)-L-fucose. Biotechnol Bioeng 2016, 113, 2443–2452.

[43] Vidra A, Németh A. Bio-produced Acetic Acid: A review. Period Polytech Chem Eng 2018, 63, 245–256.

[44] Le Berre C, Serp P, Kalck P, Torrence G P. Acetic Acid. In: Elvers, ed. Ullmann's Encyclopedia of Industrial Chemistry. Online ed., Weinheim, Germany, Wiley-VCH Verlag GmbH & Co, 2014.

[45] Ragsdale S W, Pierce E. Acetogenesis and the Wood-Ljungdahl pathway of CO2 fixation. Biochim Biophys Acta Proteins Proteomics 2008, 1784, 1873–1898.

[46] Fontaine F E, Peterson W H, McCoy E, Johnson M J, Ritter G J. A new type of glucose fermentation by clostridium thermoaceticum. J Bacteriol 1942, 43, 701–715.

[47] Anastassiadis S, Morgunov I G, Kamzolova S V, Finogenova T V. Citric acid production patent review. Recent Pat Biotechnol 2008, 2, 107–123.

[48] Majumder L, Khalil I, Munshi M K, Alam K, Rashid H, Begum R, Alam N. Citric acid production by Aspergillus niger using molasses and pumpkin as substrates. Eur J Biolog Sci 2001, 2, 1–8.

[49] Lotfy W A, Ghanem K M, El-Helow E R. Citric acid production by a novel Aspergillus niger isolate: II. Optimization of process parameters through statistical experimental designs. Bioresour Technol 2007, 98, 3470–3477.

[50] Schlegel H G. Allgemeine Mikrobiologie. Thieme. Vol. 8. Stuttgart/New York, 2008.

[51] Schlegel H G, Zaborosch C. Allgemeine Mikrobiologie. 7th ed., Stuttgart/ New York, Thieme, 1992, 296–304.

[52] Syldatk C Organische Säuren. Essigsäure (Acetat). In: Garabed Antranikian: Angewandte Mikrobiologie, Berlin/ Heidelberg, Springer-Verlag, 2006.

[53] Starr J N, Westhoff G. Lactic Acid. In: Elvers, ed. Ullmann's Encyclopedia of Industrial Chemistry. Online ed, Weinheim, Germany, Wiley-VCH Verlag GmbH & Co, 2014.

[54] Okabe M, Lies D, Kanamasa S, Park E Y. Biotechnological production of itaconic acid and its biosynthesis from aspergillus terreus. Appl Microbiol Biotechnol 2009, 84, 597–606.

[55] Vroemen A J, Beverini M. Enzymatic Production of Gluconic Acid or Its Salts. WO 96/35800, Gist Brocades, 1996.

[56] Levy J K, Crawford P C, Appel L D, Clifford E L Comparison of intraventricular injection of zinc gluconate versus surgical castration to sterilize male dogs. Am J Vet Res 2008, 69, 140–143.

[57] Reznikoff P, Goebel W F. The preparation of ferrous gluconate and its use in the treatment of hypochromic anelia in rats. J Pharmacol Exp Ther 1937, 59, 182–192.

[58] Freddi G, Allera G, Cadiani G. Degumming of silk fabrics with tartaric acid. Color Technol 2008, 112, 191–195.

[59] Gal J. The discovery of biological enantioselectivity: Louis pasteur and the fermentation of tartaric acid, 1857 – A review and analysis 150 yr later. Chirality 2008, 20, 5–19.

[60] Ress-Löschke M, Friedrich T, Bauer B, Mattes R, Engels D. New Nucleic Acid Sequence Encoding Alcaligenes Faecalis Nitrilase Polypeptide Useful for Converting Racemic Nitriles to Chiral Carboxylic Acids. DE 19848129, BASF, 2000.

[61] Putten P L. Mandelic acid and urinary tract infections. Antonie van Leeuwenhoek 1979, 45, 622–623.

[62] Roth H J, Müller C E, Folkers G. Stereochemie & Arzneistoffe, Wissenschaftliche Verlagsgesellschaft Stuttgart, 1998, 164–165.

[63] Izumi Y, Chibata I, Itoh T. Herstellung und Verwendung von Aminosäuren. Angew Chem 1978, 90, 187–194.

[64] Hoppe B, Martens J. Aminosäuren – Herstellung und Gewinnung. Chemie in Unserer Zeit 1984, 3, 73–86.

[65] Waite J H, Tanzer M L. The bio adhesive of Mytilus byssus: A protein containing L-dopa. Biochem Biophys Res Commun 1980, 96, 1554–1561.

[66] Lopez V M, Decatur C L, Stamer W D, Lynch M R, McKay B S. L-DOPA is an endogenous ligand for OA1. PLoS Biol 2008, 6, 236.

[67] Kumaja H. Production of L-3,4-dihydroxyphenylalanine. Adv Eng Biotechnol 2000, 69, 71–85.

[68] Tsuchida T, Nishimoto Y, Kotani T, Iiizumi K. Hydrophilized Curable Silicone Impression Materials with Improved Storage Behavior. JP 5123177, Ajinomoto, 1993.

[69] Ager D J. Handbook of Chiral Chemicals. Marcel Dekker, New York, 1999.

[70] Wandrey C, Wichmann R, Leuchtenberger W, Kula M R, Bueckmann A. Process for the Continuous Enzymatic Change of Water Soluble Alpha -ketocarboxylic Acids into the Corresponding Amino Acids. US 4304858, Degussa, 1981.

[71] Leuchtenberger W, Karrenbauer M, Plöcker U. Scale-up of an enzyme membrane reactor process for the manufacture of l-enantiomeric compounds. Ann N Y Acad Sci 1984, 434, 78–86.

[72] Schoemaker H E, Boesten W H J, Kaptain B, Hermes F F M, Sonke T, Broxterman Q B, van den Tweel W J J, Kamphuis J. Chemo-enzymatic synthesis of amino acids and derivatives. Pure Appl Chem 1992, 64, 1171–1175.

[73] van den Tweel W J J, Van Dooren T J G M, De Jonge P H, Kaptein B, Duchateau A L L, Kamphuis J. Ochrobacterium anthropic NCIMB 40321: A new biocatalyst with broad-spectrum L-specific amidase activity. Appl Microbiol 1993, 39, 296–300.

[74] Kragl U, Vasic-Racki M, Wandrey C. Continuous production of L-tert-leucine in series of two enzyme membrane reactors. Modeling and computer simulation. Bioprocess Eng 1996, 14, 291–297.

[75] Kula M-R, Wandrey C. Continuous enzymic transformation in an enzyme-membrane reactor with simultaneous NADH regeneration. Methods Enzymol 1987, 136, 9–21.

[76] Thomas K, Woodley J. Rational Selection of Co-factor Regeneration Processes. In: 2nd European Symposium on Biochemical Engineering Science, Porto. 1998, 201.

[77] Eichhorn E, Roduit J-P, Shaw N, Heinzmann K, Kiener A. Preparation of (S)-piperazine-2-carboxylic acid and (R)-piperazine carboxylic acid by kinetic resolution of the corresponding racemic carboxamides with stereoselective amidases in hole bacterial cells. Tetrahedron Asymmetry 1997, 8, 2533–2536.

[78] Sauter M, Venetz D, Henzen F, Schmidhalter D, Pfaffen G, Werbitsky O. Process for producing n-protected d-proline derivatives. WO 9733987, Lonza, 1997.

[79] Petersen M, Sauter M. Biotechnology in the fine chemical industry. Cyclic amino acids by enantioselective biocatalysis. Chimia 1999, 53, 608–612.

[80] Tenud L. Verfahren Zur Herstellung von Carnitin. De 2518813, Lonza, 1975.

[81] Tenud L. Verfahren zur Herstellung von Carnitin. DE 2542196, Lonza, 1976.

[82] Tenud L. Verfahren zur Herstellung von Carnitin. DE 2542227, Lonza, 1976.

[83] Meyer H-P, Kiener A, Imwinkelried R, Shaw N. Biotransformations for fine chemical productions. Chimia 1997, 51, 287–289.

[84] Chmiel H, Takors R, Weuster-Botz D. Bioprozesstechnik. 4th ed., Springer, 2018, 127–128.

[85] Jung H, Jung K, Kleber H P. Synthesis of (L)-carnitine by microorganism and isolated enzymes. Adv Biochem Eng Biotechnol 1993, 50, 21–44.

[86] Kula M-R, Joeres U, Stelkes-Ritter U. New microbial amidases. Ann N Y Acad Sci 1996, 799, 725–728.

[87] Kleber H-P. Bacterial carnitine metabolism. FEMS Microbiol Lett 1997, 147, 1–9.

[88] Shaw N M, Robins K T, Kiener A. Lonza: 20 Years of Biotransformations. Adv Synth Catal 2003, 345, 425–435.

[89] Drauz K, Grayson I, Kleemann A, Krimmer H-P, Leuchtenberger W, Weckbecker C. Amino Acids. In: Elvers, ed. Ullmann's Encyclopedia of Industrial Chemistry. Online ed., Weinheim, Germany, Wiley-VCH Verlag GmbH & Co, 2007.

[90] Maier T, Winterhalter C. Method for Production of L-Cysteine or L-Cysteine Derivatives By Fermentation. WO 0127307, Consortium für Electrochemische Industrie, April 19, 2007.

[91] Maier T H P. Semisynthetic production of unnatural L-α-amino acids by metabolic engineering of the cysteine-biosynthetic pathway. Nat Biotechnol 2003, 21, 422–427.

[92] Martens J, Offermanns H, Scherberich P. Facile synthesis of racemic cysteine. Angew Chem Int Ed Engl 1981, 20, 668.

[93] Sano K, Mitsugi K. Enzymatic production of L-cysteine from D,L-2-amino-D^2-thiazoline-4-carboxylic acid by pseudomonas thiazolinnophilim: Optimal conditions for the enzym formation and enzyme reaction. Agric Biol Chem 1978, 42, 2315–2321.

[94] Plant-Based L-Cysteine for Dough Softening (Accessed August 08 2020, at www.wacker.com).

[95] Pfaller R, Koch J. Improved Cysteine Production Strains. WO 2021259491, Wacker Chemie AG, December 30, 2021.

[96] Hanatani Y, Imura M, Taniguchi H, Okano K, Toya Y, Iwakin R, Honda K. In vivo production of cysteine from glucose. Appl Microbiol 2019, 103, 8009–8013.

[97] Seidler J, Bernhard R, Haufe S, Neff C, Gärtner T, Waldvogel S R. From screening to scale-up: The DoE-based optimization of electrochemical reduction of L-cystine at metal cathodes. Org Process Res Dev 2021, 25, 2622–2630.

[98] Wüstenberg B, Müller M-A, Schütz J, Wyss A, Schiefer G, Litta G, John M, Hähnlein. W. Vitamin A (Retinoids). In: Elvers, ed. Ullmann's Encyclopedia of Industrial Chemistry. Online ed., Weinheim, Germany, Wiley-VCH Verlag GmbH & Co, 2020.

[99] Bonrath W, Bruins M, Mair P, Medlock J, Netscher T, Schütz J, Wüstenberg B. Kirk-Othmer encyclopedia of chemical technology. Vitamin A 2015.

[100] Orsat B, Wirz B, Bischof S. A continuous lipase-catalyzed acylation process for the large-scale production of vitamin a precursors. Chimia 1999, 53, 579–584.

[101] Orsat B, Spurr P, Wirz B. Enzymatic Acylation of a Retinol Derivative. EP 0802261, Hoffmann LaRoche, 1997.

[102] Wüstenberg B, Stemmler R T, Letinois U, Bonrath W, Hugentobler M, Netscher T. Large-scale production of bioactive ingredients as supplements for healthy human and animal nutrition. Chimia 2011, 65, 420–428.

[103] Mantzouridou F, Naziri E, Tsimidou M Z. Industrial glycerol as a supplementary carbon source in the production of β-carotene by Blakeslea trispora. J Agric Food Chem 2008, 56, 2575–2668.

[104] Malisorn C, Suntornsuk W. Optimization of β-carotene production by Rhodotorula glutinis DM28 in fermented radish brine. Bioresour Technol 2008, 99, 2281–2287.

[105] Silva C, Cabral J M S, Keulen F V. Isolation of a β-carotene over-producing soil bacterium, Sphingomonas sp. Biotechnol Lett 2004, 26, 257–262.

[106] Murillo F J, Calderón I L, López-Díaz I, Cerdá-Olmedo E. Carotene-super producing strains of phycomyces. Appl Environ Microbiol 1978, 36, 639–642.

[107] Hu X, Li H, Tang P, Sun J, Yuan Q, Li C. GC–MS-based metabolomics study of the responses to arachidonic acid in Blakeslea trispora. Fungal Genet Biol 2013, 57, 33–41.

[108] Rodríguez-Sáiz M, Paz B, De La Fuente J L, López-Nieto M J, Cabri W, Barredo J L. Blakeslea trispora genes for carotene biosynthesis. Appl Environ Microbiol 2004, 70, 5589–5594.

[109] Sun J, Sun X X, Tang P W, Yuan Q P. Molecular cloning and functional expression of two key carotene synthetic genes derived from Blakeslea trispora into E. coli for increased bate-carotene production. Biotechnol Lett 2012, 34, 2077–2082.

[110] Wang G Y, Keasling J D. Amplification of HMG-CoA reductase production enhances carotenoid accumulation in Neurospora crass. Metab Eng 2002, 4, 193–201.

[111] Anderson M S, Muehlbacher M, Street I P, Proffitt J, Poulter C D Isopentenyldiphosphate dimethylallyl diphosphate isomerase. J Biol Chem 1989, 264, 19269–19275.

[112] Wang H B, Luo J, Huang X Y, Lu M B, Yu L J. Oxidative stress response of Blakeslea trispora induced by H2O2 during β-carotene biosynthesis. J IndMicrobiol Biotechnol 2014, 41, 555–561.

[113] Eggersdorfer M, Laudert D, Letinois U, McClymond T, Medlock J, Netscher T, Bonrath W. One hundred years of Vitamins – A success story of the natural sciences. Angew Chem Int Ed 2012, 51, 12960–12990.

[114] Philippsen P, Pompejus M, Seulberger H. Partial Sequences of Purine Biosynthesis Genes from Ashbya Gossypii and Their Use in the Microbial Riboflavin Synthesis, WO 98/29539, BASF, 1998.

[115] Van Loon A P G M, Hohmann H-P, Bretzel W, Hümbelin M, Pfister M. Development of a fermentation process for the manufacture of riboflavin. Chimia 1996, 50, 410–412.

[116] Han M, Laudert D, Hohmann H-P, Lehmann M. Modified Transketolase and Use Thereof. US 20080311617, DSM, 2008.

[117] Hohmann H-P, Litta G, Hans M, Friedel A, Bretzel W, Lehmann M, Kaesler B. Vitamin B_{12} (Cobalamins). In: Elvers, ed. Ullmann's Encyclopedia of Industrial Chemistry. Online ed., Weinheim, Germany, Wiley-VCH Verlag GmbH & Co, 2020.

[118] Friedmann H C, Cagen L M. Microbial biosynthesis of B12-like compounds. Annu Rev Microbiol 1970, 24, 159–208.

[119] Elste V, Peng K, Kleefeldt A, Litta G, Medlock J, Pappenberger G, Oster B, Fechtel U. Vitamin C. In: Elvers, ed. Ullmann's Encyclopedia of Industrial Chemistry. Online ed., Weinheim, Germany, Wiley-VCH Verlag GmbH & Co, 2020.

[120] Alsters P, Aubry J-M, Bonrath W, Daguenet C, Hans M, Jary W, Letinois U, Nardello-Rataj V, Netscher T, Parton R, Schütz J, V.soolingen J, Tinge J, Wüstenberg B. Selective Oxidation in DSM: Innovative Catalysts and Technologies. In: Duprez D, Cavani F, ed. Oxidation Catalysis, 2014, 382–419.

[121] Deppenmeier U, Ehrenreich A Physiology of acetic acid bacteria in light of the genome sequence of Gluconobacter oxydants. J Mol Microbiol Biotechnol 2009, 16, 69–80.

[122] Yang W, Xu H. Industrial Fermentation of Vitamin C. In: Vandamme E, Revuelta J L, eds. Industrial Biotechnology of Vitamins, Biopigments and Antioxidants. 1st ed., Wiley-VCH, 2016, 161–191.

[123] Takagi Y, Sugisawa T, Hoshino T. Continuous keto-L-gulonic acid fermentation by mixed culture of Ketogulonicum vulgare DSM 4025 and Bacillus megaterium or Xanthomas maltophilia. Appl Microbiol Biotechnol 2010, 86, 469–480.

[124] Müller M-A, Medlock J, Pragai Z, Warnke I, Litta G, Kleefeldt A, Kaiser K, De Potzolli B. Vitamin B_5. In: Elvers, ed. Ullmann's Encyclopedia of Industrial Chemistry. Online ed., Weinheim, Germany, Wiley-VCH Verlag GmbH & Co, 2019.

[125] Bonrath W, Karge R, Netscher T. Lipase-catalysed transformations as key-steps in the large-scale preparation of vitamins. J Mol Catal B: Enzymatic 2002, 19–20, 67–72.

[126] He L, Mi Z, Li B. Method for preparing calcium D-pantothenate. Jingjing Pharmaceutical Co. Ltd., 2018.

[127] Sakamoto K, Honda K, Wada K, Kita S, Tsuzaki K, Nose H, Kataoka M, Shimizu S. Practical resolution system for DL-pantoyl lactone using the lactonase from Fusarium oxysporum. J Biotechnol 2005, 118, 99–106.

[128] Sun Z. Process for Preparing D-pantoyl Lactone by Microbe Enzyme Method. Cn 1111604, Xinfu Biochemical Co. Ltd., 2001.

[129] Sun Z, Guo X. Method for Manufacturing D-pantoic Acid with High Optical Activity from Racemic D, L-pantolactone with Filamentous Fungi Capable of Producing D-pantolactone Hydrolase. CN 100351369, Zhejiang Hangzhou Xinfu Pharmaceutical CO. Ltf, Southern Yangtze University, Jiangnan, 2005.

[130] Effenberger F. Synthese und Reaktionen optisch aktiver Cyanhydrine. Angew Chem 1994, 106, 1609–1619.

[131] Pscheidt B, Liu Z, Gaisberger R, Avi M, Skranc W, Gruber K, Griengl H, Glieder A. Efficient biocatalytic synthesis of (R)-Pantolactone. Adv Syn Catal 2008, 350, 1943–1948.

[132] Fouts P J, Helmer O M, Lepkovsky S, Jukes T J. Production of microcytic hypochromic anemia in puppies on synthetic diet deficient in rat antidermatitis factor (Vitamin B6). J of Nutr 1938, 16, 197–207.

[133] Blum R. Vitamin B₃ (Niacin). In: Elvers, ed. Ullmann's Encyclopedia of Industrial Chemistry. Online ed., Weinheim, Germany, Wiley-VCH Verlag GmbH & Co, 2020.

[134] Nagasawa T, Mathew C D, Mauger J, Yamada H. Nitrile hydratase-catalyzed production of nicotinamide from 3-cyanopyridine in Rhodococcus rhodochrous. J Appl Environ Microbiol 1988, 54, 1766–1769.

[135] Nowaczyk M J M, Waye J S. The Smith-Lemli-Opitz syndrome: A novel metabolic way of understanding developmental biology, embryogenesis, and dysmorphology. Clin Genet 2001, 59, 375–386.

[136] Saunders C A, Wolf F R, Mukharji I. Method and Composition for Increasing the Accumulation of Squalene and Specific Sterols in Yeast. US 5460949, AMOCO, 1995.

[137] Bendik I, Höller U, Marty M, Schütz J, Labler L. Vitamin D. In: Elvers, ed. Ullmann's Encyclopedia of Industrial Chemistry. Online ed., Weinheim, Germany, Wiley-VCH Verlag GmbH Co, 2019.

[138] Basson M E, Thorsness M, Rine J Saccharomyces cerevisiae contains two functional genes encoding 3-hydroxy-3-methylglutaryl coenzyme A reductase. Proc Natl Acad Sci USA 1986, 83, 5563–5567.

[139] Balkenhohl F, Ditrich K, Hauer B, Ladner W. Optically active amines via lipase-catalyzed methoxylation. J Prakt Chem 1997, 339, 381–884.

[140] Balkenhohl F, Hauer B, Ladner W, Schnell U, Pressler U, Staudenmaier H R. Lipase Katalysierte Acylierung von Alkoholen Mit Diketenen, De 4329293, BASF, 1995.

[141] Matsumae H, Furui M, Shibatani T, Sosa T. Lipase-catalyzed asymmetric hydrolysis of 3-phenylglycide acid ester, the key intermediate in the synthesis of diltiazem hydrochloride. J Fermet Bioeng 1993, 75, 93–98.

[142] Taoka N, Ppoanda M, Mizuho K, Suga K. Manufacture of Optically Active Alkoxycyclohexanol Derivative by Enzymic Resolution. JP 09047298, Kanegafuchi Chemical Ind, 1997.

[143] Barros Aguirre D, Bates R H, Gonzales Del Rio R, Mendoza Losana A. Sanfetrinem or a salt or ester thereof for use in treating mycobacterial infection. GlaxoSmithKline, WO 201/206466, November 15, 2018.

[144] Gao S, Zhu S, Huang R, Lu Y, Zheng G. Efficient synthesis of the intermediate of abacavir and carbovir using a novel (+)-c-lactamase as a catalyst. Bioorg Med Chem Lett 2015, 25, 3878–3881.

[145] King A O, Corley E G, Anderson R K, Larsen R D, Verhoeven T R, Reider P J, Xiang Y B, Belley M, Leblanc Y, Labelle M, Prasit P, Zamboni R J. An efficient synthesis of LTD4 antagonist L-699,392. J Org Chem 1993, 58, 3731–3735.

[146] Belley M, Leger S, Labelle M, Roy P, Xinag Y B, Guay D. US 5565473, Merck, 1995.

[147] Liang J, Lalonde J, Borup B, Mitchell V, Mundorff E, Trinh N, Kochrekar D A, Cherat R N, Pai G G. Development of a biocatalytic process as an alternative to the (-)-DIP-Cl-Mediated asymmetric reduction of a key intermediate of montelukast. Org Process Res Dev 2010, 14, 193–198.

[148] Petersen M, Birch O, Shimizu S, Kiener A, Hischer M-L, Thoni S. Method for Producing trifluoro-3(R)-hydroxybutyric Acid Derivatives. WO 99/42590, Lonza, 1999.

[149] Dooseop K, Wang L, Beconi M, Eiermann G J, Fisher M H, He H, Hickey G J, Kowalchick J E, Leiting B, Lyons K, Marsilio F, McCann M E, Patel R A, Petrov A, Scapin G, Patel S B, Roy R S, Wu J K, Wyvratt M J, Zhang B B, Zhu L, Thornberry N A, Weber A E. (2R)-4-Oxo-4-[3-(Trifluoromethyl)-5,6-dihydro[1,2,4]triazolo[4,3-a]pyrazin- 7(8H)-yl]-1-(2,4,5-trifluorophenyl)butan-2-amine: A potent, orally active dipeptidyl peptidase IV inhibitor for the treatment of type 2 diabetes. J Med Chem 2005, 48, 141–151.

[150] Desai A A. Sitagliptin manufacture: A compelling tale of green chemistry, process intensification, and industrial asymmetric catalysis. Angew Chem Int Ed 2011, 50, 1974–1976.

[151] Savile C K, Janey J M, Mundorff E C, Moore J C, Tam S, Jarvis W R, Colbeck J C, Krebber A, Fleitz F J, Brands J, Devine P N, Huisman G W, Hughes G J. Biocatalytic asymmetric synthesis of chiral amines from ketones applied to Sitagliptin manufacture. Science 2010, 329, 305–309.

[152] Erickson R C, Bennett R E. Penicillin acylase activity of penicillium chrysogenum. Appl Microbiol 1965, 13, 738–742.

[153] Bruggink A, Roos E C, De Vroom E. Penicillin acylase in the industrial production of β-Lactam antibiotics. Org Process Res Dev 1998, 2, 128–133.

[154] Elander R P. Industrial production of β-lactam antibiotics. Appl Microbiol Biotechnol 2003, 61, 385–392.

[155] Kiener A. Microbiological oxidation of ethyl groups in heterocycles. EP 466115, Lonza, 1992.

[156] Kiener A. Enzymic oxidation of methyl groups in heteroarenes: A versatile method for the preparation of heteroaromatic carboxylic acids. Angew Chem Int Ed Engl 1992, 31, 774–775.

[157] Kiener A. Biosynthesis of functionalized aromatic heterocycles. Chemtech 1995, 9, 31–35.

[158] Frampton G A C, Zavareh H S. Process for Preparing Levobupivacaine and Analogues Thereof. WO 9612700, Lonza, 1996.

[159] Kiener A, Roduit J-P, Kohr J, Shaw N. Biotechnological Process for the Production of Cyclic S-alpha-aminocarboxylic Acids and R-alpha-aminocarboxylic Acid Amides. EP 0686698, Lonza, 1995.

[160] Eichhorn E, Roduit J-P, Shaw N, Heinzmann K, Kiener A. Preparation of (S)-piperazine-2-carboxylic acid and (R)-piperazine carboxylic acid by kinetic resolution of the corresponding racemic carboxamides with stereoselective amidases in hole bacterial cells. Tetrahedron Asymmetry 1997, 8, 2533–2536.

[161] Martiney C A, Hu S, Dumond Y, Tao J, Kelleher P, Tully L. Development of a chemoenzymatic manufacturing process for pregabalin. Org Process Res Dev 2008, 12, 392–398.

[162] Paddon C J, Westfall P J, Pitera D J, Benjamin K, Fisher K, McPhee D, Leavell M D, Tai A, Main A, Eng D, Polichuk D R, Teoh K H, Reed D W, Treynor T, Lenihan J, Fleck M, Bajad S, Dang G, Dengrove D, Diola D, Dorin G, Ellens K W, Fickes S, Galazzo J, Gaucher S P, Geistlinger T, Henry R, Hepp M, Horning T, Iqbal T, Jiang H, Kizer L, Lieu B, Melis D, Moss N, Regentin R, Secrest S, Tsuruta H, Vazquez R, Westblade L F, Xu L, Yu M, Zhang Y, Zhao L, Lievense J, Covello P S, Keasling J D, Reiling K K, Renninger N S, Newman J D. High-level semi-synthetic production of the potent antimalarial artemisinin. Nature 2013, 496, 528–532.

[163] Benjamin K R, Silva I R, Cherubim J P, McPhee D, Paddon C J. Developing commercial production of semi-synthetic artemisinin, and of β-Farnesene, an isoprenoid produced by fermentation of Brazilian Sugar. J Braz Chem Soc 2016, 27, 1339–1345.

[164] Paddon C J, Keasling J D. Semi-synthetic artemisinin: A model for the use of synthetic biology in pharmaceutical development. Nat Rev Microbiol 2014, 12, 355–367.

[165] Turconi J, Griolet F, Guevel R, Oddon G, Villa R, Geatti A, Hvala M, Rossen K, Göller R, Burgard A. Semisynthetic artemisinin, the chemical path to industrial production. Org Process Res Dev 2014, 18, 417–422.

[166] Schober M, MacDermaid C, Ollis A A, Chang S, Khan D, Hosford J, Latham J, Ihnken L A F, Brown M J B, Fuerst D, Sanganee M J, Roiban G-D. Chiral synthesis of LSD1 inhibitor GSK2879552 enabled by directed evolution of an imine reductase. Nat Catal 2019, 2, 909–915.

[167] Kumar R, Karmilowicz M J, Burke D, Burns M P, Clark L A, Connor C G, Cordi E, Do N M, Doyle K M, Hoagland S, Lewis C A, Mangan D, Martinez C A, McInturff E L, Meldrum K, Pearson R, Steflik J, Rane A, Weaver J. Biocatalytic reductive amination from discovery to commercial manufacturing applied to abrocitinib JAK1 inhibitor. Nat Catal 2021, 4, 775–782.

[168] Fehr C, Galino J. A new variant of the Claisen rearrangement from malonate-derived allylic trimethylsilyl ketene acetals: Efficient, highly enantio- and diastereoselective syntheses of (+)-methyl dihydroepijasmonate and (+)-methyl epijasmonate. Angew Chem Int Ed 2000, 39, 569–573.

[169] Fehr C, Galino J, Etter O. A new variant of the claisen rearrangement from malonate-derived allylic silyl ketene acetals. efficient highly enatio- and diastereoselective synthesis of (+)-Methyl Dihydroepijasmonate and (1-)-methyl Epijasmonate. Chimla 1999, 53, 376.

[170] Harada T, Shinnanyo-shi Y, Irino S, Kunisawa Y, Oyama K. Improved Enzymatic Coupling Reaction of N-protected-L-aspartic Acid and Phenylalanine Methyl Ester. EP 0768384, Holland Sweetener Company, 1997.

[171] Gröger H. Enzymatic routes to enantiomerically pure aromatic α-Hydroxy carboxylic acids: A further example for diversity of biocatalysis. Adv Synth Catal 2001, 343, 547–558.

[172] Yamada H, Kobayashi M. Nitrile hydratase and its application to industrial production of acrylamide. Biosci Biotech Biochem 1996, 60, 1391–1400.

[173] Yomada H. Neat Frontiers in Screening for Microbial Biocatalysis, Kieslich K, von der Beck C P, De Bunt A M, Van den Tweel W I J, eds. Amsterdam, Elsevier, 1998, 13–17.

[174] Kaltschmitt M, Streicher W, Wiese A. Erneuerbare Energien. Systemtechnik, Wirtschaftlichkeit, Umweltaspekte. Springer Vieweg, Berlin / Heidelberg, 2013.

[175] Huelin F E, Murray K E. α-Farnesene in the natural coating of apples. Nature 1966, 210, 1260–1261.

[176] Gibson R W, Pickett J A. Wild potato repels aphids by release of aphid alarm pheromone. Nature 1983, 302, 608–609.

[177] Avé D A, Gregory P, Tingey W M. Aphid repellent sesquiterpenes in glandular trichomes of Solanum berthaultii and S. tuberosum. Entomolo Exp Appl 1987, 44, 131–138.

[178] Renninger N, McPhee D. Fuel Compositions Comprising Farnesane and Method of Making and Using Same. US 20080098645, Amyris, 2008.

[179] Özaydın B, Burd H, Lee T S, Keasling J D. Carotenoid-based phenotypic screen of the yeast deletion collection reveals new genes with roles in isoprenoid production. Metab Eng 2013, 15, 174–183.

[180] Peralta-Yahya P P, Ouellet M, Chan R, Mukhopadhyay A, Keasling J D, Lee T S. Identification and microbial production of a terpene-based advanced biofuel. Nat Commun 2011, 2, 483.

[181] Pray T. Biomass R&D Technical Advisory Committee: Drop-in Fuels Panel. Emeryville, CA, USA, Amyris, 2010; https://biomassboard.gov/pdfs/biomass_tac_todd_pray_09_29_2010.pdf (21.01.2020).

11 Electrochemistry

11.1 Introduction

An electrochemical reaction is a heterogeneous reaction involving the transfer of charge from or to an electrode. The electrode is typically made of carbon, a semiconductor or a metal. The charge transfer can be a cathodic process in which a species is reduced by electron transfer, or an anodic process in which an oxidation of a species takes place. In general, an electrochemical reaction is possible if a cell (reactor) contains both a cathode and an anode. In electrolysis reactions, the electrons must pass from the anode to the cathode through an external electrical circuit connecting the electrodes, and a mechanism for charge transport between the electrodes in the cell must be given [1].

To achieve chemical reactions, various transformations can occur on an electrode, and some selected examples such as electron transfer, gas evolution, gas reduction in a porous diffusion electrode or electron transfer coupled with chemical reaction are depicted in Figure 11.1.

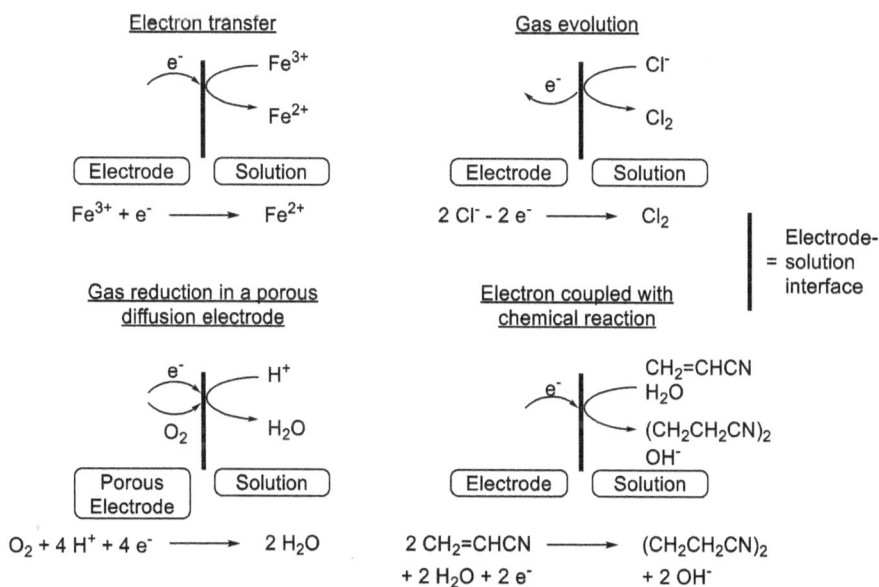

Figure 11.1: Examples of electrode processes at a single electrode.

Electrochemical reactions are redox processes carried out in an electrochemical cell, that is, the reduction and oxidation half-reactions are separated. For reversible processes, the electrical energy is equivalent to the reaction enthalpy. The redox process

https://doi.org/10.1515/9783111102672-011

itself occurs on the interface between electrode and electrolyte. If the electrochemical reaction is spontaneous, the set-up is defined as a galvanic cell (named after *L. Galvani* and *A. Volta*). At the anode an oxidation reaction occurs, whereas at the cathode, reduction occurs; for example, metal ions from the solution are reduced to metal (electrode-posits). Between the connected electrodes, the electrolyte ions can freely move [CVII]. In galvanic cells, chemical energy is transformed into electrical energy, whereas in electrolytic cells the inverse process occurs.

A common example for a galvanic cell is a set-up with a Zn anode and a Cu cathode; wherein the electronic charge can be transferred by a salt bridge between the two chambers (Figure 11.2).

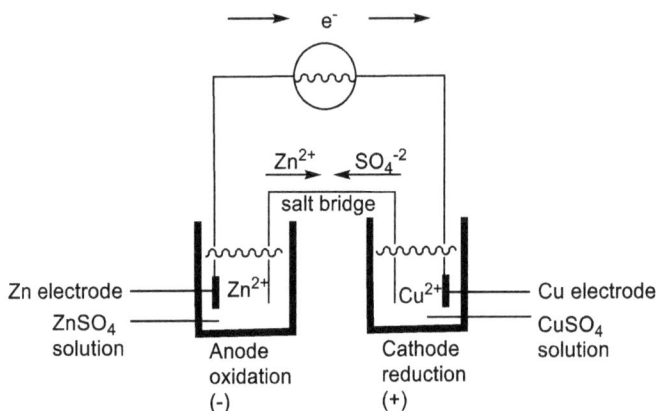

Figure 11.2: Galvanic system (*Daniell* cell).

In a galvanic cell, there are two different metal electrodes, and each is in an electrolyte. One metal is oxidised (anode) and a cation is formed, whereas a metal ion is reduced and deposited on the cathode (electrodeposition). The *Daniell* cell, invented by *J. F. Daniell* in 1836, is the historical basis for electromotive force which is given in volt (V). The anode and cathode are electrically connected, allowing a flow of electrons from the anode to the cathode surface. The redox potentials $E_o Cu = 0.34$ V and $E_o Zn = -0.76$ V result in an electromotive force of 1.1 V [1, 2]. The redox reaction in the *Daniell* element is usually described by the redox pair Zn/Zn^{2+} and Cu/Cu^{2+}, and in the following equations:

$$Zn_s \longrightarrow Zn^{2+}_{aq} + 2e^- \quad (-0.76\,V)$$

$$Cu^{2+}_{aq} + 2e^- \longrightarrow Cu_s \quad (0.34\,V)$$

in total:

$$Zn_s + Cu^{2+}{}_{aq} \longrightarrow Zn^{2+}{}_{aq} + Cu_s \qquad \begin{matrix} s = \text{solid} \\ aq = \text{aqueous} \end{matrix}$$

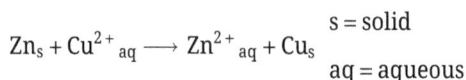

The Cu^{2+} ions move, driven by the attraction of opposite charges, towards the positive electrode. When the two processes take place in containers, each is named a "half-cell".

In general, in idealised electrochemical processes, the mass (m) of products produced on an electrode is proportional to the charge (Q) (*Faraday*'s first law of electrolysis) [3]:

$$m \propto Q$$

It was also found and described by *M. Faraday* that when the same amount of electric current is passed through different electrolytes connected in series, the mass of the substance deposited at the electrodes is directly proportional to their chemical equivalent, or equivalent weight (*Faraday*'s second law of electrolysis) [3]:

$$m \propto E \text{ and } E = \text{molar mass/valence}$$

where m is the mass of the substance produced at the electrode (in g), Q is the total electric charge that passed through the solution (in C (coulomb)), n is the valence number of the substance as an ion in solution (electrons per ion) and M is the molar mass of the substance (in grams per mol).

The *Faraday* laws can be combined to:

$$Q = nF, \text{ with } F = \text{Faraday's constant [3]}.$$

The Faraday constant is $96,485\,C/mol(1\,C = 1\,As)$.

The cell potential is obtained by subtracting the equilibrium potentials of anode and cathode and is related to the free energy of the overall reaction in the cell:

$$G = -nFE^e{}_{cell} \quad (E^e cell = \text{equilibrium cell potential})$$

Electrochemistry is applied in the fields of: the manufacture of metals, for example Al and Mg, and alkali metals; the oxidation of anions such as Cl^-; the synthesis of ozone; electrolysis of water (production of H_2 and O_2); Galvano technique (plating of metals); Galvano cells; or organic electrochemistry, which is further discussed below.

From a practical point of view, in electrocatalytic reactions, the catalyst can adsorb on the electrode and slow down the reaction by an overpotential. Such behaviour occurs, for example, in the Br^--catalysed epoxidation of propene and can be described as an indirect electrode reaction.

Hydrogen evolution is historically important for the understanding of electrode reactions and corrosion phenomena as well as water electrolysis and fuel cells. The adsorbed hydrogen plays a key role in the hydrogen evolution mechanism. For the transformation of the adsorbed hydrogen into the hydrogen evolution, the mechanisms A and B exist (Scheme 11.1). The electrode material (cathode) plays an important role,

$$2\,H^+ + 2\,e^- \longrightarrow H_2 \qquad \text{Acidic conditions}$$

$$2\,H_2O + 2\,e^- \longrightarrow H_2 + 2\,{}^-OH \qquad \text{Basic conditions}$$

$$H^+ + e^- \longrightarrow H_{ads} \qquad \text{ads = adsorbed}$$

$$H^+ + e^- + M \longrightarrow M\text{-}H$$
$$2\,M\text{-}H \longrightarrow 2\,M + H_2$$
Mechanism A

$$H^+ + e^- + M \longrightarrow M\text{-}H$$
$$M\text{-}H + H^+ + e\text{-} \longrightarrow M + H_2$$
Mechanism B

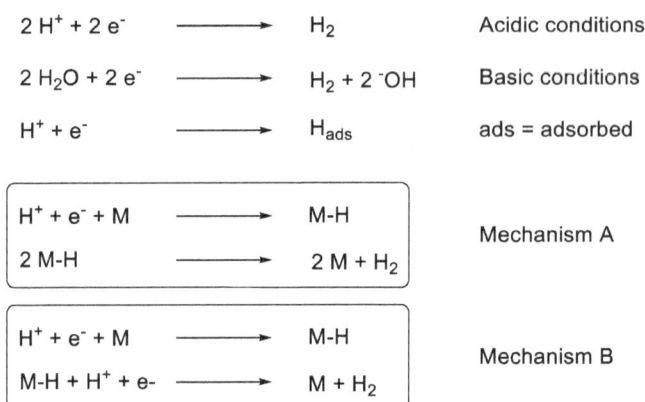

Scheme 11.1: Fundamental reactions of hydrogen evolution reaction.

because the free energy of adsorbed species favours the formation of adsorbed species. The rate of hydrogen evolution depends on the strength of the M–H (M = metal) bond formed in the electrochemical reaction. More information can be found in [4].

Electro-organic reactions depend on the electrode material. For example, the oxidation of carboxylic acid ions in aqueous solution in the presence of a Pt electrode results in the formation of dimers. If a Pb electrode is applied, the corresponding alcohol containing one carbon less is obtained (Scheme 11.2) [5].

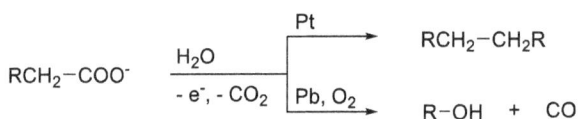

$$RCH_2\text{-}COO^- \xrightarrow[\;-\,e^-,\,-\,CO_2\;]{H_2O}$$

Pt → $RCH_2\text{-}CH_2R$

Pb, O_2 → $R\text{-}OH\ +\ CO$

Scheme 11.2: Anodic oxidation of carboxylic acid ions depending on the electrode material.

In cathodic reduction reactions, a similar behaviour is observed. For example, adsorbed hydrogen on electrodes with low overvoltage, such as Ni-, Pd- and Pt-based electrodes, induces the reduction of acrylonitrile into propionitrile, whereas on Hg or Pb electrodes adiponitrile and other organic nitriles are formed (Scheme 11.3) [5].

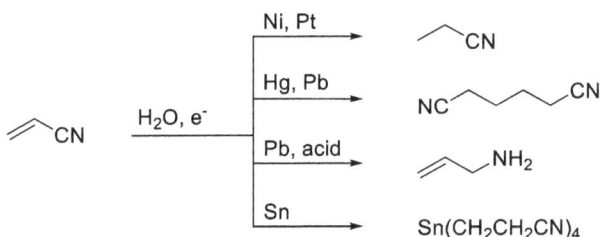

Scheme 11.3: Influence of the electrode material on the type of product formed in cathodic reduction.

From an industrial point of view, electrochemical processes require a high energy efficiency. The needed electrical energy for an electro-organic process can be calculated by the following equation:

$$E_S = \frac{z \cdot F \cdot U_z \cdot 100}{M_p \cdot \beta} \text{ in kWh per g product}$$

where z is the number of electrons, F is Faraday's constant (96,485 C/mol), U_z is the cell voltage, β is the current efficiency and M_p is the mole mass of product (g/mol).

If E_s is multiplied with the current electrical energy costs, the contribution of the electricity to the product-specific costs is obtained. It can be derived from the above equation that for an increasing value of z/M_p and U_z as well as a decreasing value of β, E_s increases. In general, an electrochemical reactor is a reactor in which two different reactions are carried out in parallel forced by electric current.

At a fixed current density, the reaction rate is proportional to the electroactive area. The active area depends on the flow conditions, that is flooded or trickle flow, which may be in different gas evolution or metal deposits in an electrochemical process. Usually, a cell with high active surface area per unit volume has to be designed. For the desired cell, over the whole electrode area, the potential and current density must be uniform. Several different cell designs are known. In a rotating cylinder cell and a parallel plate cell, they are uniform, whereas in a plate in a tank reactor or a plate cell with non-parallel electrodes of a reaction, a non-uniform situation occurs (Figure 11.3).

Organic electrochemistry can often be carried out in non-divided cells. The main advantage of such a set-up is the reduction of complexity since anode and cathode electrolyte handling is combined, and most processes are irreversible. The main challenge of a non-divided cell is avoiding the degradation of the product and/or starting material on the counter-electrode. In anodic processes, this can generally be easily achieved, especially if there is a hydrogen over-voltage on the cathode. An example of a reaction on a cathode in an undivided cell is the adiponitrile synthesis [6]. This reaction can occur if the oxidation of the electrolyte is avoided (or is very low), in the presence of a co-solvent such as methanol (Figure 11.1).

However, it is sometimes necessary to work in divided cells to avoid side reactions. In this situation, a diaphragm is used to separate the cell parts. Diaphragm materials are usually based on ceramics (for porous materials), aluminium oxide (for applications at high pH value) or polymers, such as PVC, polyamides or Nafion.

Examples of typical reactor types are shown in Figure 11.4 [7, 8].

11.2 Electrochemical oxidations

In the field of carotenoid chemistry, an important building block for the centre of the molecule is (E)-1,1,4,5-tetramethoxybut-2-ene (TMB). This compound then undergoes a Lewis acid-catalysed enol ether condensation (Section 8.4, Scheme 8.27) to form the

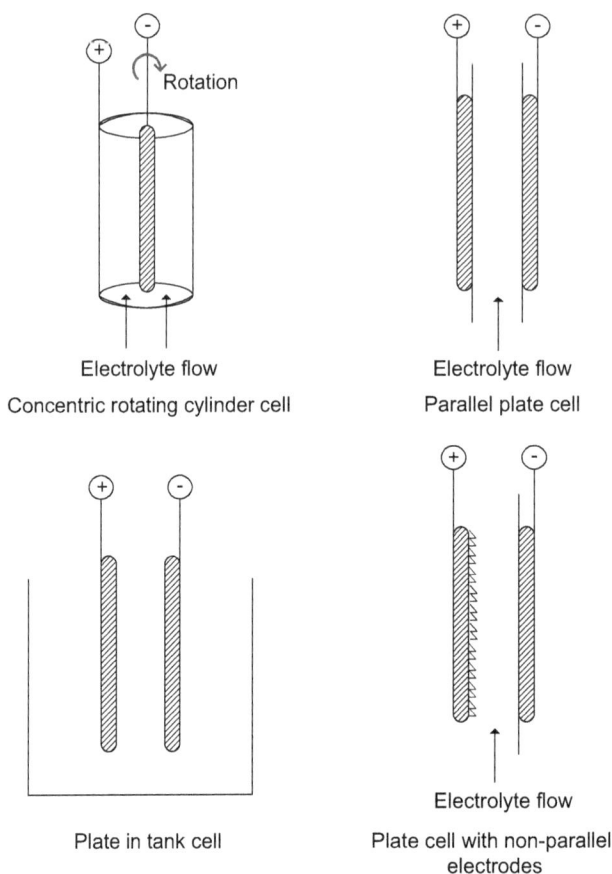

Figure 11.3: Examples of different electrochemical reactor geometries.

key dialdehyde building block. TMB is produced from furan via 2,5-dimethoxy-2,5-dihydrofuran (DMDF) using bromine. An interesting alternative approach for the synthesis of DMDF is the transformation of (Z)-2-butene-1,4-diol to DMDF under electrochemical conditions, followed by the ion-exchange resin-catalysed reaction to TMB (Scheme 11.4) [9].

The main advantages of the electrochemical approach are the reduced amount of waste, the avoidance of mineral acid formation, the selective production of DMDF and simplified down-stream processing (the product does not need to be extracted from the salt–product mixture). Furthermore, where available, "green/renewable" electricity can be used to further reduce the carbon footprint.

Vanillin, 4-hydroxy-3-methoxybenzaldehyde, is produced in several thousand tons per year and is used in cosmetics, in the food industry, and as a starting material for pharmaceutical production (see Chapter 5: Gas-phase reactions and Chapter 8: Acid–base catalysed reactions) [10]. In addition to chemical synthesis, vanillin can be produced from

Divided cell

Plate cell with
diaphragm

Undivided cell

Fluidized bed reactor
Fleischmann - Goodridge

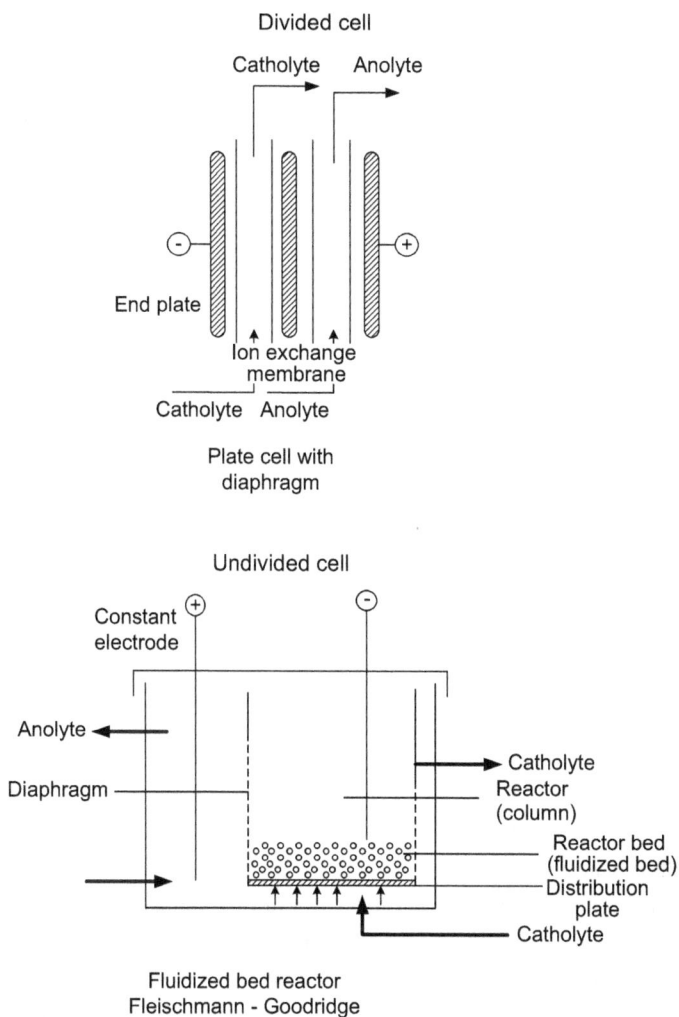

Figure 11.4: Examples of reactor types.

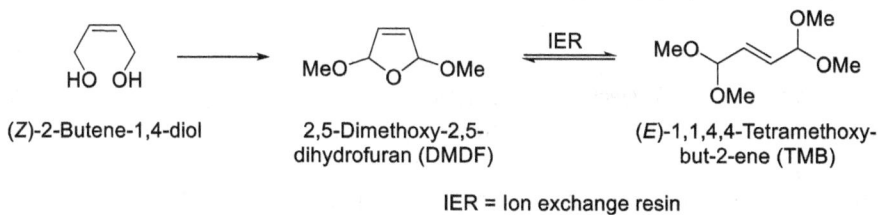

(Z)-2-Butene-1,4-diol

2,5-Dimethoxy-2,5-
dihydrofuran (DMDF)

(E)-1,1,4,4-Tetramethoxy-
but-2-ene (TMB)

IER = Ion exchange resin

Scheme 11.4: Electrochemical approach to DMDF and TMB.

lignin using an electrochemical procedure [11]. One main advantage of this process is to avoid problems based on the high sulphur content in the lignin feed stream (waste formation), which is a result of the *Kraft process* for pulp manufacture [10, 12]. In the *Kraft process*, lignocellulose is treated with sodium hydroxide and sodium sulphide at temperatures of around 170 °C at around 10 bar for up to 2 h. These harsh conditions split the b-O-4′ bonds of lignin and form new C–C bonds [13]. One of the main problems in the application of lignin is the stability of lignin itself, which can mean it is difficult to convert it into chemicals such as vanillin (Figure 11.5).

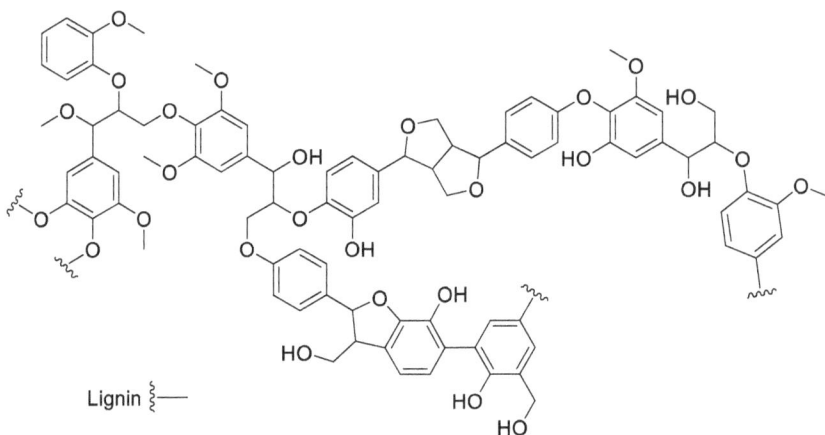

Figure 11.5: Representative structure of lignin.

The production of vanillin from bio-based resources (including lignin) is carried out by Borregaard (Sweden). Lignosulfonate from the waste stream of pulp industry is oxidised at around 15 bar with oxygen in the presence of a Cu catalyst. The yield of this process is around 1 wt% and the starting material must first be purified [14]. The electrochemical approach to vanillin by-passes the problems of the current process, especially the formation of by-product formation or the requirement to run the reaction under pressure. "Green" (renewable) electricity can be applied, resulting in a sustainable process. The electrochemical procedure for the transformation of lignin into vanillin is carried out in the presence of NaOH at 160 °C on a Ni anode (Scheme 11.5). The active layer is an in situ formed Ni-oxide-hydroxide layer, which converts the lignin by oxidative splitting. An additional advantage of this electrochemical procedure is the increased yield of 4% [11, 15]. The separation of vanillin is carried out by adsorption on a basic ion-exchange resin containing a polystyrene backbone. This has the advantage of a minimised waste stream since the reaction medium does not need to be neutralised [16].

Another area where electrochemical processes have been developed utilising naturally derived starting materials is the conversion of furfural and/or furfuryl alcohol,

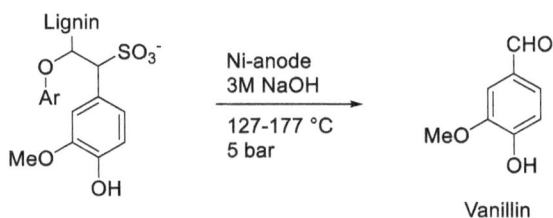

Scheme 11.5: Electrochemical approach from lignosulfonate to vanillin.

which has been proposed as future platform chemicals for the production of C_4- and C_5-building blocks [17]. Furfural is mainly produced from xylan and xylose which is readily available from hemicellulose. Both oxidation and reduction processes can be performed using electrochemistry producing a number of useful compounds and starting materials for fine chemicals [18].

Electrochemical reduction results in methylfuran or dimethylfuran which can be applied as biofuels. Alternative catalytic processes (not electrochemical) can result in levulinic acid (LA), γ-valerolactone (GVL), 2-methyltetrahydrofuran (2-MeTHF) and various diols such as 1,4-pentanediol (Scheme 11.6).

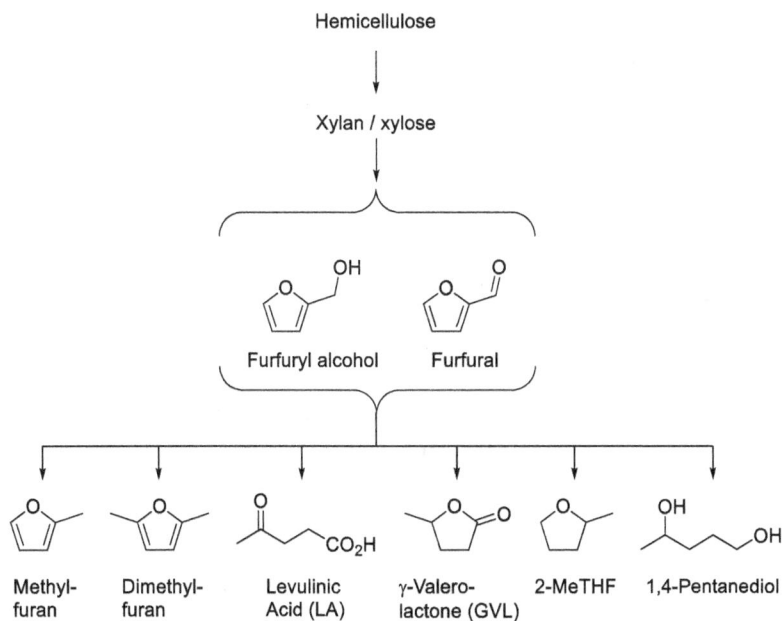

Scheme 11.6: Electrochemical reduction products of furfuryl alcohol and furfural.

Electrochemical oxidation can be used to convert furfuryl alcohol and furfural into bio-based building blocks such as furancarboxylic acid (furoic acid) or 2,5-furandicarboxylic acid (Scheme 11.7). Furan carboxylic acid can be produced in water on a Ni-anode in 80% yield [19]. 2,5-Furandicarboxylic acid can be used as replacement of terephthalic acid [20, 21]. Terephthalic acid is used in the polyester chemistry and produced in >10^6 t/a. The electrochemical oxidation reactions of furans can also be carried out in continuous mode, and electrochemical flow cells have been developed for such a process [22].

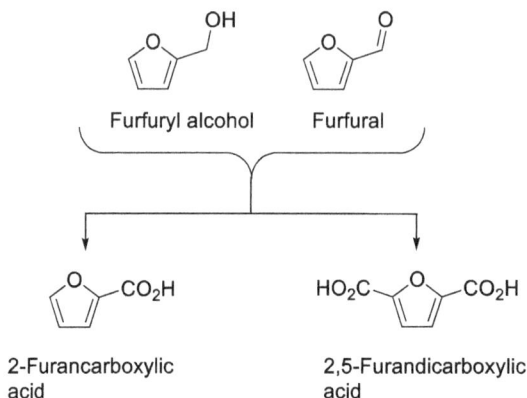

Scheme 11.7: Electrochemical oxidation products of furfuryl alcohol and furfural.

The electrochemical methoxylation of α-methylfuranol and further treatment into maltol is carried out on industrial scale in >150 t/a at Otsuka Ltd in Japan. Maltol is applied in the flavour and fragrance industry, mainly as flavour enhancer, and the starting α-methylfuranol is available from furfural (Scheme 11.8). In the Otsuka process, a 20% solution of the furan derivative in MeOH containing sodium bromide is used. The current density is 0.1–0.2 A/cm^2 and a yield of 97% with a current efficiency of up to 95% at 10 °C can be achieved. In this process, a steel cathode and a carbon anode are used. The α-methylfurfuryl alcohol is oxidised into maltol in a water/MeOH mixture at −10 °C with chlorine gas, or bromine, and heated up to around 95 °C for 3 h. After concentration, cooling to room temperature and separation of the by-products by a decanter, the product was isolated by a pH-controlled (pH 2.2) granulation at 5 °C. Yields of around 70% were achieved [23–30].

The method of anodic oxidation can be used for the synthesis of a vitamin B$_6$ precursor (Scheme 11.9) [31]. Pyridoxine was obtained in 31% yield after hydrolysis of the electrochemically oxidised product.

Electrochemistry on industrial scale finds only a limited number of applications in the fine chemical industry; the main area of application is in the production of organic molecules with a reduced formation of waste. Typically, a current density of 10–30 mA/cm^2 is required for an efficient chemical process [30]. Furthermore, the choice of the electrolyte is extremely important as it must be recycled and reused with high recovery rates. Avoiding the use or formation of perchlorates is important

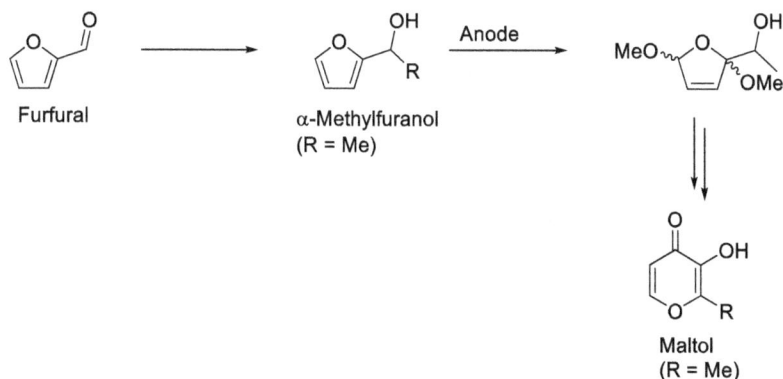

Scheme 11.8: Manufacture of maltol by anodic methoxylation (Otsuka process).

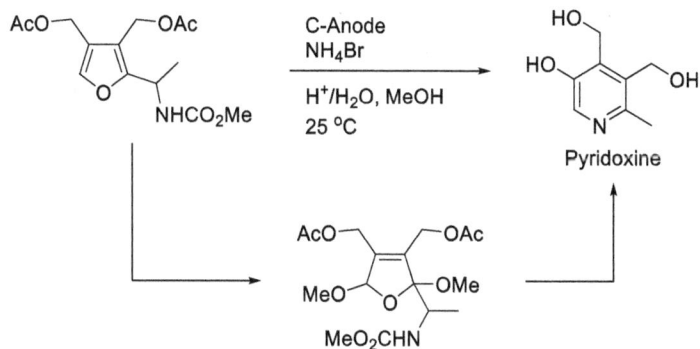

Scheme 11.9: Synthesis of pyridoxine (vitamin B$_6$), intermediates via dialkoxylation of furan derivatives [31].

since these compounds can result in explosive waste streams [30]. The down-stream processing is also a key topic for electrochemical reactions.

One of the main applications of an electrochemical reduction reaction is the manufacture of adiponitrile, for nylon 6,6, the *Baizer process* (Scheme 11.10). The reduction of acrylonitrile is carried out on copper-lead electrodes, which have replaced the formerly applied cadmium-based electrodes. Companies such as BASF, Asahi Kasei, Solutia and Monsanto have used this process on an industrial scale. After the development of cheaper processes based on propene, these processes are generally no longer in operation. However, if cheap (and potentially renewable) electricity is available, the process could be re-installed [32–35]. The main advantage of the *Baizer process* is the use of water as the hydrogen source. Roughly one third of the yearly 350,000 t production volume in 2010 of adiponitrile was produced by this process [7].

The *Baizer process* has also been carried out batch-wise using an Ag–Pb anode and a Pb cathode. A cation-exchange membrane was applied as a separator. After

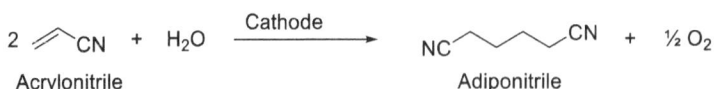

$$2 \diagup CN + H_2O \xrightarrow{\text{Cathode}} NC \diagup\diagdown\diagup CN + \tfrac{1}{2} O_2$$

Acrylonitrile   Adiponitrile

Scheme 11.10: Baizer process for the manufacture of adiponitrile.

product isolation, the non-converted starting material was recycled [36, 37]. From an industrial point of view, it was important to add toluene sulfonic acid salts, preferred tetraethyl ammonium, because this increased the solubility of acrylonitrile in water and enhanced the dimerisation reaction and decreased the formation of propionitrile.

Electrochemistry is also used to produce fragrance compounds, for example Lilial or Lysmeral. This lily of the valley-like compound is manufactured starting from 4-*tert*-butyltoluene. The electrochemical dimethoxylation is performed on several 40 kt/a at BASF [35]. The intermediate dimethoxy *tert*-butyl toluene is then condensed with propanal followed by hydrogenation to the product; the liberated methanol is recycled (Scheme 11.11). The product Lysmeral is manufactured as racemic mixture, even though the different stereoisomers do not contribute equally to the odour; the (R)-configurated enantiomer has a strong floral odour [38, 39], whereas the (S)-isomer has no strong odour. In addition, a procedure for the oxidation of 4-*tert*-butyltoluene on a Ti/PbO$_2$ or graphite electrode has been reported using the solvent–electrolyte–system (water/H$_2$SO$_4$/CH$_2$Cl$_2$) resulting in 83% conversion and 77% yield of *p-tert*-butylbenzaldehyde [40].

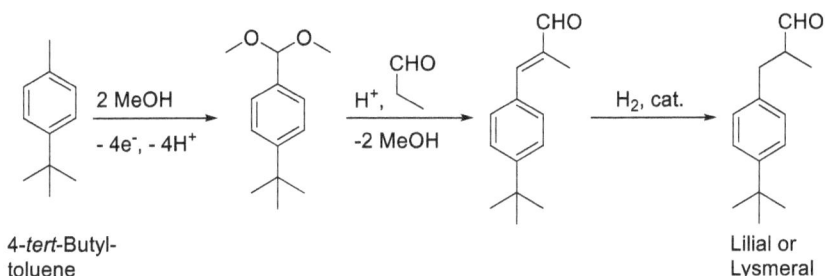

4-*tert*-Butyl-
toluene

Lilial or
Lysmeral

Scheme 11.11: Electrochemical manufacture of the aroma compound lily aldehyde.

The manufacture of *p*-anisaldehyde starting from 4-methylanisole is carried out electrochemically on industrial scale in >3,000 t/a (Scheme 11.12). The process is performed at 40–50 °C in methanol on graphite electrodes, using a voltage range of 4–6 V and a current density of 3–5 Ad/m^2. The conversion of 4-methylanisole is 90–99% and a selectivity of around 85% is achieved. Further developments include the use of cells in a voltage range of up to 30 V or at a current density of 100 mA/dm^2 [41–43]. This process is an improvement on the alternative oxidation method for the synthesis of anisaldehyde, which is the oxidation in the presence of manganese dioxide; however, this leads to significant amounts of metal waste [44].

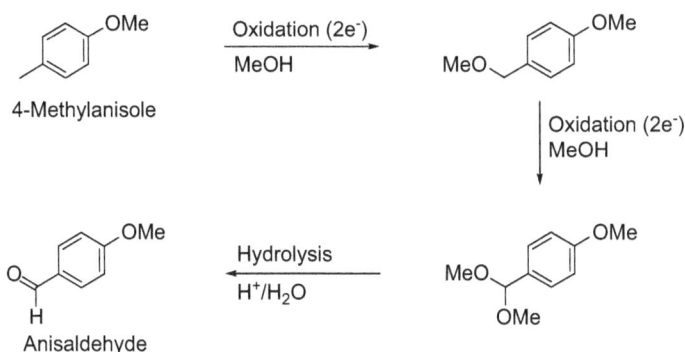

4-Methylanisole

Oxidation (2e⁻)
MeOH

Oxidation (2e⁻)
MeOH

Anisaldehyde

Hydrolysis
H⁺/H₂O

Scheme 11.12: Electrochemical approach to p-anisaldehyde.

A related industrialised process is the manufacture of p-hydroxybenzaldehyde from p-cresol (Scheme 11.13). The reaction is performed in methanol (15% solution) and 2% KF as electrolyte. Yields of 95% are achieved with a current density efficiency of 70% and a current of 0.1 A/cm [30]. The production volume based on this process is estimated to be several kt/a.

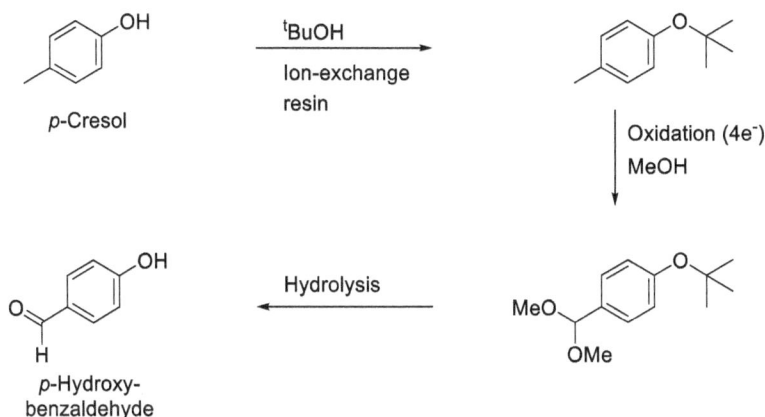

p-Cresol

ᵗBuOH
Ion-exchange
resin

Oxidation (4e⁻)
MeOH

p-Hydroxy-
benzaldehyde

Hydrolysis

Scheme 11.13: Synthesis of p-hydroxybenzaldehyde.

Rose oxide is an interesting flavour compound (see Chapter 4: Oxidations). This compound can be synthesised from citronellol (3,7-dimethyl-6-en-1-ol), on an anode that is based on carbon (Scheme 11.14) [45]. In acetonitrile and a current density of 40 mA/cm² and a working potential of 1.8 V, 26% yield of rose oxide can be obtained. An electric charge of 2 F is needed to achieve these results.

In another approach, the anodic methoxylation of citronellol is followed by a Pd-catalysed elimination and an acid-catalysed ring closure reaction resulting in the formation of rose oxide (Scheme 11.15) [46]. The methoxylation reaction is usually carried

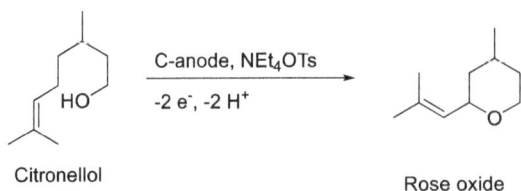

Scheme 11.14: Electrochemical approach to rose oxide.

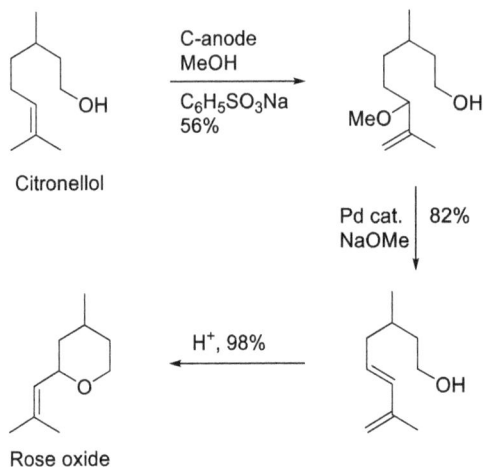

Scheme 11.15: Rose oxide synthesis via methoxylation.

out at a concentration of 40% on an anode based on palladium or more preferably carbon in a divided cell. The current density is in the range of 0.5–20 A/dm^2 at a potential of 2–10 V.

11.3 Electrochemical reductions

The selective reduction of C–C triple bonds can be performed in the presence of a *Lindlar* catalyst and hydrogen (see Chapter 2: Heterogeneous hydrogenations). In the twentieth century and also more recently, there have been a number of investigations into whether such a reduction could also be carried out electrochemically. To date, none of these processes have yet been implemented on an industrial scale, but the results suggest that the potential is there.

The electrochemical reduction of methylbutynol on Ni electrodes or Ni/Ag electrodes results in the formation of methylbutenol in up to 95% yield (Scheme 11.16) [47]. In modern trends of alkyne reduction, the formation of C–C double bonds is in focus by applying special Pd electrodes [48, 49]. The reduction of alkynols can be carried out with Pd or Pd alloy electrodes at high selectivity and yields >70%. Suitable solvent systems are water or water–solvent mixtures containing 50–99% water; the organic solvent can be chosen from the group of water-miscible solvents like acetone, polyethylene glycol, N-methyl pyrrolidone, alkyl carbonates, such as propylene carbonate, or sulfolane. The electrochemical reduction is usually performed in divided cells by applying a membrane as cell separator, and typical membranes are Nafion-based.

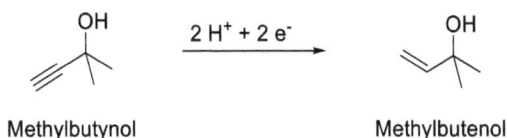

OH   2 H⁺ + 2 e⁻ →   OH

Methylbutynol   Methylbutenol

Scheme 11.16: Electrochemical reduction of methylbutynol.

Another application is found in the reduction of 1,4-butynediol to the corresponding alkene-diol. Butenediol is used in organic synthesis, for example in the preparation of vitamin B$_6$ [50]. Yields of >90% can be achieved in ethanol under basic conditions on Cu electrodes (Scheme 11.17) [51].

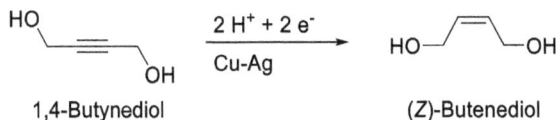

HO $\quad$ 2 H⁺ + 2 e⁻ / Cu-Ag $\quad$ HO–––––OH
OH
1,4-Butynediol $\qquad\qquad$ (Z)-Butenediol

Scheme 11.17: Electrochemical reduction of 1,4-butynediol.

In the field of xanthophyl chemistry, rhodoxanthin is of importance for colouration. Rhodoxanthin is a red pigment and has a broad range of colour shades depending upon which (E/Z)-isomer is present [52, 53]. The electrochemical alkyne reduction of a rhodoxanthin intermediate was developed by Grass and Zell [54]. The reduction occurs on a Pb cathode in 90% conversion and 54% yield, reducing both the alkyne and the tertiary C–O bond (Scheme 11.18). But to date this process has not yet been implemented on an industrial scale.

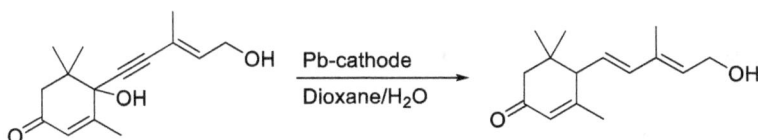

Scheme 11.18: Electrochemical approach to a rhodoxanthin building block.

D-Ribose is an important sugar molecule and it was applied in the manufacture of vitamin B_2 (see Chapter 10: Biocatalysis) [55, 56]. The electrochemical reduction of D-ribonolactone is carried out with Hg electrodes in the presence of an ammonium salt electrolyte in a divided cell (Scheme 11.19) [57, 58]. In the presence of boronic acid, a current efficiency of up to 40% and a yield of around 85% were achieved. Today, ribose is produced by fermentation starting from glucose with *Bacillus subtilis* [59].

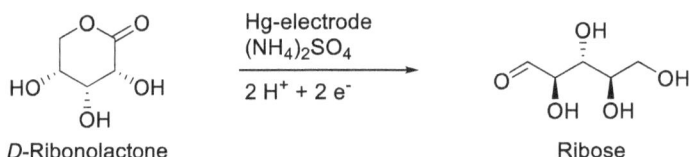

Scheme 11.19: Electrochemical synthesis of ribose from D-ribonolactone.

Another reduction of sugar molecules on the cathode is the manufacture of sorbitol from glucose with mercury or Pb/Hg electrodes in diluted H_2SO_4 as electrolyte (Scheme 11.20). During 1937–1948, the process was carried out in around 1 kt/a, with an electric density of 2 A/dm^2 and up to 70% electric efficiency. The electrochemical process was later replaced by catalytic hydrogenation (see Chapter 2: Heterogeneous hydrogenations) [60–64].

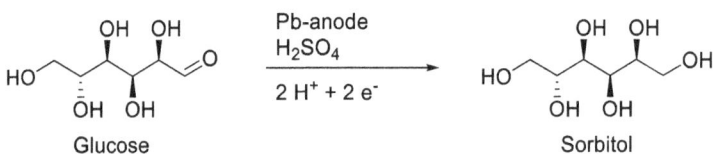

Scheme 11.20: Electrochemical reduction of glucose into sorbitol.

The electrochemical reduction of pyridine into piperidine is carried out in several kt/a at 35 °C in >90% yield applying a 5% sulphuric acid solution as electrolyte and a Pb electrode (Scheme 11.21) [65]. The reaction is performed at 15 A/dm^2 in a divided cell. The pyridine is placed in the cathode chamber, and the chambers are separated by diaphragms made from ceramics.

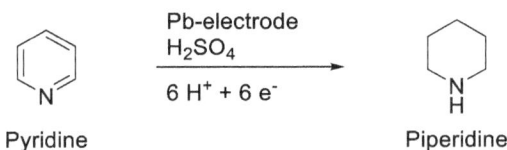

Scheme 11.21: Manufacture of piperidine by electrochemical reduction of pyridine.

11.4 Other electrochemical approaches

An interesting approach in electrochemical synthesis is the application of paired syntheses in a divided cell for the simultaneous preparation of products or intermediates on the cathode and the anode. In the field of fine chemicals, such an approach was successfully applied for vitamin B_1 [66]. On the cathode (based on Ti/Pt), the reduction reaction of a nitrile to the corresponding amine occurs in 74% conversion and 97% selectivity. This is an alternative to the heterogeneous hydrogenation of this nitrile with nickel catalysts (see Chapter 2: Heterogeneous hydrogenations). The corresponding oxidation takes place on the Pt/Ti anode to give the product vitamin B_1; a water–sulphuric acid electrolyte system is used. Conversions up to 85% and 99% selectivity can be achieved at an electrode potential of −0.2 to 0.4 V and a current density of up to 15 A/dm^2 (Scheme 11.22). The half-cells were separated by a cation-exchange membrane.

Anode

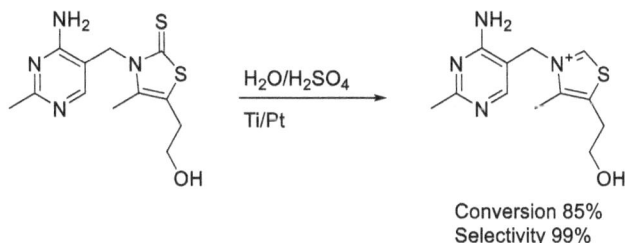

H$_2$O/H$_2$SO$_4$

Ti/Pt

Conversion 85%
Selectivity 99%

Cathode

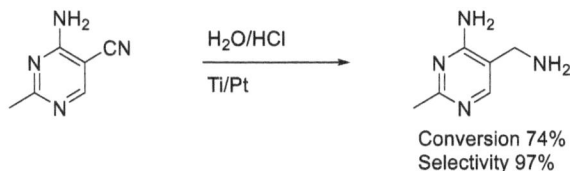

H$_2$O/HCl

Ti/Pt

Conversion 74%
Selectivity 97%

Scheme 11.22: Synthesis of vitamin B_1 and intermediates by electrochemistry.

The electrochemical manufacture of 1-naphthol is carried out in several thousand t/a at BASF. Naphthalene is transformed into 1-naphthylacetate in an undivided cell in up to 50% conversion and 85% selectivity. The current density is 65% and a conductive polymer electrode is used. The resulting naphthalene acetate is transformed into the product by hydrolysis and the co-product acetic acid is recovered (Scheme 11.23) [67–69]. 1-Naphthol is the starting material for dyes, insecticides and vitamin K_3 (see Chapter 4: Oxidations).

Naphthalene   1-Naphthol

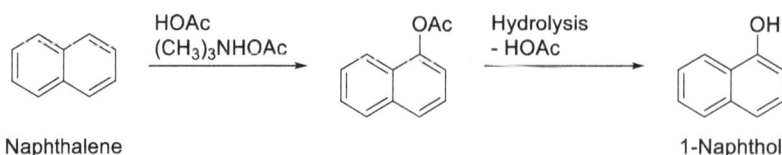

Scheme 11.23: Synthesis of 1-naphthol from naphthalene.

For many decades in the twentieth century, tetraethyl lead was used as an antiknock agent in fuels such as petrol. The electrochemical process (*Nalco* process) used a solution of an ethyl-*Grignard* reagent and an ethyl halide for its production. On the Pb anode, alkyl radicals are formed, which reacted with the electrode material (Pb) to form tetraethyl (tetraalkyl) lead (Scheme 11.24). The *Grignard* reagent was prepared in a THF/diglyme solvent mixture in >98% yield and fed in the electrolysis cell. This cell has a Pb anode and a steel cathode, which were separated by a membrane (polypropylene). The electrolyte (an ethyl magnesium halide and cyclic ether, e.g. tetrahydrofuran) is fed with ethyl chloride into the cell, which allowed the formation of the *Grignard* reagent from the deposited Mg and avoided a short circuit; the yield was 96%. In a similar set-up, tetravinyl lead can be produced [70, 71]. A representative flow sheet of such a process is depicted in Figure 11.6.

$$2 \text{ EtMgCl} \quad + \quad 2 \text{ CH}_3\text{CH}_2\text{Cl} \quad + \quad \text{Pb} \quad \longrightarrow \quad (\text{CH}_3\text{CH}_2)_4\text{Pb} \quad + \quad 2 \text{ MgCl}_2$$

Scheme 11.24: Electrochemical manufacture of tetraethyl lead.

Tetraethyl lead was produced in several kt/a based on this process [70, 73, 74]. However, the use of tetraalkyl lead compounds has been phased out several decades ago based on their toxicity and environmental aspects [75–77].

As described above, one of the fundamental reactions of electrochemistry is the hydrogen evolution (Scheme 11.1). These processes generally involve metal deposition on the cathode following a sequence of adsorption and migration of the adsorbed atoms to the grid [73]. In the presence of a ligand, this metal deposition process can be utilised to synthesise organometallic complexes. For example, nickel(0) compounds such as bis-cyclooctadiene-nickel (Ni(COD)$_2$), an important catalyst precursor (see Chapter 6: C–C-bond and C–N-bond forming reactions (metal-catalysed)), can be synthesised on an Al-cathode starting from nickel(II) acetylacetone in the presence of cyclooctadiene and pyridine in an alkylammonium salt as an electrolyte (Scheme 11.25) [78].

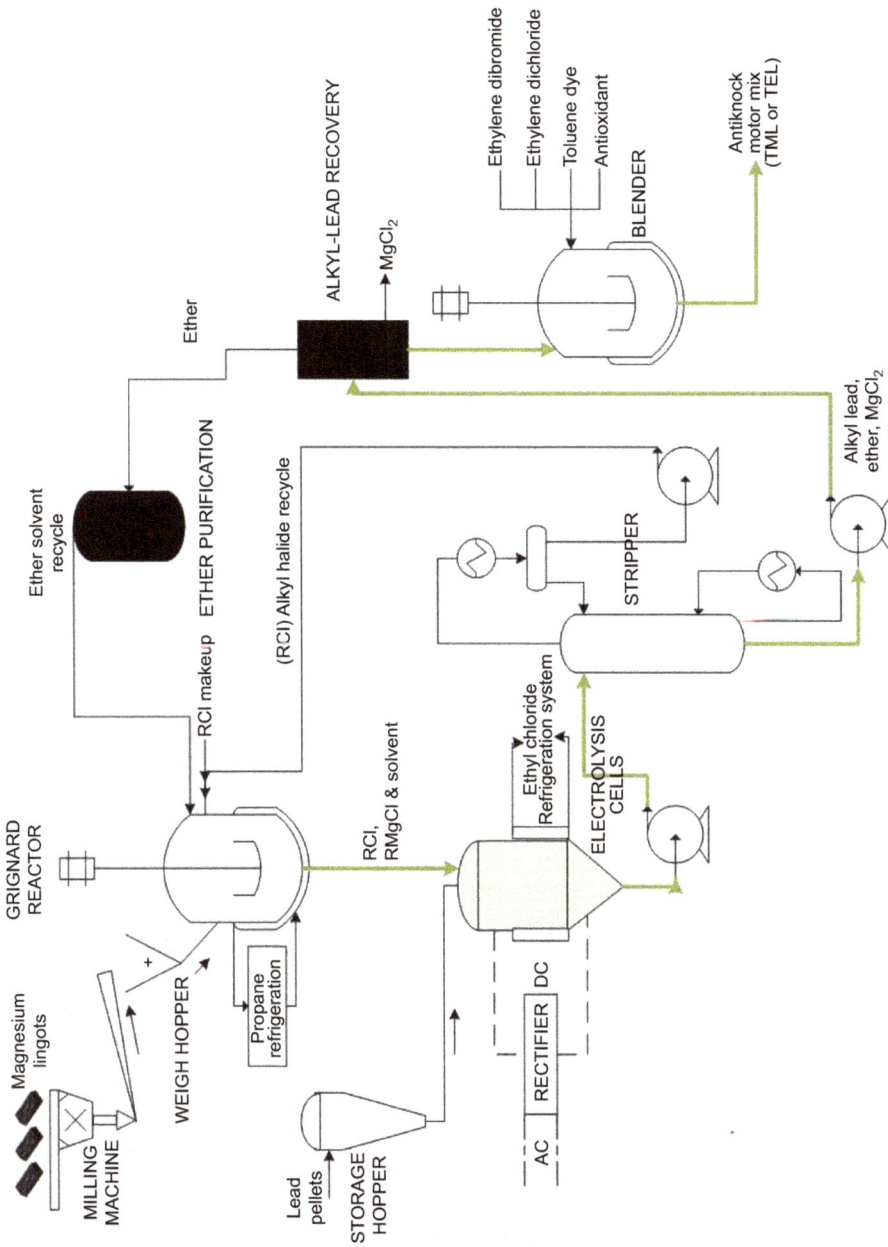

Figure 11.6: Flow sheet of Pb alkyl production [72].

Ni^{2+} + 2 (Cyclooctadiene) + 2e$^-$ $\xrightarrow[\text{Pyridine}]{\text{Al-cathode}}$ Ni(COD)$_2$

Cyclooctadiene
(COD)

Ni(COD)$_2$

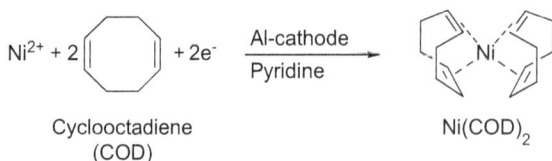

Scheme 11.25: Electrochemical synthesis of Ni(COD)$_2$.

References

[1] Kortüm G. Lehrbuch der Electrochemie, 5th ed. Weinheim, Germany, Verlag Chemie, 1972.
[2] Spencer J N, Bodner M G, Rickard H L. Oxidation–Reduction Reactions. In: Chemistry: Structure and Dynamics, 5th ed., New York, USA, John Wiley & Sons, 2010, 564.
[3] Strong F C. Faraday's laws in one equation. J Chem Educ 1961, 38, 98.
[4] Trasatti S. Work function, electronegativity, and electrochemical behavior of metals. J Electroanal Chem 1972, 39, 163–183.
[5] Beck F. Electroorganische Chemie. Weinheim, Germany, Verlag Chemie, 1974, 38–40.
[6] Noël T, Cao Y, Laudadio G. The fundamentals behind the use of flow reactors in electrochemistry. Acc Chem Res 2019, 52(10), 2858–2869.
[7] Backhurst J R, Coulson J M, Goodrige F, Plimley R E. A preliminary investigation of fluidized bed electrodes. J Electrochem Soc 1969, 116, 1600–1607.
[8] Backhurst J R, Goodrige F, South G, Fleischmann M. Elektrochemische Zelle. DE 1910286, The National Research Development Corporation, November 26, 1970.
[9] Bonrath W, Goetheer E L V, Goy R, Medlock J, Latsuzbaia R. Electrochemical hydrogenation of specific alkynes. WO 2022238349, DSM, November 17, 2022.
[10] Hocking M B. Vanillin: Synthetic flavoring from spent sulfite liquor. J Chem Educ 1997, 74, 1055–1058.
[11] Zirbes M, Waldvogel S R. Electro-conversion as sustainable method for the fine chemical production from the biopolymer lignin. Curr Opin Green Sustainable Chem 2018, 14, 19–25.
[12] Waldvogel S R, Breiner M. Biogene Produkte, Guter Geschmack aus Holz. Nachr Chem 2020, 68, 42–44.
[13] Chakar F S, Ragauskas A J. Review of current and future softwood kraft lignin process chemistry. Ind Crop Prod 2004, 20, 131–141.
[14] Bjørsvik H-R, Minisci F. Fine Chemicals from Lignosulfonates. 1. Synthesis of Vanillin by Oxidation of Lignosulfonates. Org Process Res Dev 1999, 3, 330–340.
[15] Zirbes M, Schmitt D, Beiser N, Pitton D, Hoffmann T, Waldvogel S R. Anodic degradation of lignin at active transition metal-based alloys and performance-enhanced anodes. ChemElectroChem 2019, 6, 155–161.
[16] Schmitt D, Regenbrecht C, Hartmer M, Stecker F, Waldvogel S R. Highly selective generation of vanillin by anodic degradation of lignin: A combined approach of electrochemistry and product isolation by adsorption. Beilstein J Org Chem 2015, 11, 473–480.
[17] Li X, Jia P, Wang T. Furfural: A promising platform compound for sustainable production of C4 and C5 chemicals. ACS Catal 2016, 6, 7621–7640.
[18] Kwon Y, Schouten K J P, van der Waal J C, de Jong E, Koper M T. Electrocatalytically conversion of furanic compounds. ACS Catal 2016, 6, 6704–6717.

[19]   Parpot P, Bettencourt A P, Chamoulaud G, Koloh K B, Belgsir E M. Electrochemical investigations of the oxidation-reduction of furfural in aqueous medium: Application of electrosynthesis. Electrochim Acta 2004, 49, 397–403.

[20]   Kokoh K B, Belgsir E M. Electrosynthesis of furan-2,5-dicarboxaldehyde by programmed potential electrolysis. Tetrahedron Lett 2002, 43, 229–231.

[21]   Degner D. Organic Electrosyntheses in Industry. In: Steckhan, ed. Topics in Current Chemistry. Electrochemistry III, Berlin, Germany, Springer, 1988, 1–95.

[22]   Syntrivas L-D, del Campo F J, Robertson J. An electrochemical flow cell for the convenient oxidation of Furfuryl alcohols. J Flow Chem 2018, 8, 123–128.

[23]   Shono T, Matsumura Y. Novel synthesis of maltol and related compounds. Tetahedron Lett 1976, 17, 1363–1364.

[24]   Baizer M M. Electrochemical conversion of biomass-derived materials to specialty chemicals. Proc BIO EXPO 1986, 86, 341–349.

[25]   Harima R, Kase K, Masatomi T, Misawa T, Nakacho Y, Nakanaga T, Shimizu Y, Takao H. 2,2′-Bis(3,4-Epoxy-5-oxotetrahydropyran)ethers. US 4150039, Otsuka Kagaku Yakuhin Kabushiki, April 17, 1979.

[26]   Weeks P D, Brennan T M, Brannegan D P, Kuhla D E, Elliott M L, Watson H A, Wlodecki B, Breitenbach R. Conversion of secondary furfuryl alcohols and isomaltol into maltol and related γ-pyrones. J Org Chem 1980, 45, 1109–1113.

[27]   Weeks P D, Allingham R P, Pears R. Verfahren zur Herstellung von γ-Pronen. DE 2630837, Pfizer, March 03, 1977.

[28]   Brennan T M, Brannegan D P, Week P D, Kuhla D E, Gales F. Verfahren zur Herstellung von γ-Pronen. DE 2750553, Pfizer, May 18, 1978.

[29]   Weeks P D, Kuhla D E, Allingham R A, Watson H A, Wlodecki B. The facile preparation of 6-alkoxy-2H-pyran-3(6H)-ones, and their subsequent conversion into maltol and analogous g-pyrones. Carbohydrate Res 1077, 56, 195–199.

[30]   Pletcher D, Walsh F C. Industrial Electrochemistry, 2nd ed. Springer Science + Business Media, 1990.

[31]   Shono T, Matsumura Y, Tsubata K, Takata J. Electroorganic chemistry. 55. One step synthesis of α-aminoalkylfurans and its application to a facile synthesis of pyridoxine (vitamin B6). Chem Lett 1981, 10, 1121–1124.

[32]   Steckhan E. Electrochemistry 3. Organic Electrochemistry. In: Elvers, ed. Ullmann's Encyclopedia of Industrial Chemistry, Weinheim, Germany, Wiley-VCH, 2012, 316–348.

[33]   Pütter H. Industrial Electroorganic Chemistry. In: Lund, Hammerich, eds. Organic Electrochemistry, 4th ed., New York, USA, Marcel Dekker, 2001, 1259–1307.

[34]   Knunyants I L, Vyazankin N S. Hydrodimerization of acrylonitrile. Bull Acad Sci. USSR Div Chem Sci 1957, 6, 253–256.

[35]   Hoormann D, Jörisson J, Pütter H. Elektrochemische Verfahren – Neuentwicklungen und Tendenzen. Chem Ing Tech Lab 2005, 77, 363–1376.

[36]   Weinberg N L, Tilak B V. Technique of Electroorganic Synthesis: Scale-up and Engineering Aspects, Part III. New York, USA, John Wiley & Sons, 1982.

[37]   Scott K, Hayati B. The influence of mass transfer on the electrochemical synthesis of adiponitrile. Chem Eng Process 1993, 32, 253–260.

[38]   Bartschat D, Börner S, Mosandl A, Bats J W. Stereoisomeric flavour compounds LXXVI: direct enantioseparation, structure elucidation and structure-function relationship of 4-tert-butyl-α-methyldihydrocinnamaldehyde. Zeitschrift für Lebensmitteluntersuchung und -Forschung A 1997, 205, 76–79.

[39]   Kingston C, Palkowitz M D, Takahira Y, Vantourout J C, Peters B K. A survival guide for the "electro-curious". Acc Chem Res 2020, 53, 72–83.

[40]   Seiler P. Verfahren zur Herstellung von p-tert.butylbenzaldehyde. EP 0030588, F. Hoffmann-La Roche, June 24, 1981.

[41]   Bard A J, Inzelt G, Scholz F. Electrochemical Dictionary, 2nd ed. Berlin Heidelberg, Germany, Springer-Verlag Berlin Heidelberg, 2012.

[42]   Malloy T P, Halter M A, House D W. Electrochemical oxidation of alkyl aromatic compounds. US 4459186, UOP, July 10, 1984.

[43]   Degner D, Barl M, Siegel H. In 4-Stellung substituierte Benzaldehyddialkylacetale und ihre Herstellung. EP 0011712, BASF, June 11, 1980.

[44]   Fahlbusch K-G, Hammerschmidt F-J, Paten J, Pickenhagen W, Schatkowski D, Bauer K, Garbe D, Suburg H. Flavors and Fragrances. In: Elvers, ed. Ullmann's Encyclopedia of Industrial Chemistry, Online ed., Weinheim, Germany, Wiley-VCH Verlag GmbH & Co, 2012.

[45]   Shono T, Ikeda A, Kimura Y. Anodic cyclization of olefinic alcohols to cyclic ethers. Tetrahedron Lett 1971, 39, 3599–3602.

[46]   Suzukamo G, Takano T, Tamura M, Ikimi K. Process for producing 2,6-dimethyl-3-alkoxy-oct-1-en-8-ol, dehydrocitronellol and 2-(2′-methyl-1′-propenyl)-4-methyltetrahydropyran. EP21769, Sumitomo Chemical, July 01, 1981.

[47]   Lebedewa A I. Electrolytic hydrogenation of dimethylethynylcarbinol and dimethylvinylcarbinol. I. Effect of the cathode material. Zhur Obshch Khim 1948, 18, 1161.

[48]   Bonrath W, Goetheer E L, Goy R, Latsuzbaia R, Medlock J. Electrochemical hydrogenation of specific alkynes. WO 2022238349, DSM, November 17, 2022.

[49]   Capobianco G, Vianello E, Giacometti G. Riduzione del legame acetilenico nelle aldeidi propargilica e fenilpropargilica. Gazz Chim Ital 1967, 16, 243–253.

[50]   Eggersdorfer M, Laudert D, Letinois U, McClymont T, Medlock J, Netscher T, Bonrath W. One hundred years of vitamins – A success story of natural science. Angew Chem Int Ed Engl 2012, 52, 12960–12990.

[51]   Kato J, Sakuma M, Yamada T. Electrolytic partial hydrogenation of 2-butyn-1–4-diol with Ag-Cu-Cathode. J Electrochem Soc Jpn 1957, 25, 126.

[52]   Schex R, Bonrath W, Schäfer C, Schweiggert R. The Impact of (E/Z)-isomerization and aggregation of the color of rhodoxanthin formulations for food and beverages. Food Chem 2020, 332, 12370–12378.

[53]   Schex R, Schweiggert F, Wüstenberg B, Bonrath W, Schäfer C, Schweiggert R. Kinetic and thermodynamic study of the thermally induced (E/Z)-isomerization or retro-carotenoid rhodoxanthin. J Agrico Food Chem 2020, 68, 5259–5269.

[54]   Grass H, Zell R. Process for the preparation of cyclohexane derivatives. EP85763, F. Hoffmann-La Roche, August 17, 1983.

[55]   Karrer P, Becker B, Frei P, Salomo H, Schopp K. Zur Synthese des Lactoflavins. Helv Chim Acta 1935, 18, 1435–1448.

[56]   Steiger M. Preparation of D-ribose. Helv Chim Acta 1936, 19, 189–195.

[57]   Matsumoto M, Miyazaki M. Studies on the constant-current electrolytic reduction of D-ribonolactone to D-ribose. I. Application of ammonium salts as electrolyte. J Pharm Soc Jpn 1967, 87, 627–630.

[58]   Sugasawa S, Matsumoto S. Electrolytic process for preparing D-ribose. US 3312608, Tanabe Saiyaku, April 04, 1967.

[59]   de Wulf P, Vandamme E J. Production of D-ribose by fermentation. Appl Microbiol Biotech 1997, 48, 141–148.

[60]   Creighton H J. Process for electrolytically reducing sugars to alcohol. US 1612361, Atlas Powder, December 28, 1926.

[61]   Parker E A, Swann S. Electrolytic reduction of hexoses. Trans Electrochem Soc 1947, 92, 343–348.

[62]   Hefti H R, Kolb W. Electrolytic reduction of sugars. US 2507973, May 16, 1950.

[63]   Creighton H J. The electrochemical reduction of sugars. Trans Electrochem Soc 1939, 75, 289–307.

[64]   Gräfe G. Herstellung Eigenschaften und Verwendung von Sorbit. Stärke 1969, 21, 183–192.

[65]   Parkes D W. Improvements in the electrolytic manufacture of piperidine. GB 395741, Robinson Bros, July 24, 1933.

[66] Niinobe T, Yoshida K, Yokoyama M. Verfahren zur Herstellung von Vitamin B_1 und seinem Zwischenprodukt und Vorrichtung hierzu. DE 2948343, Takeda Chemical Industries, June 19, 1980.

[67] Haufe J, Rentzea C, Degner D. Elektrochemische Herstellung aromatischer Ester. DE 2434845, BASF, February 05, 1976.

[68] Degner D, Boehlke K, Treptow W. Elektrochemische Herstellung aromatischer oder aromatischheterocyclischer Alkansäureester. DE 2659148, BASF, July 06, 1978.

[69] Schuster L, Seid B. Verfahren zur Herstellung von α-Naphthol. DE 2706682, BASF, August 24, 1978.

[70] Braithwaite D G. Manufacture of tetramethyl lead. US 3256161, Nalco Chemical, June 14, 1966.

[71] Robinson G C. Production of hydrocarbon lead compounds. US 3522156, Ethyl corporation, July 28, 1970.

[72] Guccione E. Electrolysis: New route to alkyl lead compounds. Chem Eng 1965, 72, 102–104.

[73] Beck F. Electroorganische Chemie. Weinheim, Germany, Verlag Chemie, 1974, 277–279.

[74] Lutz W-D, Zirngiebl E. Die Zukunft der Elektrochemie. Chem unserer Zeit 1989, 23, 151–160.

[75] Perino J, Ernhart C B. The reaction of subclinical lead level to cognitive and sensorimotor impairment in black preschoolers. J Learn Disabil 2016, 7, 616–620.

[76] Needleman H L. The removal of lead from gasoline: Historical and personal reflections. Environ Res 1990, 84, 20–35.

[77] Nriagu J O. The rise and fall of leaded gasoline. Sci Total Environ 1990, 92, 13–28.

[78] Lehmkuhl H, Leuchte W. Electrochemical synthesis of organometallic nickel compounds. J Organomet Chem 1970, 23, C30–C32.

12 New trends

In the preceding chapters, the crucial role of catalysis in industrial chemical processes was demonstrated in general, with a special emphasis in the field of fine chemicals. The selected examples highlight and explain the basic principles and techniques, and show the catalytic transformations that are possible and economically viable on a large scale. Whilst some examples have only been demonstrated in the laboratory or pilot plant, all would be viable in production.

Given the sheer number and variety of examples, some people might argue that in the field of catalysis, all problems have already been solved and there is nothing new to discover or develop. In fact, the opposite is true! Whilst a number of chemical methodologies are at a mature stage, there are still specific transformations that are unsuccessful with current technologies due to a lack of reactivity or selectivity, and new types of reactions are waiting to be discovered. In addition, especially in areas with high product margins, for example the pharmaceutical industry, there are many chemical transformations that are carried out stoichiometrically because an equivalent catalytic version does not exist or is not efficient enough. In the future, the importance of catalysis will continue to increase with the increasing cost and environmental pressure throughout all chemical fields. The utopia, which is still a long way in the future, is that all chemical production processes will be catalytic, with a high atom economy and minimal waste generation.

The drive for an increased number of catalytic processes with higher efficiency is many-fold, all directed at fulfilling the current and future demands. The world is changing at an incredible rate; however, these future demands can be classified into a number of over-arching trends. These include:
- increasing world population and increasing urbanisation. In addition, an increased average age all over the world; therefore, the need to sustainably produce enough food for all and maintain the health of the population,
- limiting the magnitude of global warming and our reliance on fossil-based materials as fuels resulting in efficient energy production methods and storage, including for transportation,
- related to the above, the change of the current chemical building blocks from an oil-based feedstock to a renewable feedstock and the development and use of new materials,
- digitalisation and increasing connectedness.

In all these areas, catalytic processes have a role to play.

In the next decades, the importance of catalysis will increase, especially in manufacturing processes in all the areas of pharmaceuticals, fine chemicals, monomers and polymers, and the application of catalysts in environmental technologies. New technologies, often first developed at universities, will be further developed and scaled-up

https://doi.org/10.1515/9783111102672-012

to commercial production. This will require even closer collaboration both between universities and industries, but also between different scientific disciplines. Many examples already exist of such successful collaborations including C. Bosch (*Haber–Bosch* process), I. Langmuir (surface chemistry), W. Knowles (asymmetric hydrogenation reactions), R. Schrock (olefin metathesis), R. Heck (Pd-catalysed coupling reactions) or A. Yoshino (lithium ion battery). Chemistry and catalysis sit in the centre of many scientific fields and have a critical role to play.

Another change, already observed, is the growing importance of natural science (including chemistry and catalysis) in countries in the far East. High technology companies are no longer located only in the USA, Europe and Japan; India, China, South Korea and Indonesia are already very important, and this will continue to grow. It is not inconceivable that the balance of power in these technologies will completely switch to Eastern countries in the coming decades.

In the short term, a number of general trends in the field of catalysis can be expected:

- New or improved metal-catalysed processes with high selectivity (including enantioselectivity and diastereoselectivity).
- Replacement of scarce and expensive platinum group metals (PGMs) by more abundant and less toxic alternatives.
- Increased importance of bio-transformations (using isolated enzymes or whole cells) for highly selective and complex chemical conversions, including the direct fermentation of the desired compounds.
- Fermentation processes combined with one or two subsequent chemical/catalytic steps for producing complex molecules such as pharmaceuticals and agrochemicals.
- The wider application of technologies such as organocatalysis and redox (especially photo-redox) reactions in chemical production.
- Combining catalytic process development with purification and reactor design at an early stage leading to increased efficiency and recovery of energy and heat.
- Continuing the on-going shift from batch processes to flow chemistry and continuous processing.
- Application of alternative, environmentally friendly solvents in all types of processes.
- Use of "big data" to obtain increases in process efficiency including computer-assisted design of new processes, reactors and even synthetic routes.
- Sustainable raw materials, even from waste materials such as lignin, to avoid using oil- and coal-based feedstocks.
- Development of highly efficient, reversible chemical storage of electrical energy.
- Use of (sustainable) electricity for chemical transformations, including the use of electrodes made of base-metals or organic conductors.

The different chemical industries have different needs and will focus on the development and implementation of new and improved catalytic processes in different, but related, fields. In addition, as technologies are further developed, processes that initially, for cost reasons, could only be applied to "expensive" molecules become cheaper and more efficient and will be applied to larger volumes and cheaper compounds.

In the field of pharmaceutical chemistry, there is a continuing need for new drugs, for example in the field of antibiotics to bypass the challenge of resistance in bacteria, new anti-viral treatments, cheap medicines for developing countries and personalized medicine (tailoring the medical treatment to a person's specific physiology). For small molecules, the main developments will be in the field of chemocatalysis with an emphasis on more selective catalysis and novel transformations that currently cannot be achieved (or only achieved with stoichiometric reagents), whereas large molecules with complex structures and functional groups are the domain of bio-transformations, either by direct fermentation or by a semi-synthetic approach with a late-stage derivatisation (by catalysis) of a fermentation product.

In the fine and bulk chemical industries, new catalysts for new processes and improved catalysts for current processes are the focus. Cost, availability of raw materials and sustainability are the key driving forces. New processes will be mainly, or completely, catalytic, and for existing processes, stoichiometric processes will be superseded by catalytic processes. Existing catalytic processes will evolve to become more productive and selective with reduced waste production. For the manufacture of small molecules (<20 atoms), gas-phase technology is a valid option and new applications can be expected. Furthermore, the feedstocks will change from oil- and coal-based chemistry to renewables, mainly sugar-based materials. The so-called platform chemicals are a game changer in future chemistry, and consequently, a different toolbox of catalytic methods is needed, both for the preparation of these starting materials and their subsequent transformation.

In the area of metal catalysis, the trends are new transformations and the replacement of one metal with another for cost or environmental reasons. For example, the replacement of Pd or Pt in hydrogenation and coupling reactions by metals such as Fe or Ni fulfils both criteria. Another example is the avoidance of toxic metals such as Cr in oxidation reactions, or the replacement of Pb in the traditionally used *Lindlar* catalyst by Pb-free versions [1]. This approach is directly connected with the reactor-type development, for example tubular reactors, to allow continuous processing [2, 3]. An interesting developing field is the catalytic reduction of esters and amides (especially using cheap metals), which is currently performed using stoichiometric hydride reagents.

In addition, there will be a trend towards more functionalised catalysts, bi- or multi-functionalised catalysts, which combine several functionalities and chemical steps (analogous to enzymes). An example of such a catalyst group is a solid acid doped with metals which allow acid-catalysed reactions and hydrogenation reactions; see, for example, the synthesis of dimedone (Chapter 8: Acid–base-catalysed reactions). Another example is the combination of different techniques such as transition metal catalysis and organocatalysis, and the repurposing of catalysts and/or ligands;

this has been recently demonstrated in the combination of an organocatalytic stereo-selective aldol reaction, followed by transfer hydrogenation using the organocatalyst as the ligand on the metal [4].

In the field of acid and base catalysis, a major ongoing development is the re-placement of processes containing mineral acids and also partially *Brønsted* acids with those utilising solid acids. The goal is to reduce waste formation, eliminate the need for later neutralisation with basic materials and simplify downstream process-ing. Also desired are catalysts that result in increased selectivity, especially enantiose-lectivity and diastereoselectivity. New catalysts based on new solid materials such as polymers or zeolites are likely to feature more in the future, but strong contributions from material science will be required and heterogenisation of homogeneous acids can be an option. In the field of homogeneous acid catalysis, new types of acids are needed. An interesting area of current research is the application of organocatalyst-based acids, for example proline or organo-phosphorous-based catalysts.

Whereas the acid-catalysed reactions are widely used, the field of base catalysis is in need of strong improvement. Therefore, new catalysts with higher stability, espe-cially thermal stability, and also higher selectivity are needed. Here organocatalysts or enzyme catalysts, for example aldolases for aldol condensations, can be a solution.

Whilst there is always a drive to develop new catalysts and new catalytic pro-cesses, the scale-up of catalytic processes and transfer into production is a permanent, ongoing activity throughout the chemical industry. One aspect of this is the increasing development and use of tools for parallel screening or high-throughput experimenta-tion platforms to map out the performance of a catalyst under a wide range of param-eters to ensure a large window of optimal performance. The change from batch operations to continuous processes begun many years ago, and the switch is continuing to accelerate. The catalyst, process and reactor will not necessarily be developed sepa-rately, but there will be increased interaction between the disciplines. New production modes applying new methods (technologies) and the combination of unit operations will be a breakthrough. An example of such an approach was published recently and demonstrates the advantages of this approach in view of safer process chemistry and an efficient production [5]. The motto – *The future is continuously in processing* – should be in the focus.

Where solvents must be used in a process, the type of solvent selected will change significantly. For example, the replacement of halogen-containing solvents by halogen-free ones or the use of new types of solvent such as supercritical fluids (e.g. sc-CO_2), or 2-methoxy-2-methyl butane or 2-MeTHF, will find an increased number of industrial ap-plications. The advantages of alternative ether solvents such as 2-methoxy-2-methyl bu-tane and 2-MeTHF are their stability in the presence of oxygen and minimal to zero peroxide formation. This property makes them suitable for a broad range of applications in synthetic chemistry and for extraction and/or purification processes. The industrial application of supercritical fluids, especially sc-CO_2 on a large scale, for example in the food industry (for instance the decaffeination of coffee), is well documented and is likely to be extended into chemical production.

Computers and computation chemistry will continue to become more important in catalysis in a number of areas. Computer-assisted approaches in understanding and designing new catalysts are already applied in homogeneous and enzymatic catalysis. With the increase in available computer power and techniques, this is being extended into heterogeneous catalysis. DFT calculations will lead to a new understanding of catalysis, not only in the fundamental reaction and/or rate-determining step, but also in complex, multi-pathway systems leading to improved catalysts and processes. Computation tools to design complete synthetic sequences are still in their infancy but are expected to rapidly develop. Quantum computing tools and machine-learning methods will continue to advance and play a more important role [6]. Many current production plants are fully automated with a wide variety of data being logged. In the future, this data will be managed better and more use will be made of it. Analysis applying new software tools (big data, industry 4.0) with the aim to establish safer and more stable processes is currently in progress and will continue in the future.

Energy is a further topic, in generation, storage and use. In the past decade, there has been an enormous increase in renewable electricity generation from solar panels, wind and hydro-electric projects. With low to zero CO_2 emissions, the switch from steam to electrical heating in chemical production would have a huge benefit in terms of reduction of greenhouse gas emissions. However, the storage of energy between generation and use is an area in need of improvement and catalysis could play a part, for example in novel battery technology or in the production, storage and use of hydrogen.

In addition, catalytic processes using alternative energy sources are already known and their use will increase. These include photochemistry, electrochemistry, plasma, ultrasound and microwave technologies. Photocatalysis, especially photoredox chemistry, has seen a significant increase in recent years, partially driven by the development of novel catalysts and methods. Photochemistry can also be an interesting tool in the synthesis of heterocycles, for example Co- or Rh-catalysed [2 + 2 + 2]-cycloadditions or hetero-*Diels–Alder* reactions. Electrochemistry is currently undergoing a renaissance with the development of new electrode materials for chemical and technical applications. Areas of interest include electrochemical oxidations and reductions to convert bio-based feedstocks into fine-chemical building blocks and reduction of CO_2 for energy storage and as C_1-building blocks such as CO and formic acid.

Overall, catalysis has provided huge benefits to chemical production and the world. Without it, we would not have the wide variety of chemical products (e.g. pharmaceuticals, agrochemicals, vitamins and other nutritional supplements, and polymeric materials) available at cost-effective prices and in such large volumes. Catalysis is an exciting and growing field and still has much more to offer and many more challenges to solve.

References

[1] Bonrath W, Mueller T, Kiwi-Minsker L, Renken A, Iouranov I. Hydrogenation Process. US 8592636, DSM IP Assets BV, 2010.
[2] Altheimer M, Bonrath W, Goy R, Medlock J A, Vernuccio S, Rohr V, R P. Tubular Catalytic Reactor Device for Processing and Conditioning of Material Transported through the Device. WO 20182021108, DSM IP Assets BV, 2018.
[3] Bonrath W, Goy R, Medlock J A. Metal powderdous catalyst for hydrogenation processes. WO 2018202639, DSM IP Assets BV, 2018.
[4] Bourgeois F, Medlock J A, Bonrath W, Sparr C. Catalyst repurposing sequential catalysis by harnessing regenerated prolinamide organocatalysts as transfer hydrogenation ligands. Org Lett 2020, 22, 110–115.
[5] Godineau E, Battilocchio C, Lal M. Building up a continuous flow platform as an enabler to the preparation of intermediates on kilogram scale. Chimia 2019, 73, 828–831.
[6] Fabrizio A, Meyer B, Fabregat R, Corminboef C. Quantum chemistry meets machine learning. Chimia 2019, 73, 983–989.

Index

https://doi.org/10.1515/9783111102672-013

www.ingramcontent.com/pod-product-compliance
Lightning Source LLC
Chambersburg PA
CBHW080135220326
41598CB00032B/5076